Bear and Human

The Archaeology of Northern Europe
Volume 3

General Editors
Paul S. Johnson, *University of Nottingham*
Sam Turner, *Newcastle University*

Editorial Board
Ian Armit, *University of York*
Manuel Fernández-Götz, *University of Edinburgh*
Julie Lund, *Universitetet i Oslo*
Karen Milek, *Durham University*
Neil Price, *Uppsala universitet*
Magali Watteaux, *Université Rennes 2*
Mark White, *Durham University*

Advanced Studies on the Archaeology and History of Hunting, edited by the Centre for Baltic and Scandinavian Archaeology (ZBSA), Volume 3

Bear and Human

Facets of a Multi-Layered Relationship
from Past to Recent Times, with Emphasis
on Northern Europe

Edited by Oliver Grimm,
in cooperation with Daniel Groß, Alexandra Pesch, Olof Sundqvist,
and Andreas Zedrosser

A volume based on papers presented at a conference at Orsa
Predator Park, Dalarna, Sweden, Oct. 16th to 18th, 2019

Volume 3.1

BREPOLS

British Library Cataloguing in Publication Data
A catalogue record for this book is available from the British Library.

Copy editor: Gundula Lidke

Layout, typesetting, and image editing: Matthias Bolte and Cornelia Lux-Kannenberg. Cover design by TopicA.

The editor has made every possible effort to identify all copyright owners. In the case that copyrights have not been cleared, please contact the editor.

© 2023, Brepols Publishers n.v., Turnhout, Belgium.
This is an open access publication made available under a CC BY-NC 4.0 International License: https://creativecommons.org/licenses/by-nc/4.0/. No part of this publication may be reproduced, stored in a retrieval system, or transmitted, in any form or by any means, for commercial purposes, without the prior permission of the publisher, or as expressly permitted by law, by licence or under terms agreed with the appropriate reprographics rights organization.
D/2023/0095/164
ISBN: 978-2-503-60611-8 (3 vols)
e-ISBN: 978-2-503-60613-2 (3 vols)
Volume DOI: 10.1484.M.TANE-EB.5.134319
Three-volume set : DOI: 10.1484/M.TANE-EB.5.133678

Printed in the EU on acid-free paper.

Conference sponsors

Vetenskapsrådet (Swedish Research Council)

Kungl. Gustav Adolfs Akademien för svensk folkkultur (The Royal Gustavus Adolphus Academy for Swedish Folk Culture)

Stockholms universitet (Stockholm University)

Zentrum für Baltische und Skandinavische Archäologie (Centre for Baltic and Scandinavian Archaeology)

Book sponsors

Kungl. Gustav Adolfs Akademien för svensk folkkultur (The Royal Gustavus Adolphus Academy for Swedish Folk Culture)

Länsstyrelsen Västerbotten (County Administration of Västerbotten, Sweden)

International Council for Game and Wildlife Conservation (CIC)

List of contents

Book 1

List of contents 7

Foreword by Oliver Grimm 13

Chapter 1 – Bear and human: Facets of a multi-layered relationship – introduction, discussion and synthesis 17

"Bear and human" – introduction, discussion and synthesis
Oliver Grimm 19

Chapter 2 – Bears in biology (Europe) 49

Conservation status and distribution of the brown bear in Europe
Andreas Zedrosser and Jon E. Swenson 51

The history of the Scandinavian Brown Bear Research Project – a formidable success story
Jon E. Swenson and Sven Brunberg 63

The management of brown bears in Sweden, Norway and Finland
Michael Schneider, Andreas Zedrosser, Ilpo Kojola and Jon E. Swenson 77

Genetics of brown bears in northern Europe
Alexander Kopatz 99

Hibernation ecology of brown bears in Sweden
Andrea Friebe, Jon E. Swenson and Andreas Zedrosser 109

The social system of a "nonsocial" species, the brown bear
Andreas Zedrosser, Shane C. Frank, Jennifer E. Hansen, Sam M. J. G. Steyaert, J. E. Swenson . . 121

Sexually selected infanticide as a mating strategy in brown bears
Andreas Zedrosser, Sam M. J. G. Steyaert and Jon E. Swenson 129

Bears – fact and fiction about bear hunting and intelligence
Oliver Grimm, Andreas Zedrosser and Jon E. Swenson 137

Chapter 3 – Bear hunting (Europe)..........147

Bear hunting in the later Middle Ages and early modern period, viewed from the perspective of art history and contemporary textual sources
Richard Almond..........149

Chapter 4 – Animal agency (northern Europe)..........173

Posthuman bears: Sight, agency, and baiting in Early Modern England
Liam Lewis..........175

Chapter 5 – Bears in long-term archaeo(zoo)logical studies (northern Europe)..........185

Brown bears in burials and entertainment in later prehistoric to modern Britain (*c.* 2400 BC – AD 1900s)
Hannah J. O'Regan..........187

Bears and humans in Sweden – 10,000 years of interactions from the Mesolithic to the Middle Ages
Ola Magnell..........209

Zooarchaeological brown bear (*Ursus arctos*) finds in eastern Fennoscandia
Kristiina Mannermaa, Tuija Kirkinen and Suvi Viranta-Kovanen..........235

The history of the brown bear (*Ursus arctos* L.) in the northern German lowlands
Ulrich Schmölcke..........265

In the company of bears: The role and significance of the bear from the perspective of the Holocene hunter-gatherer-fishers of the East European Plain forest zone (10th–3rd millennium BC)
Ekaterina A. Kashina and Anastasia A. Khramtsova..........291

Chapter 6 – Bears in archaeo(zoo)logical, focused analysis (northern Europe)..........315

The White One: How to frame the narrative of the world's oldest intact polar bear skeleton, specimen S10673 from Finnøy, southwestern Norway, in a museum display
Kristin Armstrong Oma and Elna Siv Kristoffersen..........317

The bear minimum. Reconsidering ursine remains and depictions at Pitted Ware culture (*c.* 3200–2300 BC) sites in Sweden
Tobias Lindström..........335

The Kainsbakke bears and changing patterns in the human-bear relationship through the Danish Mesolithic and Neolithic
Lutz Klassen and Kristian Murphy Gregersen..........351

Bears and the Viking Age transition in Sweden
John Ljungkvist and Karl-Johan Lindholm 387

Book 2

The occurrence of *Ursus arctos* in relation to other faunal remains in burials during the
Late Iron Age (560/70–1050 CE) in Uppland, Sweden
Hannah Strehlau 417

Bear bones from the Viking Age cult place at Frösö church – the unifying factor in
bear-human relationships in Viking Age Jämtland, northern Sweden
Ola Magnell 429

Bear claws in Iron Age burials on Gotland, Sweden – a first survey
Jane Jordahl, John Ljungkvist and Sabine Sten 453

Claws in Late Iron Age graves (*c.* 550–1100 CE) and bones in a castle (post 1500) –
Ursus arctos in the Åland archipelago
Rudolf Gustavsson and John Ljungkvist 469

The power of the paw. Multi-species perspectives on the bear claw burial tradition in
a long-time perspective in South Norway
Anja Mansrud 489

Bear skin burials revisited: Norway and Sweden, mainly Migration Period
Oliver Grimm 533

Sámi bear graves – results from archaeological and zooarchaeological excavations and
analyses in the Swedish part of Sápmi
Elisabeth Iregren 547

Sámi bear graves in Norway – hidden sites and rituals
Ingrid Sommerseth 587

Bear bones at Saami offering sites
Marte Spangen, Anna-Kaisa Salmi, Tiina Äikäs and Markus Fjellström 603

Bear skin trade in the late 1st/early 2nd millennium AD – what do we know from Russian sources?
Andrei V. Zinoviev 619

The bear cult in medieval Novgorod, based on archaeological finds
Elena A. Tianina 629

Evidence of bear remains in a cremation burial in the Moscow region (Burial 5, Kremenye burial
ground on the upper river Oka, 12th century)
Alexander S. Syrovatko, Natalia Svirkina and Liudmila Plekhanova 639

Chapter 7 – Bears in the history of religion (northern Europe) 653

Bears in Old Norse religion with specific references to the *berserkir*
Olof Sundqvist 655

"The Bear Ceremonial" and bear rituals among the Khanty and the Sami
Håkan Rydving 677

The songs and rituals of the Finno-Karelian bear hunt: Gifts, seduction and mimesis in the forest
Vesa Matteo Piludu 693

The Finno-Karelian bear feast and wedding: The bruin as a guest of honour of the village
Vesa Matteo Piludu 723

The Finno-Karelian bear skull rituals: Bringing the bruin home to ensure its regeneration
Vesa Matteo Piludu 745

The human-bear relationship among swidden cultivators and forest peasants in Savonia, Finland, and central Scandinavia
Marja-Liisa Keinänen 761

Karhurokka – traditional bear meat soup and other bear meat recipes from Finland
Tuija Kirkinen 777

Bear skins as a church offering
Teppo Korhonen 781

Bears in churches: Skins, paws, and claws from Norway
Jahn Børe Jahnsen 795

Book 3

Chapter 8 – Bears in literary studies and the history of ideas (northern Europe) 825

Bears, kennings and skaldic poetry
Maria Cristina Lombardi 827

The role of bears in Old Norse literature – a bestiary concept?
Agneta Ney 839

The bear in popular belief, legend and fairy tale
Klaus Böldl 851

Killer bears and bear killers in 19th-century Sweden
Karin Dirke 863

From monster to endangered animal: Three bear stories by Selma Lagerlöf
Claudia Lindén 875

Bears as *pares*: Some notes on bear stories in Zapinejie (Arkhangelskaya oblast, northern part of the Russian Federation) and the tendency to equality in human-bear relations
Andrey V. Tutorski 887

Chapter 9 – Bears in philology (northern, central and eastern Europe) . . 901

Bjørnestad, *Bjørnbåsen*, and *Godfardalen*: Bear/human relations as referred to in place names from southwestern Norway
Inge Særheim 903

Germanic "bear" and Germanic personal names before *c*. AD 1000 with elements referring to "bear"
Robert Nedoma 921

The Slavic word for "bear"
Jürgen Udolph 933

Chapter 10 – Bears in image science (northern Europe) 941

Stone Age amber bear figurines from the Baltic Sea area
Daniel Groß and Peter Vang Petersen 943

The bear in Late Iron Age and Viking Period Scandinavian art – a survey
Sigmund Oehrl 961

Bears in Swedish imagery, AD 1000–2000
Åsa Ahrland and Gert Magnusson 991

Chapter 11 – Bears in Classical Antiquity 1017

Bear und human in Greco-Roman antiquity
Florian Hurka 1019

Bears in Early and Middle Byzantine art (330–1204)
Martina Horn 1031

Chapter 12 – Further reading: Bears in a broader perspective 1063

The role of bears in the Late Bronze and Early Iron Ages in southern Germany, with a focus on the Hallstatt period
Melanie Augstein 1065

Tracking former royal dignity: The bear in medieval German literature
Sabine Obermaier 1091

"The Bear's Son Tale": Traces of an ursine genealogy and bear ceremonialism in a pan-European oral tradition
Roslyn M. Frank 1107

The bear in European folktales – with a special focus on Scandinavian variants
Angelika Hirsch 1121

The role of the bear in the Russian folk tale: Personage, plot type, and behavioural scenarios
Inna Veselova 1133

Bears bring spring: An anthropological view on the role of the bear in middle European winter feasts
Jet Bakels and Anne Marie Boer 1147

What are those bears doing there? On a painting from early Italian art
Henk van Os 1159

Bear-human interactions: Archaeological and ethnographic investigations in North American indigenous cultures
Kerry Hull 1163

Bears in the starry sky
Ernst Künzl 1185

Foreword by Oliver Grimm

The present book is the outcome of discussions that took place during the formation period of the Centre for Baltic and Scandinavian Archaeology (ZBSA, Schleswig, Germany) between late 2008 and 2010. Together with my dear colleague, Alexandra Pesch, who is co-editor of this book, the idea was coined to set in motion a workshop cycle, whose aim is the interdisciplinary analysis of high-profile 1st-millennium-AD topics that are in need of modern study and synthesis. Up to the present day, eight workshops have resulted in extended proceedings (the present book included), and more are in preparation.

The meetings themselves are characterised by lectures of restricted length (20 minutes), followed by questions and answers (10 minutes). Further, the speakers are expected to send abstracts before the actual event in order to allow orientation against the background of talks from different scientific fields. Due to the structure of the meeting, with its emphasis on discussions, we prefer to use the term "workshop".

In the meantime, two subcycles have seen the light of day. The first one addresses image analysis and is in the hands of Alexandra, in cooperation with others, whereas the second is about the archaeology and history of hunting, but in fact it moves more and more towards Human-Animal Studies in a broader understanding and for periods of time that are substantially longer than the 1st millennium AD.

The idea to look upon hunting (Human-Animal Studies) also goes back to the formation period of the ZBSA and reflects communication between another dear ZBSA colleague of mine, the archaeozoologist Ulrich Schmölcke, and me, an archaeologist with a focus on the 1st millennium AD in northern Europe.

The pilot workshop (2011) and its extended proceedings (2013) on "Hunting in northern Europe until 1500 AD" were handled by Ulrich and me, whereas the following workshops and large-scale publications – two of them, globally, on raptor and human (with a focus on falconry, books printed in 2018 and 2020), and now a third one on bear and human for northern Europe – were mostly in my hands. I was very glad, however, to have considerable back-up from others, namely, the learned falconer Karl-Heinz Gersmann for the raptor volumes and, in the present book, four scholars: Daniel Groß (once ZBSA, now Museum Lolland-Falster in Denmark) for Stone Age archaeology, Alexandra Pesch (ZBSA) for 1st-millennium-AD archaeology, Olof Sundqvist (Stockholm University, Sweden) for the history of religion and, not least, Andreas Zedrosser (University of South-Eastern Norway) for biology. Thus, we see here five parties from four different countries and with three different academic backgrounds who were responsible for the book.

The consideration of hunting (Human-Animal Studies) is meant to fill a gap in research: While the period of hunters and gatherers also finds much attention in the ZBSA, and rightly so, hunting in the era of sedentary life-style, from the Neolithic onwards, seems to get much less coverage, which is a mistake. For example, in Norway and Sweden we see a burial rite in the middle of the 1st millennium AD that included the deposition of bear remains, mostly claws but sometimes skin or teeth, in hundreds of graves. There was no such burial rite earlier or afterwards, or at least not to that extent, with the remains of bear or other wild animals.

This observation has been the starting point for launching research into the subject of "bear and human", which is also a topic that cannot be grasped without the inclusion of scholars from different fields in the Natural Sciences and the Humanities. The present book will address another gap, too; it is our feeling that Human-Animal Studies in their endeavour to describe human and animal relationships in alternative ways, detached from a purely utilitarian view, still yield unused potential for archaeology. Here, this approach is projected back in time, and it will also look at hunting, which may not be that common in Human-Animal Studies. Furthermore, the issue of Eurocentrism is faced, and the answer is one contribution that sheds light on the so-called Mound Builders of ancient America (1000 BC to AD 1600), in the present-day eastern United States of America, and their relations with bears.

I am very indebted to my co-editors for all their help: Alexandra Pesch for image analysis and Iron Age archaeology more broadly, and Andreas Zedrosser for coordinating all the worthy contributions on bear biology and for instructive talks. The conference was planned with Olof Sundqvist from Stockholm University, who also wrote successful letters of application to raise a conference budget. The workshop and book bear his mark, too. In turn, it was Daniel Groß, a former ZBSA colleague of mine, who drew attention to Orsa Predator Park and its impressive conference facilities in inner Sweden.

It turned out to be a brilliant choice and, at the same time, a most harmonic conference ("conference yoga") – an event to be remembered. We talked about bears at the conference location while bears actually moved around it. We were lucky twice. The conference in autumn 2019 took place in the late pre-virus period, and Orsa Predator Park has since been closed down.

The authors have been patient with us in the long period of book production. Thank you for this and for making such colourful manuscripts available. The present book is the result of an international and interdisciplinary cooperation, as can be seen in the book structure itself with authors from nine countries and different academic fields: archaeo(zoo)logy, art history (image science), biology, history of ideas, history of religion, literary studies, philology.

The present publication has gone through a peer-review process. The delivered manuscripts were read by at least one of the book (co-)editors and by one internal (from among the book authors) or external expert. Authors were asked to deliver a final manuscript that takes into account suggested amendments. The ongoing contact with authors proved to be very insightful; this includes John Ljungkvist and Elisabeth Iregren, to mention only two of them.

Translators (David Barick, Irina Bittner, Larissa Birrer, Wilson Huntley, Julian Jain, Sabine Lutkat, Dirk Steinforth), a language proof-reader (Sharon Shellock) and illustrators (Cornelia Golze, Anna Carina Lange, and Lars Foged Thomsen) made significant contributions to the book. This is very much appreciated.

Gratitude reaches its maximum when it comes to Gundula Lidke, the copy editor, and also the graphic designers Matthias Bolte and Cornelia Lux-Kannenberg (the former attached to the ZBSA, the latter two to the State Museum Foundation Schleswig-Holstein, Schleswig). Their work turned texts into articles and, finally, resulted in this fine book. Perfect working conditions were in place after years of collaboration on other publications, and the work proceeded in a way that was concentrated, but which also included many smiles: "book yoga".

Book production often has a sad side, too. While preparing the present one, the very gifted graphic designer and very nice fellow, Lars Foged Thomsen (Denmark), passed away suddenly. He was deeply involved in the creation of both falconry books from the ZBSA (2018, 2020), which were truly enriched and adorned by his work. This, however, was not only about work, it also included many good laughs and talks. Cooperation with Lars goes as far back as the late 1990s. Lars, you will be missed.

The book you hold in your hand has been made possible only by the benevolence of the leading persons at the ZBSA (then director Claus von Carnap-Bornheim, head of research Berit Eriksen, and head of administration Doris Rohwäder), and there was also help with the practical, economic side from Babett Winkelmann and Jutta Carstensen.

Finally, we are very glad about the contact with publishing manager Rosie Bonté from Brepols Publishers, and this includes Paul Johnson and Sam Turner, the editors of the book series "The Archaeology of Northern Europe" under the umbrella of the mentioned publishing house. Thank you very much for accepting the present book for print; this is a much-welcome opportunity to address an international audience.

Prologue: A little story as an attunement. As legend has it, there was once a grim landowner on northern Gotland. He could not find peace in the afterlife and became a revenant. No incantation could bind him. Finally, his grave was opened and he was re-buried at some distance, with a bear skin placed on top of him that was intended to make him rest. And so it did. The legend was attached to one burial mound in a local cemetery with altogether 150 graves. When that mound was excavated in the 1920s, a weapon burial was found, including a sword, that dates back to the 1st or 2nd century AD. However, bear claws also came to light, one group close to the head and another at the feet of the interred man, which may actually suggest that a bear skin was once placed in the burial (!) (see H. Hansson, *Fornvännen* 1923, 225–229; grave nr. 18, Backhagen cemetery, SHM 16492; personal communication J. Ljungkvist).

Enjoy the book!

Schleswig, March 2023

Oliver Grimm (with best regards from the co-editors Daniel Groß, Alexandra Pesch, Olof Sundqvist and Andreas Zedrosser)

Chapter 1

Bear and human: Facets of a multi-layered relationship – introduction, discussion and synthesis

*It was believed, from antiquity onwards, that bears were born as shapeless lumps which had to be licked by their mother into the proper shape. In a Christian understanding, this licking was a symbol of how Christ converted the gentiles (see V*AN *O*S *and other texts, this volume; image after the Bestiary of the University of Aberdeen, United Kingdom. End of the 12th century. Manuscript 24, folio 15r. Creative Commons Attribution 4.0 International Licence [CC BY 4.0]).*

"Bear and human" – introduction, discussion and synthesis

By Oliver Grimm

Introduction

The Centre for Baltic and Scandinavian Archaeology (ZBSA, *Zentrum für Baltische und Skandinavische Archäologie*) in Schleswig, northern Germany, has a history of instigating research into the archaeology and history of hunting or – more broadly understood – Human-Animal Studies (more on this below). This research focus goes back to the formation period of the ZBSA in the years 2008 to 2010. Right from the beginning, there was an awareness that a modern synthesis on premodern hunting, either at a country level or beyond, is missing for the areas of concern covered by the Centre. Here, the voluminous treatments of the topic by the well-known German hunting historian, K. LINDNER (1937; 1940), stand alone and there has been no effort to update his research. This does not come as any surprise, however, since the relevant source materials are so extensive and originate from such a diversity of academic fields (especially in the case of more recent times) that they are beyond the capacity of one scholar. In contrast, the only way to approach the topic is by means of case studies, upon which overall syntheses can be increasingly elaborated.

Raptor and human, bear and human, beast and human – research at the Centre for Baltic and Scandinavian Archaeology

Apart from articles and monographs from scholars of the Centre for Baltic and Scandinavian Archaeology (ZBSA) that focus on chosen aspects of hunting and gathering for parts of the Stone Age, conferences with extended proceedings are the main vehicle for gathering knowledge for the period from the Stone Age to medieval or even younger times. In the present context, the intention is to include scholars from all relevant fields in the Natural Sciences and Humanities. This approach was implemented in a pilot conference in 2011, followed by conference proceedings on "Hunting in northern Europe until 1500 AD" (GRIMM/SCHMÖLCKE 2013). This particular book covered many different aspects of the topic and formed the basis for further research.

From then onwards, work has focused upon particular animal-and-human relationships. It started with conferences and extended conference proceedings on "raptor and human" (GERSMANN/GRIMM 2018) and "raptor on the fist" – the latter of which also contributes to image analysis (GRIMM 2020). The given topic, with a focus on the archaeology and history of falconry, was considered from a global perspective. In the present case, the multi-layered bear-and-human relationhip, analysis is restricted to northern Europe (see below). Apart from that, in a session at the annual meeting of the European Association of Archaeologists (EAA) in 2021 (Kiel), the topic of "beast and human" was studied, based on archaeology and with a look at different animals. Now being prepared for publication (GRIMM in prep.), this subject can be understood as a first but small step to introduce studies that go beyond the focus on one animal species, but this is not yet an expression of a multi-species approach (see for archaeology Overton/Hamilakis 2013, 159–163; Harris/Cipolla 2017, 152–169; Pilaar Birch 2018).

Book: Scope and limitation

The present book stands in the tradition of ZBSA workshops and extended proceedings, which follow a certain scheme that was established in the aforementioned pilot conference (2011) and book (Grimm/Schmölcke 2013; see also more broadly the foreword by O. Grimm, this volume, for the interdisciplinary workshop agenda by the ZBSA).

Archaeo(zoo)logical long-term studies will stand side by side with more focused archaeo(zoo)logical analyses of restricted periods of time. The most substantial remains of actual bears from northern Europe come from Sámi bear graves, which were reserved for the animals themselves (Iregren;[1] Sommerseth), followed by skins from burials that have been preserved (O'Regan on England; Grimm on Norway and Sweden; see also Grimm 2013; Mannermaa et al. on Finland). Bear skins have also been documented, either preserved or in written records, in front of church altars (see Jahnsen on Norway and Korhonen on Finland). Furthermore, there are bear images and bear figurines from the realms of archaeology.

For mainly younger times, all kinds of non-archaeological sources are taken into account, e.g. pictorial and written, or more detailed art-historical, historical, literary and philological ones. In addition, the history of religion also plays an important role. Often, different kinds of regions have been chosen in order to analyse whether there were common or only regional developments. Ultimately, such an approach will lead to a synthesis that is substantially broader than what archaeo(zoo)logy can achieve alone. This is particularly true for the consideration of "raptor and human" and "bear and human" themes, for which there is a wealth of material from different academic fields.

In the context of the present book, the decision was taken to focus attention on northern Europe in a broad sense, from England in the west to parts of Russia in the east but, even for this restricted area, it is a sheer impossibililty to provide a complete record of all the relevant sources. This cannot be realised, no matter how extensive the book. At the same time, surprises come into play, and new topics emerge. Hence, all that can be hoped for is to achieve a "critical mass" on the basis of which common and loose threads can be identified and future research be promoted.

Decisions had to be taken about the scope of the book, and it was considered necessary to include all different populations from northern Europe, which were distinguished by their languages: north Germanic and Finno-Ugric (Sámi and Finnish/Karelian). In turn, one very important find material has been deliberately omitted due to its extensiveness and geographic location – caves in France with Late Palaeolithic bones and paintings of cave and brown bears (see for example L'ours dans l'art 2018). Another basic decision was to focus on brown bears, with the exception of one Later Upper Palaeolithic polar bear found in southwestern Norwegian Rogaland (see below). Finally, in order to avoid an Eurocentric view, one of the papers in the book (Hull) rightly considers bear-human interactions in North American indigenous cultures from an archaeological and ethnographic point of view, with a focus on the so-called Mound Builders in the area of the present-day eastern United States of America – the Adena, Hopewell and Mississipian cultures (800 BC to AD 1600).

---

1 Names in small caps without year relate to articles in the present book.

MILESTONES OF RESEARCH

Bear-human interaction has left multiple traces from the Stone Age onwards and it cannot come as any surprise that this has attracted the attention of scholars for many decades. Naming only two scholars and their influential publications cannot do justice to the subject but, in the present context, it will have to suffice.

A first milestone, the Ph.D. thesis by the anthropogist and archaeologist, I. HALLOWELL (1926), is still widely used and quoted. The controversy over his main thesis, according to which there was one type of bear ceremonialism in the northern hemisphere, but with variations, also resonates in this volume.

A second milestone of research is the book published by the French historian, M. Pastoureau, with the title that says it all (in English translation) "The bear – history of a fallen king" (PASTOUREAU 2008; see also PENTIKÄINEN 2007, but with a focus on Finland). In short, the bear is supposed to have been regarded as the king of the forest and its animals, and it was deified by humans in the olden days before it became dethroned and humiliated in Christian times and was replaced, for example in heraldry, by the lion, an exotic animal. This is supposed to have happened in the period from the 11th to the 13th centuries AD, with examples given for western, central and northern Europe.

In addition, according to M. Pastoureau, the bear assumes three roles in mythology:
- transformation,
- relationship between a male bear and a human female ("the beauty and the beast"),
- mother bear takes up a human child and nourishes it.

In its outline, the present book has a focus on northern Europe and its own agenda (see above), detached from the works of I. Hallowell and M. Pastoureau. However, both are addressed repeatedly in articles, and we shall come back to this at the end of the present text.

BIOLOGY

Worldwide, there are eight different bear species, which share a lot of characteristics (BIEDER 2005). In the present book, the focus is on the the brown bear and its European populations (ZEDROSSER/ SWENSON). Those in Sweden and Finland are among the largest ones in Europe, whereas Norway only hosts a marginal population at the border with Sweden. In comparison, in Denmark (KLASSEN/ GREGERSEN), England (O'REGAN) and Germany (SCHMÖLCKE) bears have been extinct for hundreds of years, if not (much) longer.

Modern day bear managemement either deals with the recovery of small and endangered populations, or the control of bears via regulated hunting (SCHNEIDER et al.). The Scandinavian Brown Bear Research Project (SBBRP, see SWENSON/BRUNBERG) has carried out large-scale and long-term research, with the help of many doctoral and post-doctoral studies that cover a multitude of topics, such as bear genetics (KOPATZ), social behaviour (ZEDROSSER et al.) and hibernation (FRIEBE et al.) as well as mating systems and sexually induced infanticide (ZEDROSSER et al.).

We can learn from this, amongst other things, that the brown bears on the Scandinavian Peninsula can be ascribed to two larger groups based on distinctive lineages (or clades) with characteristic mitochondrial DNA haplotypes (KOPATZ; see also LINDHOLM/LJUNGKVIST). These genetic clades reflect the Scandinavian bears' phylogeography, which is shaped by the faunal history and colonisation of the area from the south and the northeast after the last glaciation. A contact zone between the clades has been identified in northern Trøndelag (Norway) and northern Jämtland (Sweden). The question remains if this contact zone is a result of long-term developments or rather a product of recent hunting pressure and habitation loss due to human expansion.

Funeral hymns have been sung for the bear and its inevitable extinction, in Europe but also globally (Bieder 2005; Pastoureau 2008). The present book provides a status quo for brown bears in Europe (Zedrosser/Swenson), which today are more abundant than they were in the 1960s and 1970s. This is part of a more general trend towards large carnivore recovery (personal communication, Andreas Zedrosser).

Bears: Fact and fiction

In the framework of the present book, two questions have re-occurred, which is why a check on what is fact or fiction has been made from a biological point of view (see Grimm et al.).

Firstly, bears do not attack on their hindlegs but may fight that way. Standing enables bears to sniff the air, listen to sounds from a higher vantage point, observe, reach for food or an object, or to scent mark – a behaviour that is wrongly considered as aggressive by human observers. When bears are truly aggressive and intend to attack a prey or opponent, they come in low and fast on all four legs. However, the animals will sometimes fight in a standing position. For example, a bear may slightly rise on its hindlegs in order to have its forelegs available for fighting off attacking dogs.

Secondly, brown bears have been commonly portrayed in fairy tales and historic western literature as slow, both mentally and physically, and they are commonly outwitted by the smart and quick fox in fairy tales (Böldl; Hirsch; to some extent, Veselova, on fairy tales; but see also Hirsch for positive roles played by bears). This is, to some extent, based on observations in nature which, however, are misleading. Sometimes, bears move slowly and seemingly clumsily, and other animals follow them to scavenge on food resources found by them, for example a deer carcass. However, observers should not be fooled as bears are most definitely highly intelligent animals with an incredible adaptability and ingenuity, especially when it comes to gaining access to food. A striking example: According to hunters, some bears backtrack in their own footprints to throw off pursuers (!) (Bieder 2005, 24). The idea about the smartness of the bear has likely also been triggered by its similarities with humans and its supposed ability to understand human communication, which resulted in taboo names for the animal (see below). The negative views about bears go back to Christianity, in which animals were perceived as mere objects and even, as in the case of the bear, as Satanic creatures (Pastoureau 2008, 145). However, Plinius (*Naturalis historia*, book VIII) had a similarly negative opinion in antiquity.

Furthermore, a wonderful yarn has been spun with regards to bear cubs which, from antiquity onwards (Plinius, as before, amongst others), were wrongly considered as mere lumps that had to be licked by their mother into proper shape (Hurka; cf. Pastoureau 2008, 92–94). This found resonance even in early Italian art (1250–1400), in the *Dittico di Santa Chiara*, exhibited in the Pinacoteca Nazionale in Siena (Italy). In this context, the licking of the mother symbolised how Christ converted the gentiles (Van Os).

Brown bears / polar bears

As already stated, this book is about brown bears in northern Europe. However, one article describes the remains of a polar bear that came to light in southwestern Norway (Armstrong Oma/Kristoffersen). Found beneath a laundry on an island (Finnøy) to the north of Stavanger, its bones date back around 12,000 years and represent the oldest such intact specimen. The new display of the animal in the Archaeological Museum of Stavanger is discussed on different levels, such as the polar bear's status as a charismatic animal, today threatened by climate change, but also as a toy: Finn from Finnøy.

As a matter of fact, polar bears are represented in this book more often. As we can learn from written records, their furs were placed in front of Norwegian church altars in order to warm the priest's feet (Jahnsen). Another find from a church context is a figurine – a polar bear – that came to light under an older stone pavement near the altar during excavations in the 12th-century Norderö church in Jämtland, northern Sweden (Noderman 2009, 142–144).

Polar bears also play a role in the Iceland-based fairy tale "The man from Grimsö and the bear" (Hirsch). The medieval German tale *Schrätel und Wasserbär* (late 13th century) has a polar bear as an actor, too, in a period in which such bears had replaced brown ones as royal gifts; their white furs made polar bears light, pure and "tame", following Christian colour symbolism (Obermaier; see also Ahrland/Magnusson on a polar bear, a gift from Tsar Peter I of Russia, which was kept at the royal court of Stockholm in the late 17th century).

Similarities: Bear and human

Bears were not just "another kind of beast". On the contrary, they share a lot of traits with humans – the same stature, same silhouette, walking on the soles of their feet (rare among animals), standing upright, the hand-like use of the forepaws, and so on, which turn the animal almost into a "disguised human" (personal communication, A. Zedrosser; see also Pastoureau 2008, 83–84). It is also easy for a human to "become" a bear by wearing a fur, pulling the shoulders forward and walking wide-legged. Because of these similitarities, as well as their brute strength, bears were both admired and feared by humans, and, either living or dead, treated with great respect, if not as equals.

In this context, the villagers, farmers and hunters of Zapinejie, Arkhangelskaya oblast, in the northern part of European Russia deserve a mention (Tutorski). As a matter of fact, this village of 500 persons is many kilometres away from the next settlement. The inhabitants are used to encounters with bears as they spend a lot of time in nature and the forest, gathering berries, hunting and fishing. In their tales, bears occur as forest owners. They very much do what humans do and are thus equal, walking through the forest, gathering berries, and so on. The humans would say with a low voice: "I don't bother you, you don't bother me". However, if bears act intrusively – entering the village, entering storage houses by force to steal tinned food – this would result in their being despatched, but regretfully so by the hunter as this was considered enforced. However, if bears come face-to-face with humans close up and by surprise this may actually lead to an attack (personal communication, Andreas Zedrosser).

Hunting

We have a fairly good source at hand for bear hunting in northern Europe – the writings of the Catholic cleric and bishop of Uppsala, Olaus Magnus (1490–1557), who had to go into exile in 1524, following the Reformation. He is also well-known as a cartographer and geographer who left the most informative resource on pre-modern concepts of the bear in northern Europe (Oehrl 2013; see also Almond). In this respect, ancient and medieval learned literature, but also popular beliefs, were integrated (Böldl).
Three different kinds of bear hunting are distinguished by Olaus Magnus and illustrated by woodcuts in his work: pit trapping using honey as a lure, sneaking up and using a crossbow while the bear eats fruit from a tree, and, finally, the surrounding of the bear by hounds and huntsmen, with the beast stabbed by using a hunting spear with "stoppers". This stabbing was very dangerous and it made the hunting and the despatching of the bear a "heroic act" (on the killing of animals see more generally The

Animal Study Group 2006; Bronner 2008; Ulrich/Ulrich 2014). Until rather recent times, bears and other wild animals were driven into lakes in Sweden during battues (see Ahrland/Magnusson).

As to Classical Antiquity, triumph over the strong bear symbolised the human virtue of fortitude but, even more so, the heroic power of the emperor (Horn). One of these emperors, Commodus (180–192), is said to have killed 100 bears on one day in the arena (!), but bears were also kept as pets in antiquity (Hurka; see also Ahrland/Magnusson on royal and heroic bear hunting by Swedish kings). The hunting of bears occurred as a royal or aristocratic pleasure on the one hand and as pest control on the other, which decimated populations, to say the least. In Oppian's hunting book, *Cynegetica,* which dates from the late 2nd century AD, trapping by flushing the bear out of its den and capturing it in nets, or by using traps, is recorded (Almond).

There are more sources from Antiquity. The outstanding, excellent bronze she-bear in Aachen cathedral was originally part of a Greek sculpture group that showed a hunting party of the 3rd century BC (Künzl). Bear hunting imagery is also known from a Late (East) Hallstatt burial (Kröllkogel, Kleinklein, Austria) that dates to the middle of the 1st millennium BC (Augstein). Furthermore, the early Byzantine exquisite brass jug from a burial in Budakalász, Hungary, with its hunting imagery, also deserves attention (Horn).

If one looks back a long way, there is yet another pictorial source for the technicalities of bear-hunting: rock art, known not least from the UNESCO world heritage site in Alta, northern Norway, with a dating to the Mesolithic/Younger Stone Age, *c.* 5000 to 1700 BC (Helskog 2012). The bear imagery from Alta amounts to not less than 105 items, unique for northern Fennoscandia, and to some extent it belongs to narration cycles. As the rock art shows, amongst other things, this early hunting took place at the bear's den, with the animal being awakened and killed when leaving the den by a hunter with a spear. This kind of hunting is known from the Sámi in more recent times, with entire hunting parties being involved, and oversize spears, longer than the hunters and without "stoppers", but also axes were used (see Iregren).

Art history yields an abundance of images that relate to the bear hunt, too (Almond). To some extent, the depictions originate from the iconic book, *Livre de chasse,* written by Count Gaston Fébus (1331–1391) of Foix (southern France). It is important to keep in mind that the given imagery might be misleading and idealised because, in the field, the venator will have to adjust to the given situation, which cannot be planned beforehand. One exquisite and beautiful find material from the realms of art history are hunting horns, but it will have to remain open as to if and how often depictions of bears/bear hunting occur on these (on olifants see Shalem 2014).

Place names will not be forgotten. As is demonstrated for southwestern Norway, bears are the wild animal species most frequently mentioned in place names. These names are often difficult to date but, in fact, a considerable number of such denotations goes as far back as the 1st millennium AD. Not only do such names allude to places where bears were present but also to trapping, either by using natural rock formations, such as clefts, or by wooden or stone constructions (Særheim). Apart from the bear name as such, circumlocutions also occur, such as *Godfardalen,* the valley of the grandfather/good father (= bear; see below on taboo names).

In modern times, beaters and dogs drive the bear to locations where hunters are waiting. These then shoot at the first sight of the bear. This can lead to danger if the bear is only some meters away and launches a fake attack, in which case the hunter is well-advised not to shoot but to run (personal communication, Andreas Zedrosser; see also Dirke on 19th-century bear hunting in Sweden). Demographic effects on bears from hunting are well documented, whereas behavioural ones are not (Ordiz et al. 2012). In fact, hunting influences the movement patterns of brown bears since they reduce their activity in the daytime during the period of hunting, which is against their nature. So, here is confirmation: Bears know that they are hunted (see also Dirke).

The well-known anthropologist G. Marvin (2006) has considered hunting to be a cultural and social pursuit, not – as hunters sometimes argue – as predation on other animals by humans. It is different from nature since planning and premeditation is needed, whereas hunting among animals only comes into play when prey becomes visible. Hunting involves the killing of animals but, notably, not all animal killing is hunting. Two types of despatching are distinguished by G. Marvin; a domestic type by professionals in the context of the meat industry ("cold killing") or pest extermination ("hot killing"). In the latter case, this presupposes some knowledge of the landscape and animal. In contrast, wild killing is that of non-professional, knowledgeable sportsmen engaged in a "quarrel" between the hunter and the hunted. As has been noted (Dirke, see below), in 19th-century Sweden, sport and extermination targeted against bears were one and the same, and thus not in line with the aforementioned categorisation. Apart from that, we simply do not know about the perception of the animal itself which, as stated already, knew it was being hunted.

Finally, from the viewpoint of a hunter, bears were a rich resource (personal commmunication, Andreas Zedrosser; see also Pastoureau 2008, 34). An average adult brown bear is 2.30 m in height and has a weight of *c.* 220 kg in spring after hibernation but weighs 20–30 % more in autumn, due to the accumulation of body fat. In historic times, bears provided meat for nutrition, fat for light-making, bones as raw material for tools, weapons and fuel, and, furthermore, skins were used for garments or as rugs/covers. A bear is a large animal that yields a considerable amount of edible meat. However, the taste of the meat is dependent on the bear's diet. For example, the consumption of large quantities of berries may result in a sweetish and unpleasant taste of the meat (personal communication, A. Zedrosser; see Kirkinen on bear meat soup from Finland in the context of bear ceremonialism and bear recipes).

Human-Animal Studies and animal agency

Be it a consequence of the domestication of animals from Neolithisation onwards, and the persecution of others in order to protect land and flocks, be it Genesis 1:28: "Be fruitful and increase in number; fill the earth and subdue it. Rule over the fish in the sea and the birds in the sky and over every living creature that moves on the ground", or be it the Industrial Age with its capitalism and factory farming, a change in attitude towards animals has not only been called for but also set in motion, with the influence of the animal liberation movement of the 1970s on science (see below). The influential English thinker John Berger asked "why look at animals?" (Berger 1980) with a description of how they were devalued in history and have disappeared more and more in most recent times, while pets have come to play an increasing role and zoos have only led to unnatural behaviour in animals.

In what is labelled the "animal turn" (Ritvo 2007, see also Andersson Cederholm et al. 2014), non-human and human animals, as they are sometimes called, are now described with the intention of granting the animals a proper place as "agents" of their own and to dispense with the anthropocentric and utilitarian view on animals (see e.g. DeMello 2012; Roscher 2012; and, as handbooks, Marvin/McHugh 2014; Ferrari/Petrus 2015; Kompatscher et al. 2017; Roscher et al. 2021). Ultimately, a post-humanist approach has been chosen (e.g. Haraway 2003; 2008; although this label is rejected by Haraway [Harris/Cipolla 2017, 162]; see here Fredengren 2021 and Jennbert 2021 from an archaeological angle). In this context, the bonmot by the French anthropologist Claude Lèvi Stress, according to which animals are not only good to eat but also good to think comes to mind, too (Bakels/Boer; cf. Lévi-Strauss 1964, 89).

It goes without saying that Human-Animal Studies are at their best with the use of written and pictorial sources – the more recent the better because of their growing number. This will allow substantially more detail than archaeo(zoo)logical research, which may be considered a newcomer in this field. Such a narration misses, however, that in archaeo(zoo)logy not only have animal bones been analysed in surprising detail for decades now, but they have also been placed in a wider context which, depending on the field of research, means more than a purely utilitarian consideration at the expense of animals (see here Harris/Cipolla 2017, 154–156 on older research to that effect and the highly influential paper by Overton/Hamilakis 2013, which is already under the influence of the "animal turn").

A very good case in the context of Human-Animal Studies which goes beyond utilitarian thinking can be made for the archaeology and history of falconry as an expression of the delicate relationship between a raptor that, strictly regarded, is not interested in becoming domesticated and only responds to positive reinforcement while a human has to try to find acceptance by the bird (MacDonald 2014). This means the day-to-day care of the bird and not only hunting, about which the falconer says, quite remarkably, that the bird goes hunting with the human and not the other way around (personal communication, Karl-Heinz Gersmann). Following a term by the biologist and ethologist Konrad Lorenz (Lorenz 1935) who, however, referred to the biologist Jakob von Uexküll, bird and human can be described as companions in the context of falconry (Bednarek 2018; see also Schroer 2018). In the framework of the present book, bears are considered in multiple layers that go far beyond the notion of the bear as prey in hunting.

Although bears – in order to return to our topic – have been extinct in Britain for a long time, they were a common sight in early modern English cities in connection with bear baiting (in the following: Lewis). Sources include travellers' accounts, the diaries of bearwards, and archaeological remains. In the arena, bears, which had often been blinded beforehand, were either attacked by dogs or whipped, while the spectators smoked tobacco, drank wine and beer and ate fruits and nuts. While it would be in the nature of bears to quit the scene, they were tied to stakes, and faced death, but they fought bravely or, rather, desperately. However, there was an ambivalence about the bears. In the early 17th century, the author, Thomas Dekker, showed compassion with the animals baited in arenas, and he humanised the bears to a certain degree, while their tormentors, those with the whips, were considered to be half human and half animal. Notably, "poor [human] wretches" were also whipped in public.

The Bible and Humanism have placed humans at the top of living beings and thus devalued animals (still Lewis). In Humanism, there also came to be fundamental divisions: human vs. animal, science vs. religion, nature vs. culture (see here also Horn on dichotomies in Classical Antiquity). Posthumanism calls for a new approach towards animals, and here the eye contact between animal and human becomes essential. It implies the chance to establish an alternative form of contact, to come together as companions and to learn more about the other. In this respect, the encounter between the naked French philosopher J. Derrida and his cat that stares at him in the bathroom is the key scene (Derrida 2002, 372; Haraway 2003; 2008, 20). Since the aforementioned bears had been blinded, however, they could not return eye contact and this excluded them from connecting with humans (see above).

A close relationship between human and bear, unique in its time (with a general preference of lions over bears as carriers of symbolism/iconography of power), is well documented for Duke Jean of Berry in France (1340–1416), also called the "bear prince" (in the following after Pastoureau 2008, 234–246; see also Horn for imperial iconography in Classical Antiquity, and likewise Ahrland/Magnusson for royal/aristocratic imagery from Sweden in the period post-1000). To cut a long story short, not only did his seals repeatedly depict bears, but the animal also served as his identification and ownership mark and his attribute, in the form of architectural sculptures and in his books, documents and small pieces of art, too. On top of this, the duke had several bears in his two menageries, in which he also kept other (including exotic) animals. Remarkably, his bears bore the names Chapelain, Martin and Valentin, and their actual name-giving may resonate with bear folklore. Furthermore,

Jean of Berry had a master of bears who cared for the animals over many years, and this turned out to be quite expensive. His favourite bear was even with him on his voyages and, when the duke was buried, its image was carved on his white marble effigy. Thus, the duke's commitment to bears is well documented, but one can also see animal agency at work, i.e. bears with their own characters, which were recognised by a human.

A view on killer bears and bear killers in 19th-century Sweden can be gained from an analysis of hunting books written by Herman Falk (a professional hunter and royal forester) and Llewellyn Lloyd (an Englishman and private hunter). Notably, in the 19th century, bear hunting was no longer an aristocratic privilege but was open to commoners (in the following: DIRKE). Hunting books had a logic of their own, with a preference for good entertainment provided by the description of violent, adventurous encounters with aggressive, attacking, fierce and very large bears. The description of battue hunting, sometimes with thousands of participants (!) is reminiscent of warfare. At the same time, however, there is a certain ambivalence (see above on T. Dekker). Falk, who knew bears well, stated that they are majestic animals which evoke both fear and awe. According to Falk, despite the danger they pose, these animals should not be hunted to extinction. Bears, in their encounters with hunters, feel anxiety and not aggressiveness. They experience a landscape of fear and know, as stated above, that they are being hunted. If one gets to know the bear, and this was the case in a day-long hunt, Falk concluded, one finds it worthy of protection. This perception of the bear grew from a shared experience of hunters and bears and, eventually, it led to the relative protection of the animal in Sweden. It also needs to be considered how the worldview of the bears themselves was changed by the 19th-century extermination policy; up to what extent can we back-project present-day behaviour of bears?

Literary texts by the famous Swedish author, Selma Lagerlöf, who was granted the Nobel Prize for Literature in 1909, represent different ways of considering the bear at the turn of the 19th to the 20th century, the *fin de siècle*, when it had almost been wiped out following a state extermination policy; from the mythical, dangerous animal, in a long northern European tradition (chapter in "The story of Gösta Berling"), to the human's neighbour in a Christian sense (Christianity being a latecomer in northern Europe [short story: *The Truce of God*]) and, finally, the endangered species, on the verge of extinction (school book: "The Wonderful Adventures of Nils") (in the following: LINDÉN). Lagerlöf describes the tension between the individual level (human and bear as equal to one another and in a mutual relationship, with hunting being out of the question) and the societal one (human beings and their achievements are meant to dominate over the animal, which implies that it should be despatched when it is considered to be an obstacle). However, Selma Lagerlöf opens up a Christian, ethical dimension (Luke 10:27 and Mark 12:31: "you shall love thy neighbour as yourself") with the inclusion of wild animals, and she problematises the human-centered paradigm.

As we have seen, medieval historical and modern literary texts allow a wealth of insight into the bear-human relationship that goes far beyond what archaeo(zoo)logy can achieve. However, there is a need to look further back in time. In the Finno-Karelian hunting cultures, bears and *haltias* (guardian spirits) were considered agents with will, intentions and desires, and the behaviour of humans was interpreted from their own perspective (PILUDU). The hunters, on their part, had to develop a complex ritual relationship with the bear and the forest *haltias*. In the present case, the attribution of animal agency could hardly have been any more elaborate, but this cannot come as a surprise since it was the result of the animistic beliefs of humans.

Taboo Words

The strongest indication that the bear played a special role for humans are the taboo words in different languages: Germanic ("the brown one"), Slavonic ("honey eater"), and Baltic (two types: "the brown one" under Germanic influence and "animal hair, fur" or "mauler, lacerator" under Slavic influence) (Udolph; see also Nedoma on the Germanic languages, and Særheim mainly from the viewpoint of place names in southwestern Norway). In the present cases, a so-called noa-name replaced the taboo word (Særheim), "noa" being a term from the Polynesian language, which is also true for the word "taboo".

If one uses the bear's real name, so the humans feared in animistic belief, this could be overheard by the animal and cause an unwanted, dangerous encounter, or scare it away from the hunter. Another motive could have been the avoidance of the real name based on the reverence for the animal.

Remarkably, this taboo is found in only some Indo-European (Indo-Germanic) languages, whereas in others, such as Greek (*árktos*) and Latin (*ursus*), the proper inherited word was used, whose etymology, however, is uncertain, but, notably, there was no taboo (see Horn and Hurka on the role of the bear in Classical Antiquity). In other cases, such as the Sámi (Sommerseth) and the Finns/Karelians (Piludu), there were real terms for the animal but, since the same kind of fear prevailed, a wide variety of circumlocutions was used. Entire worldviews could be reconstructed on these variations in languages.

However, there is a need to go into even more detail. As regards the Germanic languages, this taboo probably goes back a very long time, to when there still was a uniform language (Proto-Germanic); in later times, with the giving of the personal name, Björn (the noa-word), which used to be and still is quite common as a personal name, there was no longer any knowledge of how this name actually came to be (personal communication, Robert Nedoma). Finally, it is also worth a mention that, in 18th-century Sweden, bears were called e.g. "gold-foot", "the big old man", "the ugly guy", "rascal" and "grandfather" (Nedoma). So, there was a "secondary taboo name-giving" with the lack of recognition that the word for bear, "the brown one", was in itself a noa-name.

Germanic name-giving before AD 1000 often included references to sovereignty, power, strength and warfare, combined with the names of powerful animals (in the following: Nedoma). Wolves dominate, followed by boar, eagles, and bears. Interestingly, personal names of this kind, with their reference to animals, do not belong to the oldest layer of the Germanic language. They cannot be found before the 4th century AD and follow a general trend in Late Antiquity name-giving, as for instance in Latin (including the term *ursus* for bear). A name like the West-Frankonian *Hilde-bernus* "fight" + "bear" may allude to animal warriors (see below on berserker) but caution is due – was this really a matter of deliberate name-giving with full awareness of the name's formation and meaning?

The bear ceremonial

The aforementioned rock art from Alta (northern Norway) shows, amongst other things, hunting at the bear den, with an armed hunter waiting for the bear to come out. Sometimes, and quite remarkably, an unarmed person is connected with these hunting scenes. Who is this person? Was it a "ritual specialist" and was there already a bear ceremonialism at this time? As has been argued, the rock art from Alta might represent a clan with its totem animals, guardian spirits and spirit helpers (Helskog 2012; see also Hull on the different roles of the bear among the indigenous population of North America).

It is only a small step from the aforementioned taboo words/circumlocutions for the bear to the ceremonialism that in northern Europe is well documented for both the Sámi and the Finns/Karelians (see below). It was embedded into a religion in which ritual specialists ("shamans") played a central role, alongside the belief in the power of gods, heroes, ghosts – and bears.

Sámi bear ceremonialism had three elements – the bear hunt, the bear feast, and the bear grave (RYDVING; see also IREGREN). The bear is often depicted on the sacred Sámi drums which, amongst other things, were used by ritual specialists in order to invoke the goodwill of the god *Leibolmai* before the actual bear hunt (SOMMERSETH). This *Leibolmai* was the leading hunting god and was considered the bear's protector and leader, who had to obtain prestigious goods as offerings, such as bows and arrows, before he would bring about good hunting fortune. On one of the Sámi drums, one can actually see the bear walking towards *Leibolmai*, both on their way to the "sacred mountain".

Finnish/Karelian bear ceremonialism also had three elements: the bear hunt, the bear feast, and the bear skull ritual (in the following: PILUDU, also briefly on ceremonialism for other large wild animals, primarily elks; see also KEINÄNEN for swidden cultivators in Savonia, Finland, and central Scandinavia and KIRKINEN on bear meat soup). The heroic bear hunter was not only a model for manhood but also the ideal groom for a village woman, whereas the mistress of the house or village welcomed the bear when the hunters entered the village with their prey. During the feast, the bear's head, which as the seat of the soul was ascribed particular power, was reserved for the hunters, whereas its teeth and claws were removed from the carcass for future use as amulets. By treating the bear respectfully, its regeneration and return was hoped for, along with good luck for future hunts. An evergreen sacred pine was used for deposition of the bones, with the skull placed in the branches and the bones buried beneath the roots (see also below on the Frösö site in northern Sweden).

Among the Finns/Karelians, the forest was considered a mythic and sacred landscape owned by non-human persons; the powerful forest *haltias* (guardian spirits), foremost of which was Tapio, the master of the forest. These spirits, but also bears, could see, listen to and understand human speech and the hunters' actions (see above). This is why bear ceremonialism had to please them. The bear, regarded as a being of partly human origin (due to the similarities between the two), was particularly sacred because it had a deep relationship with the aforementioned spirits and, quite tellingly, one of the bear's circumlocutions was "forest". The bear could even share with its master, Tapio, the honourable title of "(golden) king of the forest" (see below).

Quite remarkably, three eras of Finnish/Karelian bear ceremonialism have been described: First were the hunters and gatherers, they had "shamans" who travelled to other worlds, and who practised a bear skull ritual that related to the natural environment of the bear; second were the peasants, the bear then was the enemy of agriculture and flocks of animals, and their ritual specialists, *inter alia*, counteracted sorcerers or healed sicknesses; third was the era of the countryman in Christian times, in which saints took over the role of the earlier forest spirits (PILUDU, referring to the work of M. Sarmela; see also KORHONEN). This argument has been criticised for, amongst other things, being oversimplified but, undoubtedly, the most substantial changes in lifestyle/religion have been identified correctly, when observed over the long term (see below). Interestingly, a rather late facet of this ceremonialism no longer aimed at allowing the reincarnation of the killed animal, it was just sent back to its origin (KEINÄNEN).

What remains are the differences in Sámi/Finnish-Karelian bear ceremonialism, with regards to the third element – the bear grave vs. the bear skull ritual. Remarkably, aspects of that bear ceremonialism find support in archaeology, i.e. Sámi bear graves in Norway and Sweden (see below; cf. IREGREN; SOMMERSETH), and ethnography, i.e. still-existing trees in Finland that were once connected with the final part of the ceremony (PILUDU).

Furthermore, Sámi offering places in Fennoscandia, which included bear remains placed visibly on the ground or below rock formations, must also be taken into account (SPANGEN et al.). Research on these sites, mainly in northern Sweden, is, so far, limited. Depositions of bears, mainly cranial elements, go back to the middle of the 1st millennium AD, as can be seen at the famous Unna Saiva site in Sweden; later (post-1000) reindeer came to dominate deposition sites (lesser offerings of antlers, foodstuff, and so on, continue to the present day). "Powerful" bear crania were a means of transforming non-descript spaces into sites of particular meaning, quite possibly also in the context of underlining territorial rights.

The discussion about bear ceremonialism will be returned to at the end of this text, and it will include the question of whether there was such a thing among the speakers of the north Germanic language, too.

Berserker

It goes without saying that bears have been discussed for decades in connection with the so-called berserker, members of warrior-clans clad in bear skins, as known from Old Norse written sources (Sundqvist; see also Lombardi; Nedoma; Ney; also Höfler 1934; Samson 2020). However, it is important to keep in mind that for the first name element there are in fact two interpretations, "bear" being one of them and "bare" being the other, whereas the second element undoubtedly denoted a "shirt" (Nedoma).

Remarkably, warriors in bear skins are also described in Classical Antiquity, e.g. the Arcadians wore bear skins and also wolf skins during battle, according to the 2nd-century writer, Pausanias (Hurka). More famously, however, bearers of standards and effigies in the Roman army wore bear skins or heads (Künzl). At the end of the 4th century, Ammianus Marcellinus (*Res gestae*, liber XXXI, chapter XVI) wrote that it was a custom of the probably east Germanic tribe of the Taifals that a young man had to kill a boar or a bear in order to be admitted to the warrior community, which probably reflects some sort of initiation ritual (see below on the Torslunda finds; see also Sundqvist; Oehrl 2013, 308).

For purely practical reasons, one has to take as given that such berserker wore skins only before actual battles because they were of a considerable weight (estimated 10 kg for an adult bear individual), and there would have been the problem of fastening the skin to the body (personal communication, Andreas Zedrosser). In the present context, the "master warriors" who were placed on bear skins in their graves come to mind, but they reflect a Migration Period phenomenon (see below).

If one shifts the focus from the berserker, other animals – such as raven, eagle and wolf – were in fact more important for north Germanic religion in the Viking Age, if not earlier (Sundqvist). In the case of skaldic poetry, an elaborate form of poetry that was recitated at courts and goes back, to some extent, to the late Viking Age, bears occur in such wonderful kennings (rhetorical tropes that draw heavily on figurative language) as "bear of the sea", a circumlocution for ship (Lombardi). However, the actual animal is only represented to a decent degree in the context of hunting without a link to mythology. As a sidenote, the famous Beowulf, in fact the Bee-Wolf, from which the Old English poem takes its name, is yet another circumlocution for bear, used as a by-name (see also Nedoma). It is also worth a mention that fairy tale-type ATU 361, "Bear-skin", that is also known from northern Europe, has no or very little connection to the belief in berserker (Hirsch).

Thus, it is not legitimate to ascribe a central role to the bear in Germanic religion, and the placement of bear skins in front of church altars was not a symbol of the degradation and humiliation of the once venerated animal (Jahnsen for Norway; Korhonen for Finland; as to the latter country, however, the bear was held in high esteem and Christianity came later).

Finally, it has to be recalled that it was not only bears and their skins that played a role in martial cults, but also wolves, according to Old Norse literature (Sundqvist). As just stated, they played a more prominent role in north Germanic religion than bears, but there was no name taboo. In this context, runic stones from the southern Swedish landscape of Blekinge are worth mentioning (on runology: Krause/Jahnkuhn 1966, 203–220; broadly on philology and cultural history: Schulte

2015; on archaeology: CARSTENS/GRIMM 2015). Four runic inscriptions on stones point towards a lineage of so-called "Wülfinge" (wolflings; *-wulfaz), as the first element in the names of the relatives refer to "wolf", a very powerful animal. It is a tempting thought but it will have to remain open if a deliberate name-giving was made with full awareness of the name's composition and meaning (personal communication, Robert Nedoma).

BEAR FESTIVALS

Present-day "bear festivals" in the French Pyrenees, acknowledged as Intangible Cultural Heritage by UNESCO, and elsewhere (for instance, in Romania) seem to resonate with folklore about the fertility of the bear in its yearly cycle, with hibernation in winter and resurrection in spring – "bears bring spring" (in the following: BAKELS/BOER). An anthropological field study at one such festival on the 2nd of February 2016 in Arles-sur-Tech shows the following elements: the awakening of the bear, sex and marriage with a human female, the killing and resurrection of the animal. These festivals act as identity symbols and are probably more popular than ever.

Wild animals, such as bears, seem to be far removed from everyday life in most of present-day Europe and are met with little understanding – as can be seen by the fate of Bruno, the so-called "problem bear", who lived in the border region of Germany and Austria and was despatched by an order of the authorities in 2006, without having shown the slightest sign of being a threat to humans. Was there a kind of pact earlier on between bear and human that allowed both to lead their lives by the acceptance of certain rules, and should such a pact be rewritten (see also AHRLAND/MAGNUSSON, and below)?

It is an open question as to the extent to which these festivals echo at their core earlier, pre-Christian beliefs. Notably, in parts of Norway, bear claws or paws acted as amulets or bringers of fertility until the most recent times – but here the question of the longevity of traditions, the long run, is again relevant; is this really anchored in pre-Christian times (see e.g. MANSRUD; JAHNSEN)?

Novgorod is an extraordinary case in question; no less than 89 drilled bear fangs (canine teeth) have been found, among a larger number of altogether 239 finds, which represent the (canine) teeth of other animals. Remarkably, these "amulets" with an assumed apotropaic function do not belong to early pagan Novgorod pre-1000, but to a later phase with a peak in the period from the 12th to the 14th century (TIANINA).

MALE BEAR AND HUMAN FEMALE / MOTHER BEAR AND HUMAN CHILD

According to Saxo Grammaticus, in his *Gesta Danorum* (X, XV), and Olaus Magnus (18, 30), the royal Danish family had a male bear among the ancestors (PASTOUREAU 2008, 102; see AHRLAND/MAGNUSSON and others, this volume). Further genealogies to that effect are known from central, western and northern Europe (PASTOUREAU 2008, 102–106). In these cases, bears were not only fathers but they bestowed power, virility and fertility on their sons who were to become mighty rulers. Bear daughters are recorded, too (HIRSCH; see also NEY on the role of she-bears in Old Norse literature).

This kind of genealogy leads further to the fairy tale type ATU 301, the "Bear's Son Tale", whose main character is the offspring of a male bear and a human female (FRANK; HIRSCH; see also, to some extent, VESELOVA for Russian fairy tales and, notably, IREGREN for the Sámi). A variation is included in the collection ATU 425, aka "The Beauty and the Beast". It is a widely-spread motif in European folklore, and there might even be the opportunity to identify core elements of the narrative and later additions or changes. Here, helpers from the animal realm (lion, dog, eagle, ant) and the ability to shapeshift into these animals by "Little Bear" (descended from a male bear and a human female) may represent an older and perhaps original layer that, potentially, would echo an animistic worldview of hunters and gatherers, whereas in a younger layer the animal helpers were replaced by huge male figures with extraordinary strength, but shape-shifting no longer occurred.

As was the case before with the bear ceremonialism and the bear festival, the question has to be raised of whether the tale, in its variations, goes back to one common source from which it spread, or whether it represents independent traits developed from the perception of bears, as it was commonplace among humans in different areas. Notably, it is difficult to believe that speakers of entirely different languages, such as Germanic and Slavonic (representing Indo-European [Indo-Germanic]), Basque, and Finno-Ugric shared common themes by communication (see also HIRSCH).

Remarkably, in the case of the Basques, they seem to have had a belief, passed orally from times immemorial, that they descended from bears and this had real world manifestations in their beliefs, traditions, rituals and performance art (FRANK). In this context, ursine genealogies might also be considered to be embedded in bear cult or bear ceremonials, respectively (see also HIRSCH).

Another type of narration – a mother from the animal realm fosters a human child – is best known from the she-wolf that took care of Romulus and Remus, the founders of Rome (PASTOUREAU 2008, 44–45). However, in other cases known from Old Europe, mother bear assumes that role (HURKA) whereas, quite remarkably, the motif is not represented in fairy tales, at least not in Europe (personal communication, A. Hirsch).

BEAR IMAGERY

In Classical Antiquity, mosaic floors in Byzantine churches and chapels (6th century) in the provinces of Jordan, Libya, Palestine, Syria, Arabia, and Greece displayed representations of daily life scenes, adopted from pre-Christian floor mosaics in North African Roman villas (in the following: HORN). These depict birds and different types of animals and beasts, such as bears or lions, as well as men in combat with or chasing wild animals. The mosaics give an impression of the bears' primeval ferocity and exoticism as part of the magnificent God-created cosmos, and as a symbol of the Messianic Kingdom of Peace. Under Christian influence, acrobatic shows increasingly replaced blood sports with bears; bears now made music or performed acrobatics instead of tearing humans apart in the *damnatio ad bestias*. However, *venationes* with bears as an object of fighting and hunting in the arena may have continued for some time in the Byzantine empire.

When it comes to northern Europe, two kinds of bear imagery may be labelled iconic – on the one hand, the already-mentioned rock carvings in Alta in northern Norway (*c.* 5000–1700 BC), and on the other hand the Vendel period helmet plate dies (*c.* 550/600–750/800) from Torslunda on the Swedish island of Öland.

Interestingly, for the latter, it has been argued that the four images have an ordered thematic progression that shows the steps of an initiation sequence (see the discussion by SUNDQVIST; see also OEHRL). The first two matrices are thought to show how the warrior proves himself against two bears and a monster, the third has the warrior standing in a shield-wall whereby he indicates that the members in the war-band can now rely on him, whereas the final one depicts him dancing in an animal skin together with the god, Óðinn. It is a fascinating thought to suppose such a thematic order, but the question remains whether the images are meant to depict bears, shown with some amount of artistic freedom/alienation, or perhaps rather, to some extent, wolves (see above)?

As regards Sweden, bears have long been a part of people's reality and imagination and this is mirrored in images from different contexts, from rare church adornments to royal iconographies of power and depictions in recent story-books for children (in the following: Ahrland/Magnusson). Historically, the bear has been perceived as strong, dangerous and easily angered, and Swedish royalty and nobility have identified themselves with these particular qualities associated with the wild. In contrast, when seen in interaction with children in sources from the 19th century onwards, the bear is friendly, docile and good natured. The common thread that connects all these motifs is the readiness to identify with the animal. Remarkably, many ideas and perceptions of the brown bear have remained fairly unchanged in Sweden over time, despite the Enlightenment and modern science. Even today, some of these folk beliefs linger in society and influence the view of this animal in popular culture. Today, there is a strong bond between bears, often cubs, and the emerging generation. One question remains relevant: Who has a right to live in the landscape, human and/or bear, and if there are clashes, who has priority (see above, Lindén on Selma Lagerlöf; Bakels/Boer)?

Bear figurines

Stone Age amber bear figurines in a naturalistic style have been found along the Baltic Sea area (in the following: Gross/Vang Petersen; see also Klassen/Gregersen). Figurines were made of clay, too, and other animals were also sculptured (mainly elks). Except for one Neolithic grave in the well-known Tamula cemetery in Estonia, the bear figurines lack context and reliable dating. However, they are comparable with the portable art known from the Late Palaeolithic and the Early Neolithic. For different reasons, amber was a fascinating raw material; most remarkably, when rubbed against hair or fur, a static electric charge results in green sparks that are visible in the dark. According to use-wear traces, some of the amber figurines had been attached to strings and thus might have served as dress ornaments or, more likely, amulets.

Another type of Stone Age figurine is represented by bear-headed stone axes from Finland, as found, *inter alia*, in Paltamo in the northern part of the country, but bear figurines on their own, made of flint or amber, are also known (Mannermaa et al.). Figural axes (hammer axes) with bear heads occur in the Stone Age of the East European Plain forest zone, too (Kashina/Khramtsova). The presence of elk-headed axes as well suggests that both animals may have served as representations of totem animals, but other meanings, such as that of status symbol or ceremonial object, are likewise possible (Piludu).

Generally, figurines and also images of bears are more numerous than hitherto thought for the 1st millennium AD of northern Europe, as is shown in the results of a survey (in the following: Oehrl). The bear figurine from the western Norwegian Modvo settlement, which dates to the first centuries AD (Roman Iron Age), stands out as the only one of its kind with such an early dating, whereas the majority of figurines/images belong to the time from the Migration Period to the Early Viking Age. Accurate identification is often difficult, due to the non-naturalistic mode of representation. Among the finds, there are bear figurines placed on sockets of spear-heads and sword pommels (Migration or Vendel Period). As has been suggested, in the case of the richly furnished warrior grave 12 from the well-known Vendel cemetery, the relatively broad spearhead and the socket animals, which may have served as some sort of "stopper", could indicate a use as bear spear (see above on hunting).

As to Great Britain, so-called hogbacks, longitudinal sculptured gravestones erected under Scandinavian influence, are worth a mention. Remarkably, some of these have bear figurines at their ends. Muzzles also appear, which identify the animals as being captive. As a matter of fact, these hogbacks imitate long houses or halls and date to the end of the 1st millennium AD (Oehrl; O'Regan).

Finally, bear figurines are also known from modern Sweden (Ahrland/Magnusson). Pillars at each side of the main gate of the Swedish Museum of National History, the biggest museum in the country (inaugurated in 1916), show a she-bear with cubs and a male bear, respectively, whereas

"Playing Bears" (from 1909), sculpted by the well-known Swedish artist, Carl Milles (1875–1955) adorn the Berzelii Park, a small public park in the city center of Stockholm. In both cases, remarkably, bears are no longer depicted as ferocious creatures. There is a change in the perception of the animal. In the case of the museum, the bear figurines stand for Sweden and Swedish nature and represent a bear family while, in the park, playful young bears are represented.

Bear and human: The long, middle, and short term

It is only the consideration of the bear-human relationship over the long term (*longue durée*) that allows us to see the changes in it. This approach goes back to the influential French historian, Fernand Braudel, from the second generation of the so-called French "Annales school" (Burke 1990, 43), and his publication on the Mediterranean world in the age of the Spanish king, Philip II (1527–1598) (Braudel 1949). He also wrote a short programmatic article that explains his approach (Braudel 1958). Braudel found resonance among archaeologists, too. In a book called "Archaeology as Long-Term History", edited by I. Hodder (1987), the Norwegian archaeologist, Knut Helskog, known for his research on the aforementioned rock art in Alta in the north of the country, has published an article on the thousands of years of rock art in Arctic Norway (9000–2500 BC) in relation to much younger depictions of animals on Sámi drums (Helskog 1987). It should be noted that in a book on archaeological theory in the new millennium (the 2000s) Braudel is named only once and in passing, with the critical note that long-term considerations, as a facet of processual archaeology from the late 1960s and the 1970s, were too generalising in their attempt to identify regularities through time and missed important details that could be obtained from a close study (Harris/Cipolla 2017, 197).

In the framework of the present book, different long-term studies have been carried out, spanning the period from the Stone (Bronze) Age to medieval or even younger times. These studies are interesting in themselves but also when regarded together. Largely, this research is based on the evaluation of animal bones.

Regarding Britain, bears are rare in the overall archaeological record, but two themes become apparent; the bear in the grave (Bronze and Iron Age, early medieval times) and the bear in entertainment (bear baiting, dancing) of younger times (O'Regan). As to the latter, did bears find their way to Roman amphitheatres in England and was there an unbroken tradition in bear baiting (see Horn and Hurka on Classical Antiquity)? The animal itself, however, became extinct in Britain in the middle of the 1st millennium AD at the latest.

As to Sweden, its northern part provides evidence for a considerable amount of bear hunting from the Stone Age and onwards, with a ritual handling of bones whereas, in the south, a society of farmers that practised agriculture and animal husbandry may have led to a loss of habitation for bears, which were also hunted to protect the livestock (Magnell). Ritual handling becomes evident, *inter alia*, at the late Viking Age Frösö site, "a bear sanctuary" (?) (see below). Notably, bear parts, be it skins or claws, were a very common grave furnishing, in particular in the Migration and early Vendel Period of Uppland and Södermanland in central east Sweden (see below).

In what is today northern Germany, bears attracted little attention in the Stone Age and onwards, whereas mainly around the birth of Christ bear phalanges served as grave furnishings (Schmölcke). The animals disappeared at different times in the area under consideration, as the result of an increase in the human population and loss of habitat, but also, in younger times, as a consequence of a state policy of bear extinction.

Last but not least, Finland suffers from a general lack of bear bones (Mannermaa et al.). Only *c.* 100 bones are known (!) for seven thousand years of the Finnish Stone Age, mainly from settlements, but there is a lack of information about find circumstances and dating. However, there is a new analysis at hand for the Iron Age that reveals two different kinds of human-bear culture – one under Germanic influence in which pelts were cremated along with the deceased (Late Iron Age, 9th to 14th century), and a traditional Finno-Ugric one in which the bear was venerated as a divine forefather. In the present context, the analysis of animal hairs has proven to be of essential importance, and this method of analysis should be continued for materials from Finland and beyond (see Kirkinen 2017 for Finland and Nockert 1991, 31, 36, for the extraordinary amount of animal hairs in the Migration Period master warrior's burial in mound 2 at Högom, Sweden).

To sum up, the given areas, considered mainly over the long term, from the period of hunters and gatherers until medieval or even younger times, have in common that bear remains were used as grave furnishings only during particular periods of time (see also further below). Even though this is only referred to for England in the present book, bears in entertainment (bear baiting, dancing) were, as we may assume, yet another common thread but only for younger periods of time. In this regard, bear finds from cities are worth considering as possible testimonies for dancing bears. When it comes to the longevity of actual bear hunting, substantial archaeological evidence comes from northern Sweden and Finland, contrary to England and northern Germany where bears became extinct at least partly due to human influence, if not as a result of extermination policies.

More generally, one would expect to see a dualism in the human-bear relationship; among hunters and gatherers, bears could have been considered as ancestors or guardian spirits and may have served as clan attributes (Mannermaa et al.; see also Helskog 2012), whereas among farmers or in periods with increasing social complexity these animals were symbols of power and war or could have healing effects (see here, positively, Ney on the role of bears in Old Norse literature, which also shows bears in the hands of kings). Even more, the burials with bear skins or claws reflect a period in which bears were still considered as powerful animals from which humans could benefit, whereas in the Roman Empire or under Christianity they were degraded to killing or being killed in the arena, dancing and performing acrobatics, or being baited. The somewhat surprising Teddy-bearisation is little more than 100 years old.

As to Lithuania that shall represent the Baltic area, bear remains, usually less than 0.5 % of the total amount of animal bones at given sites, have come to light in different kinds of settlements (in the following, personal communication, Giedrė Piličiauskienė). Unfortified settlements of the Subneolithic, Neolithic and Early Bronze Ages (4000–1500 BC) have yielded mixed find materials, sometimes also from much younger times. It is only by radiocarbon-dating that bones can be allocated to particular archaeological periods. In fact, there are a few well-dated sites from the Stone Age, such as Šventoji 43 (3900–3700 cal BC). Lithuanian hillforts mostly have only broad datings, with the earliest from the Late Bronze Age (*c.* 800 cal BC) and the most recent from medieval times (13th and 14th centuries), while the bones from these sites are unstratified. A few hillforts, however, are dated to the Late Bronze Age (*c.* 800–600 BC), among them two with bear remains (Garniai 1 and Kukuliškės). The most recent excavations have led to accurate datings for comparatively young sites with bear remains, e.g. Vilnius Lower (13th to 19th centuries) and Klaipėda castle (late 13th to 18th centuries).

Two more studies have a broad chronological scope, but they remain within the boundaries of the Stone Age; let us call them considerations of the middle term, as opposed to the long term, which spans over broadly understood archaeological periods (from the Stone Age to medieval times); the short term will be introduced below.

The first study deals with the role of bears among the hunter-gatherer-fishers of the East European Plain forest zone (Kashina/Khramtsova). At the outset, reference is made to the influential paper of Knut Helskog on the role of the bear, mainly in prehistoric Scandinavian rock art. Regrettably, however, as the author states, bear effigies and bear bones are very scarce finds in Scandinavia (Helskog 2012, 217; see also above). In this context, finds from Russia yield considerable potential for conducting the same kind of study, but with a broader material base that includes portable art and petroglyphs with depictions of bears, as well as bear bones in settlement and burial contexts (10th to 3rd millennia BC).

Surprisingly, bears are almost invisible in portable art (6200–2000 BC), but other animals, particularly elks, were not (still KASHINA/KHRAMTSOVA). In the case of weapons, bear figurines are more frequent, and this relates to both stone maces and axes. Finds are largely without context but some date to the 4th/3rd millennium BC. One can well imagine that warriors considered these weapons as particularly powerful. As to petroglyphs, the ones from Alta in northern Norway (see above) stand out with a substantial representation of bears, but this also relates to sites in Russia, with the oldest one dating back to *c.* 5000 BC. In the rock art, hunting may stand side by side with bears as elements of broad narrative cycles. Knowledge about bear remains in settlement contexts is so far only accessible for the narrow time span of 3500–2700 BC (Volosovo culture at the Volga-Oka interfluve). Bear remains may reflect a practical use, for instance bones as processed tools. Incisors or fangs of bears, mostly modified, were a frequent grave furnishing of the aforementioned culture, too, which again may point to the mighty bear and the power it could bestow upon humans. In sum, the brown bear definitely played an important role in the considered area but in a dual way that encompassed both spirituality and ordinary livelihood.

The second middle-term consideration is about changes in the bear-human relationship in the Danish Mesolithic and Neolithic, *c.* 9500–2400 cal BC (KLASSEN/GREGERSEN). The point of departure is the remarkable Kainsbakke settlement site in Jutland, which dates to the early 3rd millennium BC and belonged to the so-called Pitted Ware culture, which to some extent saw a return to hunting as a subsistence strategy. In the early 1980s, brown bear bones of at least ten different individuals were excavated there (by far the largest total from any Neolithic site in southern Scandinavia) in the form of different skeletal elements, including entire crania (!). The bones also display a special treatment when compared to those of other animals. All of the bones were found in one single pit – A 47 –, which was the result of deliberate single depositions. Do we see here a case of Neolithic bear ceremonialism? The pit, however, also yielded bones from other animals, such as elk and the truly exotic Dalmatian pelican. Earlier on, the bear appears to have been a rare but hunted animal, there being no or limited evidence for bear rituals, whereas the elk might have been more dominant and an object of ceremonialism, too (see also PILUDU).

Just as worthy as archaeological studies of the long and middle term are considerations that focus on a narrower time span in either the Stone, Bronze or Iron Ages, or in medieval times. In the present book, this is particularly accentuated for the Iron Age and medieval times but there is also one contribution to that effect for the Stone Age (besides the aforementioned one from ARMSTRONG OMA/KRISTOFFERSEN, on the roughly 12,000-years-old polar bear from Finnøy in southwestern Norway).

This contribution (LINDSTRÖM) highlights the Neolithic Pitted Ware culture's ursine remains and depictions in southern Sweden (*c.* 3200–2300 BC). As is argued by the author, due to the low number of finds – only eleven, which mostly represent teeth and limb bones – any interpretation must be made with caution. Remarkably, some remains received special treatment, such as being placed in a clay pot or situated adjacent to hearths/cooking pits, whereas zoomorphic figurines cannot be ascribed to the bear, due to a lack of features that would allow a reliable identification. So, a site-by-site analysis is needed, as opposed to a broad generalisation.

As we have seen, studies that cover the long, middle and short term are insightful for the reconstruction of the past. In particular, it is the long term, as has been argued by Fernand Braudel, that shows fundamental changes in mentality, as expressed, for example, by burial customs. One has to be aware, however, that considerations of the long term in archaeology are faced with millennia whereas, in historical science, where that concept was developed, it would just be centuries – and not many of these.

The bear in the church

Among the surprising discoveries during the compilation of this book are substantial testimonies for bear skins placed in front of church altars in northern Europe, as is mentioned by Olaus Magnus in the midst of the 16th century. According to him, such skins were meant to keep the priest's feet warm in the cold church (thanks to Elisabeth Iregren for directing my attention to this).

However, written records to that effect are centuries older for Norway where bear skins have been preserved up to the present day and scrutinised (radiocarbon-dating and isotope analysis) with surprising results (Jahnsen). In addition, paws and claws were found beneath church pavements, interpreted as a kind of a deposition, placed there in the hope of recharging their assumed powers, such as for healing humans and domestic animals (see also Mansrud).

As regards Finland, we only have rather recent written testimonies (Korhonen). Again, the bear skins were meant to serve the priest in the aforementioned manner, but quite possibly the bride and the groom also stood on the skin, since this was supposed to strengthen the wedding vow. When it comes to the day of the bear hunt or the day on which the skin was donated to the church, there may have been connections with the name-days of Christian saints who were believed to protect domestic animals and to help against bears (e.g. St George and St Margaret).

The bear in the grave

As mentioned, the most remarkable burial evidence are the Sámi bear graves found in northern Norway and Sweden that were reserved for the animals themselves, with a dating that covers most of the past two millennia, at least for Norway, though most of the burials are post-1000, if not post-1500 (Iregen and Sommerseth; in the following: Iregren). A Sámi bear grave ideally consists of all the bones (unburnt) from one killed and consumed brown bear. Among the bones, the skull, the lower jaws, and the shoulder blades bear no damage from the carving, fileting, cooking, and consumption of the bear. All suitable bones were, however, split for marrow extraction (in contradiction to written sources of the 18th century according to which bones had to be left unbroken; see Rydving). In the burial, the correct anatomical order of the animal was reconstructed. Grave furnishings are very rare and include lead bullets from rather recent times. It remains an open question whether hearths found close to some graves reflect a ceremonial at the burial site. There are also potential source-critical issues, for example as to bone preservation, which is much worse in northern Sweden than in northern Norway. And there are other questions – were there also *pars pro toto* bear burials, and do the many rather recent graves reflect a period of stress caused by Christianisation?

The Sámi bear graves from Norway consist of 30 find spots with the remains of 44 bears placed in natural surroundings, that is, in caves or below boulders and screes, which have come to light by modern construction work (in the following: Sommerseth). Due to the placement, preservation conditions and disturbances of the burials, one cannot be certain how complete the bear skeletons once were – according to written sources, the bear bones had to be treated with care, following the bear ceremonial. Another open question is whether sites with at least two bears should not be seen as offering places (see Spangen et al.). Chronological allocation rests on radiocarbon-dating with three finds older than AD 1000 and the majority from the period 1000–1500. Apart from one case, there were no other objects added to the burials. Grave sites may have been considered open passages to various worlds, inhabited by gods and spirits, and hunters had to show proper reverence to the bear, which made them successful in hunting and in life more generally (Sámi *lihkku*: happiness and property).

As regards central Europe and its prehistory, one important but small group of burials which included bear claws or teeth, in worked or unworked shape, has been analysed with an emphasis on the Hallstatt period (c. 800–450 BC) of Bavaria in southern Germany (AUGSTEIN). Interpretation ranges from the bear as a resource (meat, fur) to bear remains as status indicators or trophies. Assessment, however, is difficult because of the low number of burials. The well-known Late Hallstatt centre of power in Heuneburg has yielded a voluminous animal bone material, among which nine bears are represented. Remarkably, in the nearby Hohmichele burial ground there is evidence for a bear skin (!) in a Late Hallstatt burial (the central burial in tumulus 17) but this is awaiting full publication. One more grave find deserves a mention. The famous chieftain's grave in Hochdorf, southwestern Germany, c. 2500 years old, has not yielded any bear remains but there are other references to hunting among the furnishings (HANSEN 2013).

In addition, there are c. 100 younger graves in central Europe which belong to two sub-groups, based on dating but also on burial type:
- the first one (c. 40 finds) dates to the period from c. 100 BC to AD 100 and comprises cremation burials with bear claws, which often represent men of wealth (SCHÖNFELDER 1994; DROBERJAR/PEŠKA 2002, 445–450; BEERMANN 2016);
- the second one dates to the middle of the 1st millennium AD and includes some cremations with bear claws in the area that stretches from the northern part of the continent to Frankfurt (Main), Germany, in the south. These include men or women with differences in social status. In addition, 40 contemporaneous inhumations from southwestern Germany have yielded perforated bear teeth, and sometimes claws, in burials of middle-class women or children who wore the objects on long straps at the hip or thigh (e.g. ARENDS 1978; WAMERS 2015; BEERMANN 2016).

Two 1st-millennium-AD central European burials deserve a particular mention:
- the first one is the Mušov burial, Břeslav, Moravia (Czech Republic); there a petty king and high ranking warriors were interred in a grave chamber with a length of almost 6 m (!) at the end of the 2nd century; a bear claw was also found (DROBERJAR/PEŠKA 2002, 455–459);
- the second one is a bi-ritual child grave from Frankfurt (Main) in central Germany, dated to the early 8th century; in the present context, the cremation burial with eight bear claws deserves attention. As has been argued, this child was burnt on a bear skin in northern European fashion (see further below on this). If one is to follow this suggestion, the Frankfurt burial may be seen as a reflection of an old, "classical" burial custom of the north.

When it comes to the interpretation of the bear claws and the few actual skins in the burials, discussion is ongoing. It is a fact that, among the aforementioned c. 650 burials in northern Europe, there are only a little more than a dozen, mostly inhumations, that have yielded actual skin remains. As one may assume, the deceased were once placed on these skins (GRIMM). In one case, however (mound 2 from Högom in Sweden), a skin may have served as cover of the deceased (NOCKERT 1991, 31, 36).

In this respect, the actual preservation conditions have to be taken into account, too. In the case of inhumation, bear claws and skins continue to exist only in exceptional conditions, whereas in the case of cremation, skin and also claws were largely destroyed on the funeral pyre (personal communication, Ulrich Schmölcke). In the case of cremation, two variants have to be distinguished – firstly, funeral pyres were directly covered by mounds, as in central Sweden, which would allow to identify the location of the claws in the burial (information to that effect, however, is unavailable for Sweden; personal communication, John Ljungkvist), secondly, bear remains were gathered from the burnt-down funeral pyre and (a) placed into urns and buried at some other spot, or (b) poured together with loose ashes into a flat burial, again detached from the actual pyre site.

All of the claws in the burials necessitated the killing of bears, perhaps as a heroic deed (see above), but there is no way to positively confirm that each single claw, or even a few of the claws, in burials were attached to skins that served as a rug for the deceased on the funeral pyre (affirmative, amongst others, Wamers 2015 and Beermann 2016 for all graves that have only yielded bear claws).

In contrast, alternative interpretations can be put forward. One thing is for sure – drilled claws, which apart from the use as dress ornaments may hint at amulets, are very rare. In this context, the well-known cemetery of Møllegårdsmarken (southeastern Funen) makes an interesting case with ten burials that date to the first centuries AD. Five graves have yielded single claws with drillings whereas from five more, only undrilled claws are known (Henriksen 2009, vol. 1, 216, 218, bilag 12). Furthermore, one drilled claw had an iron suspension for fastening onto clothing (?), and in three more cases beads were found, which suggests the existence of some sort of necklace.

When it comes to undrilled claws, they might have been carried attached to clothing or in organic containers. A grave in Sletten in southwestern Norway has yielded fragments of a woven belt with bronze fittings and an attached animal skin and a bear claw (Mansrud; Grimm on that burial). Furthermore, grave IV in Rösta, Ås parish in Jämtland, middle Sweden, which dates to the 10th century AD, contained two undrilled claws in a bag, placed by the waist of the buried person (Jordahl et al.).

The fittings on the claw in the Sletten grave and the placement of the two claws in the Rösta burial suggest that they may have been used as amulets, even in an undrilled state, and not only as single objects (see Ney on rune-inscribed claws in poetry that bestowed wisdom; these, however, remain to be found). Furthermore, one to five claws may reflect "the power of the paw" that is associated with their addition to the burial. Such paws were, amongst other things, used for alleviating the pain of childbirth, as is documented up until the 20th century in inner western Norway (see Böldl; Jahnsen; Mansrud on the use of different bear parts connected to healing, and Hurka on the same thing for antiquity). Finally, for burials with a minimum amount of six claws, thus beyond the number of claws for a single paw, one may in fact suggest the actual use of a bear skin in a burial context, and this would include the aforementioned child burial in Frankfurt.

Finally: Farewell to the bear as a fallen king and also to bear ceremonialism, understood as the only one of its kind?

In the period under scrutiny in this book, northern Europe, in a narrow sense (Scandinavia, including Finland), has seen three different kinds of populations on linguistic grounds – north Germanic and Finno-Ugric (Sámi and Finnish/Karelian) – each with its own human-bear relationship, and there are great differences in the evidential value of bear ceremonialism.

But before we begin, there is a need to return to the aforementioned milestone of research, the work of I. Hallowell (1926), who has purported that there was one kind of bear ceremonialism for the entire northern hemisphere, but with variations. As has been criticised, the study draws upon an uncritical gathering of materials for a very broad region and is based only on secondary sources and not on actual fieldwork that would provide insight into the actual cultures and their languages (in the following: Rydving).

Differences in bear ceremonialism have been underlined in a limitative approach (as opposed to a comparative macro-analysis) for only two cultures, namely, the Sámi and the Khanty, which represent the western and easternmost speakers of the Finno-Ugric language (each with different dialects) and for which only the southern parts were considered, owing to the source situation. Notably, there are not only differences in the ceremonials between the Sámi and Khanty but also among the groups themselves, who lived in vast areas (!). And, as a matter of fact, the bear ceremonial is still alive among the Khanty, whereas the one of the Sámi is long gone, due to the growing influence of Christianity.

For the southern Sámi, the bear ceremonials had these three elements – the hunting, the bear feast, and the bear grave (RYDVING; see above on those graves), whereas in the case of the southern Khanty it consisted of the hunt, the bear festival (with no actual feasting) and remembrance rituals (on the sixth, nineteenth and thirty-eighth day after the festival). In turn, the Finns and Karelians carried out their version of bear ceremonialism in three steps – the hunt itself, the bear feast, and the bear skull ritual (see PILUDU and also KEINÄNEN). As a matter of fact, there are still local traditions attached to certain trees (see above).

When it comes to the speakers of north Germanic, there are only weak indications for a bear ceremonialism, if there was one at all. The first clue to that effect and from times immemorial is the taboo word for bear, "the brown one", as a potential element of a ceremony (see above), whereas in the case of the Viking Age Frösö site (Jämtland, northern Sweden), bear remains may have been placed at the foot of a birch tree, whose stump came to lie under the choir of a medieval stone church (MAGNELL). Furthermore, a dark layer with fire-cracked stones and bones was found near the stump, perhaps the remains of food preparation or even a ritual meal? Bones of elk and domestic animals came to light, too, whereby the site's potential meaning as a "bear sanctuary" is devalued. Remarkably, the place name itself, "the island of the god Freyr", is testimony for pre-Christian religion and, owing to an 11th-century runic inscription, the presence of a local site of power must be considered. In any case, the question remains whether all this was done by the sedentary "proto-Swedish" population in order to express gratitude for a successful hunt. Is there even a connection with bear skins donated to the church by hunters after the successful despatchings of such animals? (JAHNSEN; KORHONEN).

If we disregard the north Germanic speakers, we are left with the ceremonialism of the Finns/Karelians and the Sámi, which differs in details, and this relates even more to the Khanty from Siberia. It thus remains an open question, as above with the bear's son tale and bear festival, whether all these facets go back to one common source from which they spread and changed over time or whether there were independent developments based on the universal notion of the bear as powerful animal (see also HIRSCH).

In the present context, it is insightful to relate to bear-human interactions in North American indigenous cultures, with a focus on the so-called Mound Builders in the present-day eastern United States of America: the Adena, Hopewell, and Mississipian cultures (800 BC to AD 1600) but also including including younger populations (HULL). The attitude towards bears is aptly captured in this saying of a tribe from Arizona: "Bears are like people, except they cannot make fire". As a matter of fact, bear claws, teeth, paws, skins as well as bear headdresses occur in various ritual contexts; the strength of the bear is transmitted to humans, for healing and shamanism, but also for war and hunting. Bear ceremonialism had different facets; before the actual hunt, bear dances were meant to guarantee success in hunting, with the medicine man and the other participants dressed in bear skins and wearing bear heads. After the hunt, the bear skull gained particular importance, which was expressed, for instance, by placing it on the branch of a tree in the forest.

We shall also return to the second milestone of research, the work by the French historian M. Pastoureau. According to him, the bear may have experienced a fall from the position of king to one of disgrace at the expense of exotic lions with the introduction of Christianity in northern Europe (see above). However, there seems to be no firm proof that the bear was a venerated animal with particular importance in religion, at least as far as the northern Germanic-speaking groups were concerned. For the Sámi and the Finns/Karelians, Christianity gained ground only much later, and the bear continued to play an important role in ceremonialism.

The change from the bear to the lion in heraldry or rather in political iconography, which was assumed by M. Pastoureau to be a part of the process of bear-dethroning, would require an in depth-analysis for northern Europe. It is a fact, for Sweden and Finland, that the bear is a frequently found heraldic motif (AHRLAND/GUSTAVSSON). However, the Swedish House of Folkung, founded in the 12th century, from which came different bishops, jarls and kings, has a lion as the heraldic animal.

As regards the role of the bear as the "(golden) king of the forest", it seems that it is only in the Finnish/Karelian material that this is directly hinted at as a "title of honour" (PILUDU; PENTIKÄINEN 2007; see also VESELOVA on Russian fairy tales). Among the north Germanic speakers and the Sámi groups, this can only be speculation, on the grounds that the bear was the biggest and strongest animal in the wild. Remarkably, in German medieval literature, dragons, lions and foxes occur more regularly than bears, and the latter seem to be absent from the forests in the literature. However, a close reading shows that different texts seem to allude to the bear's earlier role in the context of "royal dignity" or "king among the animals" (OBERMAIER). Even more, in the *Vita Galli* (8th century), saint and bear come to have a peaceful existence, "animal peace", foreshadowing the Kingdom of God (Isaiah 11:6–8; see also LINDÉN).

In the present book, owing to its specific agenda, there has been no chance for a systematic discussion of the precious works and theses of I. Hallowell and M. Pastoureau. However, a wealth of material has now been made available for conducting this kind of research, which again can only be handled in an interdisciplinary approach.

EPILOGUE: BEARS IN THE SKY

Among the preserved globes from antiquity there is one from Mainz (brass, 150–220). This object has a diameter of only 11 cm but it captures an entire worldview. Names for the constellations in Europe mainly go back to Classical Antiquity or, more precisely, ancient Greece, with its star catalogues and stars linked to mythology. "The Great Bear" and "The Little Bear" were seen in a circumpolar position; they never set (KÜNZL; see also FRANK; HORN; HURKA). There is not that much knowledge about how the Germanic people named the stars in the sky. The Great Bear and The Little Bear were unknown; instead a wain was seen in both constellations (as before in Mesopotamia).

ACKNOWLEDGEMENTS

The author would like to thank Rosie Bonté, Daniel Groß, Gundula Lidke, Robert Nedoma and Andreas Zedrosser for their help with the text and Sharon Shellock for language proof-reading.

BIBLIOGRAPHY

ANDERSSON CEDERHOLM et al. 2014: E. ANDERSSON CEDERHOLM/A. BJÖRCK/K. JENNBERT/A.-S. LÖNNGREN, Exploring the animal turn. Human-Animal Relations in Science, Society and Culture (Lund 2014).

ARENDS 1978: U. ARENDS, Ausgewählte Gegenstände des Frühmittelalters mit Amulettcharakter. Unpubl. Ph.D. thesis, Heidelberg University (Heidelberg 1978).

BATELY/ENGLERT 2007: J. BATELY/A. ENGLERT (eds.), Ohthere's voyage: a late 9th-century account of voyages along the coast of Norway and Denmark and its cultural context. Maritime Culture of the North 1 (Roskilde 2007).

BEDNAREK 2018: W. BEDNAREK, Emotions and motivation of the falconer and his relationship with the trained raptor – attempt at an evolutionary-biological interpretation. In: K.-H. Gersmann/O. Grimm (eds.), Raptor and human – falconry and bird symbolism throughout the millennia on a global scale. Publication in considerable extension of the workshop at the Centre for Baltic and Scandinavian Archaeology (ZBSA) in Schleswig, March 5th to 7th 2014. Advanced studies on the archaeology and history of hunting, edited by the ZBSA 1 (Kiel, Hamburg 2018) 285–312.

BEERMANN 2016: S. BEERMANN, Bärenkrallen und Bärenfelle in Brand- und Körpergräbern der vorrömischen Eisenzeit bis Völkerwanderungszeit in Mittel- und Nordeuropa.

Universitätsforschungen zur Prähistorischen Archäologie. Aus dem Seminar für Ur- und Frühgeschichte der Universität Göttingen 279 (Bonn 2016).

Berger 1980: J. Berger, Why look at Animals? In: J. Berger (ed.), About Looking (New York 1980) 23–28.

Bieder 2005: R. E. Bieder, Bear (London 2005).

Braudel 1949: F. Braudel, La Méditerranée et le monde méditerranéen à l'époque de Philippe II. (Paris 1949).

Braudel 1958: F. Braudel, Histoires et sciences sociales. La longue durée. Annales 1958/4, 725–753.

Bronner 2008: S. J. Bronner, Killing Tradition: Inside Hunting and Animal Rights Controversies (Lexington 2008).

Burke 1990: P. Burke, The French Historical Evolution. The Annales School, 1929–89 (Cambridge 1990).

Carstens/Grimm 2015: L. Carstens/O. Grimm, Landscape and lordship – an archaeological-historical analysis of Blekinge's topography of power (1st to 15th century AD). In: O. Grimm/A. Pesch (eds.), Archäologie und Runen. Fallstudien zu Inschriften im älteren Futhark. Beiträge zum Workshop am Zentrum für Baltische und Skandinavische Archäologie (ZBSA) in Schleswig am 3./4. Februar 2011. Schriften des Archäologischen Landesmuseums, Ergänzungsreihe 11 (Schleswig 2015) 195–227.

DeMello 2012: M. DeMello, Animals and Society. An Introduction to Human-Animal Studies (New York 2012).

Derrida 2002: J. Derrida, The Animal That Therefore I am (More to Follow), transl. by David Wills. Critical Inquiry 28(2), 2002, 369–418.

Droberjar/Peška 2002: E. Droberjar/J. Peška, Varia: Messer, Schleifsteine, Bärenkralle, Eisenfragmente. In: J. Peška/ J. Tejral (eds.), Das germanische Königsgrab von Mušsov in Mähren II. Monographien des RGZM 55 (Mainz 2002) 453–459.

Ferrari/Petrus 2015: A. Ferrari/K. Petrus (eds.), Lexikon der Mensch-Tier-Beziehungen (Bielefeld 2015).

Fredengren 2021: C. Fredengren, Beyond Entanglement. Current Swedish Archaeology 29, 2021, 11–33.

Gersmann/Grimm 2018: K.-H. Gersmann/O. Grimm (eds.), Raptor and human – falconry and bird symbolism throughout the millennia on a global scale. Publication in considerable extension of the workshop at the Centre for Baltic and Scandinavian Archaeology (ZBSA) in Schleswig, March 5th to 7th 2014. Advanced studies on the archaeology and history of hunting, edited by the ZBSA 1 (Kiel, Hamburg 2018).

Grimm 2013: O. Grimm, Bear-skins in northern European burials and some remarks on other bear-related furnishings in the north and middle of Europe in the first millenium AD. In: O. Grimm/U. Schmölcke (eds.), Hunting in northern Europe until 1500 AD – Old traditions and regional developments, continental sources and continental influences. Papers presented at a workshop organised by the Centre for Baltic and Scandinavian Archaeology, Schleswig, June 16th and 17th, 2011. Schriften des Archäologischen Landesmuseums, Ergänzungsreihe 7 (Schleswig 2013) 277–296.

Grimm 2020: O. Grimm (ed., in cooperation with K.-H. Gersmann and A.-L. Tropato), Raptor on the fist – falconry, its imagery and similar motifs throughout the millennia on a global scale. Publication in considerable extension of a conference at the New York University Abu Dhabi (NYUAD) in the United Arab Emirates, March 5th to 8th 2018. Advanced studies on the archaeology and history of hunting 2 = Advanced studies in ancient iconography II, edited by the ZBSA/Centre for Baltic and Scandinavian Archaeology in the Foundation of the Schleswig-Holstein State Museums, Schloss Gottorf, Schleswig, Germany, and NYUAD/New York University Abu Dhabi, United Arab Emirates, made possible by generous support of the International Fund for Houbara Conservation (IFHC, United Arab Emirates) and the Emirates Falconers' Club (EFC, United Arab Emirates (Kiel, Hamburg 2020).

Grimm in prep.: O. Grimm (ed.), Beast and human – Northern Europe, from prehistoric to late medieval times.

Grimm/Schmölcke 2013: O. Grimm/U. Schmölcke (eds.), Hunting in northern Europe until 1500 AD – Old traditions and regional developments, continental sources and continental influences. Papers presented at a workshop organised by the Centre for Baltic and Scandinavian Archaeology. Schleswig, June 16th and 17th 2011. Schriften des Archäologischen Landesmuseums, Ergänzungsreihe 7 (Schleswig 2013).

Hallowell 1926: I. A. Hallowell, Bear Ceremonialism in the Northern Hemisphere. American Anthropologist 28(1), 1926, 1–175.

Hansen 2013: L. Hansen, Hunting in the Hallstatt period – The example of the Eberdingen-Hochdorf "princely grave". In: O. Grimm/U. Schmölcke (eds.), Hunting in northern Europe until 1500 AD – Old traditions and regional developments, continental sources and continental influences. Papers presented at a workshop organised by the Centre for Baltic and Scandinavian Archaeology, Schleswig, June 16th and 17th, 2011. Schriften des Archäologischen Landesmuseums, Ergänzungsreihe 7 (Schleswig 2013) 239–258.

Haraway 2003: D. J. Haraway, The Companion Species Manifesto. Dogs, People, and Significant Otherness (Chicago 2003).

Haraway 2008: D. J. Haraway, When Species Meet (Minneapolis, London 2008).

Harris/Cipolla 2017: O. J. T. Harris/C. N. Cipolla, Archaeological Theory in the New Millennium. Introducing Current Perspectives (London 2017).

Helskog 1987: K. Helskog, Selective depictions. A study of 3.500 years of rock carvings from Arctic Norway and their relationship to the Sami drums. In: I. Hodder (ed.), Archaeology as Long-Term History (Cambridge 1987) 17–30.

Helskog 2012: K. Helskog, Bears and Meanings among Hunter-fisher-gatherers in Northern Fennoscandia 9000–2500 BC. Cambridge Archaeological Journal 22(2), 2012, 209–236.

Henriksen 2001: M. M. Henriksen, Bjørnen – fruktbarhetssymbol i eldre jernalder? Spor – nytt fra fortiden 1/2001, 10–13.

Henriksen 2009: M. B. Henriksen, Brudager Mark – en romertidsgravplads nær Gudme på Sydøstfyn. Fynske jernaldergrave 6.1–2 (Odense 2009).

Hodder 1987: I. Hodder (ed.), Archaeology as Long-Term History (Cambridge 1987).

Höfler 1934: O. Höfler, Kultische Geheimbünde der Germanen (Frankfurt 1934).

Ilves 2019: K. Ilves, Seals as a ritual signifier: re-evaluating the Ålandic clay paw burial rite. Fennoscandia Archaeologica XVI, 2019, 33–52.

Jennbert 2021: K. Jennbert, Post-Humanistic Approaches in Archaeology. Current Swedish Archaeology 29, 2021, 43–47.

Kirkinen 2017: T. Kirkinen, "Burning pelts" – brown bear skins in the Iron Age and Early Medieval (0–1300 AD) burials in South-East Fennoscandia. Estonian Journal of Archaeology 21(1), 2017, 3–29.

Klokkervoll in prep.: A. Klokkervoll, Animals in Iron Age mortuary rites: Tendencies and traditions, variations and transitions in Rogaland, SW Norway.

Kompatscher et al. 2017: G. Kompatscher/R. Spannring/K. Schachinger, Human-Animal Studies. Eine Einführung für Studierende und Lehrende (Münster, New York 2017).

Krause/Jankuhn 1966: W. Krause/H. Jankuhn, Die Runeninschriften im älteren Futhark. Mit Beiträgen von H. Jankuhn. Abhandlungen der Akademie der Wissenschaften in Göttingen, Philologisch-Historische Klasse 3, 65 (Göttingen 1966).

Lévi-Strauss 1964: C. Lévi-Strauss, Totemism. Transl. R. Needham (London 1964; original French edition: Le totémisme aujourd'hui, 1962).

Lindner 1937: K. Lindner, Geschichte des deutschen Weidwerk I. Die Jagd der Vorzeit (Berlin, Leizig 1937).

Lindner 1940: K. Lindner, Geschichte des deutschen Weidwerk II. Die Jagd im frühen Mittelalter (Berlin 1940).

Ljungkvist 2011: J. Ljungkvist, Mistresses of the Cult – Evidence of Female Cult Leaders from an Archaeological Perspective. In: D. Quast (ed.), Weibliche Eliten in der Frühgeschichte. Internationale Tagung vom 13. bis zum 14. Juni 2008 im RGZM im Rahmen des Forschungsschwerpunktes „Eliten" (Mainz 2011) 251–265.

Lorenz 1935: K. Lorenz, Der Kumpan in der Umwelt des Vogels – Der Artgenosse als auslösendes Moment sozialer Verhaltensweisen. Journal für Ornithologie 83, 1935, 137–213, 289–413.

L'ours dans l'art 2018: L'ours dans l'art prèhistorique, Musée d'Archéologie nationale – Domaine nationale de Saint-Germain-en-Laye 16 octobre 2016 – 30 janvier et Laténium, parc et muséed'archéologie de Nachchâtel 30 mars 2018 – 6 janvier 2019 / Musée d'Archéologique nationale, Saint-Germain-en-Laye, Laténium, parc et musée d'archéologie de Nachchâtel (Paris 2018).

MacDonald 2014: H. MacDonald, H is for hawk (London 2014).

Marvin 2006: G. Marvin, Wild killing: contesting the animal in hunting. In: The Animal Study Group (G. Marvin/C. Wilbert/D. Donald/S. Baker/E. Fudge/J. Burt/R. Mckay/C. Palmer), Killing Animals (Illinois 2006) 10–29.

Marvin/McHugh 2014: G. Marvin/S. McHugh, Routledge Handbook of Human-Animal Studies (London 2014).

Nockert 1991: M. Nockert, The Högom find and other Migration period textiles and costumes in Scandinavia. Archaeology and Environment 9. Högom II (Uddevalla 1991).

Nodermann 2009: M. Nodermann, Björnens fäll vid altaret och i graven. Saga og sed 2009, 133–159.

Oehrl 2013: S. Oehrl, Bear hunting and its ideological context (as a background for the interpretation of bear claws and other remains of bears in Germanic graves of the 1st millennium AD). In: O. Grimm/U. Schmölcke (eds.), Hunting in northern Europe until 1500 AD – Old traditions and regional developments, continental sources and continental influences. Papers presented at a workshop organised by the Centre for Baltic and Scandinavian Archaeology, Schleswig, June 16th and 17th, 2011. Schriften des Archäologischen Landesmuseums, Ergänzungsreihe 7 (Schleswig 2013) 297–332.

Ordiz et al. 2012: A. Ordiz/O.-G. Støen/S. Sæbø/J. Kindberg/M. Delibes/J. Swenson, Do bears know they are being hunted? Biological Conservation 152, 2012, 21–28.

Overton/Hamilakis 2013: N. J. Overton/Y. Hamilakis, A manifesto for a social zooarchaeology. Swans and other beings in the Mesolithic. Archaeological Dialogues 20/2, 2013, 111–136.

Pastoureau 2008: M. Pastoureau, Der Bär – Geschichte eines gefallenen Königs (Neu-Isenburg 2008: original French edition: L'ours. Histoire d'un roi déchu, 2007; English translation: The Bear – history of a fallen king, 2011).

Pentikäinen 2007: J. Pentikäinen, Golden King of the Forest. The Lore of the Northern Bear (Helsinki 2007; original Finnish edition: Karhun kannoilla, 2005).

Pilaar Birch 2018: S. Pilaar Birch (ed.), Multispecies Archaeology. Archaeological Orientations (London, New York 2018).

Price 2002: N. Price, The Viking Way: Religion and War in Late Iron Age Scandinavia (Uppsala 2002).

Ritvo 2007: H. Ritvo, On the Animal Turn. Daedalus 136(4), 2007, 118–122.

Roscher 2012: M. Roscher, Human-Animal Studies. docupedia_roscher_human-animal_studies_v1_de_2012.pdf (zeitgeschichte-digital.de).

Roscher et al. 2021: M. Roscher/A. Krebber/B. Mizelle (eds.), Handbook of Historical Animal Studies (Berlin, Boston 2021).

Samson 2020: V. Samson, Die Berserker. Die Tierkrieger des Nordens von der Vendel- bis zur Wikingerzeit. Reallexikon der Germanischen Altertumskunde 121 (Berlin 2020 [deutsche Übersetzung des französischen Originals]).

Schönfelder 1994: M. Schönfelder, Bear-claws in Germanic graves. Oxford Journal of Archaeology 13, 1994, 217–227.

Schroer 2018: S. A. Schroer, A view from anthropology: falconry, domestication and the 'animal turn'. In: K.-H.

Gersmann/O. Grimm (eds.), Raptor and human – falconry and bird symbolism throughout the millennia on a global scale. Publication in considerable extension of the workshop at the Centre for Baltic and Scandinavian Archaeology (ZBSA) in Schleswig, March 5th to 7th 2014. Advanced studies on the archaeology and history of hunting, edited by the ZBSA 1 (Kiel, Hamburg 2018) 313–321.

SCHULTE 2015: M. SCHULTE, Die Blekinger Inschriften als Status- und Machtembleme – Ein kulturhistorischer Syntheseversuch. In: O. Grimm/A. Pesch (eds.), Archäologie und Runen. Fallstudien zu Inschriften im älteren Futhark. Beiträge zum Workshop am Zentrum für Baltische und Skandinavische Archäologie (ZBSA) in Schleswig am 3./4. Februar 2011. Schriften des Archäologischen Landesmuseums, Ergänzungsreihe 11 (Schleswig 2015) 175–194.

SHALEM 2014: A. SHALEM, Die mittelalterlichen Olifante (Die Elfenbeinskulpturen Bd. 8) (Berlin 2014).

SQUIRES et al. in prep.: K. SQUIRES/H. O'REGAN/O. GRIMM, Who Gets Buried with a Bear? The Inclusion of Wild Animals in the Early Medieval Cremation Rite of Eastern England and Northern Germany.

THE ANIMAL STUDY GROUP 2006: THE ANIMAL STUDY GROUP (G. Marvin/C. Wilbert/D. Donald/S. Baker/E. Fudge/J. Burt/R. Mckay/C. Palmer), Killing Animals (Illinois 2006).

ULRICH/ULRICH 2014: J. ULRICH/A. ULRICH (eds.), Tiere und Tod. Tierstudien 05/2014 (Berlin 2014).

WAMERS 2015: E. WAMERS, Franconofurd 2. Das birituelle Kinderdoppelgrab der späten Merowingerzeit unter der Frankfurter Bartholomäuskirche ("Dom"). Archäologische und naturwissenschaftliche Untersuchungen. Schriften des Archäologischen Museums Frankfurt 22.2 (Regensburg 2015).

Dr. Oliver Grimm
Centre for Baltic and Scandinavian Archaeology (ZBSA)
Stiftung Schleswig-Holsteinische Landesmuseen, Schloss Gottorf
Schleswig
Germany
oliver.grimm@zbsa.eu

Chapter 2

Bears in biology (Europe)

Brown bears in a playful fight in Katami National Park, Alaska, USA (see G<small>RIMM</small> et al., Bears – fact or fiction; photo Th. Sbampato, image used with permission of the photographer).

Conservation status and distribution of the brown bear in Europe

By Andreas Zedrosser and Jon E. Swenson

Keywords: Europe, brown bear, Ursus arctos, *population size, conservation*

Abstract: Despite millennia of persecution, brown bears have made a remarkable comeback in Europe due to a shift in the management paradigm towards more conservation-oriented management in the 20th century. Today, the brown bear occurs in 22 countries in Europe, and the overall population size is estimated at approximately 17,000–18,000 individuals (not including Russia). Brown bears can be clustered into ten populations in Europe: Scandinavian, Karelian, Baltic, Carpathian, Dinaric–Pindos, Eastern Balkan, Alpine, Central Apennine, Cantabrian, and Pyrenean. Several of these populations are large and are sustainably managed, also by hunting (e.g. in Scandinavia and in the Carpathian mountain range). In these populations it is important to ensure that hunting quotas are kept within the limits of sustainability. However, several brown bear populations in Europe are still small and isolated, and special attention must be paid to their long-term viability in terms of the number of reproducing individuals and population genetics (e.g. the Central Apennine, Cantabrian, and Pyrenean populations). Effective mitigation measures to keep depredation rates low are crucial for conservation in Europe, as is public education and awareness of brown bears. Coexistence of humans and brown bears is fully possible, even in the human-dominated landscapes of Europe.

Introduction

The brown bear (*Ursus arctos*) is the most widespread bear, with a Holarctic distribution in Europe, Asia, and North America (Swenson et al. 2000; Zedrosser et al. 2001). The species is highly adaptable, and lives in habitats ranging from northern Arctic tundra to dry desert (Swenson et al. 2020). The subspecies generally occurring throughout Europe is *Ursus arctos arctos*, and compared to their North American cousins, for example the so-called grizzly bear (*Ursus arctos horribilis*), their claws are a bit shorter and less curved, enabling European brown bears to climb trees very well. The fur colour can vary considerably in Europe, and some individuals may seem light or dark from different angles, due to the variegated guard hairs (Swenson et al. 2000). Adult males are larger and heavier than females on average; generally, males weigh 140–320 kg, and females weigh 100–200 kg (Zedrosser et al. 2006; 2011; Swenson et al. 2007; 2020). All European populations are inland populations and do not reach the extreme body sizes found in coastal population with access to protein- and lipid-rich spawning Pacific salmon (*Oncorhynchus* spp.), as in Alaska and eastern Siberia. As a matter of fact, brown bears in Europe likely never have eaten salmon (*Salmo salar*).

People and large carnivores (i.e. brown bear, grey wolf [*Canis lupus*], Eurasian lynx [*Lynx lynx*], wolverine [*Gulo gulo*]) have been in conflict throughout their common history (Woodroffe 2000; Linnell et al. 2001), and carnivores were not killed primarily for consumption, but to prevent them from killing livestock, other wildlife, or people (Swenson et al. 2000; Zedrosser et al. 2011). Emperor Charlemagne was the first to establish a dedicated large carnivore hunting corps around AD 800 (Boitani 1995), but it took many centuries to gradually eliminate bears and other large carnivores from western Europe (Frank/Woodroffe 2001). These efforts were often encouraged with bounties paid by the state and/or local authorities for the killing of bears (Swenson et al. 2000). This was effective, because bears have a low reproductive rate and they are sensitive to high harvest rates (Swenson et al. 2000). Also, climatic variation and habitat loss may have played a role in the extinction of large carnivores. Albrecht et al. (2017) have suggested that increasing winter temperatures likely have contributed substantially to the Holocene decline of the brown bear, both directly by reducing the species' reproductive rate and indirectly by facilitating human land use. The first local extinctions occurred during the Mid-Holocene warming period (c. 7,000–5,000 years ago), but the rise of the Roman Empire 2,000 years ago marked the onset of large-scale extinctions, followed by increasingly rapid range loss and population fragmentation. These findings strongly support the hypothesis that complex interactions between climate and humans may have accelerated megafaunal extinctions (Albrecht et al. 2017).

A shift in the management paradigm towards more conservation-oriented management has occurred in the 20th century, and today many large carnivore populations are again on the increase in North America and Europe (e.g. Breitenmoser 1998; Servheen et al. 1999; Boitani 2000; Swenson et al. 2000; Woodroffe 2000; Linnell et al. 2001; Schwartz et al. 2006; Zedrosser et al. 2011). The present conservation challenges regarding large carnivores include their large area requirements and predatory behaviour (Nowell/Jackson 1996; Linnell et al. 2001), as well as their comparatively low population densities and slow life histories (Zedrosser et al. 2011). Especially small populations of large carnivores are very vulnerable to stochastic events and the loss of key individuals (Linnell et al. 2005; Zedrosser et al. 2011). Despite these somewhat pessimistic forecasts, large carnivores, including the brown bear, are making a comeback in Europe. Chapron et al. (2014) have shown that roughly one-third of mainland Europe hosts at least one large carnivore species, with stable or increasing abundance in most cases in 21st-century records. The reasons for this overall conservation success include protective legislation, supportive public opinion, and a variety of practices making coexistence between large carnivores and people possible (Chapron et al. 2014). What is so remarkable about this comeback of large carnivores in Europe is that it occurs in a man-made and human-dominated landscape, not in a wilderness area. This clearly documents that wilderness is not required, that large carnivores – including the brown bear– can adapt to the proximity of humans and human-dominated landscapes, and most importantly, that carnivores and humans can share the same landscape and coexist in Europe (Chapron et al. 2014; López-Bao et al. 2015).

Brown bears originally occurred throughout Europe (except on the largest islands such as Iceland, Gotland, Corsica, and Sardinia; cf. Swenson et al. 2000; Zedrosser et al. 2001). Today's distribution of brown bears in Europe can be defined by populations, i.e. all bears in an area that are genetically isolated, totally or substantially, from other bear populations (Swenson et al. 2000; Zedrosser et al. 2001). Overall, there are 17,000–18,000 bears spread across ten populations and 22 countries in Europe (excluding Russia) today (Kaczensky et al. 2013). The largest European populations (>1,500 individuals) are in the Carpathian Mountains, Scandinavia (see Schneider et al., this volume), Karelia, and in the Dinaric-Pindos mountain range (Fig. 1; Table 1; www.lcie.org/Large-carnivores/Brown-bear; cf. Swenson et al. 2000; Zedrosser et al. 2001; Chapron et al. 2014). These populations are generally hunted. Medium-sized populations (300–1,000) can be found in the Cantabrian Mountains, in the eastern Balkan, and in the Baltic area; very small (<100) and usually endangered populations can be found in the Alps, in the Pyrenees, and in the central Apennine Mountains (Fig. 1; Table 1; data from www.lcie.org/Large-carnivores/Brown-bear; cf. Swenson et al. 2000; Zedrosser et al. 2001; Chapron et al. 2014). Generally, these medium and small-sized populations are not hunted.

Bear populations in Europe

Alpine population

This population is located in the Alpine mountain range of Italy, Switzerland, Austria and Slovenia (Fig. 1; Table 1). Its core consists of a few brown bears in northern central Italy, originally a naturally occurring but very small population that was augmented by the release of bears from Slovenia (Swenson et al. 2000; Zedrosser et al. 2001; De Barba et al. 2010). Bears were also reintroduced into central Austria in the 1990s, again with individuals captured in Slovenia. After an initially positive development and at least 30 documented reproductions, the population in Austria has disappeared, likely due to illegal killings (Rauer 2004; Kaczensky et al. 2011). The population augmentation in Italy was successful, however, and the status of the Alpine bear population is now considered stable to increasing, and some bears have dispersed into the neighbouring countries of Switzerland, Austria, and even Germany (Kazensky et al. 2011). Unfortunately, several of these dispersers have shown problematic behaviour, i.e. some were human-habituated (i.e. not afraid of humans) and have caused damages to livestock, and were removed from the wild by the responsible authorities. Illegal killings may hamper the long-term development of the Alpine population, as can be documented by the poaching of a dispersing individual in the border area of Slovenia and Austria as well as the failed bear reintroduction in central Austria (Kaczensky et al. 2011). On the positive side, the Alpine population is starting to reconnect with the bear population in Slovenia (Kaczensky et al. 2013), especially in the triangle area of northern Italy/Slovenia/southern Austria, which is crucial for the long-term survival of this population (Jerina/Adamic 2008; Peters et al. 2015).

Baltic population

Brown bears in Estonia and Latvia are considered part of the Baltic population, which is continuous with bears in Russia and Belarus (Fig. 1; Table 1). The main distribution area in the Baltic countries is in Estonia, where bears generally are found all over the mainland; however, reproducing females are missing in the southernmost part of the country (Kaczensky et al. 2013). Bear occurrences in neighbouring Latvia are rare, however, but are most common in the eastern and northern part of the country along the Estonian and Russian border. There is no evidence of bear breeding in the territory of Latvia (Kaczensky et al. 2013). The Baltic bears are genetically connected with the Karelian and even the Scandinavian population via the very large and stable bear populations in Russia (Kopatz et al. 2012; 2014; Schregel et al. 2012; Kaczensky et al. 2013). Bears are hunted in Estonia, and the overall goal is to keep numbers at a stable level. In comparison, bears in neighbouring Latvia are strictly protected, and the management goal is to increase the number of bears.

Cantabrian population

Brown bears in the Cantabrian Mountains of northwestern Spain occur in two small, isolated, and endangered populations: one in the west (in the autonomous regions of Asturias and Castilla y León, but also in Galicia) and a smaller one in the east (mainly in the region of Castilla y León, but also in Cantabria and Asturias; cf. Palomero et al. 2007; Fig. 1; Table 1). The populations are separated by c. 50 km, however, the habitat is mainly unsuitable and also is intersected by a high-speed railway and motorway. The western segment has shown an obvious increase (from three females with cubs of the year detected in 1994 to 25 in 2010), whereas the smaller eastern segment appears to be stable or only slightly increasing. The brown bear is a strictly protected species in Spain, and there are four recovery plans (one from each autonomous region) and a National Action Plan coordinated by the Ministry of the Environment (Estrategia Nacional para la Conservación del Oso). The recovery plans and the National Plan have no quantitative population goals, but rather aim to recover the population as much as possible. The recovery plans delineate the critical areas, which are particularly protected.

They usually include the best forests and the areas with concentrations of winter dens (KACZENSKY et al. 2013). In general, brown bears in the Cantabrian Mountains are recovering, but the isolation of the two population segments jeopardises this recovery. Conservation priorities include promoting recovery of range previously occupied by breeding females and increasing contact between the populations (KACZENSKY et al. 2013).

Pyrenean population
The Pyrenean bear population occurs in Spain, France, and Andorra (Fig. 1; Table 1). It is the result of an augmentation of a very small and local occurrence of bears with bears captured in Slovenia in 2006. This population has been totally isolated from other bears for over a century, and there are no possibilities of reestablishing connectivity in the short term (KACZENSKY et al. 2013; PIEDALLU et al. 2019). In addition, the western and central cores of this very small population are spatially separated. In 2011, the minimum number of detected individuals was 22, based on genetic monitoring, camera traps, and opportunistic observations. Most of the individuals identified are both located on the Spanish and French side of the Pyrenean range (KACZENSKY et al. 2013). There is a formal cooperation agreement on brown bear management among the governments of Spain, France, and Andorra. In addition, there are informal cooperation activities between the Spanish autonomous regions and France in order to use common monitoring and genetic methods (KACZENSKY et al. 2013). The main conservation issues in this population are the acceptance of the reintroduced bears and losses mainly due to poaching (KACZENSKY et al. 2013).

Carpathian population
The Carpathian population includes the brown bears in Romania, Poland, Slovakia, Ukraine, and northern Serbia (Fig. 1; Table 1); this is the largest brown bear population in Europe outside of Russia (SWENSON et al. 2000; ZEDROSSER et al. 2001; FEDORCA et al. 2019). It increased rapidly in the second part of the last century and hunting is allowed, with the exceptions of Serbia and Poland, where bears are protected. Most of the bears in this population live in Romania (*c.* 6,000), followed by Slovakia (*c.* 1,000), whereas Poland has relatively few bears (*c.* 80), and northern Serbia has very few (*c.* 6; cf. KACZENSKY et al. 2013). There are some questions concerning internal connectivity within the Carpathian population, due to a lack of knowledge about the situation within Ukraine and the developments of bear distribution in eastern Slovakia (KACZENSKY et al. 2013). Based on genetic data, FEDORCA et al. (2019) showed that there is one large and continuous bear population across the Carpathians, which suggests that there still is sufficient suitable bear habitat to allow movement of individuals. However, results at a finer scale show some indications that highway infrastructure development may threaten to fragment regions with brown bear occurrence (FEDORCA et al. 2019). In general, brown bears thrive in this large and stable population, however, attention needs to be paid to population connectivity (ZIOLKOWSKA et al. 2016; BARTON et al. 2019).

Central Apennine population
This population is located in Abruzzo National Park and the surrounding area in the Apennine Mountains in central Italy (Fig. 1; Table 1; cf. SWENSON et al. 2000; ZEDROSSER et al. 2001). Despite its full protection, this bear population seems to be stagnant at the best, despite regular reproduction events (1–7 females with cubs have been counted annually from 2006 to 2011; CIUCCI/BOITANI 2008; KACZENSKY et al. 2013). It exists within a densely human populated area, and there are potential conflicts between bear conservation and human development and recreation activities, as also indicated by several cases of poisoning of bears in protected areas (SWENSON et al. 2000; KACZENSKY et al. 2013; TOSONI et al. 2017). The latest estimates suggest a relatively small population of around 50 bears (GERVASI et al. 2012; GERVASI/CIUCCI 2018). The population has been totally isolated for

over a century and there is no possibility of reestablishing unassisted connectivity in the short term. In the long run, this population is under severe risk of extinction (Gervasi/Ciucci 2018). Modelling suggests that efforts aimed at increasing the general food availability likely have minimal effect on population viability and extinction risk, whereas measures to decrease adult female mortality (i.e. the reproducing segment of the population) provide the best chance for the long-term survival of this critically endangered bear population (Gervasi/Ciucci 2018).

Dinaric-Pindos population
This is one of the largest European bear populations, stretching from Slovenia in the north to Albania and northwestern Greece in the south (Fig. 1; Table 1). The total number of bears in the Dinaric-Pindos area is almost 4,000 (Table 1). Especially the populations in Slovenia and Croatia are very large and viable and are also hunted, whereas the bears living in the other countries are protected. The annual harvest rate of brown bears in Slovenia is 20 % and one of the highest harvest rates reported for brown bears worldwide. Such an exceptionally high harvest rate is only possible due to the influx of especially young adult males from neighbouring Croatia in the south (Krofel et al. 2012). The suitable bear habitat in the countries of this population is less contiguous than in the Carpathian area, separating the functional habitat into more or less isolated subareas, although there are corridors. This population is also the closest to the critically endangered Alpine bear population, and bears in northern Italy and Slovenia are connected by single male dispersers. However, increasing damages and an increase in nuisance bears in Slovenia make it a challenge to maintain bear numbers at the present level, let alone allow for expanding the population into the Alps (Krofel et al. 2010; Kaczensky et al. 2013). Good information on this population is also available at its southeastern extension, in Greece, where bear numbers are slowly increasing (Karamanlidis et al. 2015). The main conservation challenge in the Dinaric-Pindos population is that it is shared by many countries and subject to widely varying monitoring methods and standards, and especially the general lack of accessible and robust data from Bosnia and Herzegovina, Montenegro, Albania, and Northern Macedonia (Kaczensky et al. 2013; Karamanlidis et al. 2014a–c).

Eastern Balkan population
The core areas of this bear population are in the Rila-Rhodope Mountains and the border regions of Bulgaria, Greece, and Serbia (Fig. 1; Table 1; cf. Kaczensky et al. 2013). The bears in these core areas are genetically connected and may potentially be linked to the very large Carpathian bear population, but this connection remains unsupported (Zlatanova et al. 2009; Zlatanova 2010). The Greek part of this population is located in northeastern Greece and is geographically relatively close to the large and viable Dinaric-Pindos population, however, no genetic connections between the populations have been documented (Kaczensky et al. 2013). In general, habitat and population fragmentation is a main conservation issue in the eastern Balkan population (Zlatanova et al. 2009; Zlatanova 2010). Brown bears are a strictly protected species in all three population range countries, and Greece and Bulgaria have developed bear management plans to aid conservation and management decisions, but especially monitoring programs to keep track of population development can still be improved in this population (Kaczensky et al. 2013).

Karelian population
The Karelian brown bear population stretches from northern Norway via the Kola peninsula and Karelia in Russia into Finland (Fig. 1; Table 1). It has genetic exchange with the Scandinavian population in the south and west and the Baltic in the east (Kopatz et al. 2012; Schregel et al. 2012). Both the Karelian and the Baltic populations also are connected to the main distribution area of Russian bears to the east and thereby with each other (Kopatz et al. 2012; Kaczensky et al. 2013).

Overall, the number of bears is increasing, mainly due to the positive trend in Finland. The density of adult females is this population (outside of Russia) is highest in southeastern and central Finland. In southwestern Finland, density is low and practically all bears are subadult males (Aspi et al. 2006; Kojola/Heikkinen 2006; Kojola et al. 2006). In the northern segment, bears are concentrated at the Finnish-Russian border (Kaczensky et al. 2013). Elsewhere in the north, only a few adult females are found (Aspi et al. 2006; Kojola/Heikkinen 2006; Kojola et al. 2006). The number of bears decreased to 150 animals, the lowest level for hundreds of years, in the late 1960s and started to increase again in the early 1970s, when local hunting associations in eastern Finland decided to stop all hunting for several years (Kaczensky et al. 2013). The return of bears to central Finland was supported by translocating two females from the east in the early 1980s. Genetic analyses indicate that these translocations have been critically important for the establishment of a reproducing bear population in central Finland (Saarma/Kojola 2007; Saarma et al. 2007; Kaczensky et al. 2013). Brown bear viewing tourism is very popular in Finland along the Russian border, and the large amounts of farmed Atlantic salmon used for bait attract bears from far within the Russian territory (Kojola/Heikkinen 2012; Penteriani et al. 2015).

Summary

Despite millennia of persecution, brown bears have made a remarkable comeback in Europe, due to a shift in the management paradigm towards more conservation-oriented management. Several populations are large and are sustainably managed, also by hunting. In these populations it is important to ensure that hunting quotas are kept within the limits of sustainability. However, several brown bear populations in Europe are still small and isolated, and special attention must be paid to their long-term viability in terms of the number of reproducing individuals and population genetics. Generally, large carnivores struggle with negative public opinions that affect conservation and management of their populations. Common management problems are usually linked to depredation of livestock, which is common in Europe. In rare cases, humans are even injured or killed by brown bears (Støen et al. 2018; Bombieri et al. 2019). Effective mitigation measures to keep depredation rates at acceptable levels are crucial for conservation in Europe, as is public education and awareness of brown bears, their ecology and behaviour. Coexistence of humans and brown bears is fully possible, even in the man-made and human-dominated landscapes of Europe.

Acknowledgements

The data and information presented in this chapter rely heavily on the work of many colleagues, researchers, and wildlife managers via the platform of the Large Carnivore Initiative for Europe – IUCN/SSC Specialist Group. More information and publications can be found at www.lcie.org.

Bibliography

Albrecht et al. 2017: J. Albrecht/K. A. Barton/N. Selva/R. S. Sommer/J. E. Swenson/R. Bischof, Humans and climate change drove the Holocene decline of the brown bear. Scientific Reports 7, 10399, 2017. https://doi.org/10.1038/s41598-017-10772-6.

Aspi et al. 2006: J. Aspi/E. Roininen/M. Ruokonen/I. Kojola/C. Vila, Genetic diversity, population structure, effective population size and demographic history of the Finnish wolf population. Molecular Ecology 15, 2006, 1561–1576.

Barton et al. 2019: K. A. Barton/T. Zwijacz-Kozica/F. Zieba/A. Sergiel/N. Selva, Bears without borders: Long-distance movement in human-dominated landscapes. Global Ecology and Conservation 17, 2019. https://doi.org/10.1016/j.gecco.2019.e00541.

Boitani 1995: L. Boitani, Ecological and cultural diversities in the evolution of wolf-human relationships. In: L. N. Carbyn/S. H. Fritts/D. R. Seip (eds.), Ecology and conservation of wolves in a changing world (Alberta 1995) 3–12.

Boitani 2000: L. Boitani, Action plan for the conservation of the wolves (*Canis lupus*) in Europe. Council of Europe Publishing (Strasbourgh 2000).

Bombieri et al. 2019: G. Bombieri/J. Naves/V. Penteriani/N. Selva/A. Fernández-Gil/J. V. López-Bao/H. Ambarli/C. Bautista/T. Bespalova/V. Bobrov/V. Bolshakov/S. Bondarchuk/J. J. Camarra/S. Chiriac/P. Ciucci/A. Dutsov/I. Dykyy/J. M. Fedriani/A. García-Rodríguez/P. J. Garrote/S. Gashev/C. Groff/B. Gutleb/M. Haring/S. Härkönen/D. Huber/M. Kaboli/Y. Kalinkin/A. A. Karamanlidis/V. Karpin/V. Kastrikin/L. Khlyap/P. Khoetsky/I. Kojola/Y. Kozlow/A. Korolev/N. Korytin/V. Kozsheechkin/M. Krofel/J. Kurhinen/I. Kuznetsova/E. Larin/A. Levykh/V. Mamontov/P. Männil/D. Melovski/Y. Mertzanis/A. Meydus/A. Mohammadi/H. Norberg/S. Palazón/L. M. Pătraşcu/K. Pavlova/P. Pedrini/P. Y. Quenette/E. Revilla/R. Rigg/Y. Rozhkov/L. F. Russo/A. Rykov/L. Saburova/V. Sahlén/A. P. Saveljev/I. V. Seryodkin/A. Shelekhov/A. Shishikin/M. Shkvyria/V. Sidorovich/V. Sopin/O.-G. Støen/J. Stofik/J. E. Swenson/D. Tirski/A. Vasin/P. Wabakken/L. Yarushina/T. Zwijacz-Kozica/M. M. Delgado, Brown bear attacks on humans: a world-wide perspective. Scientific Reports 9, 2019, 8573.

Breitenmoser 1998: U. Breitenmoser, Large predators in the Alps: the fall and rise of man's competitors. Biological Conservation 83, 1998, 279–289.

Chapron et al. 2014: G. Chapron/P. Kaczensky/J. D. C. Linnell/M. von Arx/D. Huber/H. Andren/J. V. Lopez-Bao/M. Adamec/F. Alvares/O. Anders/L. Balciauskas/V. Balys/P. Bedo/F. Bego/J. C. Blanco/U. Breitenmoser/H. Broseth/R. Bufka/R. Bunikyte/P. Ciucci/A. Dutsov/T. Engleder/C. Fuxjager/C. Groff/K. Holmala/B. Hoxha/Y. Iliopoulos/O. Ionescu/J. Jeremic/J. Kerina/G. Kluth/F. Knauer/I. Kojola/I. Kos/M. Krofel/J. Kubala/S. Kunovac/J. Kusak/M. Kutal/O. Liberg/A. Majic/P. Mannil/R. Manz/E. Marboutin/F. Marucco/D. Melovski/K. Mersini/Y. Mertzanis/R. W. Myslajek/S. Nowak/J. Odden/J. Ozolins/G. Palomero/M. Paunovic/J. Persson/H. Potocnik/P. Y. Quenette/G. Rauer/I. Reinhardt/R. Rigg/A. Ryser/V. Salvatori/T. Skrbinsek/A. Stojanov/J. E. Swenson/L. Szemethy/A. Trajce/E. Tsingarska-Sedefcheva/M. Vana/R. Veeroja/P. Wabakken/M. Wofl/S. Wolfl/F. Zimmermann/D. Zlatanova/L. Boitani, Recovery of large carnivores in Europe's modern human-dominated landscapes. Science 346, 2014, 1517–1519.

Ciucci/Boitani 2008: P. Ciucci/L. Boitani, The Appenine brown bear: a critical review of its status and conservation problems. Ursus 19, 2008, 130–145.

De Barba et al. 2010: M. de Barba/L. P. Waits/E. O. Garton/P. Genovesi/E. Randi/A. Mustoni/C. Groff, The power of genetic monitoring for studying demography, ecology and genetics of a reintroduced brown bear population. Molecular Ecology 19, 2010, 3938–3951.

Fedorca et al. 2019: A. Fedorca/I. R. M. Russo/O. Ionescu/G. Ionescu/M. Popa/M. Fedorca/A. L. Curtu/N. Sofletea/G. M. Tabor/M. W. Bruford, Inferring fine-scale spatial structure of the brown bear (*Ursus arctos*) population in the Carpathians prior to infrastructure development. Scientific Reports 9, 9494, 2019. https://doi.org/10.1038/s41598-019-45999-y.

Frank/Woodroffe 2001: L. G. Frank/R. Woodroffe, Behavior of carnivores in exploited and controlled populations. In: J. L. Gittleman/S. M. Funk/D. MacDonald/R. K. Wayne (eds.), Carnivore Conservation (Cambridge 2001) 419–442.

Gervasi/Ciucci 2018: V. Gervasi/P. Ciucci, Demographic projections of the Apennine brown bear population *Ursus arctos marsicanus* (Mammalia: Ursidae) under alternative management scenarios. European Zoological Journal 85, 2018, 243–253.

Gervasi et al. 2012: V. Gervasi/P. Ciucci/J. Boulanger/E. Randi/L. Boitani, A multiple data source approach to improve abundance estimates of small populations: The brown bear in the Apennines, Italy. Biological Conservation 152, 2012, 10–20.

Jerina/Adamic 2008: K. Jerina/M. Adamic, Fifty years of brown bear population expansion: effects of sex-biased dispersal on rate of expansion and population structure. Journal of Mammalogy 89, 2008, 1491–1501.

Kaczensky et al. 2011: P. Kaczensky/K. Jerina/M. Jonozovic/M. Krofel/T. Skrbinsek/G. Rauer/I. Kos/B. Gutleb, Illegal killings may hamper brown bear recovery in the Eastern Alps. Ursus 22, 2011, 37–46.

Kaczensky et al. 2013: P. Kaczensky/G. Chapron/M. von Arx/D. Huber/H. Andren/J. Linnell, Status, management and distribution of large carnivores – bear, lynx, wolf & wolverine – in Europe. Report prepared for the European Commission (2013).

Karamanlidis et al. 2014a: A. A. Karamanlidis/M. Paunovic/D. Cirovic/B. Karapandza/T. Skrbinsek/A. Zedrosser, Population genetic parameters of brown bears in western Serbia: implications for research and conservation. Ursus 25, 2014, 34–43.

Karamanlidis et al. 2014b: A. A. Karamanlidis/S. Pllaha/L. Krambokoukis/K. Shore/A. Zedrosser, Preliminary brown bear survey in southeastern Albania. Ursus 25, 2014, 1–7.

Karamanlidis et al. 2014c: A. A. Karamanlidis/A. Stojanov/M. D. Hernando/G. Ivanov/I. Kocijan/D. Melovski/T. Skrbinsek/A. Zedrosser, Distribution and genetic status of brown bears in FYR Macedonia: implications for conservation. Acta Theriologica 59, 2014, 119–128.

Karamanlidis et al. 2015: A. A. Karamanlidis/M. D. G. Hernando/L. Krambokoukis/O. Gimenez, Evidence of a large carnivore population recovery: Counting bears in Greece. Journal for Nature Conservation 27, 2015, 10–17.

Kojola/Heikkinen 2006: I. Kojola/S. Heikkinen, The structure of the expanded brown bear population at the edge of the Finnish range. Annales Zoologici Fennici 43, 2006, 258–262.

Kojola/Heikkinen 2012: I. Kojola/S. Heikkinen, Problem brown bears *Ursus arctos* in Finland in relation to bear feeding for tourism purposes and the density of bears and humans. Wildlife Biology 18, 2012, 258–263.

Kojola et al. 2006: I. Kojoloa/J. Aspi/A. Hakala/S. Heikkinen/C. Ilmoni/S. Ronkainen, Dispersal in an expanding wolf population in Finland. Journal of Mammalogy 87, 2006, 281–286.

Kopatz et al. 2012: A. Kopatz/H. K. Eiken/S. B. Hagen/M. Ruokonen/R. Esparza-Salas/J. Schregel/I. Kojola/M. E. Smith/I. Wartiainen/P. E. Aspholm/S. Wikan/A. M. Rykov/O. Makarova/N. Polikarpova/K. F. Tirronen/P. I. Danilov/J. Aspi, Connectivity and population subdivision at the fringe of a large brown bear (*Ursus arctos*) population in North Western Europe. Conservation Genetics 13, 2012, 681–692.

Kopatz et al. 2014: A. Kopatz/H. G. Eiken/J. Aspi/I. Kojola/C. Tobiassen/K. F. Tirronen/P. I. Danilov/S. B. Hagen, Admixture and Gene Flow from Russia in the Recovering Northern European Brown Bear (*Ursus arctos*). PLoS ONE 9(5), 2014. https://doi.org/10.1371/journal.pone.0097558.

Krofel et al. 2010: M. Krofel/S. Filacorda/K. Jerina, Mating-related movements of male brown bears on the periphery of an expanding population. Ursus 21, 2010, 23–29.

Krofel et al. 2012: M. Krofel/M. Jonozovic/K. Jerina, Demography and mortality patterns of removed brown bears in a heavily exploited population. Ursus 23, 2012, 91–103.

Linnell et al. 2001: J. D. C. Linnell/J. E. Swenson/R. Andersen, Predators and people: conservation of large carnivores is possible at high human densities if management policy is favourable. Animal Conservation 4, 2001, 345–349.

Linnell et al. 2005: J. D. C. Linnell/C. Promberger/L. Boitani/J. E. Swenson/U. Breitenmoser/R. Andersen, The linkage between conservation strategies for large carnivores and biodiversity: the view from the "half-full" forests of Europe. In: J. C. Ray/K. H. Redford/R. S. Steneck/J. Berger (eds.), Large carnivores and the conservation of biodiversity (Washington 2005) 381–399.

López-Bao et al. 2015: J. V. López-Bao/P. Kaczensky/J. D. C. Linnell/L. Boitani/G. Chapron, Carnivore coexistence: wilderness not required. Science 348, 2015, 871–872.

Nowell/Jackson 1996: K. Nowell/P. Jackson, Wild cats: status survey and action plan (Gland 1996).

Palomero et al. 2007: G. Palomero/F. Ballesteros/C. Nores/J. C. Blanco/J. Herrero/A. Garcia-Serrano, Trends in number and distribution of brown bear females with cubs-of-the-year in the Cantabrian Mountains, Spain. Ursus 18, 2007, 145–157.

Penteriani et al. 2015: V. Penteriani/J. V. López-Bao/C. Bettega/F. Dalerum/M. del Mar Delgado/K. Jerina/I. Kojol/M. Peters/W. M. Hebblewhite/M. Cavedon/L. Pedrotti/A. Mustoni/F. Zibordi/C. Groff/M. Zanin/F. F. Cagnacci, Resource selection and connectivity reveal conservation challenges for reintroduced brown bears in the Italian Alps. Biological Conservation 186, 2015, 123–133.

Peters et al. 2015: W. Peters/M. Hebblewhite/M. Cavedon/L. Pedrotti/A. Mustoni/F. Zibordi/C. Groff/M. Zanin/F. Cagniacci, Ressource selection and connectivity reveal conservation challenges for reintroduced brown bears in the Italian Alps. Biological Conservation 186, 2015, 123–133.

Piedallu et al. 2019: B. Piedallu/P. Y. Quenette/N. Bombillon/A. Gastineau/C. Miquel/O. Gimenez, Determinants and patterns of habitat use by the brown bear *Ursus arctos* in the French Pyrenees revealed by occupancy modelling. Oryx 53, 2019, 334–343.

Rauer 2004: G. Rauer, Re-introduced bears in Austria. Ecos 25, 2004, 69–72.

Saarma/Kojola 2007: U. Saarma/I. Kojola, Matrilineal genetic structure of the brown bear population in Finland. Ursus 18, 2007, 30–37.

Saarma et al. 2007: U. Saarma/S. Y. W. Ho/O. G. Pybus/M. Kaljuste/I. L. Tumanov/I. Kojola/A. A. Vorobiev/N. I. Markov/A. P. Saveljev/H. Valdmann/E. A. Lyapunova/A. V. Abramov/P. Mannil/M. Korsten/E. Vulla/S. V. Pazetnov/V. S. Pazetnov/S. V. Putchkovskiy/A. M. Rokov, Mitogenetic structure of brown bears (*Ursus arctos* L.) in northeastern Europe and a new time frame for the formation of European brown bear lineages. Molecular Ecology 16, 2007, 401–413.

Schregel et al. 2012: J. Schregel/A. Kopatz/G. Hagen/H. Broseth/M. E. Smith/S. Wikan/I. Wartiainen/P. E. Aspholm/J. Aspi/J. E. Swenson/O. Makarova/N. Polikarpova/M. Schneider/P. M. Knappskog/M. Ruokonen/I. Kojola/K. F. Tirronen/P. I. Danillov/H. G. Eiken, Limited gene flow among brown bear populations in far Northern Europe? Genetic analysis of the east–west border population in the Pasvik Valley. Molecular Ecology 21, 2012, 3474–3488.

Schwartz et al. 2006: C. C. Schwartz/M. A. Haroldson/G. C. White/R. B. Harris/S. Cherry/K. A. Keating/D. Moody/C. Servheen, Temporal, spatial, and environmental influences on the demographics of grizzly bears in the Greater Yellowstone Ecosystem (2006). https://doi.org/10.2193/0084-0173(2006)161[1:TSAEIO]2.0.CO;2.

Servheen et al. 1999: C. Servheen/S. Herrero/B. Peyton, Bears. Status survey and conservation action plan (Gland 1999).

Støen et al. 2018: O.-G. Støen/A. Ordiz/J. M. Arnemo/S. Sæbø/G. Mattsing/M. Kristofferson/S. Brunberg/J. Kindberg/J. E. Swenson, Brown bear (*Ursus arctos*) attacks resulting in human casualties in Scandinavia 1977–2016; management implications and recommendations. PLoS ONE 2018. https://doi.org/10.1371/journal.pone.0196876.

Swenson et al. 2000: J. E. Swenson/N. Gerstl/B. Dahle/ A. Zedrosser, Action plan for the conservation of the brown bear in Europe (*Ursus arctos*) (Strasbourgh 2000).

Swenson et al. 2007: J. E. Swenson/M. Adamic/D. Huber/ S. Stokke, Brown bear body mass and growth in northern and southern Europe. Oecologia 153, 2007, 37–47.

Swenson et al. 2020: J. E. Swenson/H. Ambarli/ J. M. Arnemo/L. Baskin/P. Ciucci/P. I. Danilov/ M. Delibes/M. Elfström/A. E. Evan/C. Groff/A. G. Hertel/D. Huber/K. Jerina/A. A. Karamanlidis/ J. Kindberg/I. Kojola/M. Krofel/J. Kusak/T. Mano/ A. Ordiz/S. Palazón/J. Parchizadeh/V. Penteriani/ P.-Y. Quenette/A. Sergiel/N. Selva/I. Seryodkin/ M. Skuban/S. M. J. G. Steyaert/O.-G. Støen/K. F. Tirronen/A. Zedrosser, Brown bear (*Ursus arctos*; Eurasia). In: V. Penteriani/M. Melletti (eds.), Bears of the World (Cambridge 2020) 139–161.

Tosoni et al. 2017: E. Tosoni/L. Boitani/L. Gentile/ V. Gervasi/R. Lattanie/P. Ciucci, Asessment of key reproductive traits in the Appenine brown bear population. Ursus 28(1), 2017, 105–116.

Woodroffe 2000: R. Woodroffe, Predators and people: using human densities to interpret declines of large carnivores. Animal Conservation 3, 2000, 165–173.

Zedrosser et al. 2001: A. Zedrosser/B. Dahle/J. E. Swenson/N. Gerstl, Status and managment of the brown bear in Europe. Ursus 12, 2001, 9–20.

Zedrosser et al. 2006: A. Zedrosser/B. Dahle/J. E. Swenson, Population density and food conditions determine adult female body size in brown bears. Journal of Mammalogy 87, 2006, 510–518.

Zedrosser et al. 2011: A. Zedrosser/S. M. J. G. Steyaert/ H. Gossow/J. E. Swenson, Brown bear conservation and the ghost of persecution past. Biological Conservation 144, 2011, 2163–2170.

Ziolkowska et al. 2016: E. Ziolkoswska/K. Ostapowicz/ V. C. Radeloff/T. Kuemmerle/A. Sergiel/T. Zwijacz-Kozica/F. Zieba/W. Smietana/N. Selva, Assessing differences in connectivity based on habitat versus movement models for brown bears in the Carpathians. Landscape Ecology 31, 2016, 1863–1882.

Zlatanova 2010: D. Zlatanova, Modelling the habitat suitability for the bear (*Ursus arctos* L.), the wolf (*Canis lupus* L.) and the lynx (*Lynx lynx* L.) in Bulgaria. Unpubl. PhD. thesis, Sofia University (Sofia 2010) (in Bulgarian).

Zlatanova et al. 2009: D. Zlatanova/V. Racheva/W. Fremuth, Habitatverbund für den Braunbären in Bulgarien – Grundlage für die Schaffung Transeuropäischer Wildtiernetze (TEWN) auf dem Balkan. Naturschutz und Landschaftsplanung 41, 2009, 114–122.

Prof. Andreas Zedrosser
Department of Natural Sciences and Environmental Health
University of South-Eastern Norway
Bø i Telemark
Norway

Department for Integrative Biology
Institute for Wildlife Biology and Game Management
University for Natural Resources and Life Sciences
Vienna
Austria
andreas.zedrosser@usn.no

Prof. Jon E. Swenson
Faculty of Environmental Sciences and Natural Resource Management
Norwegian University of Life Sciences
Ås
Norway
jon.swenson@nmbu.no

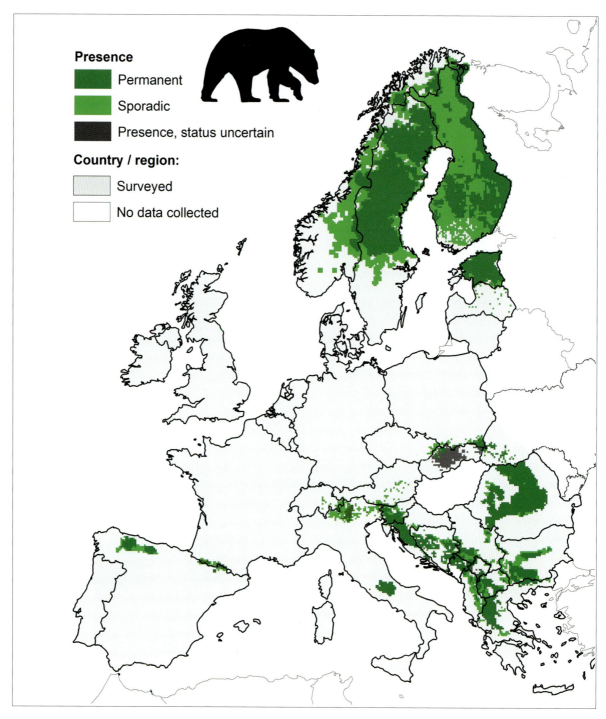

Fig. 1. Current distribution of the brown bear in Europe (data after https://www.lcie.org/Largecarnivores/Brownbear.aspx).

Table 1. Size, trend, and conservation status of brown bear populations in Europe, based on criteria by the International Union for the Conservation of Nature (data after https://www.lcie.org/Largecarnivores/Brownbear.aspx and KACZENSKY *et al. 2013).*

Population name	Countries involved	Size (2012–2016), number of individuals	Trend	Conservation status
Alpine	Italy, Switzerland, Austria, Slovenia	49–69	Stable – increasing	Critically endangered
Baltic	Estonia, Latvia	700	Stable	Least concern
Cantabrian	Spain	321–335	Increasing	Critically endangered
Carpathian	Romania, Poland, Slovakia, Serbia	7,630	Stable	Least concern
Central Apennine	Italy	45–69	Stable – increasing	Critically endangered
Dinaric-Pindos	Slovenia, Croatia, Bosnia & Herzegovina, Montenegro, Northern Macedonia, Albania, Serbia, Greece	3,950	Stable – increasing	Least concern
Eastern Balkan	Bulgaria, Greece, Serbia	468–665	Stable	Vulnerable
Karelian	Norway, Finland	1,660	Stable	Least concern
Pyrenean	France, Spain	30	Stable	Critically endangered
Scandinavian	Norway, Sweden	2,825	Decreasing	Least concern

The history of the Scandinavian Brown Bear Research Project – a formidable success story

By Jon E. Swenson and Sven Brunberg

Keywords: Brown bear, Ursus arctos, project management, international research, long-term research

Abstract: The Scandinavian Brown Bear Research Project began in northern Sweden in 1984 and central Sweden in 1985. It quickly became a cooperative Swedish study between the conservation management authority and the Hunters' Association in Sweden. This cooperative philosophy was continued by expanding to a Swedish-Norwegian cooperation in 1987 and a continually increasing internationalisation throughout the project's history. During the 35 years from 1984–2018, the period considered here, the project focused on following individual bears from birth to death. We combined research on management-relevant and general ecological topics. The project became one of the world's most successful research projects on large carnivores (actually, on birds or mammals), with 263 scientific papers in international peer-reviewed journals, 35 completed PhD degrees, and 101 completed master's degrees. We believe that one reason for this success was creating a dynamic "research family" of international experts and international students.

Introduction

Modern bear research began in North America in 1961, when a brown (grizzly) bear (*Ursus arctos*) in Yellowstone National Park was the first bear to receive a radio-collar (Craighead et al. 1995). This telemetry-based research gave the scientific community new knowledge of the ecology and behaviour of bears, which was very important to managers. Nevertheless, in the mid-1980s, it was not known how much of this knowledge about North American brown bears was applicable to brown bears in Scandinavia (Bjärvall et al. 1990), where the available knowledge was based on following tracks, and examining dens, scats, etc. (Haglund 1964; Elgmork 1979). Despite a lack of general knowledge, the major relevant questions about bears in Scandinavia in the mid-1980s were quite mundane. Hunters considered the bear population to be dense in many areas and increasingly demanded greater harvests and opening closed areas to hunting (Bjärvall/Sandegren 1987).

Research on bears in Scandinavia did not start as a part of a grand plan, but a long-standing hope among biologists to study bears led some researchers to seize a moment of opportunity while working on other wildlife projects. This is the story of the Scandinavian Brown Bear Research Project (SBBRP), focusing on the 35-year-period from its start in 1984 until 2018, as it became one of the most productive large carnivore research projects in the world. During this period, the project captured 842 brown bears and radio-marked 587 of them in two study areas (Fig. 1), trying to follow them from birth (born to a radio-marked mother) to death. This entailed recapturing them throughout their lives, involving 2,228 captures. Following so many individuals for such a long time is quite unique for studies of large carnivores and has allowed the SBBRP to make important scientific contributions and provide valuable knowledge for managers.

A COOPERATIVE PROJECT FROM THE START: THE SWEDISH BROWN BEAR PROJECT 1984–1986

The Bear Project began as an offshoot of a study of large carnivore depredation on domestic reindeer (*Rangifer rangifer*) conducted by the Research Division of the Swedish Environmental Protection Agency in the mountains of Norrbotten County, northern Sweden. The project leader was Anders Bjärvall, and the principal project personnel were Peter Segerström, Robert (Robban) Franzén, and Per Ahlqvist. The first funding was provided by the Swedish Office of the World Wildlife Fund (now World Wide Fund for Nature).

On 22 March 1984, six people followed the tracks of a yearling female bear that had been inadvertently scared out of its den 15 days earlier. They found it, hibernating or sleeping, under the branches of a Norway spruce tree (*Picea abies*), and Per Ahlqvist immobilised it using a jabstick. This started the Bear Project, which was among the first wildlife studies in Sweden using radiotelemetry. The fascinating story of this capture has been described (in Swedish) by BJÄRVALL/AHLQVIST (1985) and BJÄRVALL (2007). In 1985, three more bears were captured, two from the ground at a moose (*Alces alces*) carcass and one from a helicopter (BJÄRVALL/SANDEGREN 1987). Also in 1985, a bear study started in Dalarna County, central Sweden, on the initiative of local chapters of the Swedish Sportsmen's Association (later named the Swedish Hunters' Association, then the Swedish Association for Hunting and Wildlife Management). The association's central administration provided financing from its research fund. It started as an extension of a moose study in the area conducted by the association's Research Unit. The project leader was Finn Sandegren, and the principal project personnel were Per Ahlqvist, Lennart Petterson, and volunteer Sven Brunberg.

Three bears were captured in 1985, the first on 17 March, using a jabstick on a yearling bear in its winter den after following the mother's tracks to the den; the other yearling in the den escaped. The other two bears were immobilised from a helicopter (BJÄRVALL/SANDEGREN 1987). After this, project personnel used snowmobiles to find a track on the snow in the spring, drove in successively smaller circles to locate the area where the bear was, called in a helicopter to find it within the circle, and the bear was captured by immobilising it from the helicopter. This required manpower, which was provided by volunteers; especially important were Knut Hans Hansson and Gunnar Håkansson, Ingemar Gustavsson from Orsa Communal Forest, and helpers from the Swedish Sportsmen's Association in Gävleborg County.

Already in 1985, these two projects combined into the Swedish Brown Bear Project, a cooperation between the Swedish Environmental Protection Agency and the Swedish Sportsmen's Association. The goals of this phase of the Bear Project were to obtain information on movements, activity patterns, and population dynamics of brown bears in both the northern and southern portions of the species' range (BJÄRVALL/SANDEGREN 1987; BJÄRVALL 2007).

THE PROJECT BECOMES SCANDINAVIAN: THE SWEDISH-NORWEGIAN BROWN BEAR PROJECT 1987–1990

There had been interest in Norway to study brown bears for several years. In 1987, Petter Wabakken, a research biologist with the Norwegian Institute for Nature Research, who also was cooperating with Swedish colleagues on studies of the grey wolf (*Canis lupus*), asked to join the Swedish Bear Project. This was accepted, creating a Scandinavian cooperative project. The first two bears captured in Norway were caught in Hedmark County (later Innland County) near the Swedish border in 1988 (BJÄRVALL et al. 1990).

The bears received Very High Frequency (VHF) collars, which had to be relocated regularly by people using antennas, in contrast to the Global Positioning System (GPS) collars used today (Fig. 2). In the south, Per Ahlqvist and Sven Brunberg and other volunteers located the bears periodically from a vehicle, using the extensive network of forest roads in the rolling landscape. But the bears were often not found, apparently because they had moved out of the search area. Olle Persson, a volunteer who flew a small plane to locate radiomarked moose, also searched for radioed bears. From 1988, the Bear Project organised flights with local flying clubs and in 1990 started engaging Jämtlands Flyg, a local flying service, to have a better continuity and to locate bears not found by volunteers. The number of volunteers increased slowly to about 10–15. Their role was to locate radiomarked bears at least weekly and to help to find and capture bears. The local volunteers were intimately involved in the project and provided important transparency and informed local people about the project. Petter Wabakken expanded the use of volunteers into Norway. We also began having a meeting each winter, where the volunteers were invited to hear about our results and plans, and to discuss the volunteers' experiences and practical challenges during the past season. The field situation was much different in the north, with mountainous terrain, few roads, and travel restrictions in national parks. There, Peter Segerström and Robban Franzén conducted most of the captures and relocations from helicopters and airplanes.

This Scandinavian cooperation also brought Norwegian funding into the project. At this stage, economic support came from the Swedish Environmental Protection Agency, Norwegian Directorate for Nature Management (later Norwegian Environmental Directorate), Administrative Office of Hedmark County (later Innland County, Norway), World Wide Fund for Nature Offices in Sweden and Norway, and the Swedish companies Norma, Volvo, and Ockelbo (BJÄRVALL et al. 1990). As the project was in an early phase, it had produced only three scientific publications, one master's thesis (no fieldwork), and two bachelor's theses by 1990.

BEYOND SWEDEN AND NORWAY: THE SCANDINAVIAN BROWN BEAR RESEARCH PROJECT 1991–2018

The SBBRP experienced major changes in 1991, when Anders Bjärvall left the project for two years to work with nature conservation in Tanzania. The Swedish Environmental Protection Agency's Research Committee asked several Swedish researchers if they were interested to become a co-leader of the project, but all declined, primarily because of the project's low scientific productivity and/or the conflicts that often surround bear research and management. The financing agencies considered terminating the project, and only two new bears were marked in 1991. But the Norwegian Directorate for Nature Management was especially interested in continuing the project and offered to cover a budget deficit. Per Alqvist knew Jon Swenson, an American from Montana who worked at Grimsö Research Station, and introduced him to the Bear Project. Jon Swenson had experience with capturing and studying American black bears (*Ursus americanus*) and had helped the Bear Project estimate the number of bears in Sweden. So, he knew about the dedicated personnel in the project and their excellent database. He finished his PhD degree on hazel grouse (*Tetrastes bonasia*) in 1991 and applied to the Swedish Research Committee for a research project studying the effect of modern forestry on hazel grouse. Instead, the Committee gave him funding to be a co-leader of the Bear Project, along with Finn Sandegren.

One of the first changes Jon Swenson proposed was to consolidate the names "Swedish-Norwegian Bear Project" (used in Sweden) and "Norwegian-Swedish Bear Project" (used in Norway) to the "Scandinavian Brown Bear Research Project". The fieldwork continued as before in the north, but several changes were made in the south in 1992, where Sven Brunberg and Arne Söderberg began capturing the bears and Sven Brunberg became the field supervisor, with a 50 % position in the project (the other 50 % as supervisor of Orsa Communal Forest's machine park). Capturing and recapturing bears from a helicopter allowed us to follow individuals from birth until death. As the SBBRP became a long-term project, this individual-based dataset became increasingly valuable.

Major changes were also made in 1993. Jon Swenson accepted an offer of a permanent position as a researcher at the Norwegian Institute for Nature Research, remaining co-leader of the SBBRP. Also, Petter Wabakken accepted a position as a biologist at the County Administration in Hedmark (now Innland), and his participation in the project decreased gradually as he concentrated more on wolf management and research. The same year, the project borrowed a house for master's students in the village of Noppikoski in Dalarna County and in the center of the southern study area. Our first student with fieldwork was Bjørn Dahle. We quickly realised that it was better to have several students working together. So, from 1994, we always had several students in the field. From 1993–1997, nine students lived in Noppikoski while conducting their fieldwork. Arne Söderberg maintained the database, working full-time for the Swedish Sportsmen's Association. Sven Brunberg, who had responsibility for capturing bears and supervising the practical aspects of students' fieldwork, began to work full-time in the project in 1998. In 1999 Jon Swenson accepted a position at the Norwegian University of Life Sciences, which allowed him to keep a position at the Norwegian Institute for Nature Research and leadership of the SBBRP.

Bengt Röken was our first consulting veterinarian. Jon Martin Arnemo took over in 1997 and eventually obtained professor positions at both Hedmark University College (later Inland Norway University of Applied Sciences) and the Swedish University of Agricultural Sciences. This allowed him to lead our veterinary work, develop and improve capture protocols, support our applications for capture permits to Swedish and Norwegian ethical committees, and supervise veterinary students. Jon Arnemo's participation led to many improvements in capturing, for example by reducing the capture-caused risk of mortality from 3.8 % to 0.3 % (ARNEMO et al. 2006) and improving methods to measure and reduce stress in captured bears.

Robban Franzén had been searching for a geneticist in Sweden to determine the genetic status of the Scandinavian bear population. When that failed, Jon Swenson contacted some international experts. Pierre Taberlet at Université Joseph Fourier (CNRS) in Grenoble, France, responded positively. We met in 1994 and started an extremely productive cooperation. Pierre Taberlet had been one of the first to study the genetic structure of European bears (TABERLET/BOUVET 1994) and had developed a method to identify individuals from DNA in hair (TABERLET/BOUVET 1992). He introduced us to the multitude of ecological questions that could be answered with genetics methods (and discovered more along the way) and helped us obtain funding for PhD students and postdoctoral researchers.

We could have lost the opportunity to show how productive this cooperation was to be, because the funding authorities almost terminated the project for a second time in 1996–1997. Fortunately for us, we convinced them of the potential that the project had and we were allowed to continue.

The project had become a truly international cooperation with diverse personnel. Sven Brunberg, Peter Segerström, and Robban Franzén were excellent fieldworkers with much practical experience, and Robban Franzén worked for the Swedish Environmental Protection Agency, Peter Segerström had contacts with the Norrbotten County Administration and Sámi reindeer owners, Finn Sandegren's position connected us to Swedish wildlife managers and hunters, and Petter Wabakken brought connections and perspectives from Norway, where the attitudes towards bears and management challenges are very different from Sweden. Jon Swenson had worked as a wildlife management biologist in Montana for ten years, had research experience in conservation biology and behavioural ecology, and a Scandinavian connection (a Norwegian wife and grandparents from Norway and Sweden). Jon Arnemo had experience with many species of wildlife in both countries, and Pierre Taberlet was a world-class research geneticist and ecologist. We were united by our interest in bears and our commitment to answer management-oriented questions with solid basic research. While conducting basic research, we always remained aware of the potential uses for management that our results might have.

The project grew rapidly in number of captures, students, and volunteers (to about 25). In order to unify this growing number of people, the scientific and wildlife management questions we were asking, and the cultural and academic diversity that resulted from our internationalisation, we made an active decision to try to bond everyone into a "research family". Our plan was to design the project based on Jon Swenson's experience in the Montana Department of Fish, Wildlife and Parks, with a very flat structure, based on open discussions, cooperation, transparency, and working for a common goal. It was an enjoyable and inspiring task, although it required effort. The "research family" approach was also necessary, because basing our project on primarily external researchers and externally funded students had the advantage of channeling most of the funding from the management agencies to fieldwork (rather than wages), but the disadvantage was that the leaders had no formal supervisory authority.

Nevertheless, according to CHERUVELIL et al. (2014). this approach can lead to "high-performing collaborative research teams" who are committed to a common purpose, approach, and mutually accountable performance goals and can be created and maintained when team diversity is effectively fostered and interpersonal skills are taught and practiced. We believe that we achieved this in the SBBRP.

A major change occurred in 2002, when the Swedish Association for Hunting and Wildlife Management terminated its Research Division. Finn Sandegren retired and Arne Söderberg transferred to the National Veterinary Institute (with 50 % financing from the SBBRP through 2005). From this time, Jon Swenson and Sven Brunberg were co-leaders, Jon Swenson as the scientific leader and Sven Brunberg as the leader of practical aspects, including the fieldwork, database, and economy. In 2008, we decided to formalise the administration and cooperation of the project and formed a Steering Committee, consisting of the leaders, Genetics supervisor Pierre Taberlet, Veterinary supervisor Jon Arnemo, future Swedish project leader Jonas Kindberg (who became a co-leader of the project after he defended his PhD in 2010), and all active postdoc researchers in the project, who at the time were Andreas Zedrosser and Ole-Gunnar Støen. The reason to form the Steering Committee was to better interact with the PhD students and cooperating researchers, approve the inclusion of new researchers and PhD students into the project, help to obtain funding, and formulate policy, particularly regarding criteria for cooperating researchers, publication and authorship, media, and data ownership.

The project grew even more in 2009 as more cooperators joined the project. Physiologists Ole Fröbert from the Örebro University Hospital, Sweden, and Stepháne Blanc from the University of Strasbourg (CNRS), France, were interested in how understanding the physiology of hibernating bears might help solve some of humanity's modern health problems, related to obesity and inactivity. Our bears become obese prior to hibernation and remain inactive for six to seven months, but still remain healthy, in contrast to humans. They asked if they could cooperate with us and obtain physiological samples from individual bears while both hibernating and active. Obtaining samples from the same bears while active and hibernating required new field methods, including capturing bears in their dens (EVANS et al. 2012), which we only had done at the very start of the project. This time, we recaptured previously radioed bears in February and only used 2–4-year-old bears, for safety reasons. Then we recaptured the same bears from a helicopter in June to obtain duplicate tissue samples. The same year, Andreas Zedrosser initiated contact with Fanie Pelletier at the Université de Sherbrooke, Québec, Canada, and we started to cooperate. Fanie Pelletier provided funding for master's and PhD students, who concentrated their studies on the life history, the direct and indirect effects of hunting on population dynamics, and selection for reproductive and behavioural traits of Scandinavian bears.

We had always known that brown bears (called grizzly bears in interior North America) showed a wide range of ecological, behavioural, and life-history adaptations throughout their range, which is the largest and most diverse (deserts, deciduous and coniferous forests, alpine and tundra) for any bear. Therefore, it was important for us to understand where the Scandinavian brown bear fits into this world-wide context. We did this in two ways; by cooperating with researchers working on brown bears in other areas and by integrating PhD students working on brown bears in other ecosystems into our group. Our first such cooperation was with Djuro Huber, University of Zagreb, and Miha Adamič, University of Ljubljana, where we compared the body growth of brown bears in northern and southern Europe (SWENSON et al. 2007). In 2006, we started a cooperation with Gordon Stenhouse, the leader of the Foothills Research Institute Grizzly Bear Program in Alberta, Canada, and his collaborator Marc Cattet at the Canadian Cooperative Wildlife Health Centre, University of Saskatchewan. This cooperation was to help understand patterns of brown bear population viability in the human-dominated boreal landscapes of Alberta and Scandinavia in order to provide general applications towards conservation management. In 2013, we started a cooperation with Nuria Selva's group at the Institute of Nature Conservation, Polish Academy of Sciences, as part of the Polish-Norwegian Research Program. One product of this project was an evaluation of the factors influencing the decline of brown bears in Europe during the past 12,000 years (ALBRECHT et al. 2017). Our PhD students working on other brown bear populations in this period were Muhammad Ali Nawaz and Alice Valentini (Himalaya Mountains, Pakistan, both finished in 2008), Jodie Martin (Pyrenees Mountains, France, 2009), Andrés Ordiz (some research on bears in the Cantabrian Mountains, Spain, 2010), and Odbayar Tumendemberel (Gobi Desert, Mongolia, finished in 2020). These cooperative projects gave our students and us valuable insights into how diverse the brown bears are and how the Scandinavian brown bear is similar to, or different from, other populations.

By 2010, with the inclusion of human physiologists in the project, we needed a more formalised structure in the SBBRP. The Steering Committee decided to require that proposals for fieldwork, proposed papers, and proposed new cooperating partners would be described in "one-pagers", which we had learned from our Alberta cooperators. The fieldwork "one-pagers" were to be decided upon each September for the upcoming field season, and cooperation "one-pagers" were to be discussed at the winter Steering Committee meeting. Also, Ole Fröbert and Stepháne Blanc accepted an invitation to become Steering Committee members in 2011 (Fabrice Bertile later replaced Stepháne Blanc). It became more complicated to coordinate the many new researchers and laboratories in human physiology, as they often had different research cultures than wildlife biologists. However, this cooperation enriched the project, produced extremely interesting results, and has given us a better understanding of bear ecophysiology.

In 2011, we also decided to terminate the research in the north, because of funding restraints, the difficulty to conduct intensive studies there, and the newly adopted and aggressive bear removal policies of the government of Norrbotten County, which resulted in the killing of GPS-collared bears when they entered domestic reindeer calving areas. In 2012, for example, county managers allowed and carried out the management killing of 11 of our 29 marked bears, 20 of which had GPS collars. We were required to provide them with locations of our GPS-marked bears. We decided to keep following our two oldest females (27 and 20 years old), because we had long reproductive histories from them, and remove transmitters last from those that had not yet given birth, in order to document more ages of first reproduction. We formally terminated our research in the north in 2013. We had captured 290 bears in the north, 183 of which were radio-marked, and followed 30–50 collared bears per year there during 1994–2012. In the south, by 2018, we had captured 552 bears, followed 404 of them with radiocollars, and had 50–80 bears marked annually since 1996. As we moved to using GPS transmitters, which allowed automated locations of bears, thus making on-the-ground locations unnecessary, the number of local volunteers declined, although not before the mid-2000s. After 2010, we still had a few active local volunteers.

The Swedish Environmental Protection Agency's Wildlife Management Fund (WMF) and the Norwegian Environment Directorate have been the project's major and most stable funding sources. During the last 20 years, the project has received an average of SEK 4.5–5 million (about USD 500,000) annually. The major contributors during 2003–2014 were: Norwegian Environmental Directorate (32 %), WMF (28 %), Swedish Environmental Protection Agency's Biological Diversity Fund (16 %), regional management authorities in Sweden and Norway (8 %), funds generated by our medical physiology cooperators (4 %), and 2 % each from a Polish-Norwegian cooperative project funded by Norway, the Austrian Science Fund, the Swedish Association for Hunting and Wildlife Management, and WWF-Sweden. The project would not have been able to receive this supplemental funding without the stable, long-term support of the Swedish and Norwegian management agencies.

At the end of 2018, Jon Swenson retired, being replaced by Jonas Kindberg as sole leader. Jonas Kindberg worked for the Swedish Association for Hunting and Wildlife Management at the time, and now manages the Norwegian Large Predator Monitoring Program (Rovdata) at the Norwegian Institute for Nature Research. Sven Brunberg retired a year later and has been replaced by David Ahlqvist (the son of Per Alqvist, who captured the first bear in the project), who does most of the capturing, and by Andrea Friebe (who obtained her master's and PhD degrees in the project), who manages the other aspects, such as database, equipment, etc. The SBBRP continues and will hopefully be active and productive for many more years.

Master's students in the SBBRP

The local volunteers appreciated and were motived by being included as true members of the project, with regular feed-back from the researchers. We used the same philosophy for our students and student volunteers. We included them in our research meetings at the field station, and Sven Brunberg held two practical meetings weekly during the field period. One was to follow up on their fieldwork, ensure that they followed their planned scientific work with the expected quantity and quality of data, and give them support if some aspects did not go well. The second meeting was aimed at building a quality collective social experience; cleaning the field station, taking turns making food (serving specialties from the students' countries was very appreciated), going on group fishing trips, etc. We worked to include everyone, had private conversations to help with conflicts or other difficulties, and encouraged whenever we could. Our goal was that everyone should work hard as a part of the research family, while also obtaining a quality thesis for themselves. Many students attended the winter meetings for the local volunteers and reported on the results of their projects. Our master's students were a diverse group; 58 % were women and, although most were from Norway and Sweden, we also had master's students and university volunteers from Austria, Belgium, Canada, Chile, Croatia, Czech Republic, Denmark, Estonia, Finland, France, Germany, Iceland, Italy, Japan, Netherlands, Slovakia, Slovenia, South Africa, South Korea, Spain, Switzerland, Syria, Ukraine, UK, and USA.

In 1998, Orsa Communal Forest offered us the use of their house at Kvarnberg, close to Noppikoski, where 46 students and university volunteers lived while conducting their fieldwork until the end of the 2005 field season. When Orsa Communal Forest decided to sell this house, a donor gave the project a field station in the small village of Tackåsen, also close to Noppikoski, in 2006, and Örebro University Hospital made it possible to buy another house there (Fig. 3). More than 100 students and student volunteers stayed at the field station in Kvarnberg during 2006–2018.

How successful was our goal to include master's students in the SBBRP? Success is hard to measure, but we are convinced that our approach helped produce many successful researchers. Of the 101 students who successfully completed their master's theses by the end of 2018, 52 had published their results as a scientific paper, which is an exceptionally high percentage, and 31 were accepted

tions" at the Centre for Advanced Study at the Norwegian Academy of Science and Letters, Oslo, along with Atle Mysterud, University of Oslo, with seven other researchers participating. Jon Swenson's university awarded him a postdoctoral fellowship as a reward for obtaining this project; Andrés Ordiz was chosen for this postdoctoral position. Sven Brunberg was also recognised for his outstanding, career-long work in the SBBRP with an Honorary Doctorate Degree from the Norwegian University of Life Sciences. Andreas Zedrosser received the "Prize of the Prof. Anton Kurir Foundation 2007 for exceptional habilitation or doctoral theses" at the University for Natural Resources and Life Sciences, Vienna, for his PhD thesis.

Jon Swenson also has received several awards based on SBBRP scientific publications. Of course, the co-authors did most of the work on almost all of these publications, so they should be considered as awards to the entire SBBRP. He was listed as No. 69 in the ranking of the top 100 published researchers in Norway during 2012–2016 and was presented the "Award for Most International Publications during 2012–2014" from the Norwegian Institute for Nature Research. A paper in "Science" that he co-authored (CHAPRON et al. 2014) was ranked as No. 48 among the top 100 most influential papers for 2015 by Almetric. Richard Bischof and Andreas Zedrosser received the "Granser Research Prize for Sustainable Hunting" in 2011 for their paper BISCHOF/ZEDROSSER 2009.

Several other awards were based on public outreach. In 2011 the SBBRP was awarded the "Prize for Dissemination of Research" by the Norwegian University of Life Sciences and, in 1997, Jon Swenson was awarded the "Svein Myrberget Memorial Award for Outstanding Work in the Field of Popular Science", presented by the Norwegian Institute for Nature Research. In 1994, he was awarded the "Hugin og Munin Pris", an award presented cooperatively by the Norwegian Institute for Nature Research and Directorate for Nature Management for "bringing new influence and objectivity to a controversial area for research and management".

Epilogue

During the most active and productive period of the SBBRP, Jon Swenson and Sven Brunberg were the primary or only leaders of the project. We are grateful for this opportunity, for the trust that the management agencies showed us to fund the project for such a long time, for the friends we made with students and researchers, for the exciting and motivating research and social environment, for the knowledge the entire project produced, and that the two of us worked together so well as good, trusted friends with mutual respect and common goals. We never disagreed about our goals nor about how to reach them. This was very useful, because the project was always changing; first adding a master's student, then several, then international cooperators, then PhD students, then postdoctoral researchers, then more international co-operators. We had to adjust our group-management methods continually. Having such a diverse group caused challenges, of course, but we think everyone learned and grew from them. And we believe that the cultural and educational diversity was a major reason the project was so successful. Another reason was the project's methodology; gathering individually based data through generations over a long time, which was possible when the same basic methods were followed (with some improvements, such as adding genetic analyses and GPS technology) throughout the study, even though we investigated a wide range of topics underway. Finally, we are convinced that creating our "research family" was essential for the success of such a large and diverse project, where the leaders did not have employer responsibilities for the students and cooperators.

Acknowledgements

Leading the SBBRP was a fantastic experience, and we thank everyone who contributed. We also thank Andreas Zedrosser and Andrés Ordiz for helpful comments on the manuscript.

Bibliography

Albrecht et al. 2017: J. Albrecht/K. A. Bartoń/N. Selva/R. S. Sommer/J. E. Swenson/R. Bischof, Humans and climate change drove the Holocene decline of the brown bear. Scientific Reports 7, 2017, 10399.

Arnemo et al. 2006: J. M. Arnemo/P. Ahlqvist/R. Andersen/F. Berntsen/G. Ericsson/J. Odden/S. Brunberg/ P. Segerström/J. E. Swenson, Risk of capture-related mortality in large free-ranging mammals: experiences from Scandinavia. Wildlife Biology 12, 2006, 109–113.

Bischof/Zedrosser 2009: R. Bischof/A. Zedrosser, The educated prey – consequences for exploitation and control. Behavioral Ecology 20, 2009, 1228–1235.

Bjärvall 2007: A. Bjärvall, Tretio år med rovdjur – och människor (Möklinta 2007).

Bjärvall/Ahlqvist 1985: A. Bjärvall/P. Ahlqvist, Första björnen med radiosändare. Svensk Jakt 123, 1985, 278–281.

Bjärvall/Sandegren 1987: A. Bjärvall/F. Sandegren, Early experiences with the first radio-marked bears in Sweden. Proceedings of the International Conference on Bear Research and Management 7, 1987, 9–12.

Bjärvall et al. 1990: A. Bjärvall/F. Sandegren/P. Wabakken, Large home ranges and possible early sexual maturity in Scandinavian bears. Proceedings of the International Conference on Bear Research and Management 8, 1990, 237–241.

Chapron et al. 2014: G. Chapron/P. Kaczensky/J. D. C. Linnell/M. von Arx/D. Huber/H. Andrén/J. V. López- Bao/M. Adamec/F. Álvares/O. Anders/ L. Balciauskas/V. Balys/P. Bedo/F. Bego/J. C. Blanco/U. Breitenmoser/H. Brøseth/L. Bufka/ R. Bunikyte/P. Ciucci/A. Dutsov/T. Engleder/ C. Fuxjäger/C. Groff/M. Heltai/K. Holmala/ B. Hoxha/Y. Iliopoulos/O. Ionescu/G. Ivanov/ J. Jeremić/K. Jerina/F. Knauer/I. Kojola/I. Kos/ M. Krofel/J. Kubala/S. Kunovac/J. Kusak/M. Kutal/ P. Mannil/R. Manz/E. Marboutin/F. Marucco/ D. Melovski/K. Mersini/Y. Mertzanis/R. W. Mysłajek/ S. Nowak/J. Odden/J. Ozolins/G. Palomero/M. Paunovic/J. Persson/H. Potočnik/P.-Y. Quenette/ G. Rauer/I. Reinhardt/R. Rigg/A. Ryser/V. Salvatori/T. Skrbinšek/A. Skrbinšek-Majić/A. Stojanov/J. E. Swenson/A. Trajçe/E. Tzingarska-Sedefcheva/M. Váňa/R. Veeroja/M. Wölfl/S. Wölfl/ F. Zimmermann/D. Zlatanova/L. Boitani, Successful recovery of large carnivores in Europe's human-dominated landscapes. Science 346, 2014, 1517–1519.

Cheruvelil et al. 2014: K. S. Cheruvelil/P. A. Soranno/K. C. Weathers/P. C. Hanson/S. J. Goring/C. T. Filstrup/E. K. Read, Creating and maintaining high-performing collaborative research teams: the importance of diversity and interpersonal skills. Frontiers in Ecology and the Environment 12, 2014, 31–38.

Craighead et al. 1995: J. J. Craighead/J. S. Sumner/J. H. Mitchell, The grizzly bears of Yellowstone; their ecology in the Yellowstone ecosystem 1959–1992 (Covelo 1995).

Elgmork 1979: K. Elgmork, Bjørn i naturen (Oslo 1979).

Evans et al. 2012: A. Evans/V. Salén/O.-G. Støen/Å. Fahlman/S. Brunberg/K. Madslien/O. Fröbert/J. E. Swenson/J. M. Arnemo, Capture, anesthesia, and disturbance of free-ranging brown bears (*Ursus arctos*) during hibernation. PLOS One 7, 2012. https://doi.org/10.1371/journal.pone.0040520.

Haglund 1964: B. Haglund, Björn och lo (Stockholm 1964).

Mills et al. 2015: J. A. Mills/C. Teplitsky/B. Arroyo/ A. Charmantier/P. H. Becker/T. R. Birkhead/ P. Bize/D. T. Blumstein/C. Bonenfant/S. Boutin/ A. Bushuev/E. Cam/A. Cockburn/S. D. Côté/J. C. Coulson/F. Daunt/N. J. Dingemanse/B. Doligez/ H. Drummond/R. H. M. Espie/M. Festa-Bianchet/ F. Frentiu/J. W. Fitzpatrick/R. W. Furness/D. Garant/ G. Gauthier/P. R. Grant/M. Griesser/L. Gustafsson/B. Hansson/M. P. Harris/F. Jiguet/P. Kjellander/E. Korpimäki/C. J. Krebs/L. Lens/J. D. C. Linnell/M. Low/A. McAdam/A. Margalida/J. Merilä/A. P. Møller/S. Nakagawa/J.-Å. Nilsson/I. C. T. Nisbet/A. J. van Noordwijk/D. Oro/T. Pärt/F. Pelletier/J. Potti/B. Pujol/D. Réale/R. F. Rockwell/ Y. Ropert-Coudert/A. Roulin/J. S. Sedinger/J. E. Swenson/C. Thébaud/M. E. Visser/S. Wanless/D. F. Westneat/A. J. Wilson/A. Zedrosser, Archiving primary data: solutions for long-term studies. Trends in Ecology and Evolution 30, 2015, 581–589.

Swenson et al. 2007: J. E. Swenson/M. Adamič/D. Huber/ S. Stokke, Brown bear body mass and growth in northern and southern Europe. Oecologia 153, 2007, 37–47.

Taberlet/Bouvet 1992: P. Taberlet/J. Bouvet, Bear conservation genetics. Nature 358, 1992, 197.

Taberlet/Bouvet 1994: P. Taberlet/J. Bouvet, Mitochondrial DNA polymorphism, phylogeography, and conservation genetics of the brown bear (*Ursus arctos*) in Europe. Proceedings of the Royal Society of London, Series B 255, 1994, 195–200.

Prof. Jon E. Swenson
Faculty of Environmental Sciences and Natural Resource Management
Norwegian University of Life Sciences
Ås
Norway
jon.swenson@nmbu.no

Dr. Sven Brunberg
Orsa
Sweden
svenolov.brunberg@gmail.com

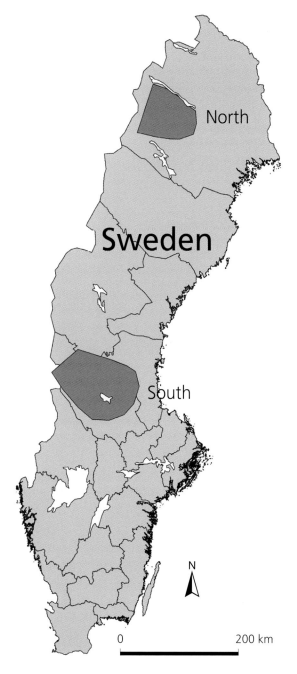

Fig. 1. The two major study areas of the Scandinavian Brown Bear Research Project. The darker areas show where most of the brown bears were captured. Note that the southern study area extends into Norway (map J. Hansen).

Fig. 2. A female brown bear is being captured using a helicopter and a tranquilising dart fired from a gas-operated dart gun as part of the research activities of the Scandinavian Brown Bear Research Project (photo A. Zedrosser).

Fig. 3. The Scandinavian Brown Bear Research Project's field station in Tackåsen, Sweden, with the main building and extra sleeping quarters (photos J. Swenson [a]/O. Fröbert [b]).

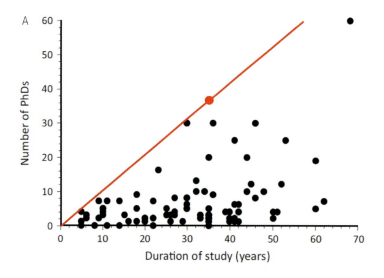
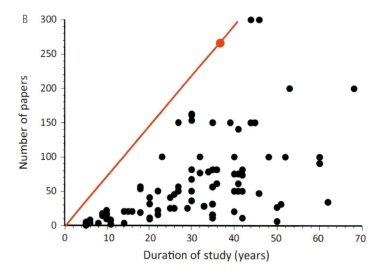

Fig. 4. Number of completed PhD theses (a) and scientific papers (b) produced by 91 long-term research projects on individual-based field studies of birds and mammals worldwide in relation to the length of the study (after MILLS et al. 2015) and those of the Scandinavian Brown Bear Research Project during 1984–2018 (red circles). The studies to the left of the red line have been more productive than the Scandinavian Brown Bear Research Project.

The management of brown bears in Sweden, Norway and Finland

By Michael Schneider, Andreas Zedrosser, Ilpo Kojola and Jon E. Swenson

Keywords: Brown bear, Ursus arctos, *population management, hunting, Sweden, Finland, Norway*

Abstract: There are about 2,700 bears in the central and northern parts of Sweden, about 150 bears in Norway (most of them along the Swedish, Finnish, and Russian borders), and about 2,400 in Finland, mostly in the eastern parts of the country. The conservation status of the brown bear is considered "Near Threatened" in Sweden and Finland and "Endangered" in Norway. All three countries have well-developed population monitoring programs, but the methods used differ widely. However, because these countries share the same population of bears, cross-border collaboration in research, management, and the sharing of information is well established. All three countries have damage compensation systems in place, however, the type of damages vary; in Sweden and Finland they are mainly due to depredation of semi-domestic reindeer in the northern parts of the countries, while damages in Norway are mainly related to the depredation of free-grazing sheep and they are concentrated in the eastern part of the country, along the border with Sweden. Bears in Sweden and Norway are managed at the regional level, while bears in Finland are managed on the national level. Hunting of bears is allowed in all three countries nowadays.

Introduction

Humans have always been interested in large carnivores, due to the threat they cause to livestock and sometimes to humans as well, but also because their strength, agility and beauty fascinate and inspire the human imagination. Five species of large carnivorous animals exist in the Nordic countries of Sweden, Finland, and Norway, the brown bear (*Ursus arctos*), grey wolf (*Canis lupus*), wolverine (*Gulo gulo*), Eurasian lynx (*Lynx lynx*), and golden eagle (*Aquila chrysaetos*). In this chapter, we focus on the management of brown bears in Sweden, by using Västerbotten County as example, but also provide information on bears and their management in Norway and Finland.

Brown bears were common in northern Europe until the middle of the 1800s, when their populations decreased rapidly due to human persecution. The species was protected in the 1900s, earliest in Sweden and latest in Norway. The populations increased after protection, especially so in Sweden and in Finland, which received immigrants from Russia. The increase in Norway was much less and came later, because the bear had been exterminated as a reproducing species and its return was dependent upon immigration from neighbouring countries (Swenson et al. 1995; see Fig. 1). Today, there are about 2,700 bears in the central and northern parts of Sweden, about 150 bears in Norway (most of them along the Swedish, Finnish, and Russian borders), and about 2,400 in Finland, mostly in the eastern parts of the country (Fig. 2; cf. Bischof et al. 2020). Bear management differs quite a lot among the Fennoscandian countries.

Historic changes in brown bear populations and distribution in Sweden, Norway, and Finland

Sweden
Swedish public policy regarding the brown bear has changed greatly through the centuries. Early on, the national policy was to exterminate the species, and as early as in the 14th century there were laws in Sweden that required people to take part to meet this goal (Danell/Bergström 2016). Bounties were introduced in 1647 as a measure to help reach the extermination objective. Thousands of bears still existed in Scandinavia in the mid-1800s, with an estimated 3,100 bears in Norway and 1,650 in Sweden. By the end of the 19th century, the numbers of bears were extremely low in both countries. The lowest population level occurred around 1930 and was estimated at 130 bears, all of them living in Sweden (Swenson et al. 1995), although genetic estimates indicate that the numbers were somewhat higher (Xenikoudakis et al. 2015).

However, changing opinions among academics, hunters, and the public resulted in a paradigm shift at the end of the 1800s in Sweden, leading to the removal of bounties in 1893. Several other measures to protect bears, such as restrictions on where they could be killed and making any dead bear the property of the state, additionally contributed to the subsequent population increase (Swenson et al. 2017). The management paradigm in Sweden changed again in 1943, when hunting seasons were introduced. Nevertheless, the bear population continued to increase and reached a size of about 3,300 bears in 2008, some 60 years later (Kindberg et al. 2009). This increase was broken when changed management objectives in the counties with bears caused a decline of the population, which was subsequently estimated to consist of about 2,700 bears in 2018 (Bischof et al. 2020).

Norway
Norway did not change its extinction policy when Sweden did, and bears were virtually eliminated by 1920–1930, even though there were about twice as many bears in Norway as Sweden in the mid-1800s (Swenson et al. 1995). Since 1975, bear observations increased again in Norway, due to immigration from neighbouring countries and coinciding with a pronounced increase in the Swedish bear population; bears reappeared sooner in areas closer to the remnant Swedish populations (Swenson et al. 1995). Large carnivores are managed intensively in Norway, based on small population goals and small management zones where resident brown bears are accepted. In combination with low dispersal abilities of female bears, this reluctant attitude towards bears has prevented the species from re-establishing a large population in the country.

Finland
According to a back-calculation, there had been approximately 1,000 bears in Finland until their decline started around 1875. Thereafter, the population declined by about 210 bears per decade until 1905. The decline continued until 1915, when an estimated population of 129 bears was left in Finland (Mykrä/Pohja-Mykrä 2015). At this time, bears had disappeared from southern, southwestern, and western Finland, with remnants of the population restricted to the northern and eastern parts of the country. By the late 1960s, the number of bears started to increase again, from about 150 bears in 1970 to 450 in 1985. The species also extended its range into the western and southern parts of the country. The first reproducing females were observed in central Finland in the late 1980s, and in the western and southern parts of the country in the early 1990s (Pulliainen 1990; Kojola/Laitala 2000). The bear population in Finland was estimated at 2,300–2,500 individuals in summer 2020 (Heikkinen et al. 2021). Most of the bears in Finland live along the Russian border, and immigration of bears from the Russian part of Karelia explains much of the growth and range expansion of the bear population in Finland, in spite of relatively high harvest rates (Pulliainen 1997; Saarma/Kojola 2007).

Management of brown bears in Sweden

The Swedish management system for brown bears is knowledge-based, rather well informed, and is required by law to be adaptive, although that does not always seem to be the case (Swenson et al. 2017). It includes monitoring of the population size and distribution, as well as of human attitudes, in addition to subsidies for measures to prevent depredation on livestock, a system for damage compensation payments, and stakeholder involvement in decision making. The elaborate system for the monitoring and management of large carnivores is in part due to the compensation system for damages to domestic reindeer (*Rangifer tarandus*). Wild reindeer became extinct in Sweden in the 1800s, but today there are about 350,000 domestic reindeer owned by native Sámi people. In accordance with the rules and regulations of the European Union, limited hunting of brown bears is allowed in Sweden as a measure to prevent agricultural damages.

The Swedish Environmental Protection Agency (EPA) is the national authority for nature conservation and environmental issues. The EPA advises the Swedish government on these issues and provides instructions, advice, and information on large carnivores to regional authorities and the public. The EPA puts much effort into the management of large carnivores, including policy tools (legislation), guidelines and strategies, funding (for regional administration, surveillance, research, and information), and issuing hunting permits.

Sweden is subdivided into 21 counties, each with its own administration. The County Administrative Board is a regional authority that is a link between the people and municipalities of the county on the one hand, and the government, Parliament, and national authorities on the other. Environmental issues, such as wildlife management, large carnivore conservation, and reindeer husbandry constitute a major field of work for the county boards. For example, about 20 people are involved in large carnivore management in Västerbotten County, working on topics such as population monitoring, managing hunting, damage compensation, and public information. Regional stakeholder involvement is achieved by so-called Delegations for Game Management at the county boards. They consist of regional politicians and representatives of different interest groups affected by or interested in large carnivores. Delegations for game management give advice to county boards and are charged with making overarching decisions regarding the management of game species in the county (Lundmark/Matti 2015; SFS 2009:1474).

Furthermore, Sweden is subdivided into three areas for the management of large carnivores (Fig. 3). These zones were established to increase and improve the cooperation among the counties and to facilitate cross-boundary management. Within a management area, counties are similar with respect to species composition and population sizes of the carnivore community. They are also comparable regarding human population density and landscape use, as well as types and extent of human-carnivore conflicts. Between management areas, differences in the aforementioned factors are relatively large. The northern management area contains a large population of brown bears and almost the entire reindeer husbandry zone. The central management area contains good numbers of bears, but few reindeer. In the southern area, bears are absent (Schneider 2017).

Monitoring size and trend of bear populations
The population size and trend of brown bears in Sweden are estimated based on a combination of genetic and observational methods. Together, these methods provide a very good overview of the dynamics and spatial distribution of brown bears in Sweden.

DNA-based methods are used to estimate population size (Bellemain et al. 2005; Bischof et al. 2020). Brown bear DNA surveys are usually carried out at five-year intervals in all counties with bears. The management authorities provide sampling kits, and volunteers, commonly hunters, are asked to collect a small sample whenever they find bear feces. These samples are then sent to a designated laboratory for analysis. The lab results are used to estimate the population size of brown bears at the county, as well as the national, level based on statistical capture-recapture models (SCHNEIDER 2006; 2015).

The trend of the brown bear population is monitored annually at the county level using effort-corrected observations of bears by moose (*Alces alces*) hunters (KINDBERG et al. 2009). Every autumn, thousands of hunters are in the Swedish forests to participate in moose hunting. Most of the hunters are organised into hunting parties, i.e. groups of people hunting together in an area that they own or lease (SCHNEIDER 2017). Hunting parties are required to keep track of bear sightings during moose hunting and to report their results to the Swedish Hunters' Organisation (KINDBERG et al. 2009). Based on these data, an annual index of bear sightings per observation effort is published for every county with a bear population. Because the visibility of bears varies among years, due to weather conditions and other factors, and among areas, due to differences in forest density, etc., this index only indicates the trend of the bear population in a given area. Together with results from DNA surveys, trends from observation indices can be used to estimate bear population size in the counties in different years (KINDBERG et al. 2011).

Framework for bear management
The brown bear is considered to be "Near Threatened" in the Swedish Red List (ARTDATABANKEN 2020), based on criterion D of the International Union for the Conservation of Nature (fewer than 2,000 reproducing individuals). Article 11 of the European Union Habitats Directive requires member states to monitor the habitats and species listed in the annexes, and Article 17 requires a report to be sent to the European Commission every six years. The core of this Article 17 report is an assessment of the conservation status of the habitats and species targeted by the directive. Conservation status is assessed using a standard methodology as being either "favourable", "unfavourable-inadequate" or "unfavourable-bad", based on four parameters as defined in Article 1 of the Directive. The parameters for species are range, population, habitat of the species, and future prospects (DG ENVIRONMENT 2017). According to the Article 17 reporting by the Swedish EPA in 2019, the reference population size for the brown bear is 1,400 individuals (NATURVÅRDSVERKET 2020). That means that at least that many bears must live in Sweden, but the population can be much larger, as long as it does not cause too many problems. Today, the brown bear has favourable conservation status in both biogeographical regions in which it occurs in Sweden (alpine and boreal).

The main goal of the management of large predators in Sweden is stated in Section 1 of the Ordinance on the Management of Bear, Wolf, Wolverine, Lynx and Golden eagle (SFS 2009:1263); these species should occur in such large numbers that they persist in the Swedish fauna in the long term and that they can spread to their natural areas of distribution. This aim is to be achieved at a pace that promotes the coexistence of humans and these species, while preventing and limiting damages and inconveniences. According to the government's proposition "A sustainable predator policy" (REGERINGEN 2013), the general and long-term objective is that large carnivores in Sweden shall achieve and maintain favourable conservation status in accordance with the European Union's Habitats Directive, at the same time as livestock husbandry is not significantly hindered and socio-economic considerations are taken into account. In 2018, in its "Strategy for Swedish Wildlife Management" (SWEDISH EPA 2018), the Swedish Environmental Protection Agency has formulated a vision for wildlife management in Sweden. The vision can be viewed as a long-term objective for Swedish wildlife management and draws on the values of wildlife in a broad sense, for nature experiences and tourism, for hunting, for the provision of game meat, and for the conservation of biodiversity. Everyone should have access to these values, regardless of background, gender, disabilities, or other conditions.

National and regional management plans
The national management plan for the brown bear (Naturvårdsverket 2016) translates the general management principles defined by the Parliament and the government into more concrete objectives and measures. According to the Swedish EPA, the following specified goals should be achieved during 2014–2019, the period covered by the plan: 1) Reach and maintain favourable conservations status; 2) Reduced damage; 3) Increase confidence in management; and 4) No illegal hunting of bears. Sweden joined the EU in 1995, which resulted in the protection of the brown bear. However, bears can still be hunted in Sweden, using the derogations (exceptions) allowed in EU's Habitats Directive, which are implemented in national Swedish legislation. The hunting law instituted by Parliament, the hunting ordinance issued by the government, and more detailed regulations from the EPA define bear hunting in Sweden today. Actual decisions on bear hunting are made by county boards at the regional scale, after input from their Delegation for Game Management.

Within the three management zones, county boards cooperate on different aspects of large carnivore management. These include common guidelines and quotas for management removals of problematic animals, justification of and quotas for license hunting, models for administrative routines, collaborative suggestions for county-wise minimum levels, and the production of regional management plans. According to section 7 in the Ordinance for the Management of the Bear, Wolf, Wolverine, Lynx, and Golden eagle (SFS 2009:1263), each county board must establish a regional management plan for large carnivores. Management plans describe the general conditions in the county and assess the basis for large carnivore occurrence. They also describe numbers and distribution of carnivores, the conservation status of the predators, their effects on prey species, the problems they cause, and their socioeconomical consequences for society. Most importantly, regional management plans include concrete objectives for population size and distribution of the species, as well as acceptable levels of problems and inconveniences that carnivores inflict.

Bears in Sweden occur mostly in the six northernmost counties, Norrbotten, Västerbotten, Jämtland, Västernorrland, Gävleborg, and Dalarna. Table 1 summarises minimum levels, management targets, and limits of management intervals for the brown bear in these six counties. Minimum levels sum up to 1,400 bears, which has been defined as the national reference value, i.e. the national minimum level for the population.

Hunting of bears in Sweden
Bears are protected by the EU Habitats Directive (Council Directive 92/43/EEC of 21 May 1992 on the conservation of natural habitats and of wild fauna and flora), but exceptions can be made, and bears can be killed if the objective of the hunt is to prevent serious damage and there is no satisfactory alternative to solve the problems. In addition, killing bears must not be detrimental to the maintenance of the population at a favourable conservation status in its natural range, at neither the regional nor the national scale.

Two types of bear hunting exist in Sweden, license hunting and protective hunting. License hunting aims at regulating populations by managing the density, size, and growth rate of these populations. It operates at large spatial scales and multi-year time frames, and it is an important and fundamental component in the management of large carnivores in Sweden. Any person who has passed a hunter's exam and purchased a general hunting permit from the EPA can take part in license hunting of bears in areas where he or she has the right to hunt and where the hunting of bears is allowed. There is no governmental fee for killing a bear, but private landowners can sell the possibility to hunt, within the hunting quota that has been set by the county board. Most of the bears are killed by specialised hunters using dogs, many are shot during still hunting (also by people sitting and waiting for moose), some bears are killed by hunters stalking them, and very few by hunters using baits (Bischof et al. 2008; Swenson et al. 2017; Zedrosser et al. 2020). In northern Sweden, license hunting is a well-founded and planned management effort and an inclusive phenomenon, where reindeer herders and local people can work together for mutual benefit, where bears can be an appreciated resource for hunting instead of being pests, and where the socio-economic and psychosocial impact on reindeer husbandry and other parts of society decreases, at the same time as hunting can generate income for local people. Managers argue that the acceptance for bears as well as large carnivore management increases because of that (see DRESSEL et al. 2021).

The other type of hunting is termed as protective hunting, i.e. management removals. County boards can grant permits for the lethal removal, i.e. killing of bears, when people who experience problems apply for it. Such applications are relatively rare in central Sweden, but quite common in the northern half of the country, where bears can cause serious damage to young reindeer on calving grounds in spring. A permit to kill a bear, if granted, is usually given to the person applying for it, but if this person does not want to hunt, he or she can ask other hunters to remove the bear. In remote mountain areas, when bears kill reindeer calves in spring, the carnivores often are removed by county board staff using helicopters. Protective hunting of problematic carnivores is an administratively extensive measure, it is usually event-driven (i.e. not planned in advance), it is usually costly, it can be controversial, and it has only a short-term effect within a limited area. In addition, management removals often exclude local hunters, and they reduce opportunities to use the bears as a resource, as the animals that are killed have usually been confiscated by the authorities.

Overall, about 350 bears are shot every year in Sweden. This is far more than were killed in the second half of the 1800s, when the bear nearly was eradicated (cf. Fig. 1). Although authorities welcome the huge interest among hunters, some recent developments in hunting practices are questionable. Hunting quotas are filled increasingly rapidly, which according to Police officials is facilitated by automatic cameras, technical equipment for tracking hunting dogs, motorised vehicles, and illegal bait sites, which some hunters use. Bear hunting has very much become a race between specialised bear hunters with trained hunting dogs, the most successful of which have killed more than 40 bears each. Also, there are indications that the selling of guided hunts is increasing rapidly in Sweden. Baiting for hunting bears was banned in Sweden in 2001 but was allowed again in 2014. Especially in Jämtland County it has become very popular, but only few bears are shot at bait sites (Fig. 4). The character of bear hunting is changing in Sweden, and the manager-caused mortality of bears has increased greatly (Fig. 5). In 2020, unusually many bears were killed on reindeer calving grounds in the spring (see the case study from Västerbotten County, below, for further details).

Bears and people in Sweden
Bear attacks on livestock
Although the bear population is relatively large in Sweden, attacks on livestock (other than reindeer) are comparatively rare, especially when compared to the number of attacks by wolves and lynx (Fig. 6). There are approximately 600,000 sheep, 350,000 dairy cattle, and 1.5 million beef cattle in Sweden. Most depredation cases by bears involve sheep. The number of attacked sheep per bear is much lower in Sweden than Norway, despite the higher numbers of bears (Table 2). There are three main reasons for the low number of bear attacks in Sweden compared to Norway. First, there are relatively fewer farms with sheep in the main area of brown bear distribution in Sweden, although most sheep in Norway also are outside of the bear range. Second, sheep are usually kept within areas protected by electric fences in Sweden. Third, animal welfare legislation in Sweden requires livestock to be visually observed at least once a day. In comparison, Norway has more than two million sheep grazing freely and unsupervised throughout the country. Unattended, free-ranging sheep are an easy prey for any bear in the area. The result is that Norway pays 210 times more in compensation for lost livestock and preventive measures to protect livestock per bear than Sweden, and 138 times more for compensation alone per bear than Sweden. Finland pays ten times more for compensation per bear than Sweden (BAUTISTA et al. 2019).

Over the last 20 years, there has been an increasing trend of bear damages to beehives in Sweden. However, bear attacks on beehives are relatively easy to prevent with electric fencing, and financial assistance is available from county boards for setting up fences to deter bears. If attacks happen anyway, monetary compensation to replace damaged equipment and bees can be paid by county boards.

Bear attacks on people
The Scandinavian brown bear is not particularly dangerous. Very few people actually meet bears in Scandinavia, and the risk of being injured by bears while engaging in outdoor activities is exceedingly low (Støen et al. 2018). Several factors influence the risk of bear attacks on people. These are, in decreasing order of importance: the presence of cubs, proximity to a carcass, proximity to a den, and the presence of a dog (Swenson et al. 1999). Bear physiology at denning, which makes bears more prone to stay than flee, may make encounters with bears riskier in the fall, when they prepare for hibernation. In Scandinavia, although attacks on humans are relatively rare, injuries from bear attacks have increased during the last decades and fatalities have been documented for the first time for more than 100 years. During a period of 40 years (1977–2016), 44 attacks occurred in Sweden and Norway, in which 42 people were injured and two were killed. During the same period, 26 attacks occurred in Finland, in which 25 people were injured and one was killed (I. Kojola, unpublished data). Victims of bear attacks in these three countries are mostly hunters, and the risk of hunters being attacked increases with bear population density in the area. Hunters are more commonly affected by bear attacks compared to other groups of outdoor users or recreationists. A fatal incident in Sweden in 2004 has resulted in information campaigns and annual hunting courses, which focus on bear behaviour and on safety issues during bear encounters (Støen et al. 2018).

People's attitudes towards bears in Sweden
The relationship between people and large carnivores is multi-faceted. Among other things, human attitudes depend on the levels of predator damage to dogs, livestock, reindeer and game animals, on actual or perceived threats to humans, and on levels of local involvement during decision making processes in relation to bears with problematic behaviour. Many people have strong feelings towards large carnivores. Often, it is not the predators that are problematic per se, but the underlying cause is a conflict between a central administrative institution and the countryside community, a conflict that may exist at several scales. People's feelings and attitudes must be taken seriously when managing large carnivores (Schneider 2008).

Thoughts and actions
A multitude of attitude surveys regarding large carnivores have been carried out in Sweden. Since 2004, these surveys have been done at a large scale every five years, encompassing mostly the northern half of Sweden (Ericsson/Sandström 2005). Results from these surveys show that an overwhelming majority of the people is supportive of both large carnivores and their current management, however, many of the people who live closest to the carnivores are negative, and over time, people's support for large carnivores and large carnivore management can fluctuate. This is true especially for wolves and bears (Sandström/Ericsson 2009; Sandström et al. 2014; Dressel et al. 2021). These are important findings for managers, as negative attitudes may result in illegal killings of large carnivores. According to Swenson et al. (2011), illegal killing of bears is low, but more common in the northern part of Sweden; annual rates of illegal mortality among adult females was estimated to be less than 1 % in the south, but 2–3 % in the north. Documented and suspected illegal deaths showed no seasonal trend in the south, but were concentrated to spring and autumn in the north. In the north, illegal mortality has been documented to be especially high in the mountain national parks (Rauset et al. 2016). Generally, illegal killing does not seem to affect population trends among brown

bears in Sweden, but may be important locally. The level of illegal killing appears to be stable and not related to the level of legal hunting. Therefore, mortality caused by illegal killing is probably additive to the mortality caused by legal hunting.

Bears both repel and attract people
Although many people in Sweden seem to be afraid of bears (DRESSEL et al. 2021), attitude surveys show that most people like the fact that bears occur in the country. After the lynx, the bear is the second most popular species of all large carnivores in Sweden. However, the popularity of the bear decreased between 2004 and 2009, but it increased again between 2014 and 2020. Most people do not like the idea of meeting a bear in the forest (DRESSEL et al. 2021). However, if bears expose themselves along roads or on fields in the spring, many people will come to watch them. Bears are the object of human curiosity, as long as observing them can be done from a safe distance, and many people can congregate around such bears.

Case study: Bear predation on reindeer calves in Västerbotten – immobile hunters, scarce herders, and climate change increase bear predation on reindeer calves
Västerbotten is the second largest (55,000 km^2) and second-most northerly county in Sweden (Fig. 7). The county stretches from the coast of the Bothnian Bay in the east up to almost 1,800 m above sea level in the mountains along the Norwegian border in the west. The climate varies considerably between different parts of Västerbotten, but generally it is characterised by cold winters with heavy snow. More than 50 % of the county is covered by forest, which is intensively used by large-scale forestry. Most of Västerbotten's about 273,000 inhabitants live along the coast in the east, where the biggest cities are located. Human population density decreases steadily from east to west, with few inhabitants in the forested inland areas and especially so in the mountain range. Most Sámi reindeer herders live in the central and western parts of the county. Today, reindeer herding is not economically rewarding, and in many Sámi families only one person works fulltime with reindeer (SJÖLANDER et al. 2009). As a consequence, there are few people guarding reindeer in the woods and mountains.

The bear population
The number of bears in Västerbotten has been estimated four times, in 2004, 2009, 2014, and 2019. The latest survey revealed a population of about 450 animals in autumn 2019. In recent years, the annual rate of harvest has been about 10 % of the population, but the number of bears has increased, nevertheless. The distribution of bears is uneven, especially so for females, which mostly occur in the southern and northwestern parts of the county. Most males occur in the central parts of Västerbotten, but densities are low in the mountains and in the coastal areas. In 2020, 73 bears are known to have died in the county, and with few exceptions, they were shot during protective or license hunting. During most years since 2005, there has been a special hunting quota for the Västerbotten mountain range, or more precisely the area between the Norwegian border and the *odlingsgränsen*, a border which was legally defined in 1890 as the western (upper) limit of new habitation in order to secure the higher altitudes for Sámi domestic reindeer herding (LUNDMARK 2006). Most of Västerbotten's calving grounds for reindeer are situated in this area.

Bears kill reindeer
For many years, reindeer herders have identified bears as a big problem on reindeer calving grounds in the spring. During 2010–2012 a study of bear predation on reindeer calves was conducted by the Swedish Wildlife Damage Center and the Scandinavian Brown Bear Research Project in cooperation with two reindeer herding cooperatives in northernmost Sweden (KARLSSON et al. 2012). The research showed that the average bear kills and eats eleven reindeer calves during the calving season

in May and June, but that there are big differences among individual bears and between years. The researchers also studied different measures to protect reindeer calves from bears, and they concluded that hunting bears is the most effective measure.

Inspired by traditional Sámi knowledge, the Västerbotten County Administration established three management zones for brown bears in 2005 (SCHNEIDER 2006). In Västerbotten, most reindeer calving grounds are situated in the mountains close to the Norwegian border. Therefore, few bears should occur in this region, and the objective was to regulate the population by targeted hunting quotas in the area. Hunting quotas were set, based on the results of bear population size surveys.

Finding bears

Bear surveys to determine population size are organised as citizen science projects by the County Board in cooperation with the regional branch of the Swedish Hunters' Organisation. Sampling equipment for bear scats is distributed to hunters, reindeer herders, hikers, and other people working or relaxing in the outdoors, but the participation is voluntary. The survey period starts on 21 August (the onset of the bear hunting season) and ends on 31 October. Most bear scat samples are collected by hunters while they are hunting moose. Many samples are collected on or close to forest roads, presumably because they are easier to find there and because many hunters and berry pickers do not go very far from roads. In most cases, the calving grounds of reindeer and surrounding lands are rather remote areas that lack a dense network of roads. In consequence, during bear population size surveys, only few scat samples are collected in these areas. Reindeer herders move often off-road, but several of them have declared that they are unwilling to collect bear fecal samples.

An unfavourable spring

When the reindeer herds reached the mountains during their traditional spring migration in 2020, they could not reach their traditional calving grounds at higher altitudes due to large amounts of snow. Instead, they had to stay in the forests of the foothills and wait for the snow higher up to melt. Consequently, the calving season started when the reindeer still were at these lower altitudes, and great numbers of pregnant reindeer and newborn calves were in the area when bears started to emerge from hibernation. Neonatal reindeer and moose calves are preferred food of Scandinavian bears in spring, and the damage inflicted to reindeer herding was massive. In response to the problematic situation, the Västerbotten County Board granted about 50 licenses for protective hunting of bears, and 43 bears were shot.

Outraged hunters

The Swedish Hunters' Organization had a media campaign against the county boards and the bear management in northern Sweden, which according to hunters is a total disaster. They claimed that bears should be shot by hunters during the ordinary hunting season in autumn, as a valuable natural resource for sport and recreation, not as vermin by state officials in spring.

In summary

According to the management plan for the brown bear in Västerbotten, there should be few bears in the mountains, to avoid that bears prey on reindeer calves. Therefore, special hunting quotas were set for this area, based on the results of bear population size surveys, which require that hunters collect samples in the entire county. However, as there are relatively few roads in the mountains, hunters only collected samples in easily accessible parts of the area. Most reindeer herders did not participate in the bear population size monitoring, except in the southern part of the Västerbotten mountains. Therefore, relatively few of the bears in the area were detected during the survey, and hunting quotas were set accordingly too low. In most years, such a situation would not have important consequences, as reindeer calve at higher altitudes, where there are fewer bears; however, the conditions with an exceptional amount of snow in spring 2020 exacerbated the situation and resulted in high bear depredation. The Västerbotten County Board did its best to help by removing many bears but was accused anyway for being utterly incompetent when it comes to bear management.

Management of brown bears in Norway

Human attitudes and political decisions, rather than natural conditions, determine the numbers and distribution of large carnivores today. This is obvious in Norway, where bear numbers have not increased very much since the reappearance of the species in the country in the 1970s. Norway is not a member of the European Union and therefore not directly affected by the EU Habitats Directive. However, Norway has signed the Bern Convention and therefore is obliged to conserve its large carnivores.

Due to climatic constraints, livestock production, especially sheep production, is important in Norwegian agriculture. Subsidies for livestock production are used to promote the goal of a "living landscape", i.e. economic activity in populated rural areas. More than two million sheep, some hundred thousand cattle and about 160,000 domestic reindeer graze freely throughout Norwegian landscapes. As an example of the conflict between farmers and bears in Norway, we quote Strand et al. (2019): "Bear attacks on grazing sheep has been prevalent in Norway [...]. Bears are large, unpredictable, and occasionally violent and therefore represent a threat that the farmers are unable to cope with. Bears can damage carnivore-repellent fences and the damage inflicted on a herd attacked by brown bear is often substantial with many animals killed. The CMZ (carnivore management zone) for brown bear is found in regions where livestock farming is particularly dependent on using outfield resources. Bears are incompatible with free-roaming sheep in the outfields and prevent the farmers from exploiting these resources. Farms in the CMZ for brown bear are small and herds cannot be sustained on their infields alone. The result is that sheep farmers are forced out of business".

Nevertheless, the number of sheep and lambs that farmers received compensation for as killed by bears was only 1,843 in 2019, and 1,054 in 2020 (Miljødirektoratet 2022). The corresponding numbers for domestic reindeer were 347 and 397 in 2018 and 2019, respectively (latest available figures; Miljødirektoratet 2022), although loss of calves is particularly difficult to document. Even if these numbers are not particularly high, the number of bears is low, resulting in Norway (outside of the northernmost province of Finnmark) having the highest number of livestock compensated and greatest compensation cost per bear in Europe, in spite of the second-highest cost of preventive measures per bear (Bautista et al. 2017; 2019). These results highlight the difficulties of trying to conserve bears in an open-range landscape with free-ranging, unguarded sheep.

Management system
Stortinget, the Norwegian parliament, has sought to establish a compromise among stakeholders in the conflict between humans and carnivores. The solution is a political consensus formalised through two parliamentary decisions, the Carnivore Settlements of 2004 and 2011, which seek to reconcile two goals: continued sustainable livestock production in the outfields (open range) and the maintenance of viable carnivore populations. Through the agreements, national objectives have been set that define how large the populations of the predators are allowed to be. These targets for population size and distribution of large carnivores are so low that all of the species are included in the Norwegian Red List; the brown bear is listed as "Endangered" in Norway (www.artsdatabanken.no).

When populations are above or below the targets, it is the task of the environmental administration to find good tools to move and maintain populations close to the targets. The relevant administration is the Norwegian Environment Agency (Miljødirektoratet), which is responsible for the management of large carnivores on the national level. It has described the Norwegian management system for large carnivores on its website (MILJØDIREKTORATET 2021).

Thus, the objectives are different than in Sweden, where minimum objectives have been set, with no maximum limits on the national level. Large carnivore management is a field where the instructions from Stortinget and the government are particularly detailed, because of the strongly conflicting interests in Norwegian society. GANGAAS et al. (2013) found that the conflict associated with large carnivores in Norway is linked to sheep farming and big game hunting and that people living in rural areas with big game hunting and sheep farming are more likely to accept illegal hunting compared to people living in areas with less rural traditions. They also found that Norwegians were four times more inclined to accept poaching than Swedes.

Stortinget decided that carnivores should be managed at a local scale. Therefore, Norway was subdivided into eight management regions for large carnivores (Fig. 8), and a predefined number of each species of predator can occur in each region. These goals are expressed in annual reproductions for each region and can be zero. Within these regions, some areas are defined to prioritise large carnivores; there the threshold for the removal of animals is higher and preventive measures to protect livestock are more relevant. Outside these areas, livestock grazing on open ranges is prioritised and large carnivores are less welcome. The management in each region is governed by a carnivore committee with members consisting of elected politicians appointed by the Ministry of Climate and the Environment and the Sámi Parliament. They are part of the environmental administration and they are subordinate to the Ministry. Within each region, it is the regional carnivore committee that is responsible for ensuring that the populations of large carnivores are maintained at the level set by Stortinget. One of the tasks of the committee is to determine management zones for lynx, wolverine, and bear as part of the regional management plan for large carnivores. However, Stortinget defined the management zone for wolves. The committees also set harvest quotas when the number of reproductions has exceeded the goal. In the regions where the target has not been met, the Norwegian Environmental Agency retains management authority.

Bears in Norway

The Norwegian brown bear population is the western edge of a larger population in Sweden, Finland, and Russia. Stortinget has decided that Norway should have 13 annual reproductions (litters) of bears in the country, and that these should occur in the four regions bordering Sweden, Finland, and Russia (Fig. 8). In the other four regions, the target number of brown bear reproductions is zero. Management zones have been established in those regions, where bear reproduction is a priority. Outside these areas, grazing animals will be given priority and there is a lower threshold for killing bears.

The national population target of 13 litters of bears per year was adopted in the carnivore settlement in Stortinget in 2011, but it has never been reached (Table 3). In 2021, the target had been reached in two regions (Table 3), so the committees in those regions had management authority over bears in their area. The bear population in Norway is increasing slowly, which in part depends on how the bear is managed in Sweden. In recent years, Swedish counties had a goal of reducing the bear population. Due to conflicts with reindeer husbandry, many bears have been killed in border areas with Norway, and many of the killed bears probably had parts of their home ranges on the Norwegian side of the border. In addition, bears can be killed in the fall by hunters who have obtained a license for bears in areas opened for bear hunting, which are primarily areas where the authorities want to reduce the number of bears and their damages outside of the areas prioritised for bears. In 2019, license hunting was allowed for nine bears in two regions and two bears were killed, one of which was a female. Also in 2019, eleven bears were killed as a response to depredation events in areas where grazing is prioritised, and two others were known to have died (one killed by a train and one died of unknown causes). Thus, 15 bears are known to have died in Norway in 2019 (www.rovbase.no).

In Norway, most brown bears live in the border areas with Sweden, Finland, and Russia. The individuals that have been detected further inland are mainly young males on the move. The female bears in Norway live mostly close to the border; there are relatively few established adult females in the country.

Monitoring

The administration is dependent on accurate data on population size and distribution to follow up the very detailed goals in the large carnivore policy. The institution Rovdata, which is part of the Norwegian Institute for Nature Research (NINA), is responsible for the national monitoring program in Norway, which is part of the joint monitoring of large carnivores in Scandinavia. The Norwegian Nature Inspectorate (Statens naturoppsyn, SNO) is responsible for monitoring in the field and delivers bear excrements and hair collected annually from bears throughout the country to Rovdata. Every year, about 1,500 samples are analysed genetically, which allows the determination of the minimum number of bears in Norway, to follow the same individuals from year to year, and to map spatial use over time, and the distribution of males and females.

Because Stortinget has set a goal of 13 litters to be born in Norway each year, the monitoring scheme is aimed at determining how many reproductions occur annually. However, it is difficult to document females with young of the year for several reasons. When the female leaves the den with her newly born cubs for the first time, the snow, which is crucial for tracking the individuals, is often already gone. It is also hard to distinguish between large young of the year and young from previous years, and many young may not be seen or reported.

Therefore, managers use a model, developed by the Scandinavian Brown Bear Research Project, to calculate the number of reproductions each year (Bischof/Swenson 2012; Table 3). The method is based on several parameters: current results from DNA analyses in Norway, age and sex distribution in the Swedish bear population, time between litters and age at first reproduction, home range sizes, and mortality risk. The parameters are then adjusted in relation to differences between Sweden and Norway, before the probable number of litters in Norway is calculated.

Management in Finland

In contrast to Norway and Sweden, where management of the brown bear population is decentralised, Finland manages the brown bear on the national level. The main goal of the management plan for brown bears in Finland (2007, updated in 2016) is to pursue ecologically, economically, and socially sustainable population management. The brown bear is considered a game species according to the Finnish legislation, and the Ministry of Agriculture and Forestry of Finland (MAFF) is responsible for its management. The population size increased greatly from ~200 bears in the 1970s to ~2,400 in 2020, i.e. by a factor of 12, and it is regulated mostly by recreational hunting. The annual bear population growth rate since the 1970s was substantially higher outside the reindeer husbandry district (0.07) than inside the district (0.02; Kojola et al., unpublished data).

The Finnish Wildlife Agency is responsible for the execution of the management strategy proposed by the MAFF as well as the handling of applications for bear hunting licenses, which are required to hunt bears outside of the reindeer husbandry district. Bear hunting quotas in the reindeer husbandry area are set on a regional level (eastern and western region), and no personal hunting licenses are required.

Finland is divided into four brown bear management zones, i.e. the reindeer husbandry zone in the north, the stable population zone in the east, the bear dispersal zone in central Finland, and the zone for a developing population in western Finland. Bear population densities decrease from the east to the west, and the proportion of females killed during bear hunting is lowest in northern Finland (Kojola et al. 2020). The MAFF sets annual harvest quotas for provinces and management zones, based on population estimates and harvest scenarios provided by the Natural Resources Institute of Finland (LUKE), also considering damages caused by bears. Stakeholder groups are provided the opportunity to comment on a proposal for the size of the annual bear hunting quota prepared by the MAFF.

The Finnish Wildlife Agency may also approve licenses to kill problem bears beyond the quota. The Finnish Police Department has a network of trained hunters that can be used to deter or remove individual problem bears that may pose a risk to human safety in residential areas. The Police may also use hunters to remove bears that have been wounded by hunting or injured in traffic accidents.

The bear harvest scenarios are developed by LUKE, based on a Bayesian model that estimates sustainable harvest rates and the associated uncertainty, based on the annual rates of growth and mortality in the population (Heikkinen et al. 2021). Separate harvest scenarios are produced for the reindeer husbandry district and for the area outside the reindeer husbandry area. The most probable sustainable harvest rate estimates in these population models for recent years have been strikingly high (14–17 % of pre-hunt estimates), which is probably possible due to a high bear immigration rate from neighbouring Russia, where population densities are high and harvest rates comparatively low (I. Kojola/K. Tirronen, unpublished data).

Population monitoring
Every spring, LUKE prepares a bear population size estimate of the pre-hunt season (autumn) of the previous and the ongoing year. Separate population size estimates are produced for 15 provinces and four bear management zones (Heikkinen et al. 2021). The data that are used as basis for these population size estimates are collected by a network of ~2,000 volunteers, most of them hunters. This process can be considered "advanced citizen science", because these volunteers have received training to prepare them for their role as citizen scientists and data collectors. The data on observations of brown bears and other large carnivores collected by these volunteers are uploaded via a link into the online observation data base "Tassu" ("Paw" in English; Kojola et al. 2018). Each observation is attributed with geographic coordinates, date, the type of observation (sighting, track, picture, etc.). More than 10,000 observations are collected annually by this volunteer network. The main focus in the annual population size estimation is on observations of females with cubs of the year (i.e. offspring born in the current year); observations are separated based on the width of footprint of a front paw, and the body size of the dependent offspring is used to differentiate between litters consisting of cubs of the year or yearlings (i.e. offspring born in the previous year). A distance criterion (Ordiz et al. 2006) is used to differentiate between different females with litters of cubs of the year. The population size estimates based on the volunteer network have been shown to correspond well with the ones based on non-invasive genetic sampling, where samples are taken without affecting the bears in any way, e.g. from scats.

Damages and compensation
About 13 % of Finland's bear population live in the area reserved for the management of domestic reindeer (Heikkinen et al. 2021). Most damages by bears are concentrated in this zone, which covers ~36 % of the area of Finland. For example, on average 650 reindeer were reported as killed by bears annually in this area during 2010–2019. Only a small fraction of these kills is actually examined by communal authorities in the field to confirm that the reindeer had been killed by a bear, but most of

these kills are compensated for by the Finnish government. Compensation is only paid for reindeer that are older than one year. Reindeer herding cooperatives (n = 56) are classified into different categories by the number of reported kills by carnivores. The category and herd size determine the sum of money paid for the compensation of calf loss.

South of the reindeer husbandry area, in total bears usually destroy only a few dozen beehives and kill a few dozen of sheep annually. In addition, there are about 100–200 reports every year of bears damaging hay bales or feeding on agricultural crops, especially oats or fruits, such as strawberries. All such types of damages are fully compensated by the Finnish government.

Current cooperative management in Fennoscandia

The management systems in Norway and Finland are rather similar to Sweden, but objectives for carnivore population size and distribution differ widely between countries. However, because these countries share the same population of bears, cross-border collaboration in research, management, and the sharing of information is well established (Table 4).

Where to go from here in Sweden

At least in Sweden, the management system for all large carnivore species is strongly affected by never-ending discussions and controversies about the wolf. During the last 20 years, several changes to the management system were made to accommodate the needs of county boards and different stakeholders in the relatively small counties with wolves in central Sweden. Changes were not always positive for the huge counties in northern Sweden, and it was difficult to have long enough phases of uniform management in between changes, to be able to thoroughly evaluate management actions and different measures that had been applied.

Currently (May 2022), the Swedish government aims at changing the system again, by removing the county-wise minimum levels for population size, at least for the wolf. If this is done, presumably it will become more difficult for county boards to defend large population sizes of carnivores against opposing views in the delegations for game management, where influential stakeholders argue intensively for increased hunting and smaller populations.

Furthermore, the Swedish EPA is currently doing a major revision of the regulations for the hunting of large carnivores. Some of the proposed changes can alter bear hunting tremendously, especially when it comes to the use of baits. Not very much, however, is done against the hurried way in which bear hunting is conducted these days.

In 2021–2022, the Swedish EPA also revises the national action plans for brown bear, wolf, wolverine, and lynx. There are several ideas of submitting assignments to the Scandinavian Brown Bear Research Project and other researchers to compile information on different topics, and of providing funding for the extension of existing studies and for the start-up of new research. The wealth of new knowledge that hopefully will be produced will inform decision making and benefit the future management of the brown bear in Sweden.

Bibliography

Artdatabanken 2020: The Swedish Species Information Centre at the Swedish University of Agriculture (SLU Artdatabanken), Rödlistade arter i Sverige 2020 (Uppsala 2020).

Bautista et al. 2017: C. Bautista/J. Naves/E. Revilla/N. Fernández/J. Albrecht/A. K. Scharf/R. Rigg/A. A. Karamanlidis/K. Jerina/D. Huber/S. Palazón/R. Kont/P. Ciucci/C. Groff/A. Dutsov/J. Seijas/P. I. Quenette/A. Olszańska/M. Shkvyria/M. Adamec/M. Jonozovič/N. Selva, Patterns and correlates of claims for brown bear damage on a continental scale. Journal of Applied Ecology 54, 2017, 282–292.

Bautista et al. 2019: C. Bautista/E. Revilla/J. Naves/J. Albrecht/N. Fernández/A. Olszańska/M. Adamec/T. Berezowska-Cnota/P. Ciucci/C. Groff/S. Härkönen/D. Huber/K. Jerina/M. Jonozovič/A. A. Karamanlidis/S. Palazón/P.-I. Quenette/R. Rigg/J. Seijas/J. E. Swenson/T. Talvi/N. Selva, Large carnivore damage in Europe: Analysis of compensation and prevention programs. Biological Conservation 235, 2019, 308–316.

Bellemain et al. 2005: E. Bellemain/J. E. Swenson/D. Tallmon/S. Brunberg/P. Taberlet, Estimating population size of elusive animals using DNA from hunter-collected feces: comparing four methods for brown bears. Conservation Biology 19, 2005, 150–161.

Bischof/Swenson 2012: R. Bischof/J. E. Swenson, Combining noninvasive genetic sampling and traditional monitoring to aid management of a trans-border carnivore population. Ecological Applications 22, 2012, 361–373.

Bischof et al. 2008: R. Bischof/R. Fujita/A. Zedrosser/A. Söderberg/J. E. Swenson, Hunting patterns, ban on baiting, and harvest demographics of brown bears in Sweden. The Journal of Wildlife Management 72, 2008, 79–88.

Bischof et al. 2020: R. Bischof/C. Milleret/P. Dupont/J. Chipperfield/M. Tourani/A. Ordiz/P. de Valpine/D. Turek/J. A. Royle/O. Gimenez/Ø. Flagstad/M. Åkesson/L. Svensson/H. Brøseth/J. Kindberg, Estimating and forecasting spatial population dynamics of apex predators using transnational genetic monitoring. Proceedings of the National Academy of Sciences of the United States of America 117, 2020, 30531–30538.

Danell/Bergström 2016: K. Danell/R. Bergström, En viltförvaltning i sin linda, 1250–1634. In: K. Danell/R. Bergström/L. Mattsson/S. Sörlin (eds.), Jaktens historia i Sverige (Stockholm 2016) 189–201.

DG Environment 2017: Directorate-General for Environment of the European Commission, Reporting under Article 17 of the Habitats Directive: Explanatory notes and guidelines for the period 2013–2018 (Brussels 2017).

Dressel et al. 2021: S. Dressel/C. Sandström/J. Bennett/G. Ericsson, En attitydundersökning om stora rovdjur och rovdjursförvaltning. Sveriges lantbruksuniversitet Rapport 2021:8 (Umeå 2021).

Ericsson/Sandström 2005: G. Ericsson/C. Sandström, Partial report on the attitudes of Swedes towards carnivore politics and management. FjällMistra-report 10 (Umeå 2005).

Fløystad et al. 2021: I. Fløystad/H. Brøseth/A. S. B. Hansen/I. H. Søvik/H. G. Eiken/S. B. Hagen, Populasjonsovervåking av brunbjørn. DNA-analyse av prøver innsamlet i Norge i 2020. NINA Rapport 1986 (Trondheim 2021).

Frank et al. 2020: J. Frank/J. Månsson/M. Levin/L. Höglund, Viltskadestatistik 2019. Skador av fredat vilt på tamdjur, hundar och gröda. Rapport från Viltskadecenter, Sveriges lantbruksuniversitet 2020:2 (Riddarhyttan 2020).

Gangaas et al. 2013: K. E. Gangaas/B. P. Kaltenborn/H. P. Andreassen, Geo-Spatial Aspects of Acceptance of Illegal Hunting of Large Carnivores in Scandinavia. PLoS ONE 8(7), 2013. https://doi.org/10.1371/journal.pone.0068849.

Heikkinen et al. 2021: S. Heikkinen/I. Kojola/S. Mäntyniemi, Karhukanta Suomessa 2020 [Bear population in Finland 2020]. Luonnonvara- ja biotalouden tutkimus 20/2021. Luonnonvarakeskus (Helsinki 2021).

Karlsson et al. 2012: J. Karlsson/O.-G. Støen/P. Segerström/R. Stokke/L.-T. Persson/L.-H. Stokke/S. Persson/N. A. Stokke/A. Persson/E. Segerström/G.-R. Rauset/J. Kindberg/R. Bischof/T. R. Sivertsen/A. Skarin/B. Åhman/I. Ängsteg/J. Swenson, Björnpredation på ren och potentiella effekter av tre förebyggande åtgärder. Rapport från Viltskadecenter 2012, 6.

Kindberg et al. 2009: J. Kindberg/G. Ericsson/J. E. Swenson, Monitoring rare or elusive large mammals using effort-corrected voluntary observers. Biological Conservation 142, 2009, 159–165.

Kindberg et al. 2011: J. Kindberg/J. E. Swenson/G. Ericsson/E. Bellemain/C. Miquel/P. Taberlet, Estimating population size and trends of the Swedish brown bear (Ursus arctos) population. Wildlife Biology 17, 2011, 114–123.

Kojola/Laitala 2000: I. Kojola/H. M. Laitala, Changes in the structure of an increasing brown bear population with distance from core areas: another example of presaturation female dispersal? Annales Zoologici Fennici 37, 2000, 59–64.

Kojola et al. 2018: I. Kojola/S. Heikkinen/K. Holmala, Balancing costs and confidence: volunteer-provided point observations, GPS telemetry and the genetic monitoring of Finland's wolves. Mammal Research 63, 2018, 415–423.

Kojola et al. 2020: I. Kojola/V. Hallikainen/S. Heikkinen/V. Nivala, Has the sex-specific structure of Finland's brown bear population changed during 21 years? Wildlife Biology 2020. https://doi.org/10.2981/wlb.00575.

Lundmark 2006: L. Lundmark, Samernas skatteland i Norr- och Västerbotten under 300 år. Rättshistoriska skrifter 3, 2006, 1650–2299.

Lundmark/Matti 2015: C. Lundmark/S. Matti, Exploring the prospects for deliberative practices as a conflict-reducing and legitimacy-enhancing tool: The case of Swedish carnivore management. Wildlife Biology 21(3), 2015, 147–156. https://doi.org/10.2981/wlb.00009.

Miljødirektoratet 2021: https://www.miljodirektoratet.no/ansvarsomrader/arter-naturtyper/vilt/rovvilt/.

Miljødirektoratet 2022: https://www.miljodirektoratet.no/ansvarsomrader/arter-naturtyper/vilt/rovvilt/husdyrtap/.

MMM/Riistavahinkorekisteri 2021: https://riistavahinko.mmm.fi/.

Mykrä/Pohja-Mykrä 2015: S. Mykrä/M. Pohja-Mykrä, Back-calculation of large carnivore populations in Finland in 1865–1915. Annales Zoologici Fennici 52, 2015, 285–300.

Naturvårdsverket 2016: Naturvårdsverket, Nationell förvaltningsplan för björn – Förvaltningsperioden 2014–2019. Revised version (Stockholm 2016).

Naturvårdsverket 2020: Naturvårdsverket, Sveriges arter och naturtyper i EU:s art- och habitatdirektiv. Resultat från rapportering 2019 till EU av bevarandestatus 2013–2018. Rapport (Stockholm 2020).

Ordiz et al. 2006: A. Ordiz/O.-G. Støen/J. E. Swenson/I. Kojola/R. Bischof, Distance-dependent effect of the nearest neighbor: spatiotemporal patterns in brown bear reproduction. Ecology 89, 2008, 3327–3335.

Pulliainen 1990: E. Pulliainen, Recolonization of Finland by the brown bear in the 1970s and 1980s. Aquilo Series Zoologica 27, 1990, 21–25.

Pulliainen 1997: E. Pulliainen, The expansion of brown bears from east into Finland. International Bear News 6(3), 1997, 10–11.

Rauset et al. 2016: G. R. Rauset/H. Andrén/J. E. Swenson/G. Samelius/P. Segerström/A. Zedrosser/J. Persson, National parks in northern Sweden as refuges for illegal killing of large carnivores. Conservation Letters 9, 2016, 334–341.

Regeringen 2013: The Swedish Government, Regeringens proposition 2012/13:191 En hållbar rovdjurspolitik (Stockholm 2013).

Rovbase 2022: www.rovbase30.se.

Saarma/Kojola 2007: E. Saarma/I. Kojola, Matrilineal genetic structure of the brown bear population in Finland. Ursus 18(1), 2007, 30–37.

Sandström/Ericsson 2009: C. Sandström/G. Ericsson, On Swedes' attitudes towards carnivores and carnivore politics. Department of Wildlife, Fish, and Environmental Studies, Swedish University of Agricultural Sciences. Report 2009:1 (Umeå 2009) (in Swedish).

Sandström et al. 2014: C. Sandström/G. Ericsson/S. Dressel/M. Eriksson/E. Kvastegård, Attitudes towards carnivores and carnivore politics. Department of Wildlife, Fish, and Environmental Studies, Swedish University of Agricultural Sciences Report 2014:1 (Umeå 2014) (in Swedish).

Schneider 2006: M. Schneider, Monitoring the Brown Bear Ursus arctos in Västerbotten County. In: C. Hurford/M. Schneider (eds.), Monitoring nature conservation in cultural habitats (Dordrecht 2006) 195–214.

Schneider 2008: M. Schneider, Managing large carnivores in Northern Sweden. In: BarentsWatch 2008. English edition: Large carnivores in the Barents region (Svanhovd 2008) 10–11.

Schneider 2015: M. Schneider, The bear dropping survey of 2014 in Västerbotten County. Report (Umeå 2015) (in Swedish).

Schneider 2017: M. Schneider, Managing large carnivores in Västerbotten. Report, Västerbotten County Administration (Umeå ³2017).

SFS 2009:1263: Svensk författningssamling/Swedish Code of Statutes, Förordning (2009:1263) om förvaltning av björn, varg, järv, lo och kungsörn (Stockholm 2009).

SFS 2009:1474: Svensk författningssamling/Swedish Code of Statutes, Förordning (2009:1474) om viltförvaltningsdelegationer (Stockholm 2009).

Sjölander et al. 2009: P. Sjölander/A. Edin-Liljegren/L. Daerga, Samernas hälsosituation i Sverige – en kunskapsöversikt. Rapport från Södra Lapplands Forskningsenhet (Vilhelmina 2009).

Stortingmelding 2016: Report to the Norwegian Parliament from the Norwegian Government, Meld. St. 21 (2015–2016). Ulv i norsk natur – Bestandsmål for ulv og ulvesone (Oslo 2016).

Strand et al. 2019: G.-H. Strand/I. Hansen/A. de Boon/C. Sandström, Carnivore Management Zones and their Impact on Sheep Farming in Norway. Environmental Management 64, 2019, 537–552. https://doi.org/10.1007/s00267-019-01212-4.

Støen et al. 2018: O.-G. Støen/A. Ordiz/V. Sahlén/J. M. Arnemo/S. Sæbø/G. Mattsing/M. Kristofferson/S. Brunberg/J. Kindberg/J. E. Swenson, Brown bear (Ursus arctos) attacks resulting in human casualties in Scandinavia 1977–2016; management implications and recommendations. PLoS ONE 13(5), 2018. https://doi.org/10.1371/journal.pone.0196876.

Swedish EPA 2018: Swedish Environmental Protection Agency, Strategy for Swedish Wildlife Management, Naturvårdsverket report (Stockholm 2018).

Swenson et al. 1995: J. E. Swenson/P. Wabakken/F. Sandegren/A. Bjärvall/R. Franzén/A. Söderberg, The near extinction and recovery of brown bears in Scandinavia in relation to the bear management policies of Norway and Sweden. Wildlife Biology 1995(1), 11–25.

Swenson et al. 1999: J. E. Swenson/F. Sandegren/A. Söderberg/M. Heim/O. J. Sørensen/A. Bjärvall/R. Franzén/S. Wikan/P. Wabakken, Interactions between brown bears and humans in Scandinavia. Biosphere Conservation 2(1), 1999, 1–9.

Swenson et al. 2011: J. E. Swenson/A. Zedrosser/S. Brunberg/P. Segerström, Causes of mortality, especially illegal killing, among Swedish brown Bears, 1984–2010. Scandinavian Brown Bear Research Project. Report 2011-3 to the World Wide Fund for Nature, WWF (Sweden) (Ås 2011).

Swenson et al. 2017: J. E. Swenson/M. Schneider/A. Zedrosser/A. Söderberg/R. Franzén/J. Kindberg, Challenges of managing a European brown bear population; lessons from Sweden, 1943–2013. Wildlife Biology 2017(4). https://doi.org/10.2981/wlb.00251.

Xenikoudakis et al. 2015: G. Xenikoudakis/E. Ersmark/J.-L. Tison/L. Waits/J. Kindberg/J. E. Swenson/L. Dalén, Consequences of a demographic bottleneck on genetic structure and variation in the Scandinavian brown bear. Molecular Biology 24, 2015, 3441–3454.

Zedrosser et al. 2020: A. Zedrosser/M. Leclerc/S. Frank/J. Kindberg/M. Schneider, Utvärdering av åteljakt efter björn i Sverige perioden 2014–2019. Scandinavian Brown Bear Research Project. Report no. 2020:1 (Bø 2020) (bilingual Swedish/English).

Dr. Michael Schneider
The County Administrative Board of Västerbotten
Länsstyrelsen Västerbotten
Umeå
Sweden
michael.schneider@lansstyrelsen.se

Prof. Andreas Zedrosser
Department of Natural Sciences and Environmental Health
University of South-Eastern Norway
Bø i Telemark
Norway

Department for Integrative Biology
Institute for Wildlife Biology and Game Management
University for Natural Resources and Life Sciences Vienna
Vienna
Austria
andreas.zedrosser@usn.no

Prof. Ilpo Kojola
Finish Game and Fisheries Research Institute
Helsinki
Finland
ilpo.kojola@luke.fi

Prof. Jon E. Swenson
Faculty of Environmental Sciences and Natural Resource Management
Norwegian University of Life Sciences
Ås
Norway
jon.swenson@nmbu.no

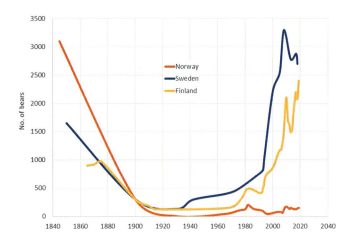

Fig. 1. Population dynamics of brown bears in Norway, Sweden, and Finland between c. 1850 and 2019, compiled from a variety of sources.

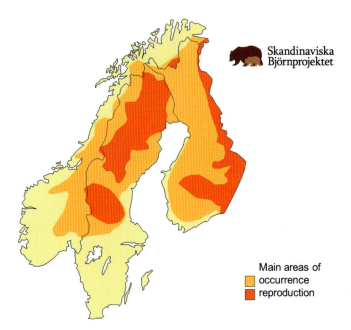

Fig. 2. Current distribution of brown bears in Norway, Sweden, and Finland (map compiled by the Scandinavian Brown Bear Research Project).

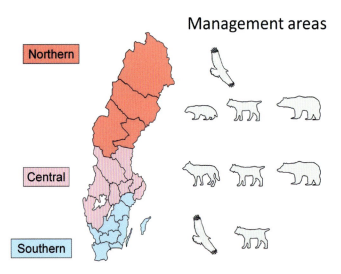

Fig. 3. Occurrence and density of large carnivores (brown bear, wolf, lynx, wolverine, and golden eagle) are different in the three management areas for the species in Sweden. Each area is subdivided into counties, the most important units for carnivore management (after SCHNEIDER 2017).

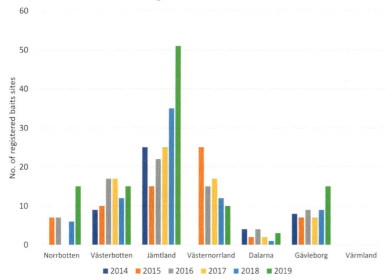

Fig. 4. Baiting for hunting of brown bears had been banned in Sweden in 2001 but was allowed again in 2014. Especially in Jämtland County it has become very popular since 2014 (after ZEDROSSER et al. 2020).

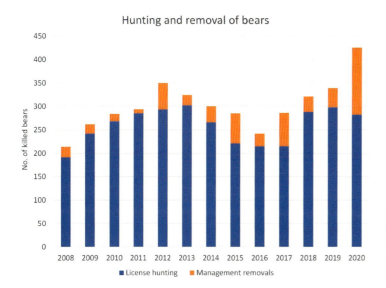

Fig. 5. The number of bears killed in Sweden during license hunting and management removals, respectively (after data from the Scandinavian Large Carnivore Database Rovbase: www.rovbase.se).

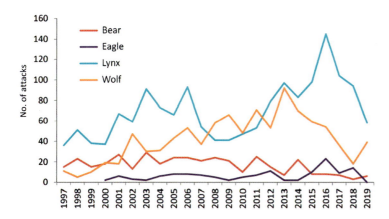

Fig. 6. Attacks by large carnivores on livestock in Sweden during the period 1997–2019. Brown bears cause relatively few problems, and the trend is decreasing (after FRANK et al. 2020, adapted).

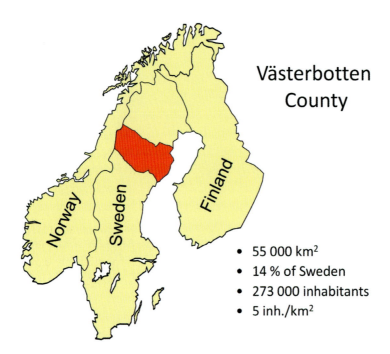

Fig. 7. Västerbotten is the second most northerly and second largest county in Sweden. The human population density is low, and most people live along the coast in the eastern parts of Västerbotten. The entire county is situated within the area of reindeer husbandry in Sweden (map M. Schneider).

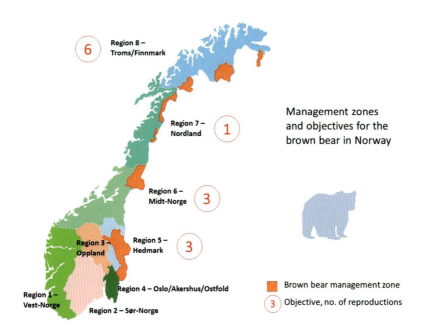

Fig. 8. Norway is subdivided into eight large carnivore management regions. The four northern-most regions have defined management areas for the brown bear and goals for the number of reproductions per year (after a map from STORTINGMELDING 2016, adapted).

Table 1. Minimum levels, management targets, and limits of management intervals for the brown bear during the period 2014–2019 in the six northernmost counties in Sweden, where bear populations are rather large. Figures express the number of individuals.

	Norrbotten	Västerbotten	Jämtland	Västernorrland	Gävleborg	Dalarna	Sum
Upper limit	910	432	800	220	460	290	3,112
Target	820	350	650	200	381	270	2,671
Lower limit	730	288	500	180	300	250	2,248
Minimum	330	110	360	100	250	250	1,400

Table 2. Comparison of the number of sheep compensated for as killed by brown bears in Norway, Sweden and Finland in 2020 (after data from the Norwegian Environment Agency [MILJØDIREKTORATET 2022], the Scandinavian Large Carnivore Database [ROVBASE 2022], and the Finnish Wildlife Damage Registry [MMM/RIISTAVAHINKOREKISTERI 2021]).

	No. of sheep compensated	Approximate no. of bears	Compensated sheep per bear
Norway	1,054	150	7.027
Sweden	11	2,700	0.004
Finland	123	2,400	0.051

Table 3. The national objective for the number of litters born in Norway is the sum of the regional objectives in the four northern large carnivore regions. The status in 2020 met the objective in two regions. See text for further explanations (after FLØYSTAD et al. 2021).

Large carnivore region	County	Objective (litters born per year)	Status 2020 (litters born)
Region 5	Hedmark	3	3.1
Region 6	Møre og Romsdal and Trøndelag	3	2.9
Region 7	Nordland	1	0
Region 8	Troms og Finnmark	6	2.5
Norway		13	8.5

Table 4. List of currently active bilateral and trilateral agreements relating to cooperative management of large carnivores in Fennoscandia, i.e. Norway, Sweden, and Finland.

Year	Agreement
2011	Agreement between the Ministry of Environment, Sweden, and the Ministry of Environment, Norway, on management of genetically important wolves in the Scandinavian wolf population. 12 August 2011.
2012	Memorandum of Understanding regarding the establishment and continuance of a public web-based database (Skandobs) for geographic information on large carnivore observations in Norway and Sweden (Norwegian Institute for Nature Research and Swedish Environmental Protection Agency). 25 March 2012.
2012	Memorandum of Understanding regarding management strategies for the Scandinavian wolf population (Norwegian Directorate for Nature Management and Swedish Environmental Protection Agency). 25 April 2012.
2015	Memorandum of Understanding regarding the establishment and continuance of a monitoring system for large carnivores in Sweden and Norway (Norwegian Environment Agency and Swedish Environmental Protection Agency). 25 March 2015.
2020	Tri-lateral framework document for transboundary cooperation on management and conservation of wolves in Fennoscandia (Ministry of Agriculture and Forestry of Finland, Norwegian Environment Agency, Swedish Environmental Protection Agency). September/October 2020.

Genetics of brown bears in northern Europe

By Alexander Kopatz

Keywords: Brown bear, genetic structure, mitochondrial DNA, noninvasive genetic sampling, phylogeography, population monitoring, Ursus arctos, *Y chromosome*

Abstract: From the dawn of wildlife genetics, brown bears in Scandinavia have been studied using molecular genetic methods and hence have been substantial in the development of these methods, which are now the gold standard in DNA-based monitoring of many wildlife species. This chapter introduces the constantly evolving field of DNA-based assessments to monitor and to study the history of brown bears. Genetic studies enable us to understand better past processes, such as the recolonisation after the last Ice Age, and present status, such as migration, and other factors influencing brown bear populations in the north of Europe.

Genetics and the development of DNA-based population monitoring

Genetics is the study of heredity and how an organism's genes are associated with the environment it is living in. The genetic variability or genetic diversity of a species and/or a population is essential for its adaptability to e.g. changing environmental conditions and therefore its viability and long-term survival (Frankham 1995; 2005; Lacy 1997; Allendorf et al. 2010). Molecular genetic methods have been crucial for the understanding of speciation and evolution as well as historical and current processes in Ursid biology and distribution. Genetic information helps us understand what may have shaped the brown bear (*Ursus arctos*) population in northern Europe and provides us with knowledge about its past as well as current status.

The development of DNA-technology has revolutionised how we monitor our natural environment and has also helped to improve our knowledge on numerous species, including brown bears. Individual brown bears, as every sexually reproducing species, differ genetically. An individual's genetic profile is unique, except for identical twins. Individuals usually differ in appearance and behaviour, however, these often-subtle differences are challenging to assess by human observers in other mammals. A genetic profile holds valuable information for deeper analyses, especially in comparison with the profiles from other bears, whether in the same region, adjacent areas, or across regional and national borders. Having precise genetic information from individual brown bears enables the assessment of genetic relatedness, which in turn allows addressing important questions concerning their management and conservation.

To obtain genetic information from an individual wild animal, a biological sample, such as tissue, blood, bones, hairs, or feces, must be collected in the field and then genetically analysed in the laboratory. In Scandinavia, brown bears are mainly monitored with the help of non-invasively collected material. Non-invasive sampling means that it is not required to directly have contact with, disturb, or capture an animal to collect a biological sample. Instead, DNA is extracted from biological samples left in the field by the individual itself, for example, in the form of feces or hairs. Such samples are routinely collected by wildlife rangers, hunters, and also hikers (BELLEMAIN et al. 2005; KINDBERG/ SWENSON 2018). Individual brown bears are then genetically "tagged" in the laboratory based on their unique DNA-profile; new individuals are identified and already known individuals are genetically "recaptured". Especially the recapture of known individuals is crucial to obtain reliable population size estimates of brown bears. Over the last decade, DNA has become the population monitoring method of choice for wildlife management authorities in Sweden and Norway, and Finland is currently in the process of establishing the same method. Data based on non-invasive collection methods create a solid scientific foundation to estimate the population size of bears in a given area, but also provide additional highly valuable information, such as population distribution, population density, and the sex ratio, i.e. the proportion between males and females in the population, which is important to understand reproduction and the effect of hunting quotas (BISCHOF et al. 2016; 2020; KINDBERG/ SWENSON 2018).

Feces and hair samples make up the largest part of biological materials collected for DNA analysis of brown bears in Scandinavia. The individuals identified, the date their sample was collected in the field, as well as their location and additional information, are stored and made accessible for the public in the Scandinavian monitoring database "Rovbase" (www.rovbase.no). Besides samples and information on brown bears, this database also contains similar information from other large carnivores collected and analysed via the same methods in Scandinavia (see e.g. TABERLET et al. 1999; SCHWARTZ et al. 2007). The Scandinavian brown bear was the main model species at the dawn of molecular methods for the development, design and quality of the genetic methods used today in the population monitoring of wildlife species on a global scale (BELLEMAIN et al. 2005; KINDBERG et al. 2011; SWENSON et al. 2011). The improvement of the genetic methods is an ever-ongoing process (NORMAN et al. 2013; NORMAN/SPONG 2015; DE BARBA et al. 2017).

PHYLOGEOGRAPHY OF BROWN BEARS IN NORTHERN EUROPE

In 1994, TABERLET/BOUVET (1994) published a groundbreaking study on the phylogeography of brown bears in Europe based on mitochondrial DNA-lineages of northern European brown bears. The authors identified two different brown bear DNA-lineages, indicating that Scandinavia was re-colonised by bears from two different population refugia coming to Scandinavia from two different directions after the last glacial maximum: from the north and the south (Fig. 1a). The Quaternary cold periods of the Pleistocene (2.6 million to 11,700 years ago) generally had great influence on life in Europe. As a result of the Scandinavian Ice Sheet, flora and fauna of the Holocene, the current geological epoche (11,700 years ago to today), experienced range contractions, expansion, and sometimes also extinctions in the northern hemisphere; most species were restricted to refuge areas in the southern parts of Europe. This Ice Age is ultimately responsible for today's composition and distribution of species in northern Europe (SIIVONEN 1982; HEWITT 1996; TABERLET et al. 1998). Once temperatures were on the rise and ice caps retreated at the end of the last glacial maximum (~22,000–17,000 years ago), animal and plant species followed the retreating ice. The leading edges of these expansion fronts can be characterised by consecutive occurrences of so-called genetic population bottlenecks. A population bottleneck describes the drastic reduction in the number of individuals in a group or population accompanied by loss of genetic variation (NEI et al. 1975). Such bottlenecks, if severe enough, can be traced back in time with modern genetic methods (HEWITT 1996). Radiocarbon-dated bone remains suggest that the brown bear was one of the earliest carnivorous mammals to re-colonise northern Europe at a comparably rapid pace, which was genetically confirmed by haplogroup data

(Hewitt 1999; 2000; Sommer/Benecke 2005; Korsten et al. 2009; Davison et al. 2011). Today, the brown bear still is one of the most important model species to illustrate Late Quaternary mammalian phylogeography and the application of genetic methods.

The first genetic assessment of brown bears in Europe used mitochondrial DNA (mtDNA) with the goal to identify potential conservation units, i.e. areas and populations valuable for the long-term conservation of the species (Moritz 1994). Mitochondrial DNA is exclusively inherited from the mother to her offspring and, in contrast to nuclear DNA, stored in the mitochondrion of all cells in an eukaryotic organism's body, i.e. animal, plant and fungal organisms with cells containing nucleus and nuclear envelope. The analysis of mtDNA is the basis for the tracking of a species' phylogeography, i.e. the historic and genetic processes and patterns across large time scales that have shaped the current distribution of a species (Hewitt 1996). The mtDNA of brown bears from the northern part of Scandinavia was shown to be part of the eastern lineage or clade, which is also characteristic for bears in Russia, Romania, and Slovakia. This suggests that northern Scandinavia was re-colonised by bears from eastern Europe via Russia and Finland. The mtDNA of brown bears from the southern part of Scandinavia was characteristic of the western lineage, which suggests that southern Scandinavia was re-colonised by brown bears which had their Ice Age refugium on the Iberian Peninsula, i.e. the so-called southern clade of bears found in central and southern Europe (Taberlet/Bouvet 1994).

The striking results of this first study inspired a number of follow-up research, which highlighted that these two clades originated from two separated Quaternary refugia in the far west and east of Europe (Fig. 1a–b; cf. Kohn et al. 1995; Taberlet et al. 1995; 1998; Hewitt 1999; 2000; Hofreiter et al. 2002; Sommer/Benecke 2005; Saarma et al. 2007; Zachos et al. 2008; Korsten et al. 2009; Davison et al. 2011; Keis et al. 2013; Xenikoudakis et al. 2015; Anijalg et al. 2018; Ersmark et al. 2019). Further analyses of mtDNA lineages pointed to a Carpathian refuge of the eastern lineage (Fig. 1b; cf. Sommer/Benecke 2005; Zachos et al. 2008). Also, a much higher matrilineal diversity was found in bears in Finland and northwestern Russia compared to the bottlenecked Scandinavian population. This is likely due to the connection of the Finnish and northwestern Russian population to the large and stable populations in Russia, while the Scandinavian population was likely isolated from potential immigration of new individuals from the east (Saarma et al. 2007). A recent study challenged the scenario of the Iberian peninsula as refugial area, as the genetic analyses of historical material of bears from Belgium and France indicate that this specific haplotype may be the origin of expansion of brown bears into southern Scandinavia (Ersmark et al. 2019).

The general conclusion is that the eastern and western genetic lineages of brown bears diverged about 0.85 million years ago and that bears re-colonised first southern Scandinavia from western Europe, followed by northern Scandinavia from eastern Europe. Here, both mtDNA-lineages possibly met about 5,000–9,000 years ago (Taberlet et al. 1995; Xenikoudakis et al. 2015), after the ice shield from the last glaciation had melted (Siivonen 1982). The eastern and western mtDNA-lineages are highly divergent, and a follow-up study further described the delination or contact zone in the middle of Sweden, where these two clades meet. The width of the contact zone was estimated to be ~130 kilometres. At the time of the study, only four individuals were identified that have crossed the contact zone from one lineage to the other (Taberlet et al. 1995). Overall, the mtDNA-haplotype diversity especially in southern Scandinavia was higher, as the genetic analysis of historical samples of brown bears archived at Swedish museums suggested (Xenikoudakis et al. 2015).

The current distribution of brown bears in Europe is shaped by centuries of severe persecution, bounty hunting, and rigorous extermination efforts (Zedrosser et al. 2011; Albrecht et al. 2017). Brown bears were also eradicated from most parts of northern Europe, and the once continuous population was split into the Scandinavian to the west and the Karelian in the east (Curry-Lindahl 1972; Swenson et al. 1995). It is assumed that the population in Scandinavia at its lowest point at the end of the 19th century was not larger than about 100 individuals (Swenson et al. 1994). Due to the beginning of modern conservation-oriented thinking and management in the early 1970s, brown bear populations, along with other large carnivores, started to recover and to re-colonise their former distribution range in northern Europe (Chapron et al. 2014). However, despite regional comebacks, such as in northern Europe, brown bears and other large carnivores are still under substantial pressure and considered as endangered in Europe and other parts of the world (Dalerum et al. 2009; Ripple et al. 2014).

Assessing contemporary status by using nuclear DNA

Shortly after the first studies of mtDNA in brown bears and other wildlife had emerged, more genetic markers as well as new techniques were developed to enable the study of diversity via nuclear genetic variation. Such studies investigate historic as well as contemporary genetic patterns by using highly variable genetic markers, such as microsatellites or short-tandem-repeats (STRs). Microsatellites have been used since the 1990s, and their application intensified until recently (Paetkau/Strobeck 1994; Paetkau et al. 1995; Taberlet et al. 1997). With the development of next-generation sequencing and genomics, even more informative genetic marker systems were developed, such as single-nucleotide-polymorphisms (SNPs; Norman et al. 2013) or restriction site-associated DNA sequencing (RADseq; Andrews et al. 2016). All genetic marker systems have their advantages but also shortcomings. Mitochondrial DNA, inherited from the mother, can only enable insights into female-mediated genes and gene flow. In comparison, nuclear DNA is inherited from both parents and can be applied to the study of the more recent history as well as the current status of populations in relation to genetic variation, inbreeding, population substructure, interpopulation connectivity via gene flow and more, all of which are important for conservation and management of a species (Frankham 1995; 2005; Waples/Gaggiotti 2006; Allendorf et al. 2010). Norman et al. (2013) published a SNP-panel to study genetics and specifically the relatedness among individuals in Scandinavian brown bear. In contrast to the first nuclear genetic markers, the higher resolution of SNPs allows even more specific assessment of the genetic relationships among bears without prior biological information on their relatedness and family groups (Norman/Spong 2015).

In 2000, the first study on brown bears in Sweden using microsatellites was published (Waits et al. 2000). It was also the first comprehensive assessment of the brown bear's contemporary genetic variation and population structure using genetic methods in northern Europe. The results of that study showed that the genetic diversity of brown bears in Sweden was comparable to brown bear populations in North America. This was surprising, considering the drastic population decline and bottleneck Scandinavian brown bears had experienced due to the intensive hunt. The study also showed that the Swedish population appeared to be subdivided into four different subpopulations, of which the southernmost was geographically located below the previously described contact zone where the two distinct mtDNA-lineages meet (Taberlet et al. 1995). The same study was re-analysed with advanced Bayesian statistical methods a few years later, and the authors concluded that the Swedish population consisted not of four but rather of three subpopulations, with the individuals from the previously identified two northernmost subpopulation showing substantial geographical overlap, leading to the conclusion that these two subpopulations comprise of just one unit (Manel et al. 2004), shaped by isolation-by-distance (Schregel et al. 2018).

The reasons for the distinct genetic structure of brown bears in Scandinavia remained unclear until Xenikoudakis et al. (2015) published a study based on the genetic analysis of archived museum samples using historical bone, skin, and tooth samples. Their findings showed that the current genetic structure is the result of historical ecological processes. The authors reported that also the historical Scandinavian population consisted of these three subpopulations and concluded that the current subdivision is not caused by anthropogenic fragmentation and a genetic bottleneck due to over-hunting. Based on the findings of different studies it is likely the combined effect of ecological processes such as e.g. the density of bears and level of relatedness among individuals in and between areas which can influence the dispersal and successful reproduction of bears outside their natal area. This has been indicated by results assessing kin-related structure (Støen et al. 2005; 2006; Frank et al. 2021) and gene flow among the areas in Sweden and Norway (Schregel et al. 2017; 2018). The current anthropogenic pressure, however, may sustain and manifest the current fragmentation further, as it has also been observed in other hunted wildlife species (Jerina/Adamic 2008; Krofel et al. 2012; Frank et al. 2021). Overall, these results warrant further studies into the history and biology of the Scandinavian brown bear population.

The disconnect of the Scandinavian and Karelian brown bear populations had strong effects on the genetics of both populations (Schregel et al. 2012; Kopatz et al. 2014; 2021). The genetic differences between these populations become obvious when bears from both populations are genotyped and compared with so-called Y-chromosomal markers (Bidon et al. 2014; Schregel et al. 2015; Hirata et al. 2017). Analogous to the mtDNA, the non-recombining fragment of the Y-chromosome enables the assessment of the paternal diversity and male lineages. This is of high interest, because such analyses can reveal dispersal patterns as males generally are the dispersing sex in brown bears, while females remain philopatric and tend to stay near their natal area (Støen et al. 2005; Zedrosser et al. 2007). Two recent studies analysed brown bears from the Scandinavian and the Karelian populations and documented strong genetic differentiation between these bear subpopulations as well as that the Y-haplotype diversity was distributed unevenly (Schregel et al. 2015; Kopatz et al. 2021). As reported by Schregel et al. (2015), Y-haplotype data of bears from Finland, northern Norway and northwestern Russia were high, with 32 haplotypes identified. In contrast, only four haplotypes could be found in the Scandinavian population in Sweden and in western and southern Norway. The haplotype distribution underlined the substantial subdivision of the bear populations in Scandinavia and Karelia (Schregel et al. 2012; 2017; Kopatz et al. 2014; 2021). Overall, these results suggested that the genetic recovery processes of these two populations have likely been very different (Kopatz et al. 2021). It is assumed that the Scandinavian bear population recovered to a large extent on its own, while in comparison the Karelian brown bear population in Finland and northern Norway experienced influx from the east, i.e. Russia (Tammeleht et al. 2010; Keis et al. 2013; Kopatz et al. 2014), which resulted in a population divided into two subpopulations, a northern and a southern one (Saarma/Kojola 2007; Tammeleht et al. 2010; Kopatz et al. 2014). The southern subpopulation showed a strong connection to bear populations in northwestern Russia, supporting the assumption of substantial migration and gene flow from Russia into Finland during the last decades (Keis et al. 2013; Kopatz et al. 2014). These high immigration rates, in contrast to the situation in Scandinavia, likely led also to the gradual mixing of these two subpopulations, so that more and more brown bears were showing the genetic signatures from both populations (Hagen et al. 2015; Kopatz et al. 2017).

In general, brown bear populations in Scandinavia and in Finland have recovered from population lows at the end of the 19th century (Chapron et al. 2014). Based on latest research, the population expansion front with bears from Scandinavia has entered Finland, i.e. bears from Scandinavia have dispersed into Finland (Kopatz et al. 2021). Also a few individuals from the Karelian population have been detected in Sweden, albeit considerably fewer, suggesting asymmetric migration (Fig. 2). This suggests that the Scandinavian bear population seems to have recovered earlier and therefore started expanding earlier compared to the Karelian population. These results highlight that conservation-oriented policies and wildlife management can lead to successful population recovery as well as the restoration of genetic connectivity between fragmented populations of brown bears (Kopatz et al. 2021).

Genetic information enabled a large body of scientific literature on the brown bears of northern Europe and has considerably increased the scientific understanding of population genetics as well as underscored the importance of genetics for conservation. Many of the studies presented would not have been possible without the strong collaborative spirit of national authorities and research groups. Especially Norway and Sweden have harmonised the monitoring and their methods and are now equipped with a unique common population monitoring system and database. The brown bear is one of the best-studied mammals in the world (BROOKE et al. 2014), and the Scandinavian brown bear in particular has been of crucial importance in the development of the genetic monitoring of wildlife in general.

BIBLIOGRAPHY

ALBRECHT et al. 2017: J. ALBRECHT/K. A. BARTON/N. SELVA/R. S. SOMMER/J. E. SWENSON/R. BISCHOF, Humans and climate change drove the Holocene decline of the brown bear. Scientific Reports 7, 10399, 2017.

ALLENDORF et al. 2010: F. W. ALLENDORF/P. A. HOHENLOHE/G. LUIKART, Genomics and the future of conservation genetics. Nature Reviews Genetics 11, 2010, 697–709.

ANDREWS et al. 2016: K. R. ANDREWS/J. M. GOOD/M. R. MILLER/G. LUIKART/P. A. HOHENLOHE, Harnessing the power of RADseq for ecological and evolutionary genomics. Nature Reviews Genetics 17, 2016, 81–92.

ANIJALG et al. 2018: P. ANIJALG/S. Y. W. HO/J. DAVISON/M. KEIS/E. TAMMELEHT/K. BOBOWIK/I. L. TUMANOV/A. P. SAVELJEV/E. A. LYAPUNOVA/A. A.VOROBIEV/N. I. MARKOV/A. P. KRYUKOV/I. KOJOLA/J. E. SWENSON/S. B. HAGEN/H. G. EIKEN/L. PAULE/U. SAARMA, Large-scale migrations of brown bears in Eurasia and to North America during the Late Pleistocene. Journal of Biogeography 45, 2018, 394–405.

BELLEMAIN et al. 2005: E. BELLEMAIN/J. E. SWENSON/O. TALLMON/S. BRUNBERG/P. TABERLET, Estimating population size of elusive animals with DNA from hunter-collected feces: Four methods for brown bears. Conservation Biology 19, 2005, 150–161.

BIDON et al. 2014: T. BIDON/A. JANKE/S. R. FAIN/H. G. EIKEN/S. B. HAGEN/U. SAARMA/B. M. HALLSTROM/N. LECOMTE/F. HAILER, Brown and polar bear Y chromosomes reveal extensive male-biased gene flow within brother lineages. Molecular Biology and Evolution 31, 2014, 1353–1363.

BISCHOF et al. 2016: R. BISCHOF/H. BRØSETH/O. GIMENEZ, Wildlife in a Politically Divided World: Insularism Inflates Estimates of Brown Bear Abundance. Conservation Letters 9, 2016, 122–130.

BISCHOF et al. 2020: R. BISCHOF/C. MILLERET/P. DUPONT/J. CHIPPERFIELD/M. TOURANI/A. ORDIZ/P. DE VALPINE/D. TUREK/J. A. ROYLE/O. GIMENEZ/Ø. FLAGSTAD/M. AKESSON/L. SVENSSON/H. BROSETH/J. KINDBERG, Estimating and forecasting spatial population dynamics of apex predators using transnational genetic monitoring. Proceedings of the National Academy of Sciences 117, 2020, 30531–30538.

BROOKE et al. 2014: Z. M. BROOKE/J. BIELBY/K. NAMBIAR/C. CARBONE, Correlates of research effort in carnivores: body size, range size and diet matter. PLoS One 9, 2014. https://doi.org/10.1371/journal.pone.0093195.

CHAPRON et al. 2014: G. CHAPRON/P. KACZENSKY/J. D. C. LINNELL/M. VON ARX/D. HUBER/H. ANDRÉN/J. V. LÓPEZ-BAO/M. ADAMEC/F. ÁLVARES/O. ANDERS/L. BALČIAUSKAS/V. BALYS/P. BEDŐ/F. BEGO/J. C. BLANCO/U. BREITENMOSER/H. BRØSETH/L. BUFKA/R. BUNIKYTE/P. CIUCCI/A. DUTSOV/T. ENGLEDER/C. FUXJÄGER/C. GROFF/K. HOLMALA/B. HOXHA/Y. ILIOPOULOS/O. IONESCU/J. JEREMIĆ/K. JERINA/G. KLUTH/F. KNAUER/I. KOJOLA/I. KOS/M. KROFEL/J. KUBALA/S. KUNOVAC/J. KUSAK/M. KUTAL/O. LIBERG/A. MAJIĆ/P. MÄNNIL/R. MANZ/E. MARBOUTIN/F. MARUCCO/D. MELOVSKI/K. MERSINI/Y. MERTZANIS/R. W. MYSŁAJEK/S. NOWAK/J. ODDEN/J. OZOLINS/G. PALOMERO/M. PAUNOVIĆ/J. PERSSON/H. POTOČNIK/P.-Y. QUENETTE/G. RAUER/I. REINHARDT/R. RIGG/A. RYSER/V. SALVATORI/T. SKRBINŠEK/A. STOJANOV/J. E. SWENSON/L. SZEMETHY/A. TRAJÇE/F. TSINGARSKA-SEDEFCHEVA/M. VÁŇA/R. VEEROJA/P. WABAKKEN/M. WÖLFL/S. WÖLFL/F. ZIMMERMANN/D. ZLATANOVA/L. BOITANI, Recovery of large carnivores in Europe's modern human-dominated landscapes. Science 346, 2014, 1517–1519.

CURRY-LINDAHL 1972: K. CURRY-LINDAHL, The Brown Bear (*Ursus arctos*) in Europe: Decline, Present Distribution, Biology and Ecology. In: International Association for Bear Research and Management (eds.), Bears: Their biology and management. IUCN Publications New Series 23 (Calgary 1972) 74–80.

DALERUM et al. 2009: F. DALERUM/E. Z. CAMERON/K. KUNKEL/M. J. SOMERS, Diversity and depletions in continental carnivore guilds: implications for prioritizing global carnivore conservation. Biology Letters 5, 2009, 35–38.

DAVISON et al. 2011: J. DAVISON/S. Y. W. HO/S. C. BRAY/M. KORSTEN/E. TAMMELEHT/M. HINDRIKSON/K. ØSTBYE/E. ØSTBYE/S.-E. LAURITZEN/J. AUSTIN, Late-Quaternary biogeographic scenarios for the brown bear (*Ursus arctos*), a wild mammal model species. Quaternary Science Reviews 30, 2011, 418–430.

DE BARBA et al. 2017: M. DE BARBA/C. MIQUEL/S. LOBREAUX/P. Y. QUENETTE/J. E. SWENSON/P. TABERLET, High-throughput microsatellite genotyping in ecology: improved accuracy, efficiency, standardization and success with low-quantity and degraded DNA. Molecular Ecology Resources 17, 2017, 492–507.

ERSMARK et al. 2019: E. ERSMARK/G. BARYSHNIKOV/T. HIGHAM/A. ARGANT/P. CASTAÑOS/D. DÖPPES/

M. Gasparik/M. Germonpré/K. Lidén/G. Lipecki/ A. Marciszak/R. Miller/M. Moreno-García/ M. Pacher/M. Robu/R. Rodriguez-Varela/M. Rojo Guerra/M. Sabol/N. Spassov/J. Storå/C. Valdiosera/A. Villaluenga/J. R. Stewart/L. Dalén, Genetic turnovers and northern survival during the last glacial maximum in European brown bears. Ecology and Evolution 9, 2019, 5891–5905.

Frank et al. 2021: S. C. Frank/F. Pelletier/A. Kopatz/ A. Bourret/D. Garant/J. E. Swenson/H. G. Eiken/ S. B. Hagen/A. Zedrosser, Harvest is associated with the disruption of social and fine-scale genetic structure among matrilines of a solitary large carnivore. Evolutionary Applications 14, 2021, 1023–1035.

Frankham 1995: R. Frankham, Conservation Genetics. Annual Review of Genetics 29, 1995, 305–327.

Frankiiam 2005: R. Frankham, Genetics and extinction. Biological Conservation 126, 2005, 131–140.

Hagen et al. 2015: S. B. Hagen/A. Kopatz/J. Aspi/I. Kojola/H. G. Eiken, Evidence of rapid change in genetic structure and diversity during range expansion in a recovering large terrestrial carnivore. Proceedings of the Royal Society of London. Series B: Biological Sciences 282, 2015. https://doi.org/10.1098/rspb.2015.0092.

Hewitt 1996: G. M. Hewitt, Some genetic consequences of ice ages, and their role in divergence and speciation. Biological Journal of the Linnean Society 58, 1996, 247–276.

Hewitt 1999: G. M. Hewitt, Post-glacial re-colonization of European biota. Biological Journal of the Linnean Society 68, 1999, 87–112.

Hewitt 2000: G. Hewitt, The genetic legacy of the Quaternary ice ages. Nature 405, 2000, 907–913.

Hirata et al. 2017: D. Hirata/T. Mano/A. V. Abramov/ G. F. Baryshnikov/P. A. Kosintsev/K. Murata/ R. Masuda, Paternal phylogeographic structure of the brown bear (*Ursus arctos*) in northeastern Asia and the effect of male-mediated gene flow to insular populations. Zoological Letters 3, 2017, 21.

Hofreiter et al. 2002: M. Hofreiter/C. Capelli/ M. Krings/L. Waits/N. Conard/S. Munzel/ G. Rabeder/D. Nagel/M. Paunovic/G. Jambresic/ S. Meyer/G. Weiss/S. Pääbo, Ancient DNA analyses reveal high mitochondrial DNA sequence diversity and parallel morphological evolution of late pleistocene cave bears. Molecular Biology and Evolution 19, 2002, 1244–1250.

Jerina/Adamic 2008: K. Jerina/M. Adamic, Fifty Years of Brown Bear Population Expansion: Effects of Sex-Biased Dispersal on Rate of Expansion and Population Structure. Journal of Mammalogy 89, 2008, 1491–1501.

Keis et al. 2013: M. Keis/J. Remm/S. Y. W. Ho/J. Davison/E. Tammeleht/I. L. Tumanov/A. P. Saveljev/ P. Männil/I. Kojola/A. V. Abramov/T. Margus/ U. Saarma/A. Phillimore, Complete mitochondrial genomes and a novel spatial genetic method reveal cryptic phylogeographical structure and migration patterns among brown bears in north-western Eurasia. Journal of Biogeography 40, 2013, 915–927.

Kindberg/Swenson 2018: J. Kindberg/J. E. Swenson, Björnstammens storlek i Sverige 2017. Scandinavian Brown Bear Research Project Rapport 2018-3 (Orsa 2018).

Kindberg et al. 2011: J. Kindberg/J. E. Swenson/G. Ericsson/E. Bellemain/C. Miquel/P. Taberlet, Estimating population size and trends of the Swedish brown bear *Ursus arctos* population. Wildlife Biology 17, 2011, 114–123.

Kohn et al. 1995: M. Kohn/F. Knauer/A. Stoffella/ W. Schröder/S. Pääbo, Conservation genetics of the European brown bear – a study using excremental PCR of nuclear and mitochondrial sequences. Molecular Ecology 4, 1995, 95–104.

Kopatz et al. 2014: A. Kopatz/H. G. Eiken/J. Aspi/I. Kojola/C. Tobiassen/K. F. Tirronen/P. I. Danilov/S. B. Hagen, Admixture and gene flow from Russia in the recovering Northern European brown bear (*Ursus arctos*). PLoS One 9, 2014. https://doi.org/10.1371/journal.pone.0097558.

Kopatz et al. 2017: A. Kopatz/H. G. Eiken/J. Schregel/ J. Aspi/I. Kojola/S. B. Hagen, Genetic substructure and admixture as important factors in linkage disequilibrium-based estimation of effective number of breeders in recovering wildlife populations. Ecology and Evolution 7, 2017, 10721–10732.

Kopatz et al. 2021: A. Kopatz/O. Kleven/I. Kojola/ J. Aspi/A. J. Norman/G. Spong/N. Gyllenstrand/ L. Dalén/I. Fløystad/S. B. Hagen/J. Kindberg/ Ø. Flagstad, Restoration of transborder connectivity for Fennoscandian brown bears (*Ursus arctos*). Biological Conservation 253, 2021, 11–25.

Korsten et al. 2009: M. Korsten/S. Y. Ho/J. Davison/ B. Pahn/E. Vulla/M. Roht/I. L. Tumanov/I. Kojola/Z. Andersone-Lilley/J. Ozolins/M. Pilot/ Y. Mertzanis/A. Giannakopoulos/A. A. Vorobiev/ N. I. Markov/A. P. Saveljev/E. A. Lyapunova/A. V. Abramov/P. Mannil/H. Valdmann/S. V. Pazetnov/ V. S. Pazetnov/A. M. Rokov/U. Saarma, Sudden expansion of a single brown bear maternal lineage across northern continental Eurasia after the last ice age: a general demographic model for mammals? Molecular Ecology 18, 2009, 1963–1979.

Krofel et al. 2012: M. Krofel/M. Jonozovič/K. Jerina, Demography and mortality patterns of removed brown bears in a heavily exploited population. Ursus 23, 2021, 91–103.

Lacy 1997: R. C. Lacy, Importance of genetic variation to the viability of mammalian populations. Journal of Mammalogy 78, 1997, 320–335.

Manel et al. 2004: S. Manel/E. Bellemain/J. E. Swenson/ O. François, Assumed and inferred spatial structure of populations: the Scandinavian brown bears revisited. Molecular Ecology 13, 2004, 1327–1331.

Moritz 1994: C. Moritz, Defining Evolutionarily-Significant-Units for Conservation. Trends in Ecology & Evolution 9, 1994, 373–375.

Nei et al. 1975: M. Nei/T. Maruyama/R. Chakraborty, The Bottleneck Effect and Genetic Variability in Populations. Evolution 29, 1975, 1–10.

Norman/Spong 2015: A. J. Norman/G. Spong, Single nucleotide polymorphism-based dispersal estimates using noninvasive sampling. Ecology and Evolution 5, 2015, 3056–3065.

Norman et al. 2013: A. J. Norman/N. R. Street/G. Spong, *De novo* SNP discovery in the Scandinavian brown bear (*Ursus arctos*). PLoS One 8, 2015. https://doi.org/10.1371/journal.pone.0081012.

Paetkau/Strobeck 1994: D. Paetkau/C. Strobeck, Microsatellite analysis of genetic variation in black bear populations. Molecular Ecology 3, 1994, 489–495.

Paetkau et al. 1995: D. Paetkau/W. Calvert/I. Stirling/C. Strobeck, Microsatellite analysis of population structure in Canadian polar bears. Molecular Ecology 4, 1995, 347–354.

Ripple et al. 2014: W. J. Ripple/J. A. Estes/R. L. Beschta/C. C. Wilmers/E. G. Ritchie/M. Hebblewhite/J. Berger/B. Elmhagen/M. Letnic/M. P. Nelson/O. J. Schmitz/D. W. Smith/A. D. Wallach/A. J. Wirsing, Status and Ecological Effects of the World's Largest Carnivores. Science 343, 2014. https://doi.10.1126/science.1241484.

Saarma et al. 2007: U. Saarma/S. Y. Ho/O. G. Pybus/M. Kaljuste/I. L. Tumanov/I. Kojola/A. A. Vorobiev/N. I. Markov/A. P. Saveljev/H. Valdmann/E. A. Lyapunova/A. V. Abramov/P. Mannil/M. Korsten/E. Vulla/S. V. Pazetnov/V. S. Pazetnov/S. V. Putchkovskiy/A. M. Rokov, Mitogenetic structure of brown bears (*Ursus arctos* L.) in northeastern Europe and a new time frame for the formation of European brown bear lineages. Molecular Ecology 16, 2007, 401–413.

Saarma/Kojola 2007: U. Saarma/I. Kojola, Matrilineal genetic structure of the brown bear population in Finland. Ursus 18, 2007, 30–37.

Schregel et al. 2012: J. Schregel/A. Kopatz/S. B. Hagen/H. Broseth/M. E. Smith/S. Wikan/I. Wartiainen/P. E. Aspholm/J. Aspi/J. E. Swenson/O. Makarova/N. Polikarpova/M. Schneider/P. M. Knappskog/M. Ruokonen/I. Kojola/K. F. Tirronen/P. I. Danilov/H. G. Eiken, Limited gene flow among brown bear populations in far Northern Europe? Genetic analysis of the east-west border population in the Pasvik Valley. Molecular Ecology 21, 2012, 3474–3488.

Schregel et al. 2015: J. Schregel/H. G. Eiken/F. A. Grøndahl/F. Hailer/J.Aspi/I. Kojola/K. Tirronen/P. Danilov/A. Rykov/E. Poroshin/A. Janke/J. E. Swenson/S. B. Hagen, Y chromosome haplotype distribution of brown bears (*Ursus arctos*) in Northern Europe provides insight into population history and recovery. Molecular Ecology 24, 2015, 6041–6060.

Schregel et al. 2017: J. Schregel/A. Kopatz/H. G. Eiken/J. E. Swenson/S. B. Hagen, Sex-specific genetic analysis indicates low correlation between demographic and genetic connectivity in the Scandinavian brown bear (*Ursus arctos*). PLoS One 12, 2017. https://doi.org/10.1371/journal.pone.0180701.

Schregel et al. 2018: J. Schregel/J. Remm/H. G. Eiken/J. E. Swenson/U. Saarma/S. B. Hagen, Multi-level patterns in population genetics: Va riogram series detects a hidden isolation-by-distance-dominated structure of Scandinavian brown bears *Ursus arctos*. Molecular Biology and Evolution 9, 2018, 1324–1334.

Schwartz et al. 2007: M. K. Schwartz/G. Luikart/R. S Waples, Genetic monitoring as a promising tool for conservation and management. Trends in Ecology & Evolution 22, 2007, 25–33. https://doi.org/10.1016/j.tree.2006.08.009.

Siivonen 1982: L. Siivonen, The history of the Fennoscandian mammal fauna. Acta Zoologica Fennica 169, 1982, 7–10.

Sommer/Benecke 2005: R. S. Sommer/N. Benecke, The recolonization of Europe by brown bears *Ursus arctos* Linnaeus, 1758 after the Last Glacial Maximum. Mammal Review 35, 2005, 156–164.

Støen et al. 2005: O.-G. Støen/E. Bellemain/S. Sæbø/J. E. Swenson, Kin-related spatial structure in brown bears *Ursus arctos*. Behavioral Ecology and Sociobiology 59, 2005, 191–197.

Støen et al. 2006: O.-G. Støen/A. Zedrosser/S. Sæbø/J. E. Swenson, Inversely density-dependent natal dispersal in brown bears *Ursus arctos*. Oecologia 148, 2006, 356–364.

Swenson et al. 1994: J. E. Swenson/F. Sandegren/A. Bjärvall/A. Soderberg/P. Wabakken/R. Franzen, Size, trend, distribution and conservation of the brown bear *Ursus arctos* population in Sweden. Biological Conservation 70, 1994, 9–17.

Swenson et al. 1995: J. E. Swenson/P. Wabakken/F. Sandegren/A. Bjärvall/R. Franzén/A. Söderberg, The near extinction and recovery of brown bears in Scandinavia in relation to the bear management policies of Norway and Sweden. Wildlife Biology 1, 1995, 11–25.

Swenson et al. 2011: J. E. Swenson/P. Taberlet/E. Bellemain, Genetics and conservation of European brown bears *Ursus arctos*. Mammal Review 41, 2011, 87–98.

Taberlet/Bouvet 1994: P. Taberlet/J. Bouvet, Mitochondrial DNA polymorphism, phylogeography, and conservation genetics of the brown bear *Ursus arctos* in Europe. Proceedings of the Royal Society of London. Series B: Biological Sciences 255, 1994, 195–200.

Taberlet et al. 1995: P. Taberlet/J. E. Swenson/F. Sandegren/A. Bjärvall, Localization of a Contact Zone between Two Highly Divergent Mitochondrial DNA Lineages of the Brown Bear *Ursus arctos* in Scandinavia. Conservation Biology 9, 1995, 1255–1261.

Taberlet et al. 1997: P. Taberlet/J. J. Camarra/S. Griffin/E. Uhres/O. Hanotte/L. P. Waits/C. Dubois-Paganon/T. Burke/J. Bouvet, Noninvasive genetic tracking of the endangered Pyrenean brown bear population. Molecular Ecology 6, 1997, 869–876.

Taberlet et al. 1998: P. Taberlet/L. Fumagalli/A. G. Wust-Saucy/J. F. Cosson, Comparative phylogeography and postglacial colonization routes in Europe. Molecular Ecology 7, 1998, 453–464.

Taberlet et al. 1999: P. Taberlet/L. P. Waits/G. Luikart, Noninvasive genetic sampling: look before you leap. Trends in Ecology & Evolution 14, 1999, 323–327. https://doi.org/10.1016/S0169-5347(99)01637-7.

Tammeleht et al. 2010: E. Tammeleht/J. Remm/M. Korsten/J. Davison/I. Tumanov/A. Saveljev/P. Mannil/I. Kojola/U. Saarma, Genetic structure in large, continuous mammal populations: the example of brown bears in northwestern Eurasia. Molecular Ecology 19, 2010, 5359–5370.

Waits et al. 2000: L. Waits/P. Taberlet/J. E. Swenson/F. Sandegren/R. Franzén, Nuclear DNA microsatellite analysis of genetic diversity and gene flow in the Scandinavian brown bear (*Ursus arctos*). Molecular Ecology 9, 2000, 421–431.

WAPLES/GAGGIOTTI 2006: R. S. WAPLES/O. GAGGIOTTI, What is a population? An empirical evaluation of some genetic methods for identifying the number of gene pools and their degree of connectivity. Molecular Ecology 15, 2006, 1419–1439.

XENIKOUDAKIS et al. 2015: G. XENIKOUDAKIS/E. ERSMARK/ J. L. TISON/L. WAITS/J. KINDBERG/J. E. SWENSON/ L. DALÉN, Consequences of a demographic bottleneck on genetic structure and variation in the Scandinavian brown bear. Molecular Ecology 24, 2015, 3441–3454.

ZACHOS et al. 2008: F. E. ZACHOS/M. OTTO/R. UNICI/ R. LORENZINI/G. B. HARTL, Evidence of a phylogeographic break in the Romanian brown bear (*Ursus arctos*) population from the Carpathians. Mammalian Biology 73, 2008, 93–101.

ZEDROSSER et al. 2007: A. ZEDROSSER/O.-G. STØEN/S. SÆBØ/ J. E. SWENSON, Should I stay or should I go? Natal dispersal in the brown bear. Animal Behaviour 74, 2007, 369–376.

ZEDROSSER et al. 2011: A. ZEDROSSER/S. M. J. G. STEYAERT/ H. GOSSOW/J. E. SWENSON, Brown bear conservation and the ghost of persecution past. Biological Conservation 144, 2011, 2163–2170.

Dr. Alexander Kopatz
Norwegian Institute for Nature Research
Trondheim
Norway
alexander.kopatz@nina.no

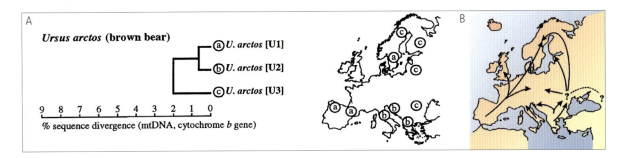

Fig. 1. a: Illustration of the phylogeography of the brown bear across Europe, based on the results by TABERLET/BOUVET 1994 (figure after TABERLET et al. 1998, fig. 2, copyright by Wiley); b: Brown bear postglacial colonisation routes with main refugial areas in Iberia, Italy, the Balkans and the Caucasus (figure after HEWITT 2000, fig. 2, copyright by Nature).

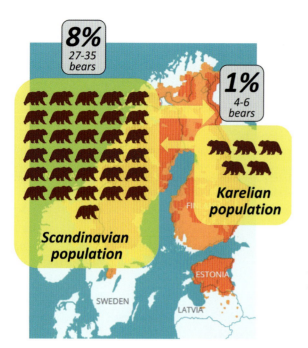

Fig. 2. Illustration of the asymmetric number of migrants per generation between the brown bear populations of Scandinavia (west) and Karelia (east), based on Kopatz et al. 2021 (background map source: Horizon [modified], distribution data source: Large Carnivore Initiative for Europe, 2012–2016; bear icon by Freepik available on flaticon.com).

Hibernation ecology of brown bears in Sweden

By Andrea Friebe, Jon E. Swenson and Andreas Zedrosser

Keywords: Scandinavia, brown bear, Ursus arctos, *den type, hibernation, physiology*

Abstract: Hibernation is an adaptive strategy to cope with unfavourable environmental conditions. Bears are the only large mammal that use this strategy during winter. They reduce their body temperature by several degrees and their metabolism by 20–50 % during hibernation. However, bears have an exceptional position among hibernators, because they are the only mammal with delayed implantation, gestation, parturition, and lactation during hibernation. Bears do not consume food but must rely on fat reserves during winter and give birth to one to three cubs during hibernation. In general, the length of hibernation is shorter at more southern latitudes and increases towards the north, and bears lose 20–45 % of their body weight during this time. Three types of winter dens are used by brown bears in Scandinavia: excavated anthill and soil dens (74 % of all dens), natural cavity dens (11 %), and open nest dens (15 %). Reproductive success in female brown bears is affected by the choice of den type, and females hibernating in better insulated dens have a higher probability of producing offspring. Brown bears prefer to den in forested areas that provide shelter from wind and cold temperatures, as well as at higher altitudes and areas with steeper slopes. Bears are especially sensitive to human disturbances during hibernation and prefer to den far from human infrastructure and settlements.

INTRODUCTION

Hibernation in mammals, i.e. the time period that an animal spends in a dormant or "sleeping" state, is an adaptive strategy to cope with unfavourable environmental conditions, such as winter (Nelson 1973; Nelson et al. 1973). It most commonly occurs in small mammals, such as ground squirrels, marmots, or bats, and can last from several days up to several months, depending on the species, ambient temperature, time of year, and an individual's body condition (Nelson 1973; Geiser 1998; Humphries et al. 2003). The physiology of a hibernating mammal in winter is profoundly different from its active state during the other times of the year. Hibernation is in generally characterised by physical inactivity, a reduction of physiological functions, low metabolic rate, and a reduction of the body temperature to as low as 0 °C (Barnes 1989; Geiser 2004; Friebe 2015).

Hibernating mammals do not necessarily remain torpid, i.e. inactive, throughout the entire hibernation season. Rather, hibernation usually is characterised by bouts of torpor that last several days or weeks, but are interrupted by periodic bouts with higher activity, i.e. short-term arousals. During those active bouts, the biochemical and physiological parameters return to an almost normal level, probably to recover from the physiological costs caused by metabolic depression (Prendergast et al. 2002; Humphries et al. 2003; Astaeva/Klichkhanov 2009). These time periods generally also are used to consume food and water, as well as to urinate and defecate (Pinter 1984; Wang 1989; Pulawa/Florant 2000; Boyles et al. 2020).

Bears formerly were not considered "true hibernators", because their body temperature does not decrease as dramatically during winter as it does in other hibernating mammals, such as ground squirrels (*Sciuridae*) (Geiser 1998; Heldmaier 2011; Tøien et al. 2011). Also, hibernating bears can "wake up" very quickly from hibernation and even leave the den when disturbed, in comparison to small hibernating mammals, that need up to several days to wake up from deep hibernation. Based on advances in our understanding of physiology and biochemistry, hibernation is now more accurately defined as comprehensive metabolic suppression rather than only based on a decline in body temperature (Carey et al. 2003; Geiser 2004). Ursids reduce their metabolism by 20–50 % during winter, and therefore are considered as true hibernators (Heldmaier et al. 2004; Heldmaier 2011). However, bears have an exceptional position among hibernating mammals, because they are the only north-temperate species with delayed implantation, gestation, parturition, and lactation during hibernation (Ramsay/Dunbrack 1986; Spady et al. 2007; Robbins et al. 2012). The physiological and biochemical processes that regulate hibernation in bears are of great interest in human medicine and thus have become an emerging field in ecological biomimicry studies (e.g. Stenvinkel et al. 2012; Berg von Linde et al. 2015; Fröbert et al. 2020).

Hibernation ecology

Depending on the climatic zone and environmental conditions, brown bears (*Ursus arctos*) can spend more than half of their life in winter dens and may hibernate between five to seven months in a given year (Friebe et al. 2001; Manchi/Swenson 2005; Bojarska/Selva 2011). In general, the length of hibernation is shorter at more southern latitudes and increases towards the north. For example, some individual bears in southern European areas, such as Croatia or Spain, may not hibernate at all (Huber/Roth 1997; Nores et al. 2010), whereas all bears hibernate in northern Europe, for example in Sweden (Friebe et al. 2001; Manchi/Swenson 2005).

Brown bears do not consume food during hibernation and therefore must build up fat reserves for the long period of caloric deprivation during winter (Hellgren 1998; López-Alfaro et al. 2013). In late summer and autumn, bears enter a physiological state called hyperphagia, which is defined as period of highly increased caloric intake and highly increased search for food. Hyperphagia can result in a caloric intake up to 20,000 kcal/day (Nelson 1987), and brown bears may add up to 40 % of their body mass in fat before hibernation (Fig. 1; Nelson 1980; Nelson et al. 1983). Activity, heart rate and body temperature start to drop slowly several weeks before den entry, and bears usually enter the den upon arrival of the first snow and when ambient temperature reaches 0 °C (Evans et al. 2016). The first demographic group of bears to start hibernating is pregnant females, followed by females with older dependent offspring and subadult bears (Friebe et al. 2001; Manchi/Swenson 2005). The demographic group that starts hibernating latest and leaves the den earliest in the spring is adult males, followed by subadult individuals and females with older dependent offspring. Females that have given birth during hibernation are the demographic group that leave the den latest (Friebe et al. 2001; Manchi/Swenson 2005). For example, pregnant females start hibernating as early as late September in Sweden and leave their den with their young as late as early May the following spring (Friebe et al. 2001; Manchi/Swenson 2005).

Bears lose 20–45 % of their body weight while in the den during hibernation. However, the amount of weight loss during winter depends on several factors, such as the duration of denning, the bear's sex and age, and the reproductive status, whether a female is solitary, pregnant, or accompanied by cubs (Nelson et al. 1973; Kingsley et al. 1983; López-Alfaro et al. 2013).

In the hibernation state, a bear's oxygen demand is reduced to approximately 25–50 % of the active state, and the respiration rate is decreased to 1–2 times per minute (Hellgren 1998; Tøien et al. 2011). The low respiratory quotient of bears during hibernation indicates a pure fat combustion, i.e. they mainly use body fat for maintaining physiological functions, while conserving lean body mass like carbohydrates and proteins (Nelson et al. 1973; Nelson 1980; Boyer/Barnes 1999). Fat serves as an efficient caloric storage medium, as it features high energy density. To regain the stored energy, fat needs to be metabolised. Burning body fat produces water and carbon dioxide; the carbon dioxide is exhaled, and the water stays in the blood to keep the animal hydrated (Nelson et al. 1973). In contrast to all other hibernating mammals, bears do not urinate during hibernation. However, azotemia (increased blood levels of nitrogen that can induce kidney failure) does not develop in hibernating bears (Stenvinkel et al. 2012). Instead, the urea production is decreased, and any urea produced is recycled and resynthesised into skeletal muscles and other body proteins, to preserve lean body mass (Nelson 1989; Harlow et al. 2001; Stenvinkel et al. 2018).

Another measure to conserve energy and adapt to the reduced respiratory rate is decreased heart rate during hibernation. The bear's heart rate can be as low as 10 beats per minute (bpm) during hibernation, compared to the heart rate of approximately 50 bpm of sleeping bears in summer (Nelson et al. 2003; Folk et al. 2008; Laske et al. 2018). Simultaneously with the lowered heart and respiratory rate, the bear's body temperature drops, but only 2–6 °C below the summer core temperature of 37–38 °C. Bears maintain this body temperature by undergoing periodic muscular shivering (French 1986; Hissa et al. 1994; Tøien et al. 2011).

Thus, compared to other hibernators, the body temperature remains above the critical values for brain and muscle activity. Continuous electroencephalogram EEG recordings of American black bears (*Ursus americanus*) during the hibernation season revealed that they are predominantly asleep but are cycling between rapid-eye-movement and non-rapid-eye-movement sleep with very few phases of wakefulness (Folk et al. 2008; Rogers et al. 2020). Nevertheless, periodical short-term bouts of activity with frequent, small movements have been reported during hibernation (Friebe et al. 2013; 2014). The relatively high body temperature enables bears to wake up spontaneously from hibernation, and they are then capable of rapid responses with high mobility and endurance (Evans et al. 2012).

Although bears continue to produce some feces during hibernation, they do not defecate, which again is in contrast with all other hibernating mammals. Therefore, the feces accumulate in the intestine. The intestinal walls absorb water from the feces and, thus, a dry and hard fecal plug develops. Bears also groom during active hibernation periods, and some of these materials are swallowed during the self-grooming process, such as plant materials, hair, and skin. These indigestible materials pass through the digestive tract and become part of the fecal plug (Rogers 1981). Additionally, hair and claws from cubs can sometimes be found in fecal plugs, which indicates that female bears consume the body of cubs that died in the den (A. Friebe, personal observation). This fecal plug is the first scat produced after hibernation and is commonly found close to den sites (A. Friebe, personal observation).

Pregnancy and birth during hibernation

Brown bears in Scandinavia mate during spring and early summer, with the peak of the mating season around the 1st of June (Steyaert et al. 2012). Brown bears exhibit delayed implantation, i.e. the embryo (blastocyst) does not implant immediately, but is maintained free-floating in the uterus in a state of suspended dormancy for four to five months until implantation occurs in November/December, after the onset of hibernation (Spady et al. 2007; Friebe et al. 2014).

For successful implantation and reproduction, females require a minimum amount of body mass and fat (~19 %) prior to hibernation. This determines whether the implantation of blastocysts happens or not (Elowe/Dodge 1989; Atkinson/Ramsay 1995; López-Alfaro et al. 2013). Gestation in bears lasts approximately 56 days (Friebe et al. 2013). This short period limits the energetic costs of reproduction by truncating embryonic development, which in turn reduces the size of offspring and thus the initial costs of lactation. Brown bears commonly give birth to one to three cubs, which are born in the den in January/February (Zedrosser et al. 2011; Friebe et al. 2014). Bear cubs are naked at birth and weigh about 300–500 g (Robbins et al. 2012). The milk of bears is very rich in protein and fat (Derocher et al. 1993; McDonald/Fuller 2005; López-Alfaro et al. 2013), and brown bear cubs in Sweden weigh approximately 5 kg when leaving the den for the first time in late April to early May (A. Friebe, personal observation).

Den types

Dens are an essential part of brown bear ecology and reproduction, because pregnant females give birth to cubs while hibernating in their dens during winter. Brown bears normally hibernate alone in their dens, except for females that give birth or that are still accompanied by older dependent offspring. In general, most brown bears hibernate in a new den every winter and there is little or no reuse of the same den over successive years (Ciarniello et al. 2005; Elfström/Swenson 2009).

The main functions of a den are reduction of energy loss during winter, as well as protection against disturbance from conspecifics or other species, including humans (Petram et al. 2004; Sahlén et al. 2011; Shiratsuru et al. 2020). The construction of the winter den seems to be an innate behaviour. Cubs of the year that had lost their mother and thus never had the possibility to learn how to construct a den, have built dens and successfully hibernated there on their own (Swenson et al. 1998). However, bears certainly gain experience in building dens and the construction improves over the years (Petram et al. 2004; González-Bernardo et al. 2020). Several studies have shown that den construction and the habitat where a den is located affect individual fitness. The insulation properties of snow contribute to constant temperature and additionally decrease energy costs for the bear (Nelson et al. 1983; Servheen/Klaver 1983; Elfström et al. 2008).

Bear dens are normally small in relation to the size of the bear. They often have an inner radius less than 1 m and a small entrance (A. Friebe, unpublished data), which can be completely covered with snow during winter. The snow seals the inner part of the den effectively from the weather conditions on the outside, contributing to increased thermal insulation (Craighead/Craighead 1972). To improve insulation properties, bears normally gather vegetation as insulating bedding materials, which helps to minimise energy loss (Reynolds et al. 1976; Tietje/Ruff 1980).

Generally, three types of dens are used by brown bears in Scandinavia: excavated anthill and soil dens (comprising 74 % of all dens; A. Friebe, unpublished data), natural cavity dens (11 %), and open nest dens (15 %; see Fig. 2). Excavated dens are preferred in many landscapes, including Scandinavia (Linnell et al. 2000). The space around an animal's body is kept tight to increase energy efficiency in excavated dens, which likely decreases the energetic costs of hibernation (Tietje/Ruff 1980; Shiratsuru et al. 2020). In comparison, natural cavity dens cannot be adjusted, and a suitable cavity may be difficult to find. In Scandinavia, the excavated anthill den is the most widely used den type, followed by dens dug in soil (Manchi/Swenson 2005). An anthill den is excavated in an abandoned ant mound that is overgrown with vegetation. The main difference between an anthill den and the other den types is the composition of the walls. The walls of anthill dens are composed by loose organic material interspersed with roots that provide stability and create an additional air layer for insulation. In comparison, soil dens usually miss this insulating air layer and may also contain more moisture (SCHOEN et al. 1987). Anthill dens are significantly more often used by females than by males, likely due to the higher degree of insulation compared to other den types, which is especially important for pregnant females (ELFSTRÖM/SWENSON 2009). Reproductive success in female brown bears is affected by the choice of den type, and female brown bears hibernating in better insulated dens have a higher probability of producing offspring (KLENZENDORF et al. 2002). NOWACK (2015) observed in Sweden that females hibernating and giving birth in anthill dens on average produced larger litters and had more male offspring, compared to females that hibernated and gave birth in other den types.

Hibernation in natural cavity dens is comparatively rare in Scandinavia, but more common in southern European populations, such as in Slovenia and Croatia (HUBER/ROTH 1997; PETRAM et al. 2004). Natural cavities are less insulated than excavated dens, however, they do not require as much energetic investment prior to use.

The least insulated den type is open dens, often also referred to as nest dens or basket dens. These dens usually contain only a thick layer of bedding material and are commonly placed close to the base of a tree or rocks. The construction lacks walls or a roof, and the only insulation is provided by the bedding material and the snow that covers the hibernating bear during winter (MARTORELLO/PELTON 2003; SERYODKIN et al. 2003; ELFSTRÖM et al. 2008). These types of dens primarily are used by large adult males in Scandinavia, which also makes them more vulnerable to disturbance during hibernation (ELFSTRÖM/SWENSON 2009; SAHLÉN et al. 2015). Males have a larger body and more fat reserves than females and smaller bears and are therefore probably better suited to bear the energetic costs of a den without protective walls or roof. However, bears using open dens hibernate for a shorter period compared to bears using other den types, which indicates that the open structure probably is less efficient when it comes to thermal insulation and energy loss, especially in periods where there is little or no snow (ELFSTRÖM/SWENSON 2009). In southern European countries with less severe winters, also female bears use open dens more frequently (GONZÁLEZ-BERNARDO et al. 2020).

Denning habitat

Numerous studies have described the denning habitats of brown bears worldwide (e.g. LINNELL et al. 2000; CIARNIELLO et al. 2005; ELFSTRÖM/SWENSON 2009; SAHLÉN et al. 2011; SMEREKA et al. 2017; MANGIPANE et al. 2018; GONZÁLEZ-BERNARDO et al. 2020). In general, the habitat surrounding a den site is highly variable and its selection depends on the available habitat types within a bear's home range. The specific ecological factors affecting a bear's den site selection are local climatic and habitat characteristics, as well as ground cover and terrain characteristics. Bears seem to select denning locations carefully in good time before actual hibernation. For example, female bears in Sweden visit the area where they finally build their dens more than once a month during the active season, probably because a suitable and safe den site is important for the survival of the cubs (FRIEBE et al. 2001; SAHLÉN et al. 2011). A study of denning behaviour in Sweden has shown that bears that had not visited their denning area prior to hibernation were more likely to abandon their den due to disturbances (SAHLÉN et al. 2015).

Brown bears generally prefer to den in forested areas that provide shelter from wind and cold temperatures, as well as at higher altitudes and in areas with steeper slopes (CIARNIELLO et al. 2005; ELFSTRÖM et al. 2008; LIBAL et al. 2011). Dens are better insulated at higher altitudes due to higher snow cover, which has a positive impact on the energetic efficiency during hibernation (LIBAL et al. 2011). Den construction also seems to be easier on a slope compared to flat ground, and the den is likely better protected from flooding due to rain or melting snow. Additionally, a den on a steep slope probably provides better protection against disturbance, as it is more difficult to access (HAROLDSON et al. 2002).

Brown bears are sensitive to disturbances in the denning area and during hibernation, especially to human activities during the time of the initiation of hibernation. This is likely also an important reason why bears in general choose den sites far from infrastructure and human settlements (SWENSON et al. 1997; SAHLEN et al. 2015). In addition, a thick vegetation cover that favours the concealment of the den entrance seems to be an important factor to avoid human-caused disturbance (ELFSTRÖM et al. 2008; ORDIZ et al. 2012; ERIKSEN et al. 2018). Bears may abandon their winter dens if disturbed. Den abandonment rates are high in Sweden, on average 22 % of dens are abandoned on an annual basis, and most abandonments occur during the start of hibernation (SAHLEN et al. 2015). Approaching dens on foot has been documented to cause den abandonment (GOODRICH/BERGER 1994; LINNELL et al. 2000; EVANS et al. 2012); other causes seem especially related to forestry activities carried out with heavy harvester vehicles during winter or proximity to roads that are cleared from snow on a regular basis (ELFSTRÖM et al. 2008; ELFSTRÖM/SWENSON 2009). Den abandonment carries a high metabolic cost. Body temperature and heart rate increase when bears are forced to change their den sites, and they require about two to three weeks to return to the physiological hibernation level after disturbance (EVANS et al. 2016). A study on American black bears (*U. americanus*) showed that bears that changed dens during winter had a greater weight loss than undisturbed bears (TIETJE/RUFF 1980). In addition, several studies have demonstrated that disturbance of pregnant female brown bears during winter can lower their reproductive success, especially if disturbance occurs during the middle or end of the hibernations season (SMITH 1986; ELOWE/DODGE 1989; GOODRICH/BERGER 1994; SWENSON et al. 1997). Mid-winter or late winter den abandonments probably cause an increased energy cost, as the bear is deeper in hibernation and the snow cover makes locating new suitable dens difficult (EVANS et al. 2012).

Bibliography

ASTAEVA/KLICHKHANOV 2009: M. D. ASTAEVA/N. K. KLICHKHANOV, Oxidative modification of proteins and blood antioxidant activity of ground squirrels during induced awakening from hibernation. Biology Bulletin 36, 2009, 562–567.

ATKINSON/RAMSEY 1995: S. N. ATKINSON/M. A. RAMSAY, The effects of prolonged fasting of the body composition and reproductive success of female polar bears (*Ursus maritimus*). Functional Ecology 9, 1995, 559–567.

BARNES 1989: B. M. BARNES, Freeze avoidance in a mammal: body temperatures below 0 degree C in an Arctic hibernator. Science 244, 1989, 1593–1595.

BERG VON LINDE et al. 2015: M. BERG VON LINDE/L. AREVSTRÖM/O. FRÖBERT, Insights from the Den: How Hibernating Bears May Help Us Understand and Treat Human Disease. Clinical and Translational Science 8, 2015, 601–605.

BOJARSKA/SELVA 2011: K. BOJARSKA/N. SELVA, Spatial patterns in brown bear (*Ursus arctos*) diet: the role of geographic and environmental factors. Mammal Review 42(2), 2012, 120 – 143.

BOYER/BARNES 1999: B. B. BOYER/B. M. BARNES, Molecular and Metabolic Aspects of Mammalian Hibernation: Expression of the hibernation phenotype results from the coordinated regulation of multiple physiological and molecular events during preparation for and entry into torpor. BioScience 49, 1999, 713–724.

BOYLES et al. 2020: J. G. BOYLES/J. S. JOHNSON/A. BLOMBERG/T. M. LILLEY, Optimal hibernation theory. Mammal Review 50, 2020, 91–100.

CAREY et al. 2003: H. V. CAREY/M. T. ANDREWS/S. L. MARTIN, Mammalian hibernation: Cellular and molecular responses to depressed metabolism and low temperature. Physiological Reviews 83, 2003, 1153–1181.

CIARNIELLO et al. 2005: L. M. CIARNIELLO/M. S. BOYCE/D. C. HEARD/D. R. SEIP, Denning behavior and den site selection of grizzly bears along the Parsnip River, British Columbia, Canada. Ursus 16, 2005, 47–58.

CRAIGHEAD/CRAIGHEAD 1972: F. C. CRAIGHEAD/J. J. CRAIGHEAD, Data on grizzly bear denning activities and behavior obtained by using wildlife telemetry. Bears: Their Biology and Management – Proceedings of the International Conference on Bear Research and Management 2, 1972, 84–106.

DEROCHER et al. 1993: A. E. DEROCHER/D. ANDRIASHEK/J. P. Y. ARNOULD, Aspects of milk composition and lactation in polar bears. Canadian Journal of Zoology 71, 1993, 561–567.

ELFSTRÖM/SWENSON 2009: M. ELFSTRÖM/J. E. SWENSON, Effects of sex and age on den site use by Scandinavian brown bears. Ursus 20, 2009, 85–93.

Elfström et al. 2008: M. Elfström/J. E. Swenson/J. P. Ball, Selection of denning habitats by Scandinavian brown bears *Ursus arctos*. Wildlife Biology 14, 2008, 176–187.

Elowe/Dodge 1989: K. D. Elowe/W. E. Dodge, Factors affecting black bear reproductive success and cub survival. The Journal of Wildlife Management 53, 1989, 962–968.

Eriksen et al. 2018: A. Eriksen/P. Wabakken/E. Maartmann/B. Zimmermann, Den site selection by male brown bears at the population's expansion front. PLoS ONE 13, 2018. https://doi.org/10.1371/journal.pone.0202653.

Evans et al. 2012: A. Evans/V. Sahlén/O.-G. Støen/Å. Fahlman/S. Brunberg/K. Madslien/O. Fröbert/J. E. Swenson/J. M. Arnemo, Capture, Anesthesia, and Disturbance of Free-Ranging Brown Bears (*Ursus arctos*) during Hibernation. PLoS ONE 7, 2012. https://doi.org/10.1371/journal.pone.0040520.

Evans et al. 2016: A. Evans/N. J. Singh/A. Friebe/J. M. Arnemo/T. Laske/O. Fröbert/J. E. Swenson/S. Blanc, Drivers of hibernation in the brown bear. Frontiers in Zoology 13, 2016, 1–14.

Folk et al. 2008: G. E. Folk/E. W. Dickson/J. M. Hunt/E. J. Nilles/D. L. Thrift, QT intervals compared in small and large hibernators and humans. Biological Rhythm Research 39, 2008, 427–438.

French 1986: A. R. French, Patterns of thermoregulation during hibernation. In: H. C. Heller (ed.), Living in the Cold. Physiology and Biochemical Adaptations (New York 1986) 393–402.

Friebe 2015: A. Friebe, Winter Ecology of Free-Ranging Brown Bears (*Ursus arctos*) in Central Sweden. Unpubl. PhD thesis, Goethe-University, Frankfurt am Main (Frankfurt a. M. 2015).

Friebe et al. 2001: A. Friebe/J. E. Swenson/F. Sandegren, Denning chronology of female brown bears in central Sweden. Ursus 12, 2001, 37–46.

Friebe et al. 2013: A. Friebe/A. Zedrosser/J. E. Swenson, Detection of pregnancy in a hibernator based on activity data. European Journal of Wildlife Research 59, 2013, 731–741.

Friebe et al. 2014: A. Friebe/A. L. Evans/J. M. Arnemo/S. Blanc/S. Brunberg/G. Fleissner/J. E. Swenson/A. Zedrosser, Factors Affecting Date of Implantation, Parturition, and Den Entry Estimated from Activity and Body Temperature in Free-Ranging Brown Bears. PLoS ONE 9, 2014. https://doi.org/10.1371/journal.pone.0101410.

Fröbert et al. 2020: O. Fröbert/A. M. Frøbert/J. Kindberg/J. Arnemo/M. Overgaard, The brown bear as a translational model for sedentary lifestyle related diseases. Journal of Internal Medicine 287(3), 2020, 263–270.

Geiser 1998: F. Geiser, Evolution of daily torpor and hibernation in birds and mammals: Importance of body size. Clinical and Experimental Pharmacology and Physiology 25, 1998, 736–740.

Geiser 2004: F. Geiser, Metabolic rate and body temperature reduction during hibernation and daily torpor. Annual Rewiev of Physiology 66, 2004, 239–274.

González-Bernardo et al. 2020: E. González-Bernardo/L. F. Russo/E. Valderrábano/Á. Fernández/V. Penteriani, Denning in brown bears. Ecology and Evolution 10, 2020, 6844–6862.

Goodrich/Berger 1994: J. M. Goodrich/J. Berger, Winter recreation and hibernating black bears *Ursus americanus*. Biological Conservation 67, 1994, 105–110.

Harlow et al. 2001: H. J. Harlow/T. Lohuis/T. D. I. Beck/P. A. Iaizzo, Muscle strength in overwintering bears. Nature 409, 2001, 997.

Haroldson et al. 2002: M. A. Haroldson/M. A. Ternent/K. A. Gunther/C. C. Schwartz, Grizzly bear denning chronology and movements in the Greater Yellowstone Ecosystem. Ursus 13, 2002, 29–37.

Heldmaier 2011: G. Heldmaier, Life on Low Flame in Hibernation. Science 331, 2011, 866–867.

Heldmaier et al. 2004: G. Heldmaier/S. Ortmann/R. Elvert, Natural hypometabolism during hibernation and daily torpor in mammals. Respiratory Physiology & Neurobiology 141, 2004, 317–329.

Hellgren 1998: E. C. Hellgren, Physiology of hibernation in bears. Ursus 10, 1998, 467–477.

Hissa et al. 1994: R. Hissa/J. Siekkinen/E. Hohtola/S. Saarela/A. Hakala/J. Pudas, Seasonal patterns in the physiology of the European brown bear (*Ursus arctos arctos*) in Finland. Comparative Biochemistry and Physiology Part A: Physiology 109 (3), 1994, 781–791.

Huber/Roth 1997: D. Huber/H. U. Roth, Denning of brown bears in Croatia. Bears: Their Biology and Management – Proceedings of the International Conference on Bear Research and Management 9(2), 1997, 79–83.

Humphries et al. 2003: M. M. Humphries/D. W. Thomas/D. L. Kramer, The role of energy availability in mammalian hibernation: A cost benefit approach. Physiological and Biochemical Zoology 76, 2003, 165–179.

Kingsley et al. 1983: M. C. S. Kingsley/J. A. Nagy/R. H. Russell, Patterns of weight gain and loss for grizzly bears in Northern Canada. Bears: Their Biology and Management – Proceedings of the International Conference on Bear Research and Management 5, 1983, 174–178.

Klenzendorf et al. 2002: S. A. Klenzendorf/M. R. Vaughan/D. D. Martin, Den-type use and fidelity of American black bears in western Virginia. Ursus 13, 2002, 39–44.

Laske et al. 2018: T. G. Laske/A. L. Evans/J. M. Arnemo/T. L. Iles/M. A. Ditmer/O. Fröbert/D. L. Garshelis/P. A. Iaizzo, Development and utilization of implantable cardiac monitors in free-ranging American black and Eurasian brown bears: system evolution and lessons learned. Animal Biotelemetry 6, 2018, 13.

Libal et al. 2011: N. S. Libal/J. L. Belant/B. D. Leopold/G. Wang/P. A. Owen, Despotism and Risk of Infanticide Influence Grizzly Bear Den-Site Selection. PLoS ONE 6, 2011. https://doi.org/10.1371/journal.pone.0024133.

Linnell et al. 2000: J. D. C. Linnell/E. S. Jon/R. Andersen/B. Barnes, How vulnerable are denning bears to disturbance? Wildlife Society Bulletin 28, 2000, 400–413.

López-Alfaro et al. 2013: C. López-Alfaro/C. T. Robbins/A. Zedrosser/S. E. Nielsen, Energetics of hibernation and reproductive trade-offs in brown bears. Ecological Modelling 270, 2013, 1–110.

Manchi/Swenson 2005: S. Manchi/J. E. Swenson, Denning behaviour of Scandinavian brown bears *Ursus arctos*. Wildlife Biology 11, 2005, 123–132.

Mangipane et al. 2018: L. S. Magipane/J. L. Belant/D. D. Gustine/G. Hilderbrand/B. A. Mangipane, Sex-specific variation in denning by brown bears. Mammalian Biology 93, 2018, 38–44.

Martorello/Pelton 2003: D. A. Martorelli/M. R. Pelton, Microhabitat characteristics of American black bear nest dens. Ursus 14, 2003, 21–26.

McDonald/Fuller 2005: J. E. McDonald/T. K. Fuller, Effects of spring acorn availability on black bear diet, milk composition, and cub survival. Journal of Mammalogy 86, 2005, 1022–1028.

Nelson 1973: R. A. Nelson, Winter sleep in the black bear: a physiologic and metabolic marvel. Mayo Clinic Proceedings 48, 1973, 733–737.

Nelson 1980: R. A. Nelson, Protein and fat-metabolism in hibernating bears. Federation Proceedings 39, 1980, 2955–2958.

Nelson 1987: R. A. Nelson, Black bears and polar bears – still metabolic marvels. Mayo Clinic Proceedings 62, 1987, 850–853.

Nelson 1989: R. Nelson, Nitrogen turnover and its conservation in hibernation. In: A. Malan/M. Canguilhem (eds.), Living in the Cold (London 1989) 299–307.

Nelson et al. 1973: R. Nelson/H. Wahner/J. Jones/R. Ellefson/P. Zollman, Metabolism of bears before, during, and after winter sleep. American Journal of Physiology – Legacy Content 224, 1973, 491–496.

Nelson et al. 1983: R. A. Nelson/G. E. Folk Jr./E. W. Pfeiffer/J. J. Craighead/C. J. Jonkel/D. L. Steiger, Behavior, biochemistry, and hibernation in black, grizzly, and polar bears. Bears: Their Biology and Management – Proceedings of the International Conference on Bear Research and Management 5, 1983, 284–290.

Nelson et al. 2003: O. L. Nelson/M. M. McEwen/C. T. Robbins/L. Felicetti/W. F. Christensen, Evaluation of cardiac function in active and hibernating grizzly bears. Journal of the American Veterinary Medical Association 223, 2004, 1170–1175.

Nores et al. 2010: C. Nores/F. Ballesteros/J. Blanco/A. García-Serrano/J. Herrero/G. Palomero, Evidence of non-hibernation in Cantabrian brown bears. Acta Theriologica 55, 2010, 203–209.

Nowack 2015: L. Nowack, Reproductive Performance of Scandinavian Female Brown Bears (*Ursus arctos*) in Relation to the Use of Den Type. Unpubl. MSc thesis, University for Natural Resources and Life Sciences Vienna (Vienna 2015).

Ordiz et al. 2012: A. Ordiz/O.-G. Støen/S. Sæbø/J. Kindberg/M. Delibes/J. E. Swenson, Do bears know they are being hunted? Biological Conservation 152, 2012, 21–28.

Petram et al. 2004: W. Petram/F. Knauer/P. Kaczensky, Human influence on the choice of winter dens by European brown bears in Slovenia. Biological Conservation 119, 2004, 129–136.

Pinter 1984: A. J. Pinter, Hibernation and torpor in mammals and birds. Journal of Mammalogy 65, 1984, 172–175.

Prendergast et al. 2002: B. J. Prendergast/D. A. Freeman/I. Zucker/R. J. Nelson, Periodic arousal from hibernation is necessary for initiation of immune responses in ground squirrels. American Journal of Physiology. Regulatory, Integrative and Comparative physiology 282, 2002, R1054–1062.

Pulawa/Florant 2000: L. K. Pulawa/G. L. Florant, The effects of caloric restriction on the body composition and hibernation of the golden-mantled ground squirrel (*Spermophilus lateralis*). Physiological and Biochemical Zoology 73, 2000, 538–546.

Ramsay/Dunbrack 1986: M. A. Ramsay/R. L. Dunbrack, Physiological constraints on life history phenomena: The example of small bear cubs at birth. The American Naturalist 127, 1986, 735–743.

Reynolds et al. 1976: H. V. Reynolds/J. A. Curatolo/R. Quimby, Denning ecology of grizzly bears in Northeastern Alaska. Bears: Their Biology and Management – Proceedings of the International Conference on Bear Research and Management 3, 1976, 403–409.

Robbins et al. 2012: C. T. Robbins/B.-D. Merav/J. K. Fortin/O. L. Nelson, Maternal condition determines birth date and growth of newborn bear cubs. Journal of Mammalogy 93, 2012, 540–546.

Rogers 1981: L. Rogers, A Bear in its Lair. Natural History 90, 1981, 64–70.

Rogers et al. 2020: L. L. Rogers/L. McColley/J. Dalton/J. Stroner/D. Hajicek/A. Partin/G. M. Burghardt, Behavior in free-living American black bear dens: parturition, maternal care, and cub behavior. Animals 10(7), 2020, 1123.

Sahlén et al. 2011: E. Sahlén/O.-G. Støen/J. E. Swenson, Brown bear den site concealment in relation to human activity in Sweden. Ursus 22, 2011, 152–158.

Sahlén et al. 2015: V. Sahlén/A. Friebe/S. Saebo/J. E. Swenson/O.-G. Støen, Den Entry Behavior in Scandinavian Brown Bears: Implications for Preventing Human Injuries. Journal of Wildlife Management 79, 2015, 274–287.

Schoen et al. 1987: J. W. Schoen/L. R. Beier/J. W. Lentfer/L. J. Johnson, Denning ecology of brown bears on Admiralty and Chichagof Islands. Bears: Their Biology and Management – Proceedings of the International Conference on Bear Research and Management 7, 1987, 293–304.

Servheen/Klaver 1983: C. Servheen/R. Klaver, Grizzly bear dens and denning activity in the Mission and Rattlesnake Mountains, Montana. Bears: Their Biology and Management – Proceedings of the International Conference on Bear Research and Management 5, 1983, 201–207.

Seryodkin et al. 2003: I. V. Seryodkin/A. V. Kostyria/J. M. Goodrich/D. G. Miquelle/E. N. Smirnov/L. L. Kerley/H. B. Quigley/M. G. Hornocker, Denning ecology of brown bears and Asiatic black bears in the Russian Far East. Ursus 14, 2003, 153–161.

Shiratsuru et al. 2020: S. Shiratsuru/A. Friebe/J. E. Swenson/A. Zedrosser, Room without a view – den construction in relation to body size in brown bears. Ecology and Evolution 10, 2020, 8044–8054.

Smereka et al. 2017: C. A. Smereka/M. A. Edwards/J. Pongracz/M. Branigan/N. W. Pilfold/A. E. Derocher, Den selection by barren-ground grizzly bears,

Mackenzie Delta, Northwest Territories. Polar Biology 40, 2017, 503–516.

SMITH 1986: T. R. SMITH, Activity and behavior of denned black bears in the lower Mississippi River Valley. Bears: Their Biology and Management – Proceedings of the International Conference on Bear Research and Management 6, 1986, 137–143.

SPADY et al. 2007: T. J. SPADY/D. G. LINDBURG/B. S. DURRANT, Evolution of reproductive seasonality in bears. Mammal Review 37, 2007, 21–53.

STENVINKEL et al. 2012: P. STENVINKEL/A. JANI/R. JOHNSON, Hibernating bears (Ursidae): Metabolic magicians of definite interest for the nephrologist. Kidney International 83(2), 2012. https://doi/10.1038/ki.2012.396.

STENVINKEL et al. 2018: P. STENVINKEL/J. PAINER/M. KUROO/M. LANASPA/W. ARNOLD/T. RUF/P. G. SHIELDS/R. J. JOHNSON, Novel treatment strategies for chronic kidney disease: insights from the animal kingdom. Nature Reviews Nephrology 14, 2018, 265–284.

STEYAERT et al. 2012: S. M. J. G. STEYAERT/A. ENDRESTØL/K. HACKLÄNDER/J. E. SWENSON/A. ZEDROSSER, The mating system of the brown bear *Ursus arctos*. Mammal Review 42, 2012, 12 –34.

SWENSON et al. 1997: J. E. SWENSON/F. SANDEGREN/S. BRUNBERG/P. WABAKKEN, Winter den abandonment by brown bears *Ursus arctos*: Causes and consequences. Wildlife Biology 3, 1997, 35–38.

SWENSON et al. 1998: J. E. SWENSON/R. FRANZEN/P. SEGERSTROM/F. SANDEGREN, On the age of self-sufficiency in Scandinavian brown bears. Acta Theriologica 43, 1998, 213–218.

TIETJE/RUFF 1980: W. D. TIETJE/R. L. RUFF, Denning behavior of black bears in boreal forest of Alberta. The Journal of Wildlife Management 1980, 858–870.

TØIEN et al. 2011: Ø. TØIEN/J. BLAKE/D. M. EDGAR/D. A. GRAHN/H. C. HELLER/B. M. BARNES, Hibernation in black bears: Independence of metabolic suppression from body temperature. Science 331, 2011, 906–909.

WANG 1989: L. WANG, Animal adaptation to cold. Advances in Contemporary and Environmental Physiology 4 (Berlin 1989).

ZEDROSSER al. 2011: A. ZEDROSSER/S. M. J. G. STEYAERT/H. GOSSOW/J. E. SWENSON, Brown bear conservation and the ghost of persecution past. Biological Conservation 144, 2011, 2163–2170.

Dr. Andrea Friebe
Scandinavian Brown Bear Research Project
Orsa
Sweden

Norwegian Institute for Nature Research
Trondheim
Norway
andrea@bearproject.info

Prof. Jon Swenson
Faculty of Environmental Sciences and Natural Resource Management
Norwegian University of Life Sciences
Ås
Norway
jon.swenson@nmbu.no

Prof. Andreas Zedrosser
Department of Natural Sciences and Environmental Health
University of South-Eastern Norway
Bø i Telemark
Norway

Department for Integrative Biology
Institute for Wildlife Biology and Game Management
University for Natural Resources and Life Sciences Vienna
Vienna
Austria
andreas.zedrosser@usn.no

Fig. 1. Researchers measure the body fat composition of a sedated five-year-old male brown bear after hibernation. Bioelectrical Impedance Analysis is used, i.e. a very weak electric current measures the body fat content of an individual (photo A. Friebe).

Fig. 2. Den types used by brown bears in Scandinavia. a: Excavated ant hill den; b: Entrance to a natural rock cavity den; c: Open nest den (photos A. Friebe).

The social system of a "nonsocial" species, the brown bear

By Andreas Zedrosser, Shane C. Frank, Jennifer E. Hansen, Sam M. J. G. Steyaert and J. E. Swenson

Keywords: Scandinavia, brown bear, Ursus arctos, *dispersal, hunting, mating system, social system*

Abstract: Brown bears are a nonterritorial and solitary species, i.e. they do not defend an exclusive territory against conspecifics. Instead, they live in overlapping home ranges – areas that are not defended and in which animals live for the majority of their lives. The brown bear has a polygamous mating system, i.e. both sexes mate with several partners. The mating season is in spring and early summer, and both sexes roam over large areas to find reproductive partners. Males commonly compete for access to females, and fights can inflict severe injuries or cause death. Male competition for reproductive success can result in sexually selected infanticide (SSI), a reproductive strategy in which males kill dependent conspecific offspring for obtaining mating opportunities. Young bears disperse at one and a half to two years of age. Males typically disperse very long distances, up to several hundred kilometres. In contrast, female dispersal distances are short, on average 10–25 km. Female brown bears commonly form matrilineal assemblages (i.e. spatial clusters of related females), with home range overlap correlated with relatedness. Hunting greatly affects the social system of brown bears in Scandinavia. The removal of conspecifics through hunting creates vacancies on the landscape, and surviving bears shift their home ranges toward these vacancies, which increases the probability of SSI and negatively affects population growth rate.

Introduction

The social system of a species describes the pattern of relationships between and among individuals and social groups and how these individuals and social groups are distributed across the landscape. Brown bears (*Ursus arctos*) are considered a nonterritorial and solitary species, i.e. they do not defend an exclusive territory against intrusion from other conspecifics, as wolverines (*Gulo gulo*) or Canada lynx (*Lynx canadensis*) do (Schwartz et al. 2003). Instead, brown bears live in overlapping home ranges – areas in which animals live and move for the majority of their lives (Fig. 1). A home range is not defended, like a territory, but it commonly overlaps with home ranges of other conspecifics of the same and opposite sex. In addition, brown bears are not group-living, such as grey wolves (*Canis lupus*), but live solitarily throughout the year, with the exception of mothers accompanied by dependent offspring (Steyaert et al. 2020). Congregations of bears at abundant food sources, such as spawning salmon (*Oncorhynchus* spp.) or garbage dumps can occur (Craighead et al. 1995; Ben-David et al. 2004), but are usually only temporary aggregations of individuals as long as the food resource lasts.

Although little is known about how this spatial reorganisation affects individual fitness, links have been made between hunting, male home-range shifts, male mating success, sexually selected infanticide, and variation in population growth (Gosselin et al. 2015; 2016; Leclerc et al. 2017; Frank et al. 2020). SSI is a male reproductive strategy, whereby males gain mating opportunities by killing dependent young (Hrdy 1979), and males should only kill offspring that they have not fathered. Swenson et al. (1997) have shown that hunting large adult males disrupts the male social organisation in an area, and 1.5 years after a large adult male was killed by hunters, SSI increases because a new male takes over the deceased male's home range (Leclerc et al. 2017) and kills the deceased males' cubs in the area to improve its own chances of reproductive success (Swenson et al. 1997; Steyaert 2012). Hunting of adult males can therefore indirectly contribute to negative population growth through increased juvenile mortality (Gosselin et al. 2015; 2016; Frank et al. 2017). Gosselin et al. (2015) found that cub survival was lowest under high hunting pressure in the Scandinavian bear population and estimated that it could explain approximately 14 % of the variation in the population growth rate. However, the effect of hunting of males on population dynamics in brown bears is a contested topic among bear biologists and managers, and effects of hunting seem to differ among populations (McLellan 2015; Steyaert et al. 2020).

Hunting may also indirectly affect reproduction and reproductive success of female bears in Scandinavia (Frank et al. 2017; 2020). Reproduction of young females remaining in their mother's home range is suppressed, likely due to the presence of the mother. In comparison, young females dispersing out of the maternal home range start reproduction earlier (Støen et al. 2006b; Ordiz et al. 2008). Competition among females for reproduction also occurs after primiparity; the probability of a female brown bear having cubs in a given year is affected by whether or not neighbouring females have cubs in the same year. Thus, dominant pregnant adult female brown bears appear to inhibit reproduction in their female neighbours, which introduces reproductive asynchrony to the population, with neighbouring females having cubs in alternating years (Ordiz et al. 2008; Frank et al. 2017).

Dispersal in large carnivores has been shown to change as a result of harvest (e.g. Cooley et al. 2009), which could affect the distribution of females across the landscape (e.g. Robinson et al. 2008); there is some evidence of this in brown bears near the Swedish-Norwegian border (Bischof et al. 2012). Hunting may also affect dispersal in Scandinavian brown bears. The hunting pressure on bears has increased in Sweden since about 2010, which has resulted in a decrease of the population, a result desired by Swedish wildlife managers (Swenson et al. 2017). The population size of brown bears in neighbouring Norway is heavily dependent upon dispersing individuals from Sweden, which has resulted in fewer bears detected in Norway in recent years due to the increased hunting quotas in Sweden (Aarnes et al. 2014; Frank et al. 2017).

Bibliography

Aarnes et al. 2014: S. V. Aarnes/C. Tobiassen/H. Brøseth/B. Banken Bakke/B. Kleppe Hansen/B. Spachmo/S. B. Hagen/H. G. Eiken, Populasjonsovervåking av brunbjørn DNA-analyse av prøver innsamlet i Norge i 2013. Bioforsk – Norwegian Institute for Environmental and Agricultural Research Report 10-46 (Svanvik 2014).

Andersson 1994: M. Andersson, Sexual Selection (Princeton 1994).

Bellemain et al. 2006a: E. Bellemain/J. E. Swenson/P. Taberlet, Mating strategies in relation to sexually selected infanticide in a non-social carnivore: The brown bear. Ethology 112, 2006, 238–246.

Bellemain et al. 2006b: E. Bellemain/A. Zedrosser/S. Manel/L. P. Waits/P. Taberlet/J. E. Swenson, The dilemma of female mate selection in the brown bear, a species with sexually selected infanticide. Proceedings of the Royal Society London – B. Biological Sciences 273, 2006, 283–291.

BEN-DAVID et al. 2004: M. BEN-DAVID/K. TITUS/L. R. BEIER, Consumption of salmon by Alaskan brown bears: A trade-off between nutritional requirements and the risk of infanticide? Oecologia 138, 2004, 465–474.

BISCHOF et al. 2012: R. BISCHOF/E. B. NILSEN/H. BROSETH/P. MANNIL/J. OZOLINS/J. D. C. LINNELL, Implementation uncertainty when using recreational hunting to manage carnivores. Journal of Applied Ecology 49, 2012, 824–832.

COOLEY et al. 2009: H. S. COOLEY/R. B. WIELGUS/G. M. KOEHLER/H. S. ROBINSON/B. T. MALETZKE, Does hunting regulate cougar populations? A test of the compensatory mortality hypothesis. Ecology 90, 2009, 2913–2921.

CRAIGHEAD et al. 1995: J. J. CRAIGHEAD/J. S. SUMNER/J. A. MITCHELL, The grizzly bears of Yellowstone: Their ecology in the Yellowstone Ecosystem, 1959–1987 (Washington 1995).

DAHLE/SWENSON 2003a: B. DAHLE/J. E. SWENSON, Family break up in brown bears: Are young forced to leave? Journal of Mammalogy 84, 2003, 536–540.

DAHLE/SWENSON 2003b: B. DAHLE/J. E. SWENSON, Seasonal range size in relation to reproductive strategies in brown bears *Ursus arctos*. Journal of the Animal Ecology 72, 2003, 660–667.

DAHLE et al. 2006: B. DAHLE/O. G. STØEN/J. E. SWENSON, Factors influencing home-range size in subadult brown bears. Journal of Mammalogy 87, 2006, 859–865.

EDWARDS/DEROCHER 2015: M. A. EDWARDS/A. E. DEROCHER, Mating-related behaviour of grizzly bears inhabiting marginal habitat at the periphery of their North American range. Behavioural Processes 111, 2015, 75–83.

EMLEN/ORING 1977: S. T. EMLEN/L. W. ORING, Ecology, sexual selection and the evolution of mating systems. Science 197, 1977, 215–223.

FRANK et al. 2017: S. C. FRANK/A. ORDIZ/J. GOSSELIN/A. HERTEL/J. KINDBERG/M. LECLERC/F. PELLETIER/S. STEYAERT/O.-G. STØEN/J. VAN DE WALLE/A. ZEDROSSER/J. E. SWENSON, Indirect effects of bear hunting: a review from Scandinavia. Ursus 28, 2017, 150–164.

FRANK et al. 2020: S. C. FRANK/F. PELLETIER/A. KOPATZ/A. BOURRET/D. GARANT/J. E. SWENSON/H. G. EIKEN/S. B. HAGEN/A. ZEDROSSER, Harvest is associated with the disruption of social and fine-scale genetic structure among matrilines of a solitary large carnivore. Evolutionary Applications 2020. https://doi.org/10.1111/eva.13178.

FRIEBE et al. 2014: A. FRIEBE/A. L. EVANS/J. M. ARNEMO/S. BLANC/S. BRUNBERG/G. FLEISSNER/J. E. SWENSON/A. ZEDROSSER, Factors Affecting Date of Implantation, Parturition, and Den Entry Estimated from Activity and Body Temperature in Free-Ranging Brown Bears. PLoS ONE 9, 2014. https://doi.org/10.1371/journal.pone.0101410.

GOSSELIN et al. 2015: J. GOSSELIN/A. ZEDROSSER/J. E. SWENSON/F. PELLETIER, The relative importance of direct and indirect effects of hunting mortality on the population dynamics of brown bears. Proceedings of the Royal Society London, Series B 282, 2015. https://doi.org/10.1098/rspb.2014.1840.

GOSSELIN et al. 2016: J. GOSSELIN/M. LECLERC/A. ZEDROSSER/S. M. J. G. STEYAERT/J. E. SWENSON/F. PELLETIER, Hunting promotes sexual conflict in brown bears. Journal of Animal Ecology 86, 2016, 35–42.

HAMER/HERRERO 1990: D. HAMER/S. HERRERO, Courtship and use of mating areas by grizzly bears in the Front Ranges of Banff National Park, Alberta. Canadian Journal of Zoology 68, 1990, 2695–2697.

HANSEN et al. 2021: J. E. HANSEN/A. G. HERTEL/S. C. FRANK/J. KINDBERG/A. ZEDROSSER, Social environment shapes female settlement decisions in a solitary carnivore. Behavioral Ecology 33, 2021, 137–146.

HRDY 1979: S. B. HRDY, Infanticide among mammals: review, classification, and examination of the implications for the reproductive strategies of females. Ethology and Sociobiology 1, 1979, 13–40.

KROFEL et al. 2010: M. KROFEL/S. FILACORDA/K. JERINA, Mating-related movements of male brown bears on the periphery of an expanding population. Ursus 21, 2010, 23–29.

LECLERC et al. 2017: M. LECLERC/S. C. FRANK/A. ZEDROSSER/J. E. SWENSON/F. PELLETIER, Hunting promotes spatial reorganization and sexually selected infanticide. Scientific Reports 7, 2017, 45222.

MCLELLAN 1994: B. MCLELLAN, Density-dependent population regulation of brown bears. In: M. Taylor (ed.), Density-Dependent Population Regulation of Black, Brown, and Polar Bears. International Conference on Bear Research and Management, Monograph Series 3 (Washington D. C. 1994) 15–24.

MCLELLAN 2015: B. N. MCLELLAN, Some mechanisms underlying variation in vital rates of grizzly bears on a multiple use landscape. Journal of Wildlife Management 79, 2015, 749–765.

MCLELLAN/HOOVEY 2001: B. N. MCLELLAN/F. W. HOOVEY, Natal dispersal of grizzly bears. Canadian Journal of Zoology 79, 2001, 838–844.

ORDIZ et al. 2008: A. ORDIZ/O.-G. STØEN/J. E. SWENSON/I. KOJOLA/R. BISCHOF, Distance-dependent effect of the nearest neighbour: spatiotemporal patterns in brown bear reproduction. Ecology 89, 2008, 3327–3335.

ROBINSON et al. 2008: H. S. ROBINSON/R. B. WIELGUS/H. S. COOLEY/S. W. COOLEY, Sink populations in carnivore management: Cougar demography and immigration in a hunted population. Ecological Applications 18, 2008, 1028–1037.

RUTLEDGE et al. 2010: L. Y. RUTLEDGE/B. R. PATTERNSON/K. J. MILLS/K. M. LOVELESS/D. L. MURRAY/B. N. WHITE, Protection from harvesting restores the natural social structure of eastern wolf packs. Biological Conservation 143, 2010, 332–339.

SCHWARTZ et al. 2003: C. C. SCHWARTZ/S. D. MILLER/M. A. HAROLDSON, Grizzly bear. In: G. A. Feldhammer/B. C. Thompson/J. A. Chapman (eds.), Wild Mammals of North America (Baltimore 2003) 556–586.

Spady et al. 2007: T. J. Spady/D. G. Lindburg/B. S. Durrant, Evolution of reproductive seasonality in bears. Mammal Review 37, 2007, 21–53.

Steyaert 2012: S. M. J. G. Steyaert, The mating system of the brown bear in relation to the sexually selected infanticide theory. Unpubl. dissertation, Norwegian University of Life Sciences (Ås 2012).

Steyaert et al. 2012: S. M. J. G. Steyaert/A. Endrestøl/K. Hackländer/J. E. Swenson/A. Zedrosser, The mating system of the brown bear Ursus arctos. Mammal Review 42, 2012, 12–34.

Steyaert et al. 2020: S. M. J. G. Steyaert/A. Zedrosser/R. Swaisgood/E. Filipczykova/B. Crudge/T. Dutta/S. Sharma/S. Ratnayeke/S. Koike/M. Leclerc/A. E. Derocher/M. Clapham/T. Spady/B. McLellan/A. Ordiz/A. Fernandez-Gil/M. Delibes/J. E. Swenson, Mating strategies. In: V. Penteriani/M. Melletti (eds.), Bears of the World. Ecology, Conservation and Management (Cambridge 2020) 21–35.

Støen et al. 2005: O.-G. Støen/E. Bellemain/S. Saebo/J. E. Swenson, Kin-related spatial structure in brown bears Ursus arctos. Behavioral Ecology and Sociobiology 59, 2005, 191–197.

Støen et al. 2006a: O.-G. Støen/A. Zedrosser/S. Sæbø/J. E. Swenson, Inversely density-dependent natal dispersal in brown bears Ursus arctos. Oecologia 148, 2006, 356–364.

Støen et al. 2006b: O.-G. Støen/A. Zedrosser/P. Wegge/J. E. Swenson, Socially induced delayed primiparity in brown bears Ursus arctos. Behavioral Ecology and Sociobiology 61, 2006, 1–8.

Swenson et al. 1997: J. E. Swenson/F. Sandegren/A. Söderberg/A. Bjärvall/R. Franzén/P. Wabakken, Infanticide caused by hunting of male bears. Nature 386, 1997, 450–451.

Swenson et al. 1998: J. E. Swenson/F. Sandegren/A. Soderberg, Geographic expansion of an increasing brown bear population: Evidence for presaturation dispersal. Journal of Animal Ecology 67, 1998, 819–826.

Swenson et al. 2017: J. E. Swenson/M. Schneider/A. Zedrosser/A. Söderberg/R. Franzén/J. Kindberg, Challenges of managing an European brown bear population – lessons from Sweden, 1943–2013. Wildlife Biology 2017(1). https://doi.org/10.2981/wlb.00251.

Trivers 1972: R. L. Trivers, Parental investment and sexual selection. In: B. G. Campbell (ed.), Sexual Selelction and the Descent of Man 1871–1971 (Chicago 1972) 163–179.

Van de Walle et al. 2018: J. Van de Walle/G. Pigeon/A. Zedrosser/J. E. Swenson/F. Pelletier, Hunting regulation favors slow life histories in a large carnivore. Nature Communications 9, 2018, 1100.

Zedrosser et al. 2007a: A. Zedrosser/E. Bellemain/P. Taberlet/J. E. Swenson, Genetic estimates of annual reproductive success in male brown bears: The effects of body size, age, internal relatedness and population density. Journal of Animal Ecology 76, 2007, 368–375.

Zedrosser et al. 2007b: A. Zedrosserr/O.-G. Støen/S. Sæbø/J. E. Swenson, Should I stay or should I go? Natal dispersal in the brown bear. Animal Behaviour 74, 2007, 369–376.

Zedrosser et al. 2009: A. Zedrosser/B. Dahle/O.-G. Støen/J. E. Swenson, The effects of primiparity on reproductive performance in the brown bear. Oecologia 160, 2009, 847–854.

Prof. Andreas Zedrosser
Department of Natural Sciences and Environmental Health
University of South-Eastern Norway
Bø i Telemark
Norway

Department for Integrative Biology
Institute for Wildlife Biology and Game Management
University for Natural Resources and Life Sciences
Vienna
Austria
andreas.zedrosser@usn.no

Dr. C. Frank
Department of Natural Sciences and Environmental Health
University of South-Eastern Norway
Bø i Telemark
Norway
shane.frank@usn.no

Dr. Jennifer E. Hansen
Department of Natural Sciences and Environmental Health
University of South-Eastern Norway
Bø i Telemark
Norway
jennifer.e.hansen@usn.no

Dr. Sam M. J. G. Steyaert
Faculty of Biosciences and Aquaculture
Nord University
Bodø
Norway
sam.steyaert@nord.no

Prof. Jon E. Swenson
Faculty of Environmental Sciences and Natural Resource Management
Norwegian University of Life Sciences
Ås
Norway
jon.swenson@nmbu.no

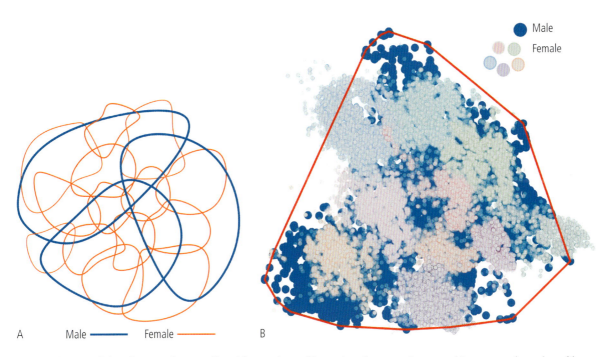

Fig. 1. The spatial distribution of GPS-collared brown bears illustrating the strong inter- and intra-sexual overlap of home ranges in Sweden. a: Home range size (based on the minimum convex polygon method) of adult male (blue) and adult female (orange) brown bears; b: Home range overlap of an adult male bear (red line depicts the outline of the home range based on the multiple convex polygon method; blue dots depict GPS radiolocations) with the home ranges of different adult female bears (dots in different colours depict GPS relocations of several females) (graphics J. E. Hansen).

Fig. 2. A female brown bear accompanied by her offspring, likely a one-year-old, in Finland (photo I. Kojola).

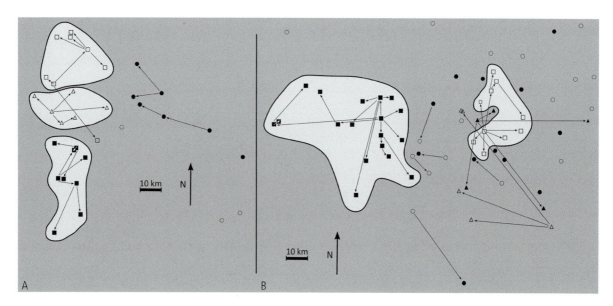

Fig. 3. Centers of home ranges and shot locations of adult female brown bears (five years and older), illustrating spatial structure of female matrilines in northern (a) and southern (b) Sweden. Home-range centers were calculated by 95 % adaptive kernel for females with >30 radiolocations and as arithmetic mean for females with <30 radiolocations. Matrilines consisting of more than three generations, or five individuals of radio-marked females are shown with open squares, closed squares, open triangles, and closed triangles. Radio-marked females belonging to other matrilines are shown with closed circles. Arrows show mother-daughter dyads, with the arrow pointing to the daughter. Matrilinear assemblages are shown by shaded areas (after STØEN et al. 2005).

Sexually selected infanticide as a mating strategy in brown bears

By Andreas Zedrosser, Sam M. J. G. Steyaert and Jon E. Swenson

Keywords: Scandinavia, brown bear, Ursus arctos, *mating strategy, reproductive success, sexually selected infanticide*

Abstract: According to the theory of sexual selection, members of one sex compete for the access to the other. The sexual selection hypothesis for infanticide proposes that a male taking over a new group of unrelated females will kill the infants. By doing so, the invader removes a competitor's offspring while speeding up the opportunity to mate with the victimised mother. However, several requirements must be met in order to categorise infanticide as sexually selected (SSI): 1) a male should not kill his own offspring; 2) killing of infants must shorten the time period to a female's next estrus and shorten the interbirth interval; and 3) the male should sire the mother's next offspring. SSI benefits the fitness of the male, but is very costly to the fitness of the mother, and thus female counterstrategies have evolved. SSI has been studied in the Scandinavian brown bear, in which it can induce up to 30 % cub loss per year. Evidence for all three requirements of the SSI hypothesis has been documented in Scandinavian brown bears, and females show evidence of behavioural counterstrategies to SSI. Hunting of especially male bears seems to increase the probability of SSI and can limit population growth, because bear cub mortality increased significantly six months and especially one and a half years after an adult male bear was killed in an area. However, this mechanism is controversial and cannot be generalised among bear species and populations.

Sexually selected infanticide as a male reproductive strategy

Reproduction and reproductive strategies are at the very core of sexual selection, and ultimately evolution (STEARNS 1992). Charles Darwin was the first to formulate the idea of the male infanticide model for reproduction (DARWIN 1871). According to the theory of sexual selection, members of one sex compete for the access to the other. Over time, the winners produce more offspring than do others. The sexual selection hypothesis for infanticide proposes that, in certain species, a male invading a new group of unrelated females will routinely kill the infants of other males. By doing so, the invader removes a competitor's offspring from the population while speeding up the opportunity to mate with the victimised mother, who stops nursing and soon again is willing to mate with the perpetrator. Together, these actions promote the invader's gene pool at the expense of his peers (BROWN 1996).

HRDY (1979) was the first to document this behaviour in Bengal Hanuman langurs (*Semnopithecus entellus*): After the usually hostile replacement of a dominant male by an invading new male, the offspring of the replaced male disappeared, and the mothers in the harem came into estrus again within a short time period and mated with the new dominant male. Based on these observations, HRDY (1979) formulated the hypothesis of sexually selected infanticide (SSI), by which males kill unrelated

Swenson et al. 1997: J. E. Swenson/F. Sandegren/A. Söderberg/A. Bjärvall/R. Franzén/P. Wabakken, Infanticide caused by hunting of male bears. Nature 386, 1997, 450–451.

Swenson et al. 2001: J. E. Swenson/F. Sandegren/S. Brunberg/P. Segerstrom, Factors associated with loss of brown bear cubs in Sweden. Ursus 12, 2001, 69–80.

Tibbets/Dale 2007: E. A. Tibbets/J. Dale, Individual recognition: it is good to be different. Trends in Ecology & Evolution 22, 2007, 529–537.

Van de Walle et al. 2018: J. van de Walle/G. Pigeon/A. Zedrosser/J. E. Swenson/F. Pelletier, Hunting regulation favors slow life histories in a large carnivore. Nature Communications 9, 2018, 1100.

Zedrosser et al. 2009: A. Zedrosser/B. Dahle/O.-G. Støen/J. E. Swenson, The effects of primiparity on reproductive performance in the brown bear. Oecologia 160, 2009, 847–854.

Zedrosser et al. 2011: A. Zedrosser/S. M. J. G. Steyaert/H. Gossow/J. E. Swenson, Brown bear conservation and the ghost of persecution past. Biological Conservation 144, 2011, 2163–2170.

Prof. Andreas Zedrosser
Department of Natural Sciences and Environmental Health
University of South-Eastern Norway
Bø i Telemark
Norway

Department for Integrative Biology
Institute for Wildlife Biology and Game Management
University for Natural Resources and Life Sciences
Vienna
Austria
andreas.zedrosser@usn.no

Assoc. Prof. Sam M. J. G. Steyaert
Faculty of Biosciences and Aquaculture
Nord University
Bodø
Norway
sam.steyaert@nord.no

Prof. Jon E. Swenson
Faculty of Environmental Sciences and Natural Resource Management
Norwegian University of Life Sciences
Ås
Norway
jon.swenson@nmbu.no

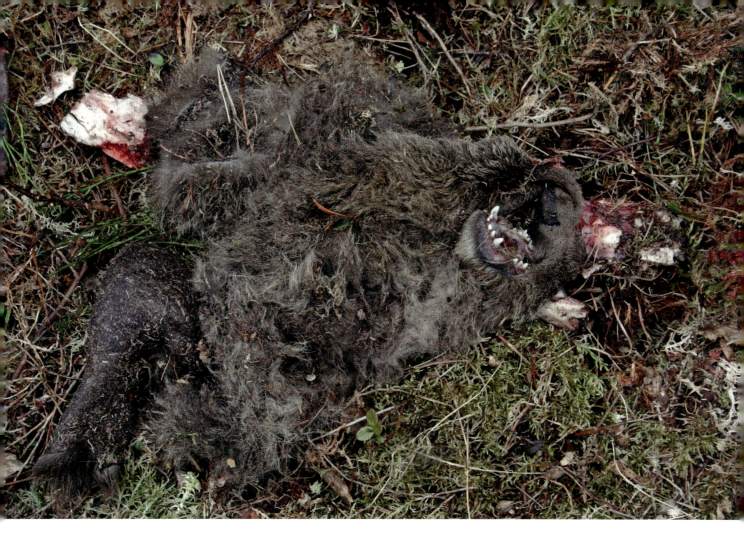

Fig. 1. Remains of a brown bear cub killed and partially consumed by an adult male brown bear during the mating season (photo S. M. J. G. Steyaert).

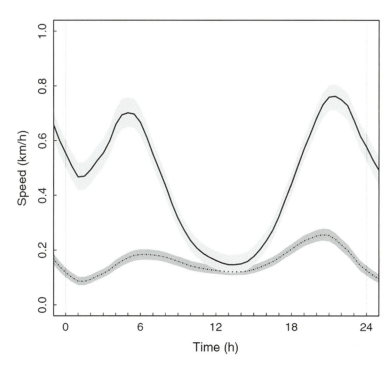

Fig. 2. Mean diurnal movement rates (km/h) fitted with a moving average spline of lone female brown bears (–) and females with cubs-of-the-year (····) during the mating season in central Sweden in the years 2006–2011. The shaded areas represent the 95 %-confidence intervals. Vertical grey lines delineate one day, from midnight to midnight. Females with cubs-of-the-year moved significantly less and more evenly distributed throughout the course of 24 hours compared to lone adult females participating in the mating season (after Steyaert et al. 2014).

Table 1. Genetic paternity analysis of Scandinavian brown bear cubs-of-the-year killed or probably killed by male bears (table adjusted and reproduced after Bellemain *et al. 2006a).*
a = no cub sample available for genetic analysis; b = mother was killed during the infanticidal attack; c = identified from DNA in hair (age unknown); d = no paternity identified among sample males; e = identified via DNA extracted from the fecal sample of an adult male bear that contained cubs' remains found in an area where almost 100 % of the adult males were radio-marked. "Unmarked" refers to a non-radio-collared male bear.

Case	Mother	Infanticidal male ID (age)	Father of the killed litter ID (age)	Paternal probability	Father of the next year's litter ID	Paternal probability
1	W9308	W8807 (12)	a		W8807	>0.98
2	W9404	W9311 (6)	a		W9311	>0.95
3	W8905	W8607 (27)	a		W8607	>0.99
4	W9307	W9301 (9)	a		W9301	<0.99
5	W8905	W9807 (11)	W8607 (28)	<0.99	b	
6	W9307	Unmarked[c]	W930 (10)	<0.99	a	
7	W9615	W9921 (9)	Unmarked[d]	<0.99	a	
8	BD007	Unmarked[e]	BD088 (8)	<0.99	b	

Bears – fact and fiction about bear hunting and intelligence

By Oliver Grimm, Andreas Zedrosser and Jon E. Swenson

Keywords: Brown bear, attack, fight, intelligence, demotion in fables and fairy tales

Abstract: Within the framework of the present book, especially two questions emerged about the bear and its natural behaviour in the wild: Do bears attack while standing upright, and are they really as slow and dumb as often described in fables and fairy tales? Bears stand upright for a variety of reasons, often simply to obtain a better overview, but never to charge at an opponent. When bears are truly aggressive and intend to attack a prey or opponent, they attack very fast by charging in great leaps and bounds on all four legs, but never on their hind legs. Upon making contact, bears then commonly rear on their hind legs and fight by biting as well as by swatting with their powerful front arms and claws at their opponent or prey. The myth of the slow and dumb bear may be rooted in their sometimes slow and seemingly clumsy gait and movements, and the fact that other animals, such as foxes or ravens, may follow a bear's trail in order to to use it as a food indicator. However, scientists working on animal intelligence as well as researchers working with bears agree that bears should be considered a highly intelligent species with an incredible adaptability and ingenuity especially when it comes to gaining access to food.

Introduction

Bears were both respected and feared in historic times. People sometimes feared that the bear would overhear when his real name was used in conversation and that this would bring about an unwanted and dangerous encounter with the animal itself. Instead, people commonly used so-called taboo names, such as "the brown one" in Germanic, "honey eater" in Slavic, and other such denotations in the Baltic languages, whereas in Finnish and Sami there were real words for the bear but also many circumscriptions (Pentikäinen 2007, 97, for Finland; cf. Nedoma, Piludu, Sommerseth, and Udolph, this volume). In turn, this suggests that people had an enourmous respect for the bear, which is also very evident in some of the honorary titles given to the animal such as the "(golden) king of the forest" in the Finnish-Karelian sphere (cf. Piludu, this volume). This respect, the bears' human-like abilities to walk upright (i.e. bipedally on its hind legs) and the danger bears presented in direct and confrontational encounters have made the bear one of the most prominent wildlife species in European history and folklore. Within the framework of the present book, especially two questions emerged about the bear and its natural behaviour in the wild:
1. When and why do bears stand upright, and are historic depictions of bears standing upright while hunted biologically correct? In premodern times, the hunting and killing of bears was considered to be a heroic deed and bears were commonly killed by hunters using a spear with a "stopper" (Fig. 1; Oehrl 2013; see Almond, this volume). In imagery, bears that are attacked by hunters are

often depicted as standing fully upright. This contradicts with observations by researchers and description by hunters that bears attack swiftly by running on all four legs. In the general public, it is commonly assumed that a bear standing upright is aggressive.

2. How intelligent are bears? Their seeming ability to understand human communication and overhear the unwanted use of their name, as well as their similarity with humans (same stature and silhouette, walking on soles, being able to stand upright and walk bipedally, etc.) may have contributed to ideas of the bear's intelligence in prehistoric and historic times. For example, hunters have claimed that bears backtrack in their own footprints to throw off pursuers or that old bears back into their dens with their rearend first, thereby leaving footprints that seem to lead out of the den (Bieder 2005, 24). In contrast, bears are commonly described as being rather dumb and outwitted by foxes in fairy tales and fables (cf. Böldl, Hirsch, and Veselova, this volume). So, how intelligent are bears? Below we attempt to answer these questions from a biological perspective.

Do bears attack standing upright and are historical images of bear hunts biologically correct?

Many mammals can stand upright. For example Alpine marmots (*Marmota marmota*), meerkats (*Surikata surikatta*), and the European ground squirrel (*Spermophilus citellus*) stand upright to look for potential dangers. Eurasian red squirrels (*Sciurus vulgaris*) and many other rodents sit on their hind legs and manipulate food with their front feet. Kangaroos propel themselves forward with their hind legs. However, besides the great apes and humans, only very few mammals share the unique ability of *Ursids* to stand and walk bipedally. Bears are plantigrade, i.e. they walk on the whole sole of their foot, similar to humans. All bear species have the ability to stand upright, and even walk bipedally for short distances (Fig. 2). This adds to their mystique, because bears can readily transform from a four-legged to a bipedal creature (Garshelis 2009, 497). "This appears to have given rise to Nepali and Tibetan legends of the Yeti: in these cultures, beings are not necessarily static, so a bear walking on four legs is a bear, but after switching to two legs becomes a Yeti. The fact that a bear's hind foot prints, which look very human-like, often cover its front tracks, strenghened the belief that this creature commonly walked bipedally for long distances (and thus might really be an ape, leaving tracks in the snow – hence the 'abominable snowman')" (ibid., 480). Standing enables bears to sniff the air or listen to sounds from a higher vantage point (ibid.). Bears can stand upright to observe, feed, reach for food or an object, to scent mark, or to fight, and they may rise up to reach or grab their prey or an opponent (Brown 2009; Garshelis 2009). However, bears do not attack, i.e. charge at an opponent, from an upright position (Brown 2009). Hererro (1985, 23) describes the behaviour of a bear when encountered by humans the following way: "A grizzly bear (i.e. a local name for the brown bear in inland North America) rearing onto its hind legs, a common stance, is trying to sense what is happening. Normally this is not an aggressive posture. On its hind legs the bear sniffs, listens and looks, trying to discover what kind of animal stands before it. Standing on its four legs a grizzly may show agitation by swaying its head from side to side, making huffing noises, or by opening and closing its mouth and making clicking noises with its teeth. Running and circling, usually to get downwind, may follow to get into a better position to sense (especially smell) the strange object. If the bear feels threatened, fleeing or charge may follow."

When bears are truly aggressive and intend to attack a prey or opponent, they attack very fast by charging in great leaps and bounds on all fours, but never on their hind legs (Brown 2009). Bear hunters in Sweden describe that bears "come in low and fast" if they truly mean to attack (Boström/ Lännbjer 2010). However, this attack or charging behaviour must be differentiated from the actual

Ursid fighting behaviour (Figs. 3–6). In a fight, bears use their powerful teeth to bite and kill, usually in the face and neck area of the prey or opponent. In addition, they use their extremely muscular forearms to swipe and swat at the opponent or prey, inflicting severe wounds with their long claws and the sheer force of their blows (Brown 2009). Bears will sometimes fight in a standing position (Garshelis 2009), i.e. to be able to fight with their forelegs, bears commonly rise to their hind legs during the fight to bite and swat at the opponents face, neck and shoulder area. This is also how bears fight with each other; they stand on their hind legs and try to bite and trash the conspecific opponent's face and shoulder area (Brown 2009). A typical injury in prey animals killed by a brown bear is a broken spine or other broken bones (Skåtan/Lorentzen 2011).

When pursued by hunters and with dogs, bears will almost always try to run and avoid the confrontation. However, once cornered or understanding that it can not escape its pursuers, a bear may turn around to fight. Bears are notorious for their ferocity and sheer strength in fights. A bear may sometimes slightly rise to its hind legs to have its fore legs availabe for fighting off attacking dogs. Hunters have used these moments to try to spear a bear that is busy fighting off a pack of aggressive and attacking dogs. Wild boars (*Sus scrofa*) have been and still are hunted in similar ways, for example in France and Australia. The hunter with a knife (pig sticker) or spear (*Saufeder*) kills a pig that is trying to fight off dogs or that may even be immobilised by several dogs that have grabbed it and pinned it to the ground. Such hunting situations are highly dramatic and heroic, and are thefore a common theme in hunting lore and art.

In summary, the secret in fully understanding when and why bears stand upright likly can be found in the distinction between a bears' observation behaviour, attacking or charging a prey or opponent, and the actual fighting behaviour with prey, conspecifics, or other species (including historic hunters with spears and accompanied by dogs). The lesson for a person going for a walk in an area with bears is to not be afraid when a bear suddenly rises to an upright position at a distance. The bear is usually trying to understand the situation and obtain a better overview. One should try to make noises that identify oneself as a human and not run away, but retreat slowly. Usually a bear will simply move away from the perceived danger.

How intelligent are bears?

Brown bears have been commonly portrayed in the historic western literature as slow, both mentally and physically (Bieder 2005, 23–24). For example, they are commonly outwitted by the smart and quick red fox (*Vulpes vulpes*) in the fables of the Greek storyteller Aesop and in fairy tales (cf. Böldl, Hirsch, and Veselova, this volume). However, in early times, bears were said to be wiser than man, because man does not know how to live all winter without eating anything (Hallowell 1926), and in the folklore of both Sweden (Tollin 2007, 256) and Norway (Schandy/Hermansen 2019, 93) a bear was said to possess the strength of twelve men and the sense of ten men.

It is exceedingly difficult to measure the intelligence of mammals other than humans, and even more so if we try to measure intelligence in human terms. No interspecific test systems are in place that enable us to directly compare the intelligence of bears and other mammals. Monkeys and apes are considered the smartest among mammals next to humans (e.g. Dicke/Roth 2016). We also consider gray wolves (*Canis lupus*) and their domestic counterpart, the dog, as very intelligent. In general, humans consider socially-living carnivores as smart, as evidenced by our close relationship with dogs and their use for a plethora of tasks in cooperation with humans. Bears have a large brain size relative to their body size, compared to other carnivores. However, they have been neglected in studies of cognition in comparison to other carnivores, such as the often socially-living canids (Vonk et al. 2012). In an experiment, American black bears (*U. americanus*) were presented with a series of natural concept discrimination tasks that varied in degree of abstraction on touchscreen computers. At least one bear performed at above-chance levels with transfer to novel images at each level of abstraction (VONK et al. 2012). Even tool use has been reported in wild brown bears (DEECKE 2012). Whereas the use of tools is comparatively common among primates and has also been documented in several species of birds, fishes, and invertebrates, tool-using behaviours have so far been observed in only four species of nonprimate mammals. DEECKE (2012) describes the observation of a subadult brown bear in southeastern Alaska that repeatedly picked up rocks in shallow water, manipulated and reoriented them in its forepaws, and then used them to rub its neck and muzzle. The behaviour probably served to relieve irritated skin or to remove food remains from the fur. Bears habitually rub against stationary objects and overturned rocks and boulders during foraging, and such rubbing behaviour could have been transferred to a freely movable object to classify as tool use (DEECKE 2012).

Researchers and bear biologists commonly consider bears as highly intelligent, although these opinions are more often based on anecdotal observations and personal experiences rather than hard data (BROWN 2009). For example, a bear biologist reports that he has observed on one occasion that an American black bear backtracked in his own footprints for about 50 meters and then jumped off the main trail, walking away in a different direction, likely to conceal its tracks (BROWN 2009; cf. HERERRO 1985). It remains unknown how common this behaviour is, however, similar observations have been reported by hunters and in hunting folklore (BIEDER 2005, 24). Bears are definitely highly trainable, as witnessed by bears in circuses that have been trained to ride bicycles, play instruments, or carry out other taks in order to entertain the audience (BROWN 2009). Although ethically questionable, this shows the ability of bears to learn and process information. Wild-living and food-conditioned bears are able to break into and raid cabins for food, as well as open garbage containers with sophisticated locking mechanisms meant to keep bears away from garbage. New "bear-proof" models of garbage containers are developed and tested with animals living in captivity almost on an annual basis, because wild bears keep figuring out how to open the lastest and newest bear-proof containers. Bears also have outstanding spatio-temporal awareness and memory and remember the exact location and temporal availability of food sources for many years (e.g. NOYCE/GARSHELIS 2011). For example, a radio-collared adult male bear in Sweden has been observed to carry out a two-week migration every year at almost exactly the same time to the same location about 150 km south of his usual home range. This bear likely fed on a resouce there, such as garbage or bait, and then returned to his usual home range (J. E. Swensson/A. Zedrosser, unpublished data). DECASIEN et al. (2017) suggested that species with a generalised or omnivorous diet, i.e. of both animal and plant origin, have larger brains and may be more intelligent than, for example, herbivores that feed just on plants. Bears have an omnivorous diet, which, together with their outstanding spatio-temporal awareness and memory, suggests that a generalised diet may be more critical than group living regarding the evolution of complex cognition in carnivores (VONK et al. 2012; DECASIEN et al. 2017). Bears are certainly able to recognise each other individually, as evident by their spatial dominance hierarchies (STØEN et al. 2005) and mating strategies (BELLEMAIN et al. 2006; STEYART et al. 2012). Mothers also spend up to three years with their young, and during this time the cubs learn the skills necessary for survival.

We speculate that part of the explanations for the myth of the slow bear may be their sometimes slow and seemingly clumsy gait and movements, but also that other animals, such as foxes, ravens (*Corvus corax*) and magpies (*Pica pica*), may follow on the trails of bears to benefit from the bears' ability to find and dig out or make accessible in other ways, food resources, such as ungulate carcasses. However, one should not be fooled by the slow gait and the sometimes clumsy or playful behaviour of bears, or their depiction of being slow and dumb by Aesop and in fairy tales: Bears are most definitly highly intelligent animals with an incredible adaptability and ingenuity especially when it comes to gaining access to food.

BIBLIOGRAPHY

Bellemain et al. 2006: E. Bellemain/A. Zedrosser/J. E. Swenson, The dilemma of female mate selection in the brown bear, a species with sexually selected infanticide. Proceedings of the Royal Society B 273, 2006, 283–291. https://doi.org/10.1098/rspb.2005.3331.

Bieder 2005: R. E. Bieder, Bear (London 2005).

Brown 2009: G. Brown, The Bear Almanach (Guilford [Connecticut] 2009).

Boström/Lännbjer 2010: R. Boström/J. Lännbjer, Svensk Björnjakt (Örkelljunga 2010).

DeCasien et al. 2017: A. R. DeCasien/S. A. Williams/J. P. Higham, Primate brain size is predicted by diet not sociality. Nature Ecology and Evolution 1, 2017. https://doi.org/10.1038/s41559-017-0112.

Deecke 2012: V. Deecke, Tool-use in the brown bear (*Ursus arctos*). Animal Cognition 15, 2012, 725–730. https://doi.org/10.1007/s10071-012-0475-0.

Dicke/Roth 2016: U. Dicke/G. Roth, Neuronal factors determining high intelligence. Philosophical Transactions of the Royal Society B 371: https://doi.org/10.1098/rstb.2015.0180.

Garshelis 2009: D. L. Garshelis, Family *Ursidae* (bears). In: D. E. Wilson/R. A. Mittermeier (eds.), Handbook of the mammals of the world 1. Carnivores (Barcelona 2009) 448–497.

Hallowell 1926: A. Hallowell, Bear ceremonialism in the Northern Hemisphere. American Anthropologist 28, 1926, 1–175.

Hererro 1985: S. Hererro, Bear Attacks. Their causes and consequences (New York 1985).

Noyce/Garshelis 2011: K. V. Noyce/D. L. Garshelis, Seasonal migrations of black bears (*Ursus americanus*): causes and consequences. Behavioral Ecology and Sociobiology 59, 2011, 823–883. https://doi.org/10.1007/s00265-010-1086-x.

Oehrl 2013: S. Oehrl, Bear hunting and its ideological context (as a background for the interpretation of bear claws and other remains of bears in Germanic graves of the 1st millennium AD). In: O. Grimm/U. Schmölcke (eds.), Hunting in northern Europe until 1500 AD. Old traditions and regional developments, continental sources and continental influences. Papers presented at a workshop organized by the Centre for Baltic and Scandinavian Archaeology (ZBSA), Schleswig, June 16th and 17th, 2011. Schriften des Archäologischen Landesmuseums, Ergänzungsreihe 7 (Neumünster 2013) 297–332.

Pentikäinen 2007: J. Pentikäinen, Golden King of the Forest: The Lore of the Northern Bear (Helsinki 2007).

Schandy/Hermansen 2019: T. Schandy/P. Hermansen, Norske pattedyr; levevis, forskning, trusler, folketro (Vestfossen 2019).

Skåtan/Lorentzen 2011: J. E. Skåtan/M. Lorentzen, Drept av rovvilt? (Trondheim 2011). https://www.miljodirektoratet.no/publikasjoner/2019/mars-2019/drept-av-rovvilt/.

Steyart et al. 2012: S. M. J. G. Steyaert/A. Endrestøl/K. Hackländer/J. E. Swenson/A. Zedrosser, The mating system of the brown bear *Ursus arctos*. Mammal Review 42, 2012, 12–34. https://doi.org/10.1111/j.1365-2907.2011.00184.x.

Støen et al. 2005: O.-G. Støen/E. Bellemain/S. Sæbø/J. E. Swenson, Kin-related spatial structure in brown bears *Ursus arctos*. Behavioral Ecology and Sociobiology 59, 2005, 191–197. https://doi.org/10.1007/s00265-005-0024-9.

Tollin 2007: C. Tollin, Djurgårder och jaktparker. In: H. Tunón/M. Iwarsson (eds.), Människan och faunan. Etnobiologi I, Sverige 3 (Stockholm 2007) 242–257.

Vonk et al. 2012: J. Vonk/S. E. Jett/K. W. Mosteller, Concept formation in American black bears, *Ursus americanus*. Animal Behaviour 84(4), 2012, 953–964. https://doi.org/10.1016/j.anbehav.2012.07.020.

Dr. Oliver Grimm
Centre for Baltic and Scandinavian Archaeology (ZBSA)
Stiftung Schleswig-Holsteinische Landesmuseen, Schloss Gottorf
Schleswig
Germany
oliver.grimm@zbsa.eu

Prof. Andreas Zedrosser
Department of Natural Sciences and Environmental Health
University of South-Eastern Norway
Bo i Telemark
Norway
Department for Integrative Biology

Institute for Wildlife Biology and Game Management
University for Natural Resources and Life Sciences
Vienna
Austria
andreas.zedrosser@usn.no

Prof. Jon E. Swenson
Faculty of Environmental Sciences and Natural Resource Management
Norwegian University of Life Sciences
As
Norway
jon.swenson@nmbu.no

Fig. 1. Hunting spear used for brown bears and wild boars. 44 cm long; Germany/Austria, 15th century (© Metropolitan Museum, Gift of William H. Riggs, 1913. Public Domain).

Fig. 2. A family group of brown bears raised on their hind legs to observe a situation in the distance. This posture is commonly used by bears to obtain a better overview, to reach for food, or manipulate a food item or object. A bear standing on its hind legs is usually not a sign of aggression. Picture taken at a bait site for photography in Finland (photo S. Klaus).

Fig. 3. Brown bears in a playful fight in Katami National Park, Alaska, USA. Both bears are on their hind legs and engage in harmless play. The same posture is commonly also taken during serious and harmful encounters with intra- or interspecific opponents or prey. However, a bear would not charge or run towards an opponent on its hind legs, but on all fours, and raise to its hind legs at the very last moment to have its front arms ready for fighting (photo Th. Sbampato).

Fig. 4. Illustration of a bear hunt in medieval times. From a biological perspective this picture is difficult to interpret other that it depicts the actual moment of killing of a bear with a bear spear. No dogs or other hunters are visible. "Herr Hawart" in the Codex Manesse, a Liederhandschrift (manuscript with songs), furnished with numerous illustrations. Made in Zurich, Switzerland, c. 1300 (© Universität Heidelberg, Cod.Pal. Germ. 848, f. 313r. Public Domain Mark 1.0).

Fig. 5. Historic print depicting a medieval bear hunt. From a biological point of view, this depiction appears highly dramatised and with an unrealistic but "heroic" touch. Brown bears are solitary living and do not occur in groups or packs. Also knights in heavy full-body armour were likley too slow to approach a bear on foot and stab it with a dagger unless the bear was cornered by dogs. Made by the Flemish artist Johannes Stradanus (1523–1605), active in Florence, with prints by the name of "Venationes Ferarum" (© Metropolitan Museum, The Elisha Whittelsey Collection, The Elisha Whittelsey Fund, 1949. Public Domain).

Fig. 6. Historic engraving depicting a bear hunt with dogs and hunters armed with bear spears. Despite a certain element of dramatisation, this picture likely reflects a realistic bear hunt. A single bear is occupied by trying to fight off a pack of dogs while a hunter approaches with a spear to stab the bear. Wild boar were and still are hunted in a similar manner. These types of hunts commonly resulted also in the injury or death of many of the dogs. Made by the German artist Virgil Solis (1514–1562) (© Metropolitan Museum, The Elisha Whittelsey Collection, The Elisha Whittelsey Fund, 1957. Public Domain).

Chapter 3

Bear hunting (Europe)

Head of a hunting spear with stoppers. Used in the bear or boar hunt. Germany/Austria, 15th century. 44 cm long (see ALMOND, this volume; photo Metropolitan Museum, 14.25.321, Gift of William H. Riggs, 1913. Public Domain. CO 1.0 Universal [CC0 1.0]).

The bear hunt. The hunter sneaks up behind the bear and shoots it in the back while it is eating autumn berries (see GRIMM, Summary, this volume; image after Olaus Magnus' Historia de gentibus septentrionalibus 1555, XVIII,25).

Bear hunting in the later Middle Ages and early modern period, viewed from the perspective of art history and contemporary textual sources

By Richard Almond

Keywords: Bear hunting, images, courtly, aristocratic, par force de chiens, *anthropomorphism, bear-baiting*

Abstract: The aim of this chapter is to select, examine, describe and interpret images of bear hunting from the later Middle Ages and early modern period,[1] supported by reference to contemporary textual sources, in order to understand Man's interaction and relationship with Ursus arctos *during the title period. Basically, this involves clarifying and assessing the consistency and validity of the evidence of three aspects of bear hunting: who did it, what practices and weapons were used, and why bear hunting was regarded as special within the aristocratic hunting culture of the time frame. Images can be misleading, as in the case of didactic hunting illustrations in the so-called "hunting books" of the period which represent the idealised or "perfect" scenario of events and practices. Hunting has always been and remains an unpredictable activity, subject to many variable elements which are partly or even largely out of the direct control of the venator. In the field, the hunter often must make ad hoc-plans to adapt to reality. Instructional texts and accompanying images sought to redress these disadvantages as far as possible by educating the learner and courtly audience in the correct behaviour, language and procedures. Of course, these didactic sources were socially biased, written or compiled by royal or aristocratic authors, and their utilisation applied specifically to the social elites who were literate and had access to such sources, as well as the wealth and leisure to indulge themselves in such activities. However, it is also clear from some sources that bears were hunted in different ways by common hunters and others who had valid reasons other than sport to kill them.*

Bears have been hunted and killed for sport and raw materials, as well as for security and protection reasons, from time immemorial. The Greeks and Romans enjoyed bear hunting. A 4th-century Roman mosaic (Fig. 1) in the Museo della Civilita Romana, Rome, Italy, features two methods of classical bear hunting: On the right a hunter on foot directs large mastiffs to drive a bear into a long net fixed to trees; on the left a hunter waits to drop the door of a box-like trap with a ramp after a bear has entered, no doubt attracted by bait. In his didactic poem on "The Chase", *Cynegetica*, compiled in the early 3rd century AD, Oppian (the so-called Pseudo-Oppian) describes the bear hunt: "For

---

[1] For the purposes of this essay, the date which divides the later medieval period/later Middle Ages from the early modern period is 1485, the beginning of the English Tudor monarchy; this period in turn closing in 1714, the end of the Stuart dynasty.

bears an exceeding glorious hunt is made by those who dwell on the Tigris and in Armenia famous for archery. A great crowd go to the shady depths of the thickets [to hunt them]". He includes the ancient practice of using a *lymerer* or scenting hound, to trace a bear to a thicket (dense cover), setting nets around it, then driving the bear into the nets (OPPIAN, 189–193). This enduring technique is repeated in the late 14th century by Gaston Fébus in Chapter 52 of his iconic book, *Livre de chasse* (Paris MS Fr 616 Ch. 52, fol. 93).

The bear as sporting quarry occupied a particular niche in the minds of royal and aristocratic medieval and early modern period hunters. As the largest predator in Europe, enormously strong with immense endurance, fast and extremely dangerous (CUMMINS 1988, 122; OEHRL 2013, 304), the bear was classified in the contemporary European instructional texts as one of the high-status beasts of the chase and was included as "noble" quarry, thus suitable for hunting by aristocratic hunters. However, the overriding reason for its "special" position in the array of noble quarry is the apparent similarity of the bear to Man in a number of its habits. These range from an upright gait and fighting stance, as shown in some contemporary illustrations (see GRIMM et al. on Bear – fact or fiction, this volume) to its human-like copulating position, as has been wrongly assumed by Gaston Fébus, author of a famous hunting book (see below). These apparently anthropomorphic aspects will emerge clearly from the visual and literary sources considered in this chapter. A 13th-century Greek School miniature entitled "Game – The Animals you can Hunt" from the didactic *De Venatione* by Oppian (Fig. 2), now in the Biblioteca Nazionale Marciana, Venice, Italy, makes the point of suitability and choice to the novice hunter. An experienced Master (right) lectures the young Apprentice (left) on what is correct game: wolf, bear, boar, deer (Fig. 2: top), lion, elephant, rhinoceros, leopard (Fig. 2: bottom).

The bear does not feature in the English medieval hunting books such as *The Art of Venerie*, the *Boke of Huntyng*, *The Tretyse off Huntyng* or *The Master of Game*, the logical reason being its complete extinction in England centuries earlier (RACKHAM 1993, 33–34). It appears that the bear had lapsed from the memory of English hunters and only remained as an icon in the Church, in heraldry, and as a fearsome, later loveable, beast in legends and children's stories. However, in his Foreword to *The Master of Game* (1904, and repeated in the 1909 edition), the President of the USA, Theodore Roosevelt, an acknowledged historian and sportsman, mentions the bear's importance on the continent: "The kings and nobles, and the freemen generally, of the regions which now make up France and Germany, followed not only wolf, boar and stag […] but [also] the bear" (BAILLIE-GROHMAN/BAILLIE-GROHMAN 1904; 1909, xxi).

An exception to the general silence on bear hunting in the English manuals is the two chapters in Turbervile's *Booke of Hunting* from 1576. These are entitled "Of the Hunting of the Beare, and first of hir nature and properties Chap. 77" and "The manner of hunting the Beare Chap. 78" (TURBERVILE 1908, 216–220). Turbervile's book of 1575, *The Noble Art of Venerie*, printed together with Turbervile's *The Book of Falconry or Hawking*, is a translation by George Gascoigne (1525–1577), English poet and soldier, of the French manual *La Venerie* (1561) by Jacques du Fouilloux. This unrevised plagiarising of a continental manual explains the inclusion of the chapters on bear hunting, relevant to French and continental *veneurs* but not English hunters in the late 16th century. The author (whoever it is) writes that he has included the wolf, the bear and the reindeer as: "I have not thought good to leave out, although they be not in use here with us in England: since they seeme by the description, to be noble chases, and much esteemed in other countreys" (TURBERVILE 1908, 220). The descriptions of bears and the bear hunt are detailed and are clearly derived from earlier continental works, particularly *Livre de chasse*. In this paper, we will not refer to this manual again but will keep to the information provided by the older treatises.

A rare comment on the bear in later English history is found in *The Description of England: the Classic Contemporary Account of Tudor Social Life* by William Harrison, discussing King Canute's (spurious) *Forest Charter* of 1019: "The beasts of the chase were commonly the buck, the roe, the fox and the [pine] marten. But those of venery in the old time were the [p. 260] hart, the hare, the boar and the wolf; but as this held not in the time of Canute [as given by the Elizabethan barrister, gamekeeper of Waltham Forest and Justice of the New Forest Eyre, John Manwood (d. 1610), in his writings on Forest Law; bears do not appear among the beasts given in cl. 23, 24 and 27] so instead of the wolf the bear is now crept in, which is a beast commonly hunted in the East [Baltic] countries and fed upon as excellent venison, although with us I know not any that feed thereon or care for it at all" (EDELEN 1968, 258, 260).

Gaston III, compte de Foix, called Fébus (1331–1391), wrote probably the most informative and technically useful text on medieval hunting, *Livre de chasse* (ALMOND 2003, 9). This canonical treatise is a personal and original work, with the exception of the chapters on dismembering deer and wild boar plagiarised from an earlier hunting book by Henri de Ferrières, *Le Livre du roy Modus et de la royne Ratio*, written between 1354 and 1376/77, and a long poem by Gace de la Buigne, a chaplain/courtier and hunting author. It was begun on 1st May, 1387 and completed in 1389 (GASTON PHÉBUS, 5, 14; ALMOND 2003, 13). 44 copies of *Livre de chasse* are known to exist, the majority from the 15th century, a few from the early 16th century (GASTON PHÉBUS, 5–6). Many of them are illuminated or illustrated, some unfinished; MS Fr 616 (dated 1405–1410, now in the Bibliothèque nationale, Paris) being the most beautifully illustrated and complete. *Livre de chasse* was widely known and regarded and as such had a profound influence on courtly hunting practices in Europe. Fébus devotes several chapters to the nature and hunting of the bear, a familiar quarry to him living on his vast estates in the Pyrenees of southern Europe. These are entitled: Chapter 8, "Of the bear and its nature" (Fig. 3), Chapter 52, "How to hunt and slay bears", and Chapter 62, "How to slay bears and other beasts with spring traps" (Paris MS Fr 616: Ch. 8, fol. 27v; Ch. 52, fol. 93; Ch. 62, 106v). An illuminated miniature in a copy of *Livre de chasse* at the Bibliothèque Mazarine, Paris, France, shows the kind of rough mountainous terrain where bears could be found (Fig. 4). Fébus respected the bear for its great strength but considered its head vulnerable to a sharp blow which would stun or even kill it. He goes into interesting detail of the bear's life cycle. The pregnant she-bear "holes up" in December after mating until she gives birth in March. Fébus says that the cubs, two at most, lie as if dead for one day while their mother warms them with her hot breath and literally "licks them into life" (GASTON PHÉBUS, 25–26). According to the medieval bestiaries, bear cubs were born in the lair as formless lumps of fat and the "licking into shape" symbolises Christianity converting the heathen (CLARK 1984, 42; cf. also VAN OS, this volume). This action can be seen in the *bas de page* illustration of folio 318v, MS Egerton 1146, British Library. Fébus continues that the she-bear nurses them for a month then feeds them pre-masticated food. Intriguingly, Fébus believed that bears mated on top of each other like humans, *aguise dôme et de femme* (GASTON PHÉBUS, 25–26), another example of the anthropomorphism applied to bears by medieval hunting authors. Ironically, Fébus died in 1391, aged 60, whilst washing his hands before the hunt supper prepared in the hall of the hospice at his castle at Orthez, after returning from a bear hunt in the forest of Sauveterre, near Pamplona. Falling backwards, he cried "Je suis mort!" before expiring (CUMMINS 1988, 128–129, taken from Gunnar Tilander's edition of *Livre de chasse* [TILANDER 1971]).

In Iberia the bear had high quarry status, both Alfonso XI of Castile (r. 1312–1350) and John I of Portugal (r. 1385–1433) regarding it as royal game. Bears were protected throughout Portugal for hunting by the king; killing one resulting in an enormous fine (CUMMINS 1988, 121). The Holy Roman Emperor Maximilian I (Roman-German King from 1486; Holy Roman Emperor r. 1508–1519) was also an ardent bear hunter; in his book *Thuerdank* there are three sections on bear hunting, and in his *Hunting Notebook* Maximilian advises "You must go hunting with a spear, and always have one – go after him with the spear […]" (NEIDERWOLFSGRUBER 1992, 35–36). His favourite method was to tackle the beast in its lair, on foot and single-handed, armed only with a short hunting spear or hunting sword (BAILLIE-GROHMAN 1904, 158). This almost suicidally brave technique reflects Maximilian's high regard for the fighting qualities and courage of the bear, making it a worthy foe to take on face-to-face. The bear is thus seen as a personal challenge to the dedicated hunter rather than as a quarry simply providing a prolonged and exciting chase like the hart or buck. However, in spite of dedicated enthusiasts like Fébus, Alfonso XI and Maximilian, opinions on the bear varied, and *The Lexicon of the Mediaeval German Hunt* says of the bear: "Amongst other heavy game [...] even the brown bear is of little importance" (DALBY 1965, xvii). German sources do include the bear, but its value as a quarry species is generally regarded as being considerably inferior to that of the stag, boar and hare. However, images of bears do feature in some German manuscripts. MS Egerton 1146, a Germanic Book of Hours in the collection of the British Library dating from 1475–1485, is a beautiful manuscript containing hundreds of life-study images of animals, birds and fish. It has three illuminated miniatures of bears and bear hunting. In the *bas de page* of fol. 11v, "October" in the Calendar, a mounted hunter thrusts a cross-hilted spear into a huge blackish-furred bear which is being harried by hounds (Fig. 5); in a marginal illustration of fol. 20r the garlanded blond-haired hunter is on foot, gleefully using a long spear to dispatch the beast (Fig. 6; cf. ALMOND 2003, 69–70).

Illustrations in the hunting manuals and treatises, frescoes, tapestries, paintings, carvings and other art media associated with socially elite spaces provide plentiful details of how bears were hunted by royal and aristocratic hunters. This activity was a highly organised and strictly managed team event, designed for guaranteeing maximum excitement and effectiveness. Many European sources show that the bear was hunted on horseback and on foot, usually assisted by large hounds including running-hounds, alaunts and mastiffs, collectively known as *Canis ursaticus* (OEHRL 2013, 299). This was the classic hunt, *par force de chiens*, described and illustrated in the hunting books compiled or written by *learned* pragmatic noble authors, presenting didactic texts for courtly audiences. It was also good training and exercise for men and horses in preparation for warfare. This elitist type of hunting large and small mammalian game remained standard and current throughout the later medieval period and into the early modern period. The chase *par force* required highly trained horses, hounds and professional assistants, proper weaponry and usually plenty of time, leisure being a prerogative of the nobility. A particular bear chase described by Alfonso XI of Castile in the first half of the 14th century took no less than five days (CUMMINS 1988, 126–128). Gaston Fébus details the classic mounted bear hunt in Chapter 52 of *Livre de chasse*, "How to hunt and slay bears" (Paris MS Fr 616, Ch. 52 fol. 93). Finding a suitable bear to hunt was the first problem, as the hunter could not judge the sort (black or brown), age or size of a bear by its dung as it is not consistent, bears being omnivores. This is unlike the regularly shaped *fewmets* and *croties* of other game animals, although the footprints of a male bear are rounder and larger than those of a female. Alfonso XI of Castile states that if a bear had urinated close to its droppings then it was a she-bear (CUMMINS 1988, 122). So Fébus advises questing with a *lymerer*, a scenting hound, in the areas where the beast is likely to be feeding according to the season: cornfields, hayfields, orchards, vineyards, oak woods and beech woods. Once traced to its lair, termed *harboured* (BAILLIE-GROHMAN/BAILLIE-GROHMAN 1909, Prologue 9), and hunted, the chase is like boar hunting, using large running hounds and strong greyhounds. To slay the beast quickly is preferable, the pack of hounds bringing it to bay to be dispatched by several hunters armed with heavy spears, supported by grooms with bows and crossbows. He emphasises the dangers of tackling an infuriated wounded bear, mounted hunters keeping their distance with lance or spear, never using the sword.

A late 14th-century fresco by Maestro Venceslao (flourished [fl.] 1390–1400) at the Castello del Buonconsilio, Torre dell'Aquila, Italy, illustrating winter activities in November (Fig. 7), shows nobles armed with cross-hilted spears on horseback with hounds and assistants hunting bears in the mountains, whilst farmers drive pigs for slaughter into the walled town, contrasting the differing societal "Labours of the Months". An early modern drawing called "The Bear Hunt" by Giovanni

Stradano (Jan van der Straet/Johannes Stradanus, 1523–1605) in the Gabinetto dei Disegni e delle Stampe degli Uffizi, Florence, Italy, captures the main elements of this aristocratic practice (Fig. 8): The bear has been driven by hounds and horsemen through woodland and has been halted by a line of spearmen on foot. The elaborately dressed nobles gallop up and kill the bear, using long lances. The basic weapon featured in many such images is the cross-hilted bear spear, a stout wooden shafted spear with a broad double-edged steel head fitted with a crosspiece to stop the beast running up the shaft and killing the user. Sometimes two men were necessary to wield such a heavy weapon effectively or they worked alternately with spears, stabbing and weakening the bear for dispatching. In *Hunting Weapons*, Howard Blackmore comments: "Spears made specifically for bear hunting were of stouter proportions than the rest [of hunting spears], the blades of some are nearly 2 ft. in length". A detail from an engraving of Hans Burgkmair (1473–1531) entitled "The Bear Hunters" in the *Triumph of Maximilian* (1526; Fig. 9) shows a group of five bear hunters carrying broad-bladed spears with crossbars which are tied to the shaft just below the socket (BLACKMORE 2000, 88). Good examples of this type of specialist spear can be seen in the Leeds Royal Armouries Collection, Leeds, England. Nobles on foot aided by professional assistants and hounds also hunted bears, particularly in difficult mountainous terrain. A miniature entitled "The [Bear] Hunt" from a 15th-century Italian manuscript, *Tractatus de Herbis*, by Dioscorides, shows a bear gripped by a mastiff, being speared by a lone hunter in a red tunic on foot, using a cross-hilted spear while another hunter in blue, probably his servant, blows *la mort* on a hunting horn. The scene takes place on a mountainous wooded hillside, a favourite habitat of bears in Italy and elsewhere in Europe (Fig. 10).

In Christian iconography the bear had a mixed reputation. It was the attribute of St Euphemia, as bears and lions refused to savage her when she was thrown to bears and lions by the Romans, but it was also regarded as a symbol of gluttony (CLARK 1984, 42, 117), sloth and clumsiness. St Ursula acquired her name from *Ursus* because of the ferocity with which she defended her 11,000 virgins (BIEDER 2005, 9). St Columba of Sens was saved from being burned alive by a she-bear, and her attribute is a chained she-bear with prominent teats. A tempora on wood panel painting, "St Columba Saved by a Bear", dated 1340, by Baronzio da Rimini (fl. 1362; Fig. 11), demonstrates that bears were not necessarily regarded as evil beasts. It is not surprising therefore that several prominent artists featured bears in their paintings and drawings. Leonardo da Vinci (1452–1519) was one of these. A small drawing of a bear's head by Leonardo was recently sold for a record £8.8m ($12.1m) at a London auction. This drawing was created using the silverpoint technique on pale pink-beige paper and is among a number of the artist's small-scale drawings of animals which date back to the early 1480s. In the late 1480s he dissected the left hind leg of a bear, an animal at that time widespread in the mountains in Italy and with which he would have been familiar. Leonardo may have been interested in looking at the bear's leg and foot anatomy because of its plantigrade gait of walking like a human with its feet flat on the ground. Access to human cadavers was restricted, but by using a bear's leg he was able to produce an objective and accurate study in metalpoint and pen and ink, to which he referred twenty years later in his dissection of the human hand and foot. "The anatomy of a bear's foot", *c.* 1488–90, is in the Royal Collection, Windsor Castle, England (CLAYTON 2019, 64). Again, this exemplifies the anthropomorphism with which the bear was regarded at this time.

Images of bears in paintings can be included in composite groups of animals, mostly made up of hunting quarry, all of which also have their own religious and social symbolism. The Italian medalist, fresco and panel painter Pisanello (1394[?]–1455) produced such a composition, "The Vision of St Eustace". The bear is shown in a dark woodland on the upper right quartile, as if trundling on all fours out of the pictorial space behind the crucifix, perhaps symbolising the former heathen background of the newly converted Christian saint or possibly the vanquishing of the Devil, a label given to bears in some medieval bestiaries (BIEDER 2005, 82). This painting is in the National Gallery, London. In the 1430s Pisanello produced "Two Bears", a study in chalks in the Codex Vallardi, 2414v, Louvre, Paris

good training for a young hound, but it also provides "an amusement" for the young lady who will not be involved in the brutality and danger of the proper bear hunt. The panel may thus symbolise noblewomen's passive roles on the margins of this particular alpha-male type of dangerous big game hunting, those roles including attendance at the climax, applause and praise for the brave hunters. Images of women portrayed at the close proximity of the demise of a bear, as in the Devonshire Hunting Tapestry example, are rare.

Visual sources can apparently provide a wealth of information on what hunters wore for bear hunting. The problem is how much reliance can be placed on illustrations, most of which are decorative and have either a didactic purpose in the case of the hunting manuals and treatises, or make a status statement especially in the case of tapestries. In his edition of William Twiti's *The Art of Hunting*, Bror Danielsson writes: "As yet (c. 1300) there did not exist any special hunting dress. Everybody moved about in their everyday dresses, varied according to social status, even though the colour might be adapted to the environment" (DANIELSSON 1977, 23). Certainly, it appears that aristocrats and courtiers dressed up increasingly beyond practicality as we progress from the later Middle Ages into the early modern period. It has already been demonstrated that bear hunting required some specialisation in weaponry, but apart from some hunters wearing protective armour for individual "bear combat" it did not necessitate particular forms of dress. Contemporary images show that *par force* hunting dress was basically the same for whatever animal was being pursued on horseback with a mixed pack of hounds, the aristocratic technique and attendant procedures dictating the clothing. The colour of clothing is debatable. John Cummins comments that there appears to be little uniformity in the colouring of hunting garments in medieval illustrations, although there can be some consistency within one manuscript, such as between mounted hunters and assistants on foot. The present author discusses the use of camouflage in *Medieval Hunting*, but there is no specific imagery evidence that green clothing was used by bear hunters as was advised for serious deer hunters (CUMMINS 1988, 179; ALMOND 2003, 100). The illustration (detail) of "The Bear Hunters" by Hans Burgkmair (1473–1531) shows five bear hunters carrying bear spears (cf. Fig. 9), dressed in elaborately-styled slashed tunics and breeches, typical of courtly hunters of the Maximilian regime, with no concession to concealment. Similar modes of courtly hunting dress can be seen in the twelve months of hunting scenes of *Les Chasses de Maximilien* tapestries, woven in Brussels by the 1530s (ALMOND 2003, 49–50). The conclusion must be that contemporary illustrations may not always have reflected real dress codes in the hunting field, including bear hunting.

Bears were also trapped in many ingenious ways by the rural peasantry protecting their crops, stock and lives. Gaston Fébus mentions that nets, snares and traps may be used to capture bears (GASTON PHÉBUS, 57). Bears love honey and were notorious for plundering beehives. A 15th-century Italian School miniature "Collecting Honey" from *Tractatus de Herbis* by Dioscorides, now in the Biblioteca Estense, Modena, Emilia-Romagna, Italy, shows a bear raiding and overturning two beehives (Fig. 20). The peasant woman at the chaotic scene looks distraught at the loss of her honey and her bees escaping. A wooden carving, probably medieval, in the cathedral at Toledo, Spain, features a bear on its hind legs plundering a cylindrical straw bee skep, being attacked by a farm dog (Fig. 21). In addition to protecting crops and livestock, a bear carcass was much prized, providing not only much needed protein but also valuable fur pelts for bedding, clothing and trimmings, claws, teeth and bones. Fébus mentions that except for its paws, bear meat is not very tasty, but its fat can be used to make ointments for gout and in a tonic for the nerves (GASTON PHÉBUS, 26). An 11th-century miniature illustrating bear hunting from *The Cynegetica* by Oppian of Syria, written in the 3rd century AD, now in the Biblioteca Marciana, Venice, shows two rustic hunters (Fig. 22), one armed with a bow and quiver, the other pointing at a bear enmeshed in a net securely tied to a tree.

Spring-traps set in fence or hedge gaps were a common method used by peasants to protect vineyards and orchards against the predations of large animal poachers, including bears. A trip-device in the gap released a spear or bolt fixed to a bent bough which pierced the animal's flanks (*Livre de chasse*: Paris MS fr 616, Ch. 62, fol. 106v; Cummins 1988, 241; Almond 2003, 107). This method is shown in an engraving after an illustration in one of the 14th-century copies of *Livre de chasse* (Fig. 23). The bear's flank has been pierced by a spear, released by triggering a bent bough in a fence gap. A woodcut "Reynke de Vos" (Reynke the Fox) from an *incunabula* (early printed book) printed by Hans van Ghetelen (before 1480 – before 31.1.1528), Lübeck, Germany, 1498, now in a private collection, shows peasants beating up a bear they have trapped in a press.

Bear-baiting was another aspect of Man's relationship with bears, connected to and dependent upon hunting/trapping and perpetuating the violent co-existence of humans with these large dangerous predators. The tradition of bear and other animal baiting probably originated in the *venationes* of the Roman arena (Blackmore 2000, xxiii; see also O'Regan, this volume). Artificial or imitation hunting, it was widely practised in England from Anglo-Saxon times and was specifically connected with the royal courts (Oehrl 2013, 309). Animals were trapped in the wild or bred in captivity and exported all over Europe. There were no wild bears in the British Isles in the later Middle Ages, so bears for enclosed park-hunting and baiting were imported, probably from Scandinavia. The unfortunate bear was chained to a pole, often handicapped by having been blinded or having had its claws cut off, and a group of strong and aggressive dogs were set on him. Betting on the outcome drove this brutal spectator sport, which remained popular in England and other European countries well into the 19th century (Oehrl 2013, 309). In the Middle Ages the specialised servant in a great household who was responsible for looking after and presenting the bears used in this activity was the bear-ward or *Ursarius*. An example from documentary sources is Master Spernellus who was the bear-ward at Richmond Castle for the Earls of Richmond, Yorkshire, England, in the mid-1200s. He had a generous salary, accommodation and after retirement a pension and house (according to Richmond Castle signage). On the *bas de page* of Psalm 88, fol. 161, of *The Luttrell Psalter*, Add. 42130, an East Anglian prayer book dated *c.* 1325–1335, housed in the British Library, a chained bear is being attacked by four dogs, encouraged by three men; one dog is seized in its forepaws (Fig. 24). The keeper of the bear, the *Ursarius*, supervises the fight with a stick and long club.

Murals on the walls of the Turret Room at Madingley Hall, near Cambridge, England, dated to between 1605 and 1633, show sporting activities in the surrounding park, including separate images for bear hunting (Fig. 25), boar hunting, and falconry. Both mammals were extinct in the wild in England, so they would have been brought in for the purpose of hunting as a high-status sport and spectator event, clearly viewed by family, friends and guests from the Turret Room. The gentleman-hunters on horseback and the more plainly dressed servants on foot are portrayed using spears to slay the beasts which are being attacked by mastiffs and greyhounds. The wall paintings were probably commissioned by Sir Edward Hynde, the owner of Madingley Hall at this time and a noted enthusiast of hawking, hunting and animal baiting (Almond 2003, 34). It may well be that Sir Edward Hynde is portrayed as one of the mounted hunters.

Bear hunting by royal and aristocratic hunters continued and flourished well into and beyond the early modern period, aided by the increasing use of sporting firearms. W. A. Baillie-Grohman comments on the vast numbers of game killed by the two Electors of Saxony, John George I (r. 1611–1656) and his son John George II (r. 1656–1680), "probably the greatest slaughterers of game known in modern history", which included 238 and 239 bears, respectively (Baillie-Grohman 1896, 169–170). The vast majority of these bears were presumably shot, having been beaten from covert, driven and funnelled past shooting stands like other game, rather than hunted *par force de chiens*, although this ancient practice also continued in the forests of Europe. These figures of game killed in one region of Europe over many years are a critical indication of the unknown but undoubtedly vast numbers of game which existed at that time and until relatively recently, a factor we tend to overlook in our understanding of game and hunting in the later Middle Ages and early modern period.

The close and critical examination and assessment of art historical sources illustrating bear hunting, supported and reinforced by reference to contemporary texts, provides valid answers to the issues posed in the opening paragraph of this paper. Firstly, bear hunting was clearly a royal and aristocratic sporting activity, carried out by male elites as part of their public and private assertion of their superior social position. Secondly, these elites hunted bears in particular ways; on horseback with hounds, classic *par force de chiens,* and on foot with hounds and attendants. These socially acknowledged and approved methods of bear hunting required some specialisation in appropriate weaponry: long sturdy cross-hilted bear spears. Bears were also driven by hounds and hunters into nets. Wounding and holding bears before dispatching them was an accepted essential practice. No special clothing was worn, fashion dictated design rather than practicality. Some hunters wore protective plate armour in close "bear-combat". Bears were also killed by other non-sporting means for different reasons; this was not regarded as "hunting" by the social elites. Thirdly, bear hunting was clearly regarded as a special activity, a test of manhood, a demonstration of fighting skills and personal courage highlighting the warrior codes of the aristocratic knightly domain. Although the red deer hart and stag, fallow buck, and boar were given "noble" status and rated as worthy opponents, no other quarry beast engendered so much respect as the bear. This was not only because of its size and ferocity but more importantly the anthropomorphic habits of its sex life (wrongly assumed), upright shambling gait and fighting stance, unique ability to use its forepaws as weapons and parry thrusts, or pluck out arrows and knives. A bear acted like a giant deranged human when cornered, a monster from ancient legend and nightmare. This beast was the ultimate opponent for the courtly knight and hunter.

BIBLIOGRAPHY

ALMOND 2003: R. ALMOND, Medieval Hunting (Stroud 2003).

ALMOND 2009: R. ALMOND, Daughters of Artemis: The Huntress in the Middle Ages and Renaissance (Cambridge 2009).

BAILLIE-GROHMAN 1896: W. A. BAILLIE-GROHMAN, Sport in the Alps (London 1896).

BAILLIE-GROHMAN 1904: W. A. BAILLIE-GROHMAN, Ancient Weapons of the Chase. Burlington Magazine 1904, Pt. II, 164.

BAILLIE-GROHMAN/BAILLIE-GROHMAN 1904: W. A. BAILLIE-GROHMAN/F. BAILLIE-GROHMAN (eds.), The Master of Game (London 1904).

BAILLIE-GROHMAN/BAILLIE-GROHMAN 1909: WM. A. BAILLIE-GROHMAN/F. BAILLIE-GROHMAN (eds.), The Master of Game (London ²1909).

BIEDER 2005: R. E. BIEDER, Bear (London 2005).

BLACKMORE 2000: H. L. BLACKMORE, Hunting Weapons, From the Middle Ages to the Twentieth Century (New York 2000).

CAMPBELL 2007: C. CAMPBELL (ed.), Temptation in Eden, Lucas Cranach's Adam and Eve (London 2007).

CLARK 1984: K. CLARK, Hall's Dictionary of Subjects & Symbols in Art (London 1984).

CLAYTON 2019: M. CLAYTON, Leonardo Da Vinci, A life in drawing (London 2019).

CUMMINS 1988: JOHN CUMMINS, The Hound and the Hawk: The Art of Medieval Hunting (London 1988).

DALBY 1965: D. DALBY, Lexicon of the Mediaeval German Hunt (Berlin 1965).

DANIELSSON 1977: B. DANIELSSON (ed.), William Twiti, The Art of Hunting, 1327 (Stockholm 1977).

EDELEN 1968: G. EDELEN (ed.), The Description of England: the Classic Contemporary Account of Tudor Social Life by William Harrison (Washington D. C. 1968).

GASTON PHÉBUS: The Hunting Book of Gaston Phébus; manuscrit français 616, Paris, Bibliothèque nationale. Introd. M. Thomas/F. Avril. Comments W. Schlag (London 1998).

KUENZEL 1973: H. KUENZEL, Lucas Cranach the Elder (Milan 1973).

NEIDERWOLFSGRUBER 1992: F. NEIDERWOLFSGRUBER, Kaiser Maximilians I. Jagd- und Fischereibücher (Innsbruck 1992).

OEHRL 2013: S. OEHRL, Bear hunting and its ideological context (as a background for the interpretation of bear claws and other remains of bears in Germanic graves of the 1st millennium AD). In: O. Grimm/U. Schmölcke (eds.), Hunting in northern Europe until 1500 AD – Old traditions and regional developments, continental sources and continental influences. Papers presented at a workshop organised by the Centre for Baltic and Scandinavian Archaeology. Schleswig, June 16th and 17th 2011. Schriften des Archäologischen Landesmuseums, Ergänzungsreihe 7 (Schleswig 2013) 297–332.

Oppian: Oppian, Colluthus and Tryphiodorus. Transl. A. W. Mair (Harvard 1928).

Rackham 1993: O. Rackham, The History of the Countryside (London 1993).

Tilander 1971: G. Tilander (ed.), Gaston Phébus, Livre de Chasse, Cynegetica XVIII. The Published Studies Series – Journal/Translation (Karlshamn 1971).

Turbervile 1908: G. Turbervile, Turbervile's Book of Hunting, 1576 (Oxford 1908).

Richard Almond
Independent Scholar
richardlalmond@yahoo.co.uk

Fig. 1. Bear Hunting, Roman mosaic, 4th century BC; Museo della Civilita Romana, Rome, Italy (© Bridgeman Images).

Fig. 2. "Game – The Animals you can Hunt", in De Venatione, by Oppian, Greek School, 13th century (© Bridgeman Images).

Fig. 3. "Of the Bear and its Nature", in Livre de chasse, French, late 14th century; Bibliotheque Mazarine, Paris, France (© Bridgeman Images).

Fig. 4. "How to Hunt and Slay Bears", in Livre de chasse, late 14th century; Bibliotheque Mazarine, Paris, France (© Bridgeman Images).

Fig. 5. Mounted hunter spearing a bear. MS Egerton 1146, fol. 11v, German, 1475–1485; British Library (© Bridgeman Images).

Fig. 6. Garlanded hunter on foot spearing a bear. MS Egerton 1146, fol. 20r, German, 1475–1485; British Library (© Bridgeman Images).

Fig. 7. Activities for the month of November. Fresco by Maestro Venceslao, Italian School, 1390–1400; Castello del Buonconsilio, Torre dell'Aquila, Italy (© Bridgeman Images).

Fig. 8. "The Bear Hunt". Drawing by Giovanni Stradano (Jan van der Straet), Flemish, mid-16th century; Gabinetto dei Disegni e delle Stampe degli Uffizi, Florence, Italy (© Bridgeman Images).

Fig. 9. "The Bear Hunters" (detail), from The Triumphal Procession, Triumphs of Maximilian I, *by Hans Burgkmair, German, 16th century (© Bridgeman Images).*

Fig. 10. "The [Bear] Hunt", from Tractatus de Herbis, *by Dioscorides, Italian School, 15th century; Biblioteca Estense, Modena, Emilia-Romagna, Italy (© Bridgeman Images).*

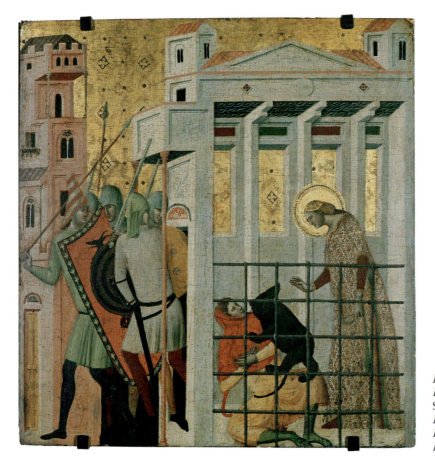

Fig. 11. "St Columba Saved by a Bear", from Episodes of the Life of St Columba, *by Baronzio da Rimini, Italian School, 1340; Pinacoteca di Brera, Milan, Italy (© Bridgeman Images).*

Fig. 12. "Two Bears", chalk study from Codex Vallardi, by Pisanello, Italian School, mid-15th century (photo © RMN-Grand Palais [musée du Louvre] / M. Urtado).

Fig. 13. "Hunt in Honour of Charles V at Torgau Castle" (detail), by Lucas Cranach the Elder, German, 1544; Museo Nacional del Prado, Spain (© Bridgeman Images).

Fig. 14. "Bear and Boar Hunting", in De Venatione, *by Oppian, Greek, 13th century; Biblioteca Nationale Marciana, Venice, Italy (© Bridgeman Images).*

Fig. 15. "Study for the Bear Hunt", for the Alcazar, Madrid, by Peter Paul Rubens, c. 1639; Cleveland Museum of Art, Ohio, USA (© Bridgeman Images).

Fig. 16. "Men in Heavy Armour Attack Bears by Stabbing them with Daggers in the Vitals", plate 26, engraved by Jan Collaert after Jan van der Straet, from Venationes Ferarum, Avium, Piscium, *c. 1600; private collection (© Bridgeman Images).*

Fig. 17. "The Boar and Bear Hunt", from The Devonshire Tapestries, Arras, 1425–1450; Victoria & Albert Museum, London, UK (© Victoria & Albert Museum).

Fig. 18. "The Otter and the Swan Hunt" (detail), from The Devonshire Tapestries, Tournai, 1425–1450; Victoria & Albert Museum, London, UK (© Victoria & Albert Museum).

Fig. 19. "The Bear Hunt", "Chasse à l'ours" tapestry, French School, Arras, France, 1420–1450; Burrell Collection, Glasgow, Scotland (© Bridgeman Images).

Fig. 20. "Collecting Honey", from Tractatus de Herbis, *by Dioscorides, Italian School, 15th century; Biblioteca Estense, Modena, Emilia-Romagna, Italy (© Bridgeman Images).*

Fig. 21. Bear plundering a bee skep, wooden carving, Spanish, undated; Toledo Cathedral, Spain (© Bridgeman Images).

Fig. 22. Bear Hunting, miniature from The Cynegetica, *by Oppian, Venetian, 11th century; Biblioteca Nationale Marciana, Venice, Italy (© Bridgeman Images).*

Fig. 23. Bear caught in a spring trap, engraving after Livre de chasse, *French, late 14th century; Bibliotheque Mazarine, Paris, France (© Bridgeman Images).*

Fig. 24. Bear-baiting, bas-de-page illumination, psalm 88, from the Luttrell Psalter, *English, 1325–1335; British Library, London, UK (© Bridgeman Images).*

Fig. 25. "They're going on a bear hunt", wall painting, English, 1605–1633; Turret Room, Madingley Hall, University of Cambridge UK (© University of Cambridge, UK).

Chapter 4

Animal agency (northern Europe)

HE BECOMES A CREDIT TO HIS MASTER

Martin the dancing bear and its bearward from the late 19th-century children's book "The Life of the bear" (Anonymous [Author of "The life of an elephant"], The life of a bear, his birth, education and adventures. New edition London 1901 [first published 1874]). This book has a strong moral undertone to instruct the young readers to obey their elders. Notably, around that time, also the Teddy-bearisation took place (see O'REGAN, this volume).

species (HARAWAY 2003). This is an encounter between species that would risk "knowing something more about cats and *how to look back*, perhaps even scientifically, biologically, and *therefore* also philosophically and intimately" – a project that is truly posthuman in scope (HARAWAY 2008, 20).

Haraway argues that Derrida failed a simple obligation of companion species because his *looking* did not fully engage with the responsivity of the other. The human part of a companion species has a responsibility to become curious about what the animal "might actually be doing, feeling, thinking, or perhaps making available to him in looking back" (HARAWAY 2008, 20). Such curiosity has the power to destabilise the modernist category of "Man", restricted to (usually) white, wealthy, educated, Western males, and therefore excluding women, the poor, people of colour, and disabled people, as well as nonhuman animals. It is precisely this notion of companionship that posthumanism seeks to explore in an era when humankind is increasingly divorced from the realities of animal suffering. As Bénédicte Boisseron remarks, "domestication has enabled creatures who act like us [...] but who may not think like us or share a similar *Umwelt* [...] to share our daily life", especially because animals cannot speak, and thus "can be thought only in terms of 'what if?' – What if they did judge us?" (BOISSERON 2018, 171). Likewise, much ink has been spilled trying to fathom what really happens in the human psyche, in our biology, and in social transactions, when we engage with animals as beings, as property, or as companions (LEDUFF 2003; HEARNE 2007; DECKHA 2021). But what would happen if the animal looking back were not a cat waiting for her daily feed, but rather, a bear? What would Derrida's response have been if he had looked not into the eyes of *Felis silvestris catus* but of *Ursus arctos*? This is not a wild, foraging bear, or even one found in a modern zoo, but rather the tamed brown bear. For this thought experiment it is necessary to go back in time to the Early Modern period when it would have been normal to see or hear a bear walking through one of the major English towns.

In modern-day Britain, we tend to think of encounters with megafauna such as bears in isolated terms. Bears either haunt the enclosures of zoos or remind us of a bygone era in which bears had considerably more space to roam away from human habitation. Despite having probably become extinct by the early Bronze Age in Britain, bears were a common sight in English towns in the Early Modern period as a consequence of the popularity of bear baiting (O'REGAN 2018; 2020). Baiting entailed setting dogs upon a bear tied to a stake; bears were thus intrinsic to a variety of social and economic interactions as baiting rivalled commercial theatre as a spectator sport. Contemporary accounts record that animals such as bears, bulls, horses, and monkeys were brought into arenas and set upon by dogs, and not always with the aim of killing the animals involved, although a great number of dogs were killed in fights with larger fauna.[2] Much of our evidence for bear baiting in this period comes from travellers' accounts, the rare diaries of bearwards (those who owned and baited bears), and the archaeological record, which provides evidence that bears were exploited from the Medieval to Victorian periods for baiting, skins, or for use as commercial products such as bear's grease. It is likely then, that walking the streets of Early Modern England one may have stumbled not only across a cat or a dog, but also a bear.

The encounters generated by bear baiting in Early Modern contexts force further reconsideration of the nature of relationships between bears and humans in urban, performative contexts. The urban ecologies in which these animals existed and fought were the end points on long routes of animal trade, which are not yet fully understood (DAVIES in print). In London the Bear Garden was a popular arena for the sport and attracted tourists in the same fashion as royal palaces, St Paul's Cathedral, or the menagerie at the Tower of London (SCOTT-WARREN 2003, 70). However, as Przemysław

---

2 See the Box Office Bears website for further details and analysis: https://boxofficebears.com; last accessed 26 September 2022. See also HÖFELE 2011, 1–2, 12. An earlier study of baiting in Early Modern England is provided by DAIGL 1997.

Pożar notes, "the stench of animals and their carcasses can all be seen as aspects of London's sinister or 'disturbing' side, especially when viewed against 'the instability of the category of the human'" (Pożar 2021, 101–102). On the reverse side of this coin, many bears would have been intimately acquainted with the sights and sounds of human habitation in Early Modern England, although some were blind and deprived of their visual senses for baiting. In terms of encounters the blind bears render Derrida's visual model for ethical enquiry with the nonhuman significantly more complex as a methodology for evaluating human/animal relations. The primacy of the visual is a particularly humanist perspective, bolstering the dominance of vision in human culture, which restricts the possibilities of encounter to normative sensory experiences. These encounters force us to think beyond Derrida's shame, especially when violence, disability, and exploitation are integral components of what it means to be posthuman or a companion species.

Companion bears

Using the word "companion" to describe the relationship between humans and bears in Early Modern England is not to suggest that these two species shared a cosy relationship similar to that enjoyed by many humans and domesticated pets today. Rather, the term is a posthumanist one that reveals our increasing understanding of the social, cultural, and biological interactions that inflect human/nonhuman relations, many of which lead to asymmetrical power relations. As Cary Wolfe notes, humans and animals "may share a vulnerability and passivity without limit as fellow living beings, but what they do *not* share equally is the power to materialise their misrecognition of their situation and to reproduce that materialisation in institutions of exploitation and oppression whose effects are far from symmetrical in species terms" (Wolfe 2010, 95). This is true of Early Modern baiting, which provided an arena in which to articulate notions of human superiority over others (Fudge 2000, 8). The decline of the bear's status as a cultural symbol, combined with the reality of their treatment in the arena, meant that the conditions in which they were kept and baited deprived them of agency in fundamental ways.[3]

Although not all Early Modern accounts of bear baiting refer to blind bears, most do convey the brutality of the sport. The mid-16th-century Italian traveller Alessandro Magno described a baiting in vivid detail. In this case the bears are not described as blind, but the account reflects bears' low status and brute strength. Writing about an arena in London, he recounted how over two hundred dogs used for baiting bulls were kept in small kennels separated from each other, while bears and bulls were kept in other houses around a central arena. Magno describes how, on Sundays, everyone watched the dog training, firstly attacks on a horse and monkey in the ring with five or six young dogs, before moving to the more experienced dogs. He offers a positive review of the activities: "In this sport it is *wonderful* to see the horses galloping along, kicking up the ground and champing at the bit, with the monkey holding very tightly to the saddle, and crying out frequently, when he is bitten by the dogs". Then they bring out the bears and bulls: "After they have entertained the audience for a while with this sport, which often results in the death of the horse, they lead him out and bring in bears – sometimes one at a time, sometimes all together. But this sport is *not very pleasant to watch*. At the end, they bring on a fierce bull and tie it with a rope about two paces long to a stake fixed in the middle of the ring. This sport is *the best one to see* and more dangerous for the dogs than the other: many of them are wounded and die. This goes on until evening" (Barron et al. 1983, 143–144; all emphases my own). Like the gladiators or wrestlers of their day, bears and bulls took the brunt force

---

3   For a book-length argument tracing the outline of the bear's decline as a cultural symbol, see Pastoureau 2011.

of this sport and would have fought repeatedly over long periods of time, even as dogs were used as cannon fodder for the spectacle.

Behind the spectacle of baiting, contemporary records suggest that special care was made to ensure that bears were ready to fight again another day. A typical bear used for baiting would have been kept well enough to enable it to fight continuously over long periods of time, reflecting its high socio-economic value. Because of the brutality of baiting, bears would have needed the constant care of their companion bearwards, especially when travelling between cities or counties. A good example of this is one contemporary inventory that records the oil used for applying to and caring for a blind bear, which featured as an outgoing cost in a bearward's "diary" from 1608 (DAVIES 2021). Studies of modern remains of captive bears in Europe and South East Asia have shown that captivity causes dental problems, perhaps due to chewing bars, and there are high levels of osteoarthritis and skeletal pathologies, which may in part be due to age or activity levels (O'REGAN/KITCHENER 2005). Consistent rounds of dogs set in the Early Modern baiting arena put bears under constant stress, especially as the instinctive reaction for bears feeling threatened (by humans at least) is to flee unless they are caring for or protecting offspring (KROFEL 2019, 190). The baiting arena removes the possibility of flight, and capitalises on bears' innate strength, which would have forged the appearance of a bottomless source of able-bodied stamina, in which the bear could choose to fight and survive. Bear baitings thus revealed an undercurrent of radical inequality that is brought into relief by a central tenant of crip theory – that "like compulsory heterosexuality [...] compulsory able-bodiedness functions by covering over, with the appearance of choice, a system in which there is actually no choice" (MCRUER 2006, 8).

If personal accounts or reflections on the nature of the relationship between bears and their bearwards existed, they have been lost to time or are yet to resurface. However, knowing that these companion relationships must have existed forces us to think deeply about the posthuman implications of such partnerships. This relationship was a form of radical inequality, beginning with capture or birth in captivity, transportation (possibly over long distances), and finally the baiting arena (CUYTEN/CONVERY 2019). Elizabeth Baldwin has suggested that one John Seckerston, the innkeeper of the Bear Inn in Nantwich, was both a bearward and a bear-breeder, who had four bears in his stable at the time of the 1583 great fire that destroyed most of the town (BALDWIN 1998, 98). Seckerston, or his associates, travelled widely with bears, perhaps even tracing routes as far as Lancashire, Bristol, and Coventry (BALDWIN 1998, 96). If these accounts provide only a snapshot of a broader practice of travelling with, and baiting bears, the bears of England's fighting pits were kept, cared for, and simultaneously brutally exploited by humans across the country.

Descriptions of bear baiting from Early Modern England clearly demonstrate the gap that exists between humans and nonhuman animals in terms of companion species. This is particularly striking for such a totemic species as the bear. Writers have equated bears with gods, princes, leaders and advisors, warriors, and even sexual icons across the centuries in ways that encourage us to believe that they are like us (PASTOUREAU 2011). This anthropomorphising is an integral component of human conceptions of bears, and thus of companion species. The anthropocentric circuit that guides much modern, Western thinking about animals, or the "mechanism underlying our current means of determining the human-animal distinction", is described by Giorgio Agamben as the "anthropological machine" (CALARCO 2008, 92). According to Agamben this machine, which is the scientific and philosophical discourses that distinguish human from animal, differs over time. He highlights a distinction between the function of this machine in the premodern and modern periods. On one hand, the modern machine is post-Darwinian. It isolates the animal aspect of the human animal to exclude select agents from the category of humanity. According to Agamben, the ape-man, and later in the 20th century the Jew, are humans who are excluded as "not (yet) human" (AGAMBEN 2004, 37). On the other hand, the premodern anthropological machine, from Aristotle to Linnaeus, works in a symmetrical way, but in inverted form: "Rather than animalising certain aspects of the human, animal life is humanised. Human beings who take an essentially animal form are used to mark the constitutive outside of humanity proper – the infant savage, the wolf-man, the werewolf, the slave, or the barbarian" (CALARCO 2008, 93). Both machines have at their centre a zone of indifference, a space of exception or caesura, which allows for the constant rearticulation of human and animal according to context.

Are the bears of Early Modern baitings an example of a modern or premodern anthropological machine? Perhaps they are both, in ways that epitomise the Early Modern period as a transformative moment in how humans and animals have shared companionship. The bear is the premodern animal humanised. This is evident not only in the custom of naming celebrity bears, such as Robin Hood, George Stone etc., which mirrors the act of colonial acquisition – but also in how baiting was set up as a spectacle in arenas not dissimilar to contemporary theatres, in which bears were the key protagonists (DE SOMOGYI 2018). However, we can also trace through Agamben's model the ways that bears signal the operation of the modern anthropological machine. Through Early Modern baiting the humanised species that trod the pages of medieval fables and stories like the *Roman de Renart* begins to be disassembled (for the Reynard cycle see OWEN 1994). Early Modern bearwards and those in the trade rebuilt bears as fighting machines, as the animal aspect of the humanised bear excludes the bear from the human realm – a shoring up of traditional distinctions between human and animal, religion and science, and nature and culture that are the key markers of Early Modern Humanism (LATOUR 1993, 1–16; DESCOLA 2013, 1–88).

The impulse of Humanism bolstered the figure of Man by contrasting him with the animal, a process that in large part mirrored the biblical model of animals being placed into the charge of humans by God's command at the time of Creation (Genesis 1:28). Both the humanist and the biblical models build a hierarchy of life in which humankind holds dominion over the nonhuman – a model exemplified in acts of animal baiting for entertainment. A recent response to the humanist model, termed posthumanism, seeks not only to reconfigure animals as a catch-all category in relation to the human, but also to question the principles that underlie models of human/nonhuman distinction, such as those outlined in Agamben's anthropological machines. Wolfe defines posthumanism as "the necessity for any discourse or critical procedure to take account of the constitutive (*and* constitutively paradoxical) nature of its own distinctions, forms, and procedures". The posthumanist lens is thus distinguished from the "reflection and introspection" associated with the critical subject of Humanism, and must therefore take account of nonhuman animals not only in terms of dominion and rationality, but also of encounter, ethical enquiry, and companion species (WOLFE 2010, 122). We have already seen how such companionships in Early Modern England might be more profitably explored through thinking in posthuman terms. As we shall discover below, this process is highlighted more explicitly in the case of blind bears who, deprived of their sense of vision, nevertheless fought back.

BLIND BEARS

The baiting of blind bears in Early Modern arenas presented a posthuman challenge to the principles of Early Modern Humanism that enabled the deprivation of bears' status and thus their subjugation by the common man. The bears of Early Modern baiting therefore articulated a problem with definitions about where to draw the line between human and nonhuman, and thus how to ascribe agency. Whereas Derrida's thought of shame was instigated by the gaze of his cat – that is through sight –, in the case of Early Modern bear baiting this field of encounter was not always possible. Deprived of vision, blind or blinded bears offered audiences an altogether posthuman spectacle in multiple ways.

Blind bears disrupted the belief that if they look back at us, bears are like us.[4] The unequal relations captured in accounts of bear baiting in which bears are reduced to blind entertainment fodder rupture well-trodden cycles of normative anthropocentrism – if they cannot look back, bears cannot be like us.

The spectacle of baiting blind bears encoded the bear in the patterns of violence instigated by the humans who owned them. In one account of Elizabethan-era bear baiting given by the German lawyer and traveller Paul Hentzner, the bear demonstrates an impressive capacity to fight back even when deprived of his field of vision. Although escape is not an option, the bear's ability to continue to fight his opponents with "force and skill" is worthy of note: "There often follows that of whipping a blinded bear, which is performed by five or six men, standing circularly with whips, which they exercise upon him without any mercy, as he cannot escape from them because of his chain. He defends himself with all his force and skill, throwing down all who come within his reach, and are not active enough to get out of it, and tearing the whips out of their hands, and breaking them" (HENTZNER 1797, 30). Hentzner's travel account sharply contrasts this turbulent scene with a description of the crowd in the arena, which is full of people leisurely smoking tobacco, drinking ale and wine, and eating fruit and nuts. Despite the contrast between the actions of human baiters, human onlookers, and the bear itself, this account nevertheless portrays the baited bear as in charge of his own agency. The bear reacts, throws down its opponents, and even tears the whip out of their hands in an unexpected reversal of baiting roles. But for viewers of the fight, and readers of Hentzner's account, the true fate of the bear lies in the juxtaposition of real peril for the humans who get too close to the bear, and an underlying familiarity with the theatricality of the fight.

The key to understanding the conceptual challenge that blind bears created is the juxtaposition of the commoner with the bear. Both are categories subjugated by the exclusive humanist Man. In a pamphlet called *Worke for Armourers* published in 1609, the writer Thomas Dekker observed with some distaste a similar spectacle of bear baiting with dogs in which a blind bear was whipped. In contrast to Hentzner's travel account, Dekker focuses on the moral dimension of the baiting by using religious metaphor to transform the bear, called Hunkes, into the figure of the sinner. Although the ethical universe in which he watches the spectacle is very different to our own, he nevertheless stresses that the people who torture Hunkes are people like ourselves – colliers, carters, or watermen – workers from the streets of London: "No sooner was I entered [the Bear Garden] but the very noise of the place put me in mind of Hell: the bear (dragged to the stake) shewed like a black rugged soul, that was damned… the dogs like so many devils inflicting torments upon it… At length a blind bear was tied to the stake, and instead of baiting him with dogs, a company of creatures that had the shapes of men, and faces of Christians (being either colliers, carters or watermen) took the office of Beadles upon them, and whipped monsieur Hunkes, till the blood ran down his old shoulders".[5] Dekker's mystifying gaze on his fellow human company registers the bear baiting in quasi-religious terms through a comparison with the devilish dogs. However, it is noteworthy that Dekker's world seems upside-down when, instead of baiting the bear, the crowds take on the role of torturers or executioners. In Agamben's terms, the animal aspect of these human workers is isolated to exclude them from the category of humanity purported by Humanism. Julie Sanders observes that the mastiff dogs in Dekker's description play the role of the devils in a morality play, with the closing vision of Dekker's statement humanising and aging the bear called Hunkes (SANDERS 2014, 59). Hunkes simultaneously embodies the trope of "disability as a metaphor for deviance or moral failing", which marginalises alterity through negation (SENIER 2017, 277). Alongside the hybrid devil-dogs, Dekker positions the Christian men who whip Hunkes as part-men, part-creature – a dehumanisation that

---

4   HÖFELE (2011, 115–170) has shown that images of cruelty produced similar reflections in the work of writers such as Montaigne, Foxe, and Shakespeare.
5   Adapted from DEKKER 1609, sig. B1v– B2r, and quoted from SANDERS 2014, 58.

animalises the crowds of human onlookers. This is a move that resembles Agamben's model for the modern, post-Darwinian anthropological machine that isolates the animal aspect of the human and excludes them from humanity. To whip a blind bear so must imply a certain animalisation of the human, as an implication of an awry morality.

The act of naming bears, which was common practice in Early Modern England, brings into question the extent to which blinding bears could have been a response to increasing humanisation. In contrast to the animalisation of the human crowd, the process of naming the bear Hunkes incorporates a humanisation that resembles Agamben's premodern anthropological machine (in which animal life is humanised). But whipping a blind bear represents a gross act of inhuman cruelty. Likewise, in a contemporary letter from Edward Barrett to Edward Alleyn about animals for the Royal Game, dated 11 June 1610, the author describes how one "Littell Besse of Bromly" fought over twenty double and single "courses" with the best dogs in the country. Some of the dogs she killed outright but "the moste parte shee sent haltting awaie".[6] This humanisation and gendering of a baited bear helps to articulate the instability of the category of the human in accounts of Early Modern baiting. In both accounts, bears are named and thus humanised through a process of mirroring human culture. But it is Dekker's account – the one that humanises Hunkes – that evokes a feeling of pity when comparing the bear at the stake with London's "poor wretches" being led to the whipping posts: "Yet me thought this whipping of the blind bear moved as much pity in my breast towards him, as y(e) leading of poor starved wretches to the whipping posts in London (when they had more need to be relieved with food) ought to move the hearts of citizens, though it be the fashion now to laugh at the punishment".[7] Dekker states that he was moved to feel pity at the sight of the blind bear being whipped. Might he have been moved further to feel shame if, like Derrida, he had been able to meet the gaze of the nonhuman animal?

The types of encounter afforded by bear baiting demonstrate the importance of sight and the gaze for defining what kind of agency is at work. For Derrida, the gaze is a trigger for ethical encounter between him and his cat, but the case of blind bears introduces a dilemma to this visual mode of enquiry into the distinction between human and nonhuman. A blind bear makes what it can of the situation in which it finds itself, yet its agency is prescribed by human baiters. The bear cannot see and perhaps has to rely on its olfactory senses, which are vastly superior to the human sense of smell, but the baiting arena did not allow humans to put their own sense of smell to the test as a form of encounter. There are, of course, many sensory ways of encountering nonhuman animals, but blindness does not allow the bear to look back, as Haraway would like it to do, to present the human onlooker with the posthuman question of what the nonhuman is feeling, thinking, or making available through that encounter. This stretches the notion of companionship to its limit, and reinforces the dark undercurrent of exploitation in the partnership between human and bear, in which the bear's agency, and ability to challenge human command, is deliberately forestalled.

Conclusion

What is posthuman about a bear that lived and fought in the arenas of Early Modern England, a bear so deprived of agency that the easiest comparisons for some observers were with London's "wretches"? Bears were integral to Early Modern theatre economics and animal trade. They were bears that lived long lives and, perhaps often, fought back. They were companion species, who relied on a cross-species

---

6   The Archive of Dulwich College (London), "MSS 2 Dulwich Letters of PH and EA as Masters of Royal Game, 1598–1626", 013. Transcription provided by Callan Davies.
7   Adapted from DEKKER 1609, sig. B2r.

partnership with bearwardens and a different species *umwelt* (perspective) to survive (Von Uexküll 2010, 45; Lewis 2022). The descriptions of bears included in this discussion demonstrate that writers in the Early Modern period did not have fixed conceptions of what it meant to be human or animal. Using Agamben's anthropological machines as tools to examine the distinctions between humans and nonhumans, it quickly becomes clear that bears straddled multiple categories in the minds of observers. They were humanised through naming, animalised alongside lower status humans, and ascribed super-human, but altogether bear-like, characteristics such as strength, stamina, and perhaps courage, which brought into question the more rigid conceptions of able-bodiedness and species difference that defined the humanist agenda.

In a study of surface encounters with the nonhuman, Ron Broglio asks whether the Humanities – a human endeavour traditionally for humans – can be "hospitable enough to give itself over and recognise our lives as entwined with other beings?" (Broglio 2021, 138). Those who work to uncover the records of humans and animals involved in bear baiting and to understand the lives of animals, know that we must recognise the entanglements in which bears were caught up – social, economic, and biological – which were at once exploitative, cruel, and violent. By looking at animals, and allowing them to look back, the Humanities can be hospitable enough to show that our entanglements are deep and entrenched. But even if bears cannot look back, understanding their lives helps to uncover the diverse ways that they were ascribed agency by bearwardens, writers, and onlookers. The Early Modern bears that we study will never benefit from the pursuit of better knowledge about their lives, but perhaps in the process we will strengthen the histories we tell about animal agency and cruelty, entertainment practice, and the social worlds in which bears have wandered.

Acknowledgements

I am grateful to Sophy Charlton, Greger Larson, and Lizzie Wright for their research on the *Box Office Bears* project, and to Callan Davies, Andy Kesson, and Hannah O'Regan for their comments on this piece.

Bibliography

Agamben 2004: G. Agamben, Man and Animal. Transl. K. Attell (Stanford 2004).

Baldwin 1998: E. Baldwin, John Seckerston: The Earl of Derby's Bearward. Medieval English Theatre 20, 1998, 95–103.

Barron et al. 1983: C. Barron/C. Coleman/C. Gobbi (eds.), The London Journal of Alessandro Magno 1562. The London Journal 9(2), 1983, 136–152.

Boisseron 2018: B. Boisseron, Afro-Dog: Blackness and the Animal Question (New York 2018).

Broglio 2021: R. Broglio, Multispecies. In: J. Cohen/S. Foote (eds.), The Cambridge Companion to Environmental Humanities (Cambridge 2021) 128–140.

Calarco 2008: M. Calarco, Zoographies: The Question of the Animal from Heidegger to Derrida (New York 2008).

Cottingham et al. 1985: J. Cottingham/R. Stoothoff/D. Murdoch (eds.), The Philosophical Writings of Descartes. Three volumes (Cambridge 1985).

Cuyten/Convery 2019: K. Cuyten/I. Convery, Bears Behind Bars: Captive Bears Throughout History. In: O. Nevin/I. Convery/P. Davis (eds.), Culture, Nature, Heritage (Woodbridge 2019) 94–105.

Daigl 1997: C. Daigl, 'All the world is but a bear-baiting': Das englische Hetztheater im 16. und 17. Jahrhundert (Berlin 1997).

Davies 2021: C. Davies, Bearward's Diary, 1608. Blog post, 16 December 2021: https://boxofficebears.com/resource/bearwards-diary-1608.

Davies in print: C. Davies, The Place of Bearwards in Early Modern England. The Historical Journal.

Deckha 2021: M. Deckha, Animals as Legal Beings: Contesting Anthropocentric Legal Orders (Toronto 2021).

Dekker 1609: T. Dekker, Worke for Armourers, or The Peace is Broken (London 1609).

Derrida 2002: J. Derrida, The Animal That Therefore I am (More to Follow). Transl. D. Wills. Critical Inquiry 28(2), 2002, 369–418.

Descola 2013: P. Descola, Beyond Nature and Culture. Transl. J. Lloyd (Chicago 2013).

Fudge 2000: E. Fudge, Perceiving Animals: Humans and Beasts in Early Modern English Culture (Basingstoke 2000).

Haraway 2003: D. Haraway, The Companion Species Manifesto: Dogs, People, and Significant Otherness (Chicago 2003).

Haraway 2008: D. Haraway, When Species Meet (Minneapolis 2008).

Hearne 2007: V. Hearne, Adam's Task: Calling Animals By Name (New York 2007).

Hentzner 1797: P. Hentzner, Paul Hentzner's Travels in England during the Reign of Queen Elizabeth (London 1797).

Höfele 2011: A. Höfele, Stage, Stake, and Scaffold: Humans and Animals in Shakespeare's Theatre (Oxford 2011).

Krofel 2019: M. Krofel, Living with Bears in Europe. In: O. T. Nevin/I. Convery/P. Davis (eds.), The Bear: Culture, Nature, Heritage (Woodbridge 2019) 187–198.

Latour 1993: B. Latour, We Have Never Been Modern. Transl. C. Porter (Cambridge 1993).

LeDuff 2003: C. LeDuff, At the Slaughterhouse, Some Things Never Die. In: C. Wolfe (ed.), Zoontologies: The Question of the Animal (Minneapolis 2003).

Lewis 2022: L. Lewis, Animal Umwelt and Sound Milieus in the Middle English Physiologus. Exemplaria 34(1), 2022, 24–39.

McRuer 2006: R. McRuer (ed.), Crip Theory: Cultural Signs of Queerness and Disability (New York 2006).

O'Regan 2018: H. O'Regan, The Presence of the Brown Bear *Ursus arctos* in Holocene Britain: A Review of the Evidence. Mammal Review 48, 2018, 229–244.

O'Regan 2020: H. O'Regan, Menageries and Bearskin Caps: Experiencing North American Bears in Postmedieval Britain. In: G. A. Waselkov/H. A. Lapham (eds.), Bears: Archaeological and Ethnohistorical Perspectives in Native Eastern North America (Gainesville 2020) 256–270.

O'Regan/Kitchener 2005: H. J. O'Regan/A. C. Kitchener, The Effects of Captivity on the Morphology of Captive, Domesticated and Feral Mammals. Mammal Review 35(3/4), 2005, 220–224.

Owen 1994: D. D. R. Owen (ed.), The Romance of Reynard the Fox, transl. from Old French (Oxford 1994).

Pastoureau 2011: M. Pastoureau, The Bear: History of a Fallen King. Transl. G. Holoch (Belknap 2011).

Pożar 2021: P. Pożar, An Unexpected Journey 'from the naves to the chops': *Macbeth,* Animal Trade, and Theatrical Experience'. Multicultural Shakespeare: Translation, Appropriation and Performance 24(39), 2021, 87–104.

Sanders 2014: J. Sanders, The Cambridge Introduction to Early Modern Drama, 1576–1642 (Cambridge 2014).

Scott-Warren 2003: J. Scott-Warren, When Theaters Were Bear-Gardens: Or, What's at Stake in the Comedy of Humors. Shakespeare Quarterly 54(1), 2003, 63–82.

Senier 2017: S. Senier, Blind Indians: Káteri Tekakwí:tha and Joseph Amos's Visions of Indigenous Resurgence. In: S. Jaquette Ray/J. Sibara (eds.), Disability Studies and the Environmental Humanities: Toward an Eco-Crip Theory (Lincoln 2017) 269–289.

De Somogyi 2018: N. De Somogyi: Shakespeare and the Naming of Bears. New Theatre Quarterly 34(3), 2018, 216–234.

Von Uexküll 2010: J. Von Uexküll, A Foray into the Worlds of Animals and Humans: with a Theory of Meaning. Transl. J. D. O'Neill (Minneapolis 2010).

Wolfe 2010: C. Wolfe, What is Posthumanism (Minneapolis 2010).

Dr. Liam Lewis
Research Fellow in Animal Studies for *Box Office Bears*
University of Nottingham
Nottingham
UK
Liam.Lewis1@nottingham.ac.uk

Chapter 5

Bears in long-term archaeo(zoo)logical studies (northern Europe)

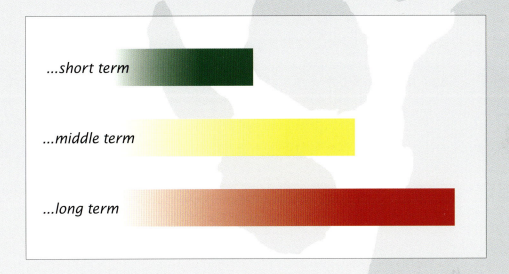

Understanding the changing role of the bear over the long term. Only such considerations will make the longevity of certain traditions, such as bear ceremonialism, visible.

Brown bears in burials and entertainment in later prehistoric to modern Britain (c. 2400 BC – AD 1900s)

By Hannah J. O'Regan

Keywords: Bear baiting, iconography, prehistory, medieval, cremations, graves, Britain, Ursus arctos, brown bear

Abstract: The brown bear, Ursus arctos, was a native British mammal, but is now extinct. This chapter briefly explores the history of the native brown bear before turning to the anthropogenic evidence for bears in prehistoric and later Britain. There are two main foci – bears in burials and bears in entertainment. Bear remains are very rare in Britain, but their phalanges are found in human cremations in the Bronze Age, late Iron Age and early medieval periods. The role of the bear in each period is discussed, concluding that there is unlikely to be a single association between bears and cremations that endured. Rather, bear remains are likely evidence of long-distance trade in high status Bronze Age and Iron Age cremations, and a possible indication of ancestral identity in the early medieval period. Roman and Viking bear iconography is also considered. In the Roman period jet bear figurines are associated with the graves of infants, while in the Viking Age bears are carved on stone "hogback" grave covers. Both may indicate some sort of protection in the afterlife. In contrast bear iconography in the medieval and post-medieval period is often associated with bear-baiting, where dogs were induced to attack tethered bears as a form of public (and royally sanctioned) entertainment. Some of the extensive documentary evidence for this "sport" is discussed, as is the practice of dancing bears, which continued in Britain into the early 20th century.

Bears in Holocene Britain

The brown bear (*Ursus arctos*) has had a long history in Britain, as a native wild mammal, and as a tamed creature imported for human entertainment. The species was present in all mid to late Pleistocene interglacials, and some glacial periods, as well as the Holocene (Turner 2009; Schreve 2019). However, brown bears are unlikely to have been present in the coldest periods, particularly as much of the British Isles was ice-covered. For example, Leonard et al. (2013) modelled brown bear survival during the Last Glacial Maximum (27,000–15,000 BP) in Ireland, and concluded that bears were unlikely to have remained *in situ*, but would have recolonised once the ice sheets had retreated. An ancient DNA study by Edwards et al. (2014) of brown bears from the Yorkshire Dales indicates that the same mitochondrial haplotypes are present before and after the later cold period of the Younger Dryas (12,900–11,700 BP; cf. Rasmussen et al. 2014), which ended at the start of the Holocene. Their results suggest that either the brown bear was able to survive the Younger Dryas in a refugium in southern England or was able to recolonise from elsewhere in Europe very shortly after it had ended.

The evidence for brown bears in the Mesolithic (9600–4000 BC, cf. Table 1) is relatively sparse, with only six localities identified so far, in Berkshire, Yorkshire, and Scotland. One is from a riverbank deposit at Eton Rowing Lake (Allen et al. 2013), and three are from caves (O'Regan 2018). None have clear evidence of anthropogenic influence. The remaining two sites are the well-known lakeside hunter-gatherer camp of Star Carr (Edwards et al. 2014) and a midden in the An Corran rock shelter on the Isle of Skye (Bartosiewicz 2012). There are more bear specimens suggested to date to the Neolithic (4000–2400 BC), but the only secure dates are for bears from archaeological sites (Yalden 1999; O'Regan 2018). One of the key issues with examining bear remains in Britain is that many have been found in cave sites, and caves are particularly prone to bioturbation, water sorting, and other taphonomic processes that can mix up material from different periods. Therefore, it is hard to take at face value the suggested dates for any bears that are not from archaeological sites, unless they have been directly radiocarbon dated. However, even with a radiocarbon date a bear specimen from a cave need not necessarily preclude an anthropogenic influence. For example, a bear canine from Greater Kelco Cave and two terminal phalanges from Sewell's Cave (both Yorkshire) were found with Roman material, but were radiocarbon dated to the Lateglacial (Edwards et al. 2014). The authors suggested that as these were the only remains of bears from these caves they could have been collected as curiosities by the Romans and deposited later (Edwards et al. 2014, 134). The securely dated Neolithic bears, always represented by a single bone or tooth, are largely from pits or ditches, and the sites cluster in East Anglia and southern England (Wiltshire and Dorset; O'Regan 2018). One of the most intriguing specimens in terms of considering human-bear interactions in this period is a single ulna from pit 11a at Down Farm on Cranborne Chase (Legge 1991). This ulna was found in a pit fill with a large number of animal bones, including a complete cattle (*Bos taurus*) cranium, which could suggest that these bones were part of a placed or structural deposit (i.e. it was not simply rubbish disposal, but had some ritual purpose or context to it; Legge 1991). However, along with other bones in the pit, the bear ulna had been gnawed, probably by dogs (*Canis familiaris*), which suggests that the bear bone was not treated with any particular reverence. Although it is only a single example, it makes an interesting contrast with ethnographic records from northern Eurasia and North America, where bear remains were often treated as important and specifically kept away from scavengers (Hallowell 1926).

A key question that is currently unanswered is when brown bears became extinct in Britain. They have been extirpated from a number of other European countries including Denmark, Ireland, Germany, and Switzerland (Pasitschniak-Arts 1993; Klassen/Gregersen, this volume; Schmölcke, this volume), likely through a combination of deliberate hunting and habitat loss. There appears to have been a staggered loss of large carnivores in Britain, with lynx (*Lynx lynx*) becoming extinct in the early medieval period (Hetherington et al. 2006) and wolf (*Canis lupus*) most likely in the medieval to early modern period (Yalden 1999), but the extirpation date of the brown bear is unknown. O'Regan (2018) examined the bear record in Britain and identified two plausible scenarios for extinction. In the first, bears became extinct in the late Neolithic/early Bronze Age (i.e. between 3000–1500 BC), while in the second, they became extinct in the first part of the early medieval period (~AD 410–700). The reason for these different scenarios is the almost total lack of bear remains in the British Bronze Age and Iron Age. The second scenario is supported by a single early medieval radiocarbon date on a bear bone from a cave in North Yorkshire, which could indicate that bears were still present in the wild at this point. However, body part analysis conducted by O'Regan (2018) found that the bones present at Bronze Age (2400–800 BC), Iron Age (800 BC – AD 43) and most early medieval sites were likely to have come from skins, but that living bears appear to have been present in the Roman (AD 43–410), medieval (AD 1066–1485) and post-medieval (AD 1485 to present) periods. This pattern was used to suggest that wild bears were either extinct or at a very low level during the Roman period, and that the bones that have been identified are likely to be from live bears that had been imported. O'Regan's (2018) review of bears in Britain also identified two topics that are worth exploring further – the potential link between bears and human burials, and the role of bears in Roman and later entertainment. The rest of this chapter focusses on these two topics.

Bears and burials

While there is very limited evidence of bears in Britain in the Bronze and Iron Age, all of the anthropogenic examples that are present have been found with human cremations (cf. Fig. 3). Bear bones are also present in human cremations in the Anglo-Saxon period (Bond/Worley 2006; Squires 2011). In contrast, there are no physical remains of bears identified from Roman period burials, but bear figurines have been found in a number of graves (Crummy 2010). I am not aware of bear bones or figurines associated with burials of any period from Scotland or Wales, although it must be noted that owing to soil chemistry the burial record for both these countries is limited, so this absence may not be an accurate reflection of original practices.

The Bronze Age
In the Bronze Age, only one grave has been identified as containing a bear specimen. This item is particularly unusual as it is a preserved pelt (Fig. 1), and is the only non-osseous bear specimen in the British archaeological record. A human cremation was found wrapped within the pelt, and placed in an early Bronze Age cist at Whitehorse Hill, Dartmoor, Devon (Jones 2016). The cremated individual (which could not be sexed) of approximately 15–25 years of age (Mays 2016) was buried with a number of high-status objects including a copper alloy pin, a composite necklace with beads of tin, shale, amber and clay (Sheridan 2016), two pairs of ear studs (labrets) made from spindle wood, as well as a woven basket and a single flint flake (Jones 2016). Dartmoor is an upland landscape, and the cist had been cut into a hummock of peat which effectively preserved the fur, but not the skin, of the animal. The fur was examined and identified as a bear (Family Ursidae), using proteomic analysis of the hair fibres (Solazzo 2016). The pelt was from the rear portion of the bear, and likely to be from one side only (i.e. it was a portion extending from the midline of the back to the belly), and would have measured some 56 x 39 cm when unfolded (Cameron/Mould 2016). Relatively few animal skins have been identified in British Bronze Age barrows, although this apparent absence may be partly due to taphonomic factors, as skins are likely to decay in the absence of very specific burial conditions (e.g. anaerobic, very cold or very dry deposits). Jones (2016, table 21.2) summarised all known Bronze Age barrow burials and cremations that contained textiles, leather/skins, matting and basketry. Eighteen sites contained animal skins or pelts, and of these, 16 were inhumations and two were cremations, indicating that even where conditions appear favourable for skin preservation, skins are less common in cremations than inhumations. Where skins have been identified, they have almost all been from cattle, with the exception of one possible sheepskin, and three sites that have yielded wild taxa – a pine marten (*Martes martes*) and a red fox (*Vulpes vulpes*) from Gristhorpe, Yorkshire, a stoat (*Mustela erminea*) or red fox from Dysgwylfa Fawr, Cardiganshire, Wales, and a possible stoat from Cuninghar, Clackmannanshire, Scotland (Cameron/Mould 2016). However, only the Gristhorpe identifications are confirmed, as the Welsh and Scottish finds were found in the late 19th or early 20th century and have not been subject to modern scientific analysis. All of the above small carnivores can be used for their pelts, but no large carnivores have been identified other than at Whitehorse Hill. Intriguingly, Dysgwylfa Fawr and Whitehorse Hill are the only cremations in the dataset, both contained the skins of non-domestic taxa, and they are of very similar date (Whitehorse Hill dated to 1740–1560 cal. BC; Dysgwylfa Fawr dated to 1760–1410 cal. BC (see Jones 2016 for full details). This could hint at an association between cremation and carnivore pelts in the early-middle

Bronze Age, but with only two sites, and such very different taxa, this cannot be explored further at present.

A key question is whether or not the individuals creating the Whitehorse Hill assemblage would have recognised the pelt as bear, and if that recognition would have had significance for them, rather than it simply being an exotic pelt instead of the more common cattle skin. It must have been clear to the mourners that the pelt was different – the fur was thicker and curlier, but unless bears were known animals, in the absence of any claws or canines to clearly mark the pelt as that of a carnivore, a piece of skin, which when unfolded would only have been 56 cm x 39 cm, may have been valued for its expense, utility and the distance it had travelled, rather than the species it was made from.

There are two other potential Bronze Age bears from burials in the archaeological literature. The first is a very fragmentary canine that was found in the tertiary fill of a ditch surrounding a late Neolithic (2900–2350 cal. BC) long barrow in Eynesbury, Cambridgeshire (ELLIS 2004). Only six other identifiable bones were found in the tertiary fills – one red deer (*Cervus elaphus*) bone and five antler fragments (SYKES 2004). Radiocarbon dating of the bear canine failed, owing to lack of collagen, so the date of this specimen remains unknown (O'REGAN/DEVIÈSE in prep.). We could assume that it is late Neolithic or later, but given the Lateglacial radiocarbon dates on specimens that were thought to be Roman (EDWARDS et al. 2014), it is always possible that it could be from a much older animal. Whatever its date we can be sure that, given the rarity of bears in Holocene Britain, the deposition of this specimen in the barrow ditch was a deliberate act by the people who placed it there. The second potentially Bronze Age example is two reported bear claws from the lower ditch fill of an early Bronze Age cremation and cairn at Sant-y-Nyll, Glamorgan, Wales (SAVORY 1960). The specimens are unfortunately missing from the site archive at the National Museum of Wales, Cardiff, but the photographs included in SAVORY (1960, plate VII) clearly show that they are not bear claws. They look much more like the canines from very young animals (just an enamel cap with little root development or dentine), and it is not possible to confidently identify them from the photographs.

In summary, there is almost no evidence for bears in Britain in the Bronze Age, and only one grave which can demonstrably be said to have a direct link between a bear and a cremation – that of Whitehorse Hill. It is therefore unlikely that bears played any significant role in Bronze Age mortuary rituals.

The Iron Age

There are only two Iron Age sites with bear remains in Britain, and they are both graves – Welwyn Garden City (STEAD 1967) and Baldock (STEAD/RIGBY 1986). They are late Iron Age high status cremation burials, with a considerable number of imported items including amphorae. SEALEY (2009) used the amphorae and other pottery to date the Baldock cremation to 100–75 BC, and Welwyn Garden City to ~25–15 BC (Fig. 2a–b). The amphorae would have come from Italy (FITZPATRICK/TIMBY 2002) and therefore demonstrate considerable trading links for these individuals and their societies. At both sites the bear remains are terminal phalanges that had been burned, and they were found mixed into the human cremation deposit (Fig. 2b); three phalanges at Baldock and six at Welwyn. The Welwyn grave, of an individual identified as "probably male, probably over 25" (POWERS 1967, 40–41), also contained evidence of a carnivore pelt, which had been preserved through association with a decaying bronze platter. Based on the presence of fine medullary hairs, Michael L. Ryder (cited in STEAD 1967) suggested this pelt might be from a stoat. The Baldock cremation, although heavily disturbed prior to formal excavation, appeared to have had only a "token" amount of the cremated individual interred, with a total weight of cremated bone of 10.5 g (STEAD/RIGBY 1986). Despite this, three bear phalanges were included in the "token" deposit, perhaps suggesting that they were of particular importance. No further age or sex information is available for the Baldock individual. There is also a record of a bear phalange from Fishbourne Roman Palace, on the south coast, a site that has produced considerable quantities of continental imports from Iron Age deposits (MANLEY et al.

2005). The proximal bear phalange is from a ditch deposit dated to the Iron Age to Romano-British transition (AD 40–79; Busby 2017). There are three cutmarks towards the distal end of the phalange, which perhaps indicate its removal during skinning, although Kirkinen's (2017) study suggests that only the terminal and part of the intermediate phalanges are likely to be retained in a skin. Busby (2017) suggested that the presence of the phalange could be evidence that a bear was skinned on site, but it may also have come from an imported skin.

Iron Age cremations with bear phalanges are rare in Britain, but Schönfelder (1994) identified a number of other late Iron Age burials with bear toe bones, largely in Germany, with outliers in Britain (n = 2, described above), Luxembourg (n = 1) and the Czech Republic (n = 2). This suggested the two British burials could be grouped with other Germanic burials, through the presence of the bear phalanges. However, a key point to consider is that bears were still extant in much of Germany in the Iron Age (Schmölcke, this volume), making the statement that their presence in graves "display[s] the ability of the dead to hunt and protect the village economy" (Schönfelder 1994, 224), much more plausible for these sites than it is for central England, where bears are likely to have been extinct for at least two thousand years (O'Regan 2018). For the inclusion of bear within these graves to have a protective meaning, people must have been aware of bears and their potential power. As briefly discussed by O'Regan (2018), while there are statuettes of wild boars and wolves from Iron Age East Anglia, there are no recorded statues of bears, and images of bears on coins are very rare and limited to the regions closest to the English Channel. Stead (1967) catalogued seven other sites with similar grave goods to those from the cremation with the bear phalanges from Welwyn Garden City, and listed another eight "Welwyn-type" sites that were found in the 18th and 19th centuries, two of which were confirmed to be burials. This suggests that there are multiple ways of characterising the Welwyn Garden City and Baldock graves – if only the bear phalanges are considered, then they can be grouped with the Germanic graves, as suggested by Schönfelder (1994). However, if the other grave goods such as amphorae and metal firedogs are taken into consideration, they are more similar to graves from England and Gaul rather than Germany (Sealey 2009).

A key point when looking at the prehistoric findings of bears is that very few graves have bear remains, but it is true to say that the three that do have bears are all high status and associated with imported goods. Therefore it seems reasonable to say that high status people in prehistoric Britain did not necessarily have bears in their graves, but the presence of bear remains is an indication of high status.

Early medieval
The early medieval record of bears in burials is rather different. Eleven cremation cemeteries from the 5th–7th centuries AD have been identified in eastern England in a region extending from Yorkshire to Norfolk (Squires 2011). While the human remains from most of these sites have received attention, there has been much less work on the non-human remains. However, there are records of bear bones from four of the eight sites that have been studied – Sancton I, Spong Hill, Elsham Wold (Bond/Worley 2006), and Cleatham (Squires 2011) (Fig. 3). A relationship between bears and the dead is widely seen in Scandinavian practices in the early Medieval period (Lindholm/Ljungvist 2016; Kirkinen 2017), with people being buried with full skins or being cremated with bear skins or isolated claws as part of the funerary rite (inferred from the presence of bear terminal phalanges in cremation deposits; Grimm 2013). The majority of phalanges from the English sites are also terminal (Bond 1996; Bond/Worley 2006; Squires 2011), but proximal and intermediate phalanges are reported from Elsham Wold and Cleatham, along with a possible long-bone diaphysis fragment (Squires 2011; cf. Fig. 4). It seems most likely, given the paucity of other bone elements, that the phalanges from the cemeteries represent bear skins that were cremated along with the humans on the pyre. At Elsham Wold, bear bones were only found in adult cremations (n = 6), one of which was

identified as male, one as female, and four could not be sexed. In terms of grave/pyre goods, the most common items were fragments of bone combs, found in five of the six cremations containing bear. The sixth cremation had no comb fragments but did have glass globules, copper alloy globules, and an ivory fragment (SQUIRES 2011, appendix 1). In contrast, at Cleatham, out of ten cremations with bear bones, no individuals could be sexed, and only eight could be aged (a child [5–12 years], two adolescents [12–18 years] and five adults [19+ years]). In terms of status, the cremated individuals with bear remains at Cleatham had pyre and grave goods ranging from an ivory ring and melted copper alloy (two cremations), gaming counters (two cremations), to no goods at all (four cremations; see SQUIRES 2011, appendix 2 for full details). The differing ages, sexes, and levels of grave goods at these two sites suggests a variety of statuses for the individuals with bear phalanges, and could potentially indicate a family relationship rather than a social rank (SQUIRES 2013). A role for bear skins rather than live bears in early medieval Britain can be supported by the fact that the only other bear bones dated to this period are also from the foot – two or more terminal phalanges from Viking Age York (O'CONNOR 1989), a metapodial from the Anglo-Saxon settlement at West Stow (CRABTREE 1989), and a single phalange from the early medieval monastery at Eynsham Abbey (HARDY et al. 2003; see Fig. 3 for locations).

However, the ethnographic and archaeological record from North America offers an alternative explanation for sites where metapodia and proximal and intermediate phalanges are found – the possibility of bear paw amulets, or of bear paw feasting (see WASELKOV/FUNKHOUSER 2020, and references therein). Given the probable extinction of bears in Britain by this time, feasting is perhaps less likely, but the possibility of paws preserved as amulets could be considered. Indeed BOND/WORLEY (2006) tentatively suggested that the inclusion of bear and other wild animal remains in the cremations could relate to shamanistic practices, or certainly symbolism of the bear, given its human-like appearance (LAPHAM 2020). They also commented on the presence of both bear and fox bones in a single cremation from Spong Hill (no. 2890) and a single cremation at Sancton I (no. MS202, which also contained a horse). Both bear and fox are rare inclusions in the cremations, leading to the suggestion that there may be some additional but unknown significance to their placement together.

Locating the bears
Assuming that, as discussed above, the bear was likely extinct (or nearly so) in Britain; where might the bear remains in burials have come from? The young adult (15–25 years old) in the Bronze Age Whitehorse Hill cremation was interred with a necklace of amber, tin, shale, and clay beads. The amber could have been picked up from beaches on the east coast of England, although it will have originated in the Baltic (SHERIDAN 2016), while the nearest shale deposit is Kimmeridge, 130 km east of Dartmoor. The finished items were probably traded to Dartmoor via Wessex, over 180 km away (JONES 2016). Analysis of the amber beads found that three out of the seven were chipped or worn, indicating that they were not new when deposited in the grave (SHERIDAN 2016). The age of the beads contrasts with the age of the bear skin which was likely to be fairly new when deposited, as the radio-carbon dates were similar to those of the purple moorgrass (*Mollina caerulea*) that had been used as matting at the base of the cist (MARSHALL et al. 2016). It is evident then, that the cremated individual and the people that created the cist burial were linked into a wide trading network, which spread well into central England and beyond. By the late Iron Age, the artefacts found in the graves of the individuals at Baldock and Welwyn Garden City demonstrate that they had links that stretched as far as the Mediterranean.

Recent overviews on the genetics and biogeography of brown bears in Holocene Europe demonstrate that brown bears were already scarce in the regions closest to southern Britain by the middle Bronze Age (CREES et al. 2016; ALBRECHT et al. 2017; ERSMARK et al. 2019). Certainly by the late Iron Age it is likely that any bear specimens would have had to have been traded over considerable distances. It is also important to note that the published distributions of European bears are models based on a database of archaeological sites (e.g. Sommer/Benecke 2005; Crees et al. 2016; Albrecht et al. 2017). Given the potential for trade in bears and bear parts, linking their distribution to anthropogenic sites may overestimate where wild bears could be found, especially in the case of islands such as Britain (O'Regan 2018). For example, the study of Albrecht et al. (2017) on the European Holocene found that bear remains were most likely to be identified in castles and burial sites. For burials, the many Scandinavian graves that contained bears in the Roman Iron Age and medieval periods may have influenced this result (e.g. Grimm 2013; Lindholm/Ljungkvist 2016; Kirkinen 2017; see different contributions, this volume), while the data from castles may indicate animals that had been killed by the occupants (see Oehrl 2013 for a discussion of aristocratic bear hunting), or skins that had been traded from elsewhere. A modern comparator would be the presence of stuffed bears and bear skins in many British aristocratic residences in the 19th and 20th century. These bears did not live in the wild in Britain, but live animals were kept in menageries and there was a huge taxidermy trade in skins and hunting trophies (O'Regan 2020).

Where the early medieval bears were traded from is a very interesting question, and will likely only be answered with ancient DNA analysis. Viking Age Coppergate in York had wide trade networks, so the origin of the bear remains cannot be determined (O'Connor 1989), while the phalanges from the cremation cemeteries in Yorkshire, Lincolnshire and Norfolk demonstrate clear similarities between the populations. Squires (2016) performed a comparison between English and German early medieval/migration period cremation cemeteries. The earlier Roman period cremation cemeteries in Germany had a very high number of cremations with animal remains (74 %), but showed a marked reduction in the migration period to only 7 %, while the early medieval cremations from England were intermediate with 29 % (Squires 2016, 125). This pattern suggested that there are links between the cremation rites of the two countries and that these customs were probably imported into England during the early migration period and continued after they had declined in Germany (Squires 2016).

Bear material culture and graves

In contrast to the Iron Age and early medieval records of bear phalanges in cremations, there are no bear bones from Romano-British burials. However there is still an association between bears and burials, as jet bear figurines (Fig. 5) have been found in five infant graves (or probable graves) and in a probable votive deposit in eastern England, and at two sites in Germany (Trier and Cologne). The bear figurines are associated with multiple other grave goods in each case, such as jet or glass beads, lunulae, coins and other items, which together indicate an intention to protect the infants (Crummy 2010). The jet bears are small (<4 cm long) and are often pierced between the legs, or in one case through the shoulder hump, to form a bead or pendant and were probably amulets. Crummy (2010) has comprehensively described each grave and figurine, and discussed their potential importance as a symbol of protection for infants in the underworld. She notes that the richness of two graves from Colchester indicates that the families of those individuals were wealthy, but that examples of damaged or worn bears (from Malton and York) may indicate the importance of them as amulets rather than being indicative of wealth (Crummy 2010, 78). As well as the overall iconography of the bear, Parker (2018) has noted a potential link between the bears and magic in the Roman period, and particularly the carvings on the backs of three of the figurines, which appear to form a star shape (Fig. 5). These markings have previously been identified as renderings of fur, but Parker (2018) suggests that they could be a genuine attempt to link the bears with the constellations of Ursa major and Ursa minor, or the great and the little bear. This, and the ability of jet to form static electricity could indicate a magical element, especially given the heavily abraded or rubbed appearance of the bears from Malton and St Stephen's, Fishergate, York (Parker 2018). A second aspect that may have been overlooked regarding the bears is that where there is information on the burials that were found, two of the three infants had been cremated (Crummy 2010). This is interesting for two reasons – firstly, as they are late Roman burials, inhumation was becoming the dominant rite, and secondly, as discussed above, there is a link between bears and cremations in both the late Iron Age and the early Anglo-Saxon period, in the same region of England that the bear figurines have been found (cf. Fig. 3). A key difference is that the bear figurines were clearly not pyre goods, while the Iron Age and Anglo-Saxon bear bones are burned.

Small copper-alloy bear figurines are also rare finds from the Roman period in Britain, Switzerland and Germany. The best-known example from near Berne shows a seated figure interpreted as the goddess Artio holding a bowl of fruit and facing a bear with a raised paw (Schmölcke et al. 2017). An isolated figurine from Lorch shows a bear with a raised right paw, which could also have formed part of an Artio group, but at present the interpretation is uncertain (Bollacher 2015). In Cologne, a figurine of a bear sitting on its hind legs was found in a grave in Luxemburger Strasse in the early 1900s (Ritter 1994), and a bear figurine with a human figure in its mouth was found by a metal detectorist at Longstanton in Cambridgeshire, England (Fig. 6; Evans/Mackay 2004). The human figure has been interpreted as a child (although it could also represent a doll), and it was tentatively suggested to be a "funerary beast" (Evans/Mackay 2004). The base of the figurine is unfinished and, like the Lorch item, it may have originally been attached to furniture or another object. Although there is an association between a human and a bear in both the Artio and Longstanton figures, the specific pose of the British example is very unusual. The positioning of a human figure in the mouth of a seated carnivore is also seen in the "wolf-god" figure from Woodeaton in Oxfordshire (Durham 2014), but in this case only the legs and feet of the figure protrude from the animal's mouth. Durham (2014, 209) states that such androphagous figurines are most often found in Gaul, and "apart from wolves or dogs can be lions, Cerberus, sphinxes, griffins or wild boars". The purpose of such juxtapositions is not known, but could be as different as the human being devoured or protected by the larger figure.

Viking Age hogbacks are the only items to link bears and graves in the later early medieval period in Britain. They are large carved recumbent stones, usually 1.5 m long and mainly found in northern England and southern Scotland (Williams 2016). They are thought to have been grave covers or markers, but none have been found *in situ* to confirm this (Williams 2016). There are a great variety of types (Cramp 1991), including those with "end-beasts" that are apparently grasping the main body of the stone. Some of these end-beasts appear to be bears, and three excellent examples are found in Brompton in Yorkshire, one of which is shown in Figure 7. The presence of bears on the hogbacks does not necessarily indicate that bears were present in the region, rather they are likely to have had some protective purpose, with Williams (2016, 505) suggesting that the dead were "protected by the static gazes and enwrapping bodies of attendant ursine or dragonesque beasts". The positioning of the bears is also intriguing, as some appear to be hugging or holding on to the stone, in a manner that is very reminiscent of a bear cub suckling its mother. This could evoke ideas of protection, or perhaps of fertility. An additional feature of these hogbacks is that many of the bears appear to be muzzled, and this contrasts with the bears from the Roman period which are often shown in their wild form, unencumbered by harnesses (e.g. Fig. 5). Such a change in iconography may indicate a change in attitudes to bears, from creatures of the wild to those that were harnessed for human purposes, as is increasingly seen in the medieval and later periods (see below). A relationship between bears and the dead is widely seen in Scandinavian practices (Grimm 2013; Lindholm/Ljungvist 2016; Kirkinen 2017), and in the cremation cemeteries of eastern England (see above for discussion). Therefore, it is tempting to speculate that the placement of a carved bear with teeth and paws around the burial stone of a high-status dead person or family may (amongst other things) have been rendering in stone a practice that occurred "in the flesh" in places where bear skins were more readily available.

Bears in entertainment

Roman period
In the Roman period, there is clear evidence of exploitation of bears for entertainment on the continent, but relatively little information for Britain. Both historical records and mosaics attest to bears and other large carnivores in the Roman arenas in Europe and North Africa (Toynbee 1973; MacKinnon 2006). These "entertainments" included the *Damnatio ad bestias* (death by beasts), when people were executed in the arena by large predators, and also bears fighting against specially trained gladiators or large animals such as bulls or lions (Auguet 1972). Intriguingly, there is one mosaic from Rades in Tunisia that gives the names of the bears (Coleman 2012), which might suggest that they had some longevity, at least in this arena.

Evidence for bears, or indeed any exotic animals, being used in entertainment in Roman Britain is limited. The only direct evidence comes from the amphitheatre in London, and although the bear bone was missing when the final publication was completed (Bateman et al. 2008), Dr Jane Sidell has confirmed that she identified it as a bear humerus, with a knifemark (possibly from skinning) and stated it was found at the edge of the arena (J. Sidell, pers. comm. Jan. 11, 2022). To my knowledge, this is the only evidence of an exotic animal from a Roman arena in Britain. At least five other Roman sites from the 3rd–4th centuries AD also have bear limb bones (e.g. femora, tibiae), which have been interpreted as indicating the presence of complete (and therefore live) animals (see O'Regan 2018, appendix 2, for full details). These sites are Courage Brewery and Tabard Square in London (Reilly 2008), Binchester (Jessop 2012), Fullerton Roman Villa (Hammon 2010), and Catterick Bridge (Meddens 1990). Live bears may also have been represented at Balkerne Lane and Butt Road in Colchester, but body part data is not given in the publication (Luff 1993). An obvious interpretation for the presence of these live animals in human settlements is entertainment, which might be supported by the "Colchester Vase" found with a human cremation, pottery jug, and a Samian dish at West Lodge, Colchester (Thompson Watkin 1877). This vase, dating to the late 2nd century AD, shows a bear in gladiatorial combat with a man wielding a whip (Fig. 8). Unlike imported Samian bowls that depict bears, this vase was made in the Colchester area (Toynbee 1963), and may therefore represent a scene that the potters and customers were familiar with. It was also inscribed after manufacture with the names of the gladiators, but the bear is not named (Thompson Watkin 1877).

Medieval and early modern periods
The early medieval (incl. Anglo-Saxon and Viking) records of bears in Britain are discussed above and appear to relate to skins and a probable trade in bear products rather than live animals. However, there is a distinct change in the record of bears from the start of the medieval period (AD 1066–1485). Out of ten sites with bear remains from medieval England (O'Regan 2018, appendix 2) three have yielded body parts of bears that are not likely to have been kept as souvenirs; a humerus from Barnard Castle (Austin 2007) and another from Gaol Street, Hereford (Hamilton Dyer 2007), and a scapula from Seal House, London (Museum of London Archaeology [MOLA], database). This again suggests live bears were present, which is supported by iconographic and documentary evidence, such as the Bayeux Tapestry (as noted by Oehrl, this volume), which shows a bear tethered to a tree and a man with a sword looking as if he is about to strike it. FitzStephen in 1174 said that in London "in the winter holidays […] bears of a large bulk are baited with dogs" (Pegge 1772). This statement clearly describes bear-baiting, a very popular entertainment of the time, where dogs were set on tethered bears, and it is likely that there was betting on the outcomes. Further medieval evidence comes from John of Gaunt's Record Book, where he is recorded as arranging for payments for baiting in Newcastle-under-Lyme in Staffordshire, in 1372 (The National Archives: DL 42/13, cited in Somerset 2017). Relatively little is known about medieval baiting, other than occasional mentions

in manuscripts and illustrations or carvings. The latter examples include a famous image from the Luttrell Psalter (AD 1325–1340). The bear is shown with a chain and muzzle, but also rather oddly with a stripey wrap or blanket around its hindlegs (Luttrell Psalter, bear baiting image f.161r.). This is unquestionably an image of baiting, as the dogs are being set on the bear, while the audience and the bearward (i.e. bear keeper) look on. A much less well-known medieval image of baiting was brought to my attention by Dr Mark Hall. This undated misericord (a ledge for perching on during church services) in the Burrell Collection in Glasgow, Scotland, shows a muzzled bear fighting a dog (Fig. 9). Stowell Phillips (2008, 159) found that in a sample of 696 misericords showing terrestrial animals, only 2 % depicted bears (n = 14), 9 % showed monkeys, and 18 % showed lions. This study also found a muzzled bear on the misericords at Durham Castle, and a bear included in a stained glass window in York Minster. The presence of bear iconography in churches is particularly intriguing, as Pastoureau (2011) suggested that the Christian church deliberately tried to suppress the role of bears as they had formed part of earlier, pagan religions. Whether or not this is the case, it is clear that bears are rarely present in medieval iconography and material culture (although they are included in heraldry, such as the famous "bear and ragged staff" emblem of the Earl of Leicester [Fig. 10]). However, archival and archaeological evidence demonstrates that live bears were present in medieval towns and cities.

It was not until the early modern period that bear-baiting reached its peak, with the building of specialised arenas, mainly in London at Bankside (Bowsher 2012), but also in rural areas, such as Little Budworth, Cheshire (Hindle 1995). The post of "Master, Guyder and Ruler of all our Bears" was established by Richard III in 1484 and continued as a royal appointment (with various name changes) until at least the 1620s (Cerasano 1991). By the late Tudor period this was a highly sought-after post, with courtiers and entrepreneurs vying for it. Evidence for this comes from the archives at Dulwich College, London, which contains several letters from the theatre impresario Philip Henslowe, attempting to use his influence to gain the job when the previous incumbent was sick (Greg 1907; Cerasano 1991). The Master of the Bears licenced bearwards in England after the Vagabond Act of Elizabeth I in 1572. This Act meant that like players (i.e. actors), bearwards now had to have a patron, such as a member of the aristocracy, or else they could be prosecuted for vagrancy. Several lords had both playing troupes and bearwards, for example in 1565/66 Lord Strange's players were performing in Lincoln and Cambridge, while Lord Strange's bearward was in Newcastle (Records of Early English Drama: REED 2020a–c). Such widespread patronage, and the determination people showed to gain the role of "Master" suggests that there was prestige, money, or both, involved in bear-baiting in England at this time. Multiple monarchs enjoyed baiting, and were able to command a performance by the Masters of the Bears at whichever palace or castle they happened to be occupying. King James I (reigned 1603–1625) went so far as to remodel the Lion Tower at the Tower of London to allow a better view of the animals being baited below (O'Regan et al. 2006).

There is also evidence for polar bears (*Ursus maritimus*) in London in both the medieval and early modern periods. For example, in 1251 a polar bear was gifted to the English monarch by the King of Norway (Grigson 2016), and in 1609 two polar bear cubs were captured on Cherry Island (reported as south of Greenland, but "Greenland" probably means Svalbard in this case, cf. Grigson 2016) and presented to James I (Ravelhofer 2002). The two cubs were sent to Bankside to be kept by the Master of the Bears (Ravelhofer 2002). There is a possibility that one of these polar bears could be the same "white bear" that was baited during the visit of the Spanish Ambassador in 1623, as John Chamberlain reported that they "then turned a white beare into the Thames where the dogs baited him swimming, which was the best sport of all" (Chamberlain Letters, cited in Ravelhofer 2002). While baiting was most formalised in London, bears and baitings were to be seen all over England, with some towns becoming particularly well known for their bear-related activities. Congleton, in Cheshire, for example, is remembered in a rhyme "Congleton rare, Congleton rare, sold the Bible to pay for the bear" which alleges that the town spent the money intended for a new bible on a bear as theirs had died just before a major public event (BALDWIN 1995). An archival record from 1613/14, again from Congleton, shows that overnight couriers were sent out to find a bear for the "Wakes" (town carnivals or holidays) after the intended animal failed to show up (BALDWIN 1995, 264). Bearwards may have travelled with single animals or groups, as records in both Cheshire and Somerset refer to multiple "beares" (BALDWIN 1995; STOKES 1996). In fact documents relating to a great fire in the town of Nantwich, Cheshire, in 1583, report that "John Seckerson who having in his stable iiijor [four] great beyres of his dyd lose theyme out in the beginning to the stretes wheroff the women were soe affrayed. They durst not carrye water". That they were "great" bears rather than cubs indicates that he owned at least four adult animals, and BALDWIN (1998, 99) suggests that he was involved in bear-baiting.

By the mid-18th century many public activities such as bear dancing and bear baiting were popular enough to be made into stoneware ornaments and jugs at potteries in Nottinghamshire and Staffordshire (BEDDOE 2015, 214–215; HALFPENNY/BEDDOE 1990; cf. Fig. 11). These items were sold to the growing middle classes, and were taking off in popularity (c. 1750s) as bear baiting is thought to have been in decline (however, I know of no study of abolition that has actually attempted to chart this). Parliamentary activity attempted to get baiting banned for a number of years in the early 1800s, finally being successful in the Cruelty to Animals Act of 1835 (TÜNAYDIN 2013). Despite this, it appears the British introduced bear-baiting into parts of the British Empire, particularly Pakistan, as a method of disrupting local hierarchies (FAKHAR-I-ABBAS 2015; KAVESH 2018; 2019).

In addition to baiting, bear-dancing was also popular. These animals must have been imported to Britain, but who travelled with them is currently unknown. TÜNAYDIN (2013) provides a fascinating insight into bear dancing, demonstrating that bear training and performing was associated with gypsies throughout Europe in the late medieval through to modern periods. A second source of dancing bear trainers (*orsanti*) was northern Italy, as detailed by SERRA (2013). The *orsanti* would leave Italy with their performing animals (which could also include other taxa such as horses, camels, monkeys or dogs) and travel across Europe during the summer months, often returning to Italy for the winter (SERRA 2013). Bears were taught multiple tricks such as pushing prams and saluting, as well as dancing, sometimes in pairs (TÜNAYDIN 2013). One intriguing early modern record of bear dancing in England comes from 1528/29 when 20 pence was paid to the servant of the Duke of Suffolk with "the dancing bear and the dancing wife" (STOKES 1996). Sadly it is not clear from the record whether they danced together or separately. A children's book from the late 19th century gives an overview of the life of a bear from its birth in the mountains, through a career as a dancing bear, to its eventual exhibition and death at the Zoo in the Jardin des Plantes in Paris (ANON. 1901). While clearly fiction, the text and illustrations give some impression of the life of the animal, and the sort of tricks dancing bears might have performed (Fig. 12).

Bear dancing continued in England into the early 20th century, with specialist bear-leaders travelling over from Italy during the summer months (SERRA 2013), finally being outlawed in 1911 (TÜNAYDIN 2013). Bears were also very popular in menageries and later in Zoological Gardens, which exhibited multiple species including brown, black and polar bears (O'REGAN 2020). The presence of bears in zoos and in the streets may have helped lead to the widespread take-up of bears as children's toys (Fig. 13), with teddy bears remaining very popular to this day.

CONCLUSIONS

Brown bears have had a varied history in England, from native mammal to source of entertainment. This paper has focussed on the presence of bears and bear iconography in archaeological sites from

Hallowell 1926: A. I. Hallowell, Bear ceremonialism in the Northern Hemisphere. American Anthropologist 28, 1926, 1–175.

Hamilton-Dyer 2007: S. Hamilton-Dyer, Gaol Street/Bath Street, Hereford. Unpubl. faunal report (2007).

Hammon 2010: A. Hammon, The brown bear. In: T. O'Connor/N. Sykes (eds.), Extinctions and Invasions: a Social History of British Fauna (Oxford 2010) 95–103.

Hardy et al. 2003: A. Hardy/A. Dodd/G. D. Keevill, Aelfric's Abbey: Excavations at Eynsham Abbey, Oxfordshire, 1989–1992. Thames Valley Landscapes Monograph 16 (Lancaster 2003).

Hetherington et al. 2006: D. A. Hetherington/T. C. Lord/J. M. Jacobi, New evidence for the occurrence of Eurasian lynx (*Lynx lynx*) in medieval Britain. Journal of Quaternary Science 21, 2006, 3-8.

Hindle 1995: S. Hindle, Custom, festival and protest in Early Modern England: the Little Budworth Wakes, St Peter's Day 1596. Rural History 6, 1995, 155–178.

Hunter/Ralston 2009: J. Hunter/I. Ralston, The Archaeology of Britain: an Introduction From Earliest Times to the Twenty-First Century (London 2009).

Jessop 2012: L. Jessop, Extinct Mammals from the Holocene. Mammals of the North East website. Natural History society of Northumbria (2012). http://www.nhsn.ncl.ac.uk/interests/mammals/mammals-north-east/9213-2/.

Jones 2016: A. M. Jones (ed.), Preserved in the Peat: An extraordinary Bronze Age burial on Whitehorse Hill, Dartmoor, and its wider context (Oxford 2016).

Kavesh 2018: M. A. Kavesh, From colony to post-colony: animal baiting and religious festivals in South Punjab, Pakistan. In: D. W. Kim (ed.), Colonial Transformation and Asian Religions in Modern History (Newcastle upon Tyne 2018) 10–29.

Kavesh 2019: M. A. Kavesh, Dog fighting: performing masculinity in rural South Punjab, Pakistan. Society and Animals 2019, 1–19.

Kirkinen 2017: T. Kirkinen, "Burning pelts" – brown bear skins in the Iron Age and early Medieval (1–1300 AD) burials in South-eastern Fennoscandia. Estonian Journal of Archaeology 21, 2017, 3–29.

Lapham 2020: H. A. Lapham, Introduction. In: H. Lapham/G. Waselkov (eds.), Bears: Archaeological and Ethnohistorical Perspectives in Native Eastern North America (Gainsville 2020) 1–15.

Legge 1991: A. J. Legge, The animal remains from six sites at Down Farm, Woodcutts. In: J. Barrett/R. Bradley/M. Hall (eds.), Papers on the Prehistoric Archaeology of Cranborne Chase. Oxbow Monograph 11 (Oxford 1991) 54–100.

Leonard et al. 2013: S. A. Leonard/ C. L. Risley/S. T. Turvey, Could brown bears (*Ursus arctos*) have survived in Ireland during the Last Glacial Maximum? Biology Letters 9, 2013. https://doi.org/10.1098/rsbl.2013.0281.

Lindholm/Ljungkvist 2016: K-J. Lindholm/J. Ljungkvist, The bear in the grave: Exploitation of top predator and herbivore resources in first millennium Sweden – first trends from a long-term research project. European Journal of Archaeology 19, 2016, 3–27.

Luff 1993: R. Luff, Animal Bones from Excavations in Colchester 1971–1985. Colchester Archaeological Report 12 (Colchester 1993).

Luttrell Psalter: Luttrell Psalter (1325–1340). The British Library MS42130. http://www.bl.uk/manuscripts/FullDisplay.aspx?index=1&ref=Add_MS_42130.

MacKinnon 2006: M. MacKinnon, Supplying exotic animals for the Roman amphitheatre games: new reconstructions combining archaeological, ancient textual, historical and ethnographic data. Mouseion Series III.6, 2006, 137–161.

Manley et al. 2005: J. Manley/D. Rudkin/N. Sykes/M. Lyne/G. Dannell/R. Scaife/L. Somerville/L. Barber/D. Allen/D. Williams/R. Pelling/S. Clegg, A pre-A.D. 42 ditch at Fishbourne Roman Palace, Chichester. Britannia 36, 2005, 55–99.

Marshall et al. 2016: P. Marshall/C. Bronk Ramsey/N. Russell/F. Brock/P. Reimer, Interpreting the chronology of the cist. In: A. M. Jones (ed.), Preserved in the Peat: An extraordinary Bronze Age burial on Whitehorse Hill, Dartmoor, and its wider context (Oxford 2016) 184–194.

Mays 2016: S. Mays, The human remains. In: A. M. Jones (ed.), Preserved in the Peat: An extraordinary Bronze Age burial on Whitehorse Hill, Dartmoor, and its wider context (Oxford 2016) 42–43.

Meddens 1990: B. Meddens, Animal bones from Catterick Bridge (CEU 240), a Roman town (North Yorkshire) excavated in 1983. Ancient Monuments Laboratory Report 31 (London 1990).

O'Connor 1989: T. P. O'Connor, Bones from Anglo-Scandinavian levels at 16-22 Coppergate. Archaeology of York, The Animal Bones 15.3 (York 1989).

Oehrl 2013: S. Oehrl, Bear hunting and its ideological context (as a background for the interpretation of bear claws and other remains of bears in Germanic graves of the 1st millennium AD. In: O. Grimm/U. Schmölke (eds.), Hunting in northern Europe until 1500 AD – Old traditions and regional developments, continental sources and continental influences. Papers presented at a workshop organised by the Centre for Baltic and Scandinavian Archaeology, Schleswig, June 16th and 17th, 2011. Schriften des Archäologischen Landesmuseums, Ergänzungsreihe 7 (Neumünster 2013) 297–332.

O'Regan 2018: H. J. O'Regan, The presence of the brown bear *Ursus arctos* in Holocene Britain: a review of the evidence. Mammal Review 48, 2018, 229–244.

O'Regan 2020: H. J. O'Regan, Menageries and bear-skin caps: experiencing North American bears in post-Medieval Britain. In: H. Lapham/G. Waselkov (eds.), Bears: Archaeological and Ethnohistorical Perspectives in Native Eastern North America (Gainsville 2020) 256–270.

O'Regan/Devièse in prep: H. J. O'Regan/T. Devièse, The decline of the bear: Tracing the extinction of Britain's largest carnivore.

O'Regan et al. 2006: H. J. O'Regan/A. Turner/R. Sabin, Medieval big cat remains from the Royal Menagerie at the Tower of London. International Journal of Osteoarchaeology 16, 2006, 385–394.

Parker 2018: A. Parker, Jet bears and magical stars (blog post). https://romanmagic.wordpress.com/2018/05/17/jet-bears-and-magical-stars/.

Pasitschniak-Arts 1993: M. Pasitschniak-Arts, *Ursus arctos*. Mammalian Species 439, 1993, 1–10.

Pastoureau 2011: M. Pastoureau, The Bear: History of a Fallen King (Cambridge [Mass.] 2011).

Pegge 1772: S. Pegge, Fitz-Stephen's description of the city of London, newly translated from the Latin original; with a necessary commentary &c. B. White, at Horace's Head, in Fleet-Street, London (London 1772).

Powers 1967: R. Powers, Report on the cremation. In: I. M. Stead, A La Tène III burial at Welwyn Garden City. Archaeologia 101, 1967, 1–62.

Rasmussen et al. 2014: S. O. Rasmussen/M. Bigler/S. P. Blockley/T. Blunier/S. L. Buchardt/H. B. Clausen/I. Cvijanovic/D. Dahl-Jensen/S. J. Johnsen/H. Fischer/V. Gkinis/M. Guillevic/W. Z. Hoek/J. J. Lowe/J. B. Pedro/T. Popp/I. K. Seierstad/J. P. Steffensen/A. M. Svensson/P. Vallelonga/B. M. Vinther/M. J. C. Walker/J. J. Wheatley/M. Winstrup, A stratigraphic framework for abrupt climatic changes during the Last Glacial period based on three synchronized Greenland ice-core records: refining and extending the INTIMATE event stratigraphy. Quaternary Science Review 106, 2014, 14–28.

Ravelhofer 2002: B. Ravelhofer, "Beasts of Recreacion": Henslowe's White Bears. English Literary Renaissance 32, 2002, 287–323.

REED 2020a: REED, Patrons and Performances index (2020). Lord Strange's Bearward, 25 Feb. 1565 – 3 Mar. 1565, Newcastle upon Tyne. https://reed.library.utoronto.ca/node/291112.

REED 2020b: REED, Patrons and Performances index (2020). Lord Strange's Players, 16 Sep. 1565 – 15 Sep. 1566, Lincolnshire. https://reed.library.utoronto.ca/node/293122.

REED 2020c: REED, Patrons and Performances index (2020). Lord Strange's Players, 29 Sep. 1565 – 28 Sep. 1566, Cambridgeshire. https://reed.library.utoronto.ca/node/289580.

Reilly 2008: K. Reilly, The Drapers Gardens bear. London Archaeologist, Spring 2008, 138.

Ritter 1994: S. Ritter, Die antiken Bronzen im Römisch-Germanischen Museum, Köln: Die Statuetten aus Köln. Kölner Jahrbuch 27, 1994, 317–403.

Savory 1960: H. N. Savory, The excavation of a Bronze Age Cairn at Sant-y-Nyll, St. Brides-Super-Ely (Glam.). Reports and Transactions of the Cardiff Naturalists' Society 89, 1960, 9–30.

Schmölcke et al. 2017: U. Schmölcke/D. Gross/E. A. Nikulina, Bears and beavers – 'The Browns' in daily life and spiritual world. In: B. V. Eriksen/A. Abegg-Wigg/R. Bleile/U. Ickerodt (eds.), Interaction without borders. Exemplary archaeological research at the beginning of the 21st century (Schleswig 2017) 901–916.

Schönfelder 1994: M. Schönfelder, Bear-claws in Germanic graves. Oxford Journal of Archaeology 13, 1994, 217–227.

Schreve 2019: D. Schreve, All is flux: the predictive power of fluctuating Quaternary mammalian faunal-climate scenarios. Philosophical Transactions of the Royal Society of London B 374, 2019. https://doi.org/10.1098/rstb.2019.0213.

Sealey 2009: P. R. Sealey, New light on the wine trade with Julio-Claudian Britain. Britannia 40, 2009, 1–40.

Serra 2013: I. Serra, On men and bears: a forgotten migration in Nineteenth-Century Italy. History Workshop Journal 76, 2013, 57–84.

Sheridan 2016: A. Sheridan, The composite necklace. In: A. M. Jones (ed.), Preserved in the Peat: An extraordinary Bronze Age burial on Whitehorse Hill, Dartmoor, and its wider context (Oxford 2016) 88–116.

Solazzo 2016: C. Solazzo, Appendix C: Report on the proteomic analysis of hairs from the basketry container, the braided band and the pelt from Whitehorse Hill Cist. In: A. M. Jones (ed.), Preserved in the Peat: An extraordinary Bronze Age burial on Whitehorse Hill, Dartmoor, and its wider context (Oxford 2016) 274–287.

Somerset 2017: J. A. B. Somerset, Staffordshire, Records of Early English Drama (REED) online. https://ereed.library.utoronto.ca/collections/staff/.

Sommer/Benecke 2005: R. S. Sommer/N. Benecke, The recolonization of Europe by brown bears *Ursus arctos* Linneaus, 1758 after the Last Glacial Maximum. Mammal Review 35, 2005, 156–164.

Squires 2011: K. E. Squires, An Osteological Analysis and Social Investigation of the Cremation Rite at the Cemeteries of Elsham and Cleatham, North Lincolnshire. Unpubl. PhD thesis, University of Sheffield (Sheffield 2011).

Squires 2013: K. E. Squires, Piecing together identity: A social investigation of early Anglo-Saxon cremation practices. Archaeological Journal 170, 2013, 154–200.

Squires 2016: K. E. Squires, Neighbours and networks: funerary trends among cremation practicing groups in early medieval England and north-western Europe. In: L. Keys/I. Riddler/J. Soulet (eds.), The Evidence of Material Culture: Studies for Professor Vera Evison (Montagnac 2016) 116–135.

Stead 1967: I. M. Stead, A La Tène III burial at Welwyn Garden City. Archaeologia 101, 1967, 1–62.

Stead/Rigby 1986: I. M. Stead/V. Rigby, Baldock: The Excavation of a Roman and Pre-Roman Settlement, 1968–72. Britannia Monograph Series 7 (London 1986).

Stokes 1996: J. Stokes, Bull and bear baiting in Somerset: The gentles' sport. In: A. F. Johnston/W. Hüsken (eds.), English Parish Drama. Ludus, Medieval and early Renaissance theatre and drama 1 (Amsterdam 1996) 65–80.

Stowell Phillips 2008: S. J. F. Stowell Phillips, Animal Visual Culture in the Middle Ages. Durham theses, Durham University (Durham 2008). http://etheses.dur.ac.uk/3664/.

Sykes 2004: N. J. Sykes, Neolithic and Saxon animal bone. In: C. J. Ellis (ed.), A Prehistoric Ritual Complex at Eynesbury, Cambridgeshire: Excavation of a Multi-Period Site in the Great Ouse Valley, 2000–2001. East Anglian Archaeology Report 17 (Salisbury 2004) 87–91.

Thompson Watkin 1877: W. Thompson Watkin, On the Roman inscriptions at Colchester. Archaeological Journal 34, 76–82.

Toynbee 1963: J. Toynbee, Art in Roman Britain (London ²1963).

Toynbee 1973: J. Toynbee, Animals in Roman Life and Art (Baltimore [Maryland] 1973 [Reprint]).

Tünaydin 2013: P. Tünaydin, Pawing through the history of bear dancing in Europe. Frühneuzeit-Info 2013, 51–60.

Turner 2009: A. Turner, The evolution of the guild of large Carnivora of the British Isles during the Middle and Late Pleistocene. Journal of Quaternary Science 24, 2009, 991–1005.

Waselkov/Funkhouser 2020: G. A. Waselkov/J. L. Funkhouser, Bear-human relationships in Native eastern North America: an overview of archaeological and ethnohistorical evidence. In: H. Lapham/G. Waselkov (eds.), Bears: Archaeological and Ethnohistorical Perspectives in Native Eastern North America (Gainsville 2020) 271–310.

Williams 2016: H. Williams, Citations in stone: the material world of hogbacks. European Journal of Archaeology 19, 2016, 497–518.

Yalden 1999: D. Yalden, A History of British Mammals (London 1999).

Prof. Hannah O'Regan
Department of Classics and Archaeology
University of Nottingham
Nottingham
United Kingdom
Hannah.Oregan@nottingham.ac.uk

Fig. 1. A post-conservation photograph of the Early Bronze Age brown bear pelt from Whitehorse Hill, Dartmoor. The grave contained the cremation of a young adult, with a number of high status artefacts, and is radiocarbon dated to 1740–1560 cal. BC. The bear pelt had been folded in half to wrap the cremation, and the two layers can clearly be seen here with packing between. Note there is no skin present, only the hair was preserved within the peat (photo courtesy of the Conservation and Museums Advisory Service, Wiltshire Council).

Fig. 2. Reconstruction of a high status Iron Age human cremation grave from the Panshanger Estate, Welwyn Garden City, England, reported in STEAD 1967, and on display in the British Museum. The amphorae and a silver cup were imported from Italy, and other items are from Gaul. Part of a unique set of glass gaming pieces can be seen above the human bones in Fig. 2b. Six bear terminal phalanges were found intermingled with the cremated human remains (bear remains not shown here). The grave has been dated between 25–15 BC (after SEALY 2009, © The British Museum).

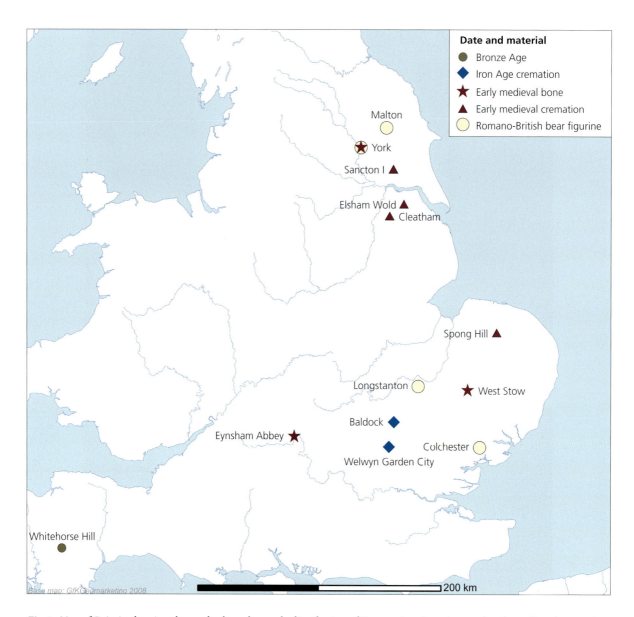

Fig. 3. Map of Britain showing the southerly and easterly distribution of Bronze Age, Iron Age, and early medieval cremations with bear remains, sites with Romano-British bear figurines, and sites with early medieval bear bones. See text for details and references (map GIS department, ZBSA).

Fig. 4. Cremated intermediate bear phalanges from burial MT89BLG (urn 983) from the Anglo-Saxon cremation cemetery at Cleatham, Lincolnshire. It was not possible to sex the human remains, but they were from an older mature adult (31–40 years). There were no pyre goods associated with this individual (after SQUIRES 2011, appendix 2; photo courtesy of Dr K. Squires).

Fig. 5. Romano-British jet bear figurine from Bootham, York. It was found in AD 1845 with a jet bead, a coin of Constantine I (AD 312–315) and a Castor ware beaker, and it is thought to have formed part of a burial group, but no bones were identified or noted at the time (CRUMMY 2010, 43). Note the incised marks on the shoulder and rump of the bear which PARKER (2018) suggested could represent stars. The figurine is in the Bateman collection, Weston Park Museum, Sheffield (© Museums Sheffield).

Fig. 6. Romano-British copper alloy bear figurine found by a metal-detectorist at Longstanton in Cambridgeshire. Note the human figure held within the mouth of the bear (after EVANS/MACKAY 2004).

Fig. 7. Hogback number 17a, from the village of Brompton, Yorkshire. The end-beasts are clearly muzzled, and possibly shown in a suckling position (photo T. Middlemass, image reproduced by permission and copyright of the Corpus of Anglo-Saxon Stone Sculpture).

Fig. 8. The "Colchester Vase", a locally made cup showing two gladiatorial scenes (one of which shows combat between a human with a whip and a bear), and a hunt scene of dogs chasing hares. It was found at West Lodge, Colchester, in 1853 with a human cremation, a small pottery jug or ewer and a Samian dish (cf. THOMPSON WATKIN 1877; © Ancient Art and Architecture Collection Ltd. / Bridgeman Images).

Fig. 9. Bear baiting scene on an undated medieval misericord from the Burrell Collection, Glasgow (cat. no. 50/2044, photo M. A. Hall, Perth, Scotland; image reproduced courtesy of Glasgow Museums).

Fig. 10. The bear and ragged staff was the emblem for the Earl of Leicester's family in the medieval and early modern periods, and is now the badge of the County of Warwickshire. This medieval lead alloy livery badge in the collections of the British Museum (item no. 1853,0502.8) would have been sewn onto the clothing of a servant. The staff has been lost, but the bear is still very clear, as is its muzzle – a very common feature of medieval and early modern bear images (© The Trustees of the British Museum).

Fig. 11. In the mid-18th century lidded jugs or mugs in brown or white-glazed stoneware depicting bear-baiting were made in Nottinghamshire and Staffordshire. The lid of the mug is formed by the head of the bear and can be removed. This is a typical example, showing a collared, chained, and muzzled bear holding a dog between its fore-paws. It is also clearly meant to represent a male bear (© The Metropolitan Museum of Art, New York, Gift of Carleton Macy, 1934. Accession number: 34.165.4a,b; public domain).

Fig. 12. Engraving of Martin the dancing bear from the late 19th century children's book "The life of a bear" (ANON. 1901). The illustrator has also shown the key accoutrements of a travelling bearward – pole, muzzle, and drum (see SERRA 2013 for more details). Also note the caption – the book has a strong moral undertone to instruct the young readers in obeying their elders.

Fig. 13. Bears as childrens' toys became increasingly popular in the 20th century. This example, made in Germany between 1910–1920, shows a dancing bear with muzzle. While grotesque to modern eyes, this toy represents a common sight in towns and villages in the 1800–1900s, as bears travelled with their owners to dance for money (Social History collection, Weston Park Museum, Sheffield, accession no. 1989.522; © Museums Sheffield).

Table 1. Names and dates of British archaeological time periods referred to in the text. Dating based on HUNTER/RALSTON 2009, except *Lateglacial which is in years BP (before present) and based on RASMUSSEN et al. 2014.

Period	Dates
Lateglacial*	14,692–11,400 BP
Mesolithic	9600–4000 BC
Neolithic	4000–2400 BC
Bronze Age	2400–800 BC
Iron Age	800 BC – AD 43
Roman	AD 43–410
Early medieval (includes Anglo-Saxon and Viking)	AD 410–1066
Medieval	AD 1066–1485
Post medieval (includes early modern and Victorian)	AD 1485–present

Bears and humans in Sweden – 10,000 years of interactions from the Mesolithic to the Middle Ages

By Ola Magnell

Keywords: Brown bears, zooarchaeology, Sweden, body part frequency, butchering

Abstract: The interactions between humans and bears from the Mesolithic to the Middle Ages in Sweden have been studied through the zooarchaeological record. The abundance of bear remains at settlement sites through time has been quantified, and a decrease in the frequency of bears from the Early Mesolithic to the Early Neolithic has been noticed. The anatomical representation of bears in the records indicates hunting nearby settlements and the transport of most types of body parts during the Mesolithic. After an increase in bear remains in the Middle Neolithic, associated with the Pitted Ware culture, a decreasing trend could be discerned for southern Sweden from the Bronze Age until the Early Iron Age, while the situation is the opposite for the northern areas. From the Late Iron Age to the Middle Ages, a slight increase in bear remains can be noticed, reflecting a development of trading networks and the circulation of bear furs from the north to the southern regions of Sweden. This is confirmed by the body part representation, which consists mainly of claws from Iron Age and medieval sites in southern Sweden. Finds of bear bones in ritual contexts indicate a symbolic significance of the bear to humans. In northern Sweden, a long tradition of ritual handling of bear bones can be traced from at least the Late Neolithic up until recent times. Finds of tooth pendants made of bear canines from the Mesolithic until the 11th century AD indicate a long use of bear fangs as amulets in Sweden.

INTRODUCTION

Bears have, through time and in different geographical regions, played various and changing roles in relation to humans, in regard to subsistence as well as the symbolic and religious aspects of different societies. The hunting of bears has varied in its importance to different groups of humans through time, and, besides being a part of their subsistence and a way to obtain food and furs, it has served to protect livestock, as well as providing wild game management and recreation. Bear hunting has, in many societies, been of large social relevance and even religious importance (BJÖRKLÖF 2010; cf. different contributions in the present volume). In most cases, the purpose of bear hunting has been a combination of several of these factors. At the same time humans have, through hunting and the alteration of environments and the landscape, had a large impact on bear populations, which even resulted in the extinction of the brown bear in several regions (SERVHEEN 1990, 3).

Through the zooarchaeological record, the interactions between humans and brown bears (*Ursus arctos*) in Sweden from the last Ice Age until the end of the Middle Ages have been studied for this paper. The aim has been to show the frequency of osteological bear remains in the faunal record from archaeological sites in different parts and time periods of Sweden, as an estimate of the importance of bear hunting over time. What was the significance of bear hunting for the subsistence during different periods? Is it possible to trace the impact of hunting on the bear populations?

The utilisation of bears has also been studied by the analysis of body part frequency in the faunal record from archaeological sites and butchery marks on bear bones. If bears were hunted near to settlements, most parts of their bodies can be expected to be found in the settlement, while, if hunting took place far from the settlements, only the most important body parts of the bears might be expected to have been transported back to the community (BINFORD 1978, 10–11; LYMAN 1994, 223–234). Higher frequencies of bones from body areas with relatively more meat, such as the trunk and proximal limbs, indicate that bear meat was important to humans. Bones from distal parts of the limbs can be taken as an indication that the bear skins, with the bones of the paws still attached to the skins, were the main reason for hunting bears. Body part frequencies of bear bones from different periods have been analysed to trace chronological developments in the utilisation of the bears. Analyses of butchering marks have also been used to trace how different body parts of the bears were utilised.

Additionally, finds of osteological bear remains in ritual contexts, and depositions of bones that suggest how they were handled, have been used as an indication of the symbolic and religious significance of bears to humans in Sweden through time. Sami bear burials in the northern part of the country will not be dealt with here, since, in Sweden, they are mainly dated to periods after the Middle Ages (*c*. AD 1500), and have been thoroughly discussed earlier (and also in this volume; ZACHRISSON/IREGREN 1974; cf. IREGREN, this volume). Similarly, finds of bear bones, mainly claws (phalanx 3), in Iron Age burials will not be included in this study, since this topic has been covered in several earlier papers, as well as in this publication (PETRÉ 1980; IREGREN 1988; LINDHOLM/LJUNGKVIST 2016; cf. different contributions in the present volume). Finds of teeth and bones from bears in burials pre-dating the Iron Age in Sweden are very rare and will be used as an example of the ritual handling of osteological bear remains.

MATERIAL AND METHODS

The abundance of brown bears through time has been quantified by the number of settlements with faunal assemblages including bear bones from different archaeological periods. To evaluate the frequencies of the brown bear, the occurrence of wolf, lynx, and red fox has also been analysed for comparisons with, and with reference to, the hunting of other species of the Order Carnivoria. Only sites with a NISP (number of identified specimens) over 100 have been included in the study, since it can be assumed that the chance of finding bear bones in smaller samples is very low, based on the relative rareness of the species. Ideally, only sites with larger samples should be included in the study, but this would result in only a few sites from large excavations that in many cases are special sites, such as central places or urban centres.

The study is based on 406 faunal assemblages and a database of zooarchaeological analyses in Sweden kept at Arkeologerna, Statens Historiska Museer. The database is based on published and unpublished zooarchaeological analyses that were compiled by the author to include as many sites as possible, but it is not the ultimate and complete database of all zooarchaeological analyses from Sweden. Due to the generally low number of bear bones in faunal remains from archaeological sites in relation to those of other animals, such as ungulates and domestic animals, the frequency of osteological bear remains in relation to other mammals has only been used for the quantification of the abundance of brown bears at Mesolithic sites with their generally large presence of bear bones.

The chronology in the long-term analysis of the frequency of bears follows the standard for (southern) Sweden. Due to the long duration of the Mesolithic, it has been divided into three parts:

Early Mesolithic (9500–6400 BC), Middle Mesolithic (6400–5400 BC), and Late Mesolithic (5400–3900 BC). The Neolithic has been divided into two parts: Early Neolithic (3900–3300 BC) and Middle Neolithic (3300–2300 BC). Due to the relatively few sites from the Late Neolithic and the Bronze Age (2300–500 BC), there is no division of these periods, but most sites date to the Late Bronze Age. The Iron Age has been divided into the Early Iron Age (500 BC–AD 400) and the Late Iron Age (AD 400–1050). The final period is the Middle Ages (AD 1050–1500). In a comparison of the frequency of bear bones at settlement sites between the northern and southern parts of Sweden, a slightly different division of the chronology into longer periods has been used, due to a generally lower chronological resolution of the northern sites.

Due to a combination of factors, namely ecology, cultural history, and bone preservation, the occurrence of bears has been studied and presented separately concerning the southern (the regions Göta- and Svealand) and northern parts (Norrland, including the county Dalarna) of Sweden. Due to different aspects concerning soil conditions, such as the predominance of podsol soils with their low pH-values, the preservation conditions for bones are poorer at most sites in the northern parts of Sweden, and the faunal remains mainly consist of burned bones in this region (Ekman/Iregren 1984, 13). Given the difficulty in identifying burned bones in comparison with well-preserved unburned bones, the faunal remains of archaeological sites from the southern regions make direct comparisons problematic. This division of Sweden into two regions is also relevant from an ecological and climatic perspective (with a temperate climate with broadleaf forests and larger agricultural areas in the southern parts, and a subarctic one with boreal forests in the northern parts).

Several sites from the northern parts of Sweden have been occupied repeatedly through time and thus have no clear stratigraphy. This means that many settlements have been dated to within long time periods, such as from the Late Neolithic to the Early Iron Age (Ekman/Iregren 1984, 7–13).

Due to the generally low number of bear bones at single sites, the body part frequency has been analysed through the compilation of several sites from different archaeological periods. The quantification has been based on the number of identified bone specimens (NISP) and the division into five main body regions; head (cranium and mandible), loose teeth, trunk (vertebrae and ribs), long bones (scapula, humerus, radius, ulna, pelvis, femur, patella, tibia, fibula), and paws (carpals, tarsals, metacarpals, metatarsals, phalanges). The anatomical distribution of single bone elements has also been compiled and is presented in the bear skeleton figures (see below).

By the analysis of types of butchering marks and their anatomical position on the bones, it is often possible to describe stages of the butchering process, such as skinning, dismembering, filleting of meat or marrow fracturing (Binford 1981, 106–135).

Prelude – Finds of polar bear remains from Sweden

Finds of polar bear bones (*Ursus maritimus*) show the presence of this species during the end of the last Ice Age in the southwestern parts of Sweden. The radiocarbon dating of bones from polar bears can be correlated with deglaciation and the retreating ice sheet, with the oldest ones dating from 12,500 BC from Scania and the most southern parts of Sweden to 9800 BC from Bohuslän on the west coast (Berglund et al. 1992). All these polar bear bones come from geological finds, such as deposits of clay or peat, and not from any archaeological sites. However, the chronology of the dated bones and the earliest archaeological finds from the Late Glacial indicate possible interactions between humans and polar bears in Sweden. Finds of lithics belonging to the Late Hamburgian culture (12,800–12,000 BC) in Scania and Denmark are approximately contemporaneous with the oldest polar bear find, a femur from Kullen in northwestern Scania (Holm 1993, 15–18; Larsson 1994, 165–168). Sites from the maritime-adapted Hensbacka culture from the west coast in Sweden, with flint material dated by shore-line replacement to about 11,000–10,000 BC, show an overlap with the dating of polar bears in this region (NORDQVIST 1997, 34–35). There is no evidence for the hunting of polar bears in the Late Palaeolithic in Sweden, but encounters between polar bears and the first humans in Sweden can be expected to have occurred.

THE FIRST BROWN BEARS IN SWEDEN

The earliest evidence of brown bears in Sweden comes from the Early Mesolithic settlement of Almeö in Västergötland by Lake Hornborgasjön, which dates to about 8700 BC (KINDGREN 1995). The earliest radiocarbon dating of osteological brown bear remains from Sweden is from a geological find of a complete skeleton discovered in Ugglarp, Scania, dating to 8600 BC (IREGREN et al. 1990, 13).

A mandible of a brown bear from Faurbo Knold on Zealand, Denmark, has been dated to the Allerød interstadial, about 11,400 BC (AARIS-SØRENSEN 2019, 19). Considering the presence of a land-bridge between Denmark and Sweden during the Allerød, it is likely that the brown bear might at this period already have spread to the southern parts of the country. During the Younger Dryas (10,600–9700 BC), when the climate was colder than during the Allerød, the return to glacial conditions likely resulted in the depletion of brown bears in the region (AARIS-SØRENSEN 2009, 19). At the onset of the Postglacial, the increasing temperatures and the establishment of vegetation during the Preboreal probably resulted in a fast spread of brown bears from the continent to the southern parts of Sweden. The brown bear is a highly adaptive animal, found in various environments; it is an omnivore with a broad diet based on the meat of scavenged cadavers, by preying on ungulates mainly in spring, along with various plants such as grasses and herbs, as well as ants in summer and berries in autumn (SAHLÉN et al. 2006, 6–7). Pollen and macrofossils of crowberries (*Empetrum nigrum*) and blueberries (*Vaccinium myrtillus*) from the Preboreal show that suitable habitats and food to sustain a brown bear population were already established by the onset of the Holocene at 9700 BC in Sweden (BJÖRCK et al. 1997, 459; PARDUCCI et al. 2019, 7).

THE DISTRIBUTION AND FREQUENCY OF THE BROWN BEAR AT SETTLEMENT SITES FROM THE MESOLITHIC UNTIL THE MIDDLE AGES

The Mesolithic
The Mesolithic finds of brown bear remains are clearly concentrated in Scania, on the west coast, and in Östergötland (Fig. 1; Table 1). This reflects the history of Stone Age archaeology and several larger excavations of Mesolithic sites in these regions, where soil conditions generally are also favourable for the preservation of bones. During the Mesolithic, due to isostatic and sea level change processes large parts of the coastal region of present-day northern and eastern middle Sweden were covered by the Baltic Sea or formed an archipelago, which also explains why no bear bones from this period have been recovered in certain parts of Sweden.

The absence of brown bears from the northern parts of Sweden is probably also a reflection of the small number of sites from this period with large bone assemblages that are securely dated to the Mesolithic. Finds of brown bear are known from Inari and Kemijärvi in the northern parts of Finland, dating to 6000–5800 BC (UKKONEN/MANNERMA 2017; cf. MANNERMA et al., this volume). This shows a bear presence in the northern parts of Fennoscandia during the Mesolithic and most likely also in northern Sweden.

Bear bones occur at 75 % of the Mesolithic sites in the southern parts of Sweden, indicating that the bear was regularly hunted during this period. Even though osteological remains of bear occur at most settlements, bear bones, as well as those of other larger predators, are of low frequency at all Mesolithic sites in comparison with bones from ungulates, such as red deer, roe deer, and wild boar. The frequency of bear in the mammalian faunal remains varies between 0.1–3.6 % at the settlements. The largest number of osteological bear bone remains from a settlement is represented by the 64 bone fragments from the site of Ageröd I:HC, while the highest frequency is documented for the site of Tågerup II (Fig. 2). The site of Kanaljorden in Motala, where 214 bear bones make up 12.6 % NISP of the osteological remains, is an exception, but this is a ritual site and not actually a settlement. This site will be discussed further below in the section that deals with ritual depositions and the handling of bear bones.

In a ranking of the most hunted mammalian prey, based on the frequency of NISP from Mesolithic settlements, the median values indicate that the bear was the ninth most hunted animal, but the values range from the third to the thirteenth most hunted animal. In comparison with other predators, the bear seems to have been hunted regularly, and 24–65 % these bones are from brown bears.

A decrease in the occurrence of brown bears can be noticed from the Early Mesolithic to the Middle and Late Mesolithic sites. On the contrary, the occurrence of other larger predators, such as wolf and red fox, increases at Middle and Late Mesolithic sites in comparison with Early Mesolithic sites (Fig. 3). Due to the low sample size, the validity of the result is a bit uncertain, but the frequency of the brown bear, based on NISP in relation to other mammals, also indicates a generally higher occurrence of bear bones at Early Mesolithic in comparison with Late Mesolithic settlements (cf. Fig. 2).

The depletion of other large game, such as elk (*Alces alces*) and aurochs (*Bos primigenius*) has been noticed over time at Mesolithic sites in south Scandinavia, and there is a lower frequency of these taxa on coastal sites in comparison with inland sites. This is suggested to have been a result of a change in habitat and vegetation, but also due to an increasing hunting pressure over time in the coastal regions (MAGNELL 2017, 127–129). On the Danish islands, the brown bear, among other larger mammals, also became extinct in this period (AARIS-SØRENSEN 1980, 131–138). The indication of a lower occurrence of bear bones in settlements during the Mesolithic is possibly a reflection of a decreasing bear population. Changes in vegetation during the Atlantic chronozone (7000–3800 BC) and the spread of broadleaf forests with a denser canopy together with less understory on the forest floor resulted in poorer habitats for light-demanding plants with berries, such as blueberries (*Vaccinium myrtillus*) and raspberries (*Rubus idaeus*; cf. NOE-NYGAARD 1995, 246–248; RICARD/MESSIER 1996, 153–159; MIINA et al. 2009, 588). This change of biotopes may have resulted in less suitable habitats for brown bears. An increasing hunting pressure over time may have affected the bear population in more densely populated areas with sedentary coastal settlements. The bear still seems to have been regularly hunted and occurs at nearly two thirds of the settlement sites from the Late Mesolithic (Fig. 3).

The Neolithic

The frequency of bears is distinctly lower for Early Neolithic sites in comparison with the Mesolithic ones (cf. Fig. 3). Most likely this does not represent a decrease in the bear population, but rather it reflects the neolithisation and the shift in subsistence strategy from hunting-gathering to farming, with less focus on hunting. A similar decrease can be noticed for the other Carnivoria, indicating that the decrease probably reflects a general diminishing in the importance of hunting for subsistence during the Early Neolithic in the southern parts of Sweden. Bear bones have been noted at only two sites belonging to the Funnelbeaker culture in Scania in southernmost Sweden (Table 2). These finds show that the brown bear was still present and was also occasionally hunted, but not as regularly as during the Mesolithic.

The occurrence of bears is clearly higher at Middle Neolithic settlements compared to the previous Early Neolithic period, but still lower than during the Late Mesolithic. An increase can also be noticed for wolf and lynx in this period, but not for red fox (cf. Fig. 3). This increase in bear hunting can be associated with sites of the Pitted Ware culture, which had a subsistence strategy mainly based on the hunting of seals and fishing, but also on the hunting of terrestrial game. If only sites belonging to the Pitted Ware culture are included in the quantification for the Middle Neolithic settlements, the frequency is higher and the bear occurs at 50 % of the sites, which is still a slightly lower occurrence than during the Late Mesolithic sites with 64 %. On Pitted Ware culture sites, the brown bear is the third to twelfth most hunted mammalian taxon, with a median as the eighth most hunted wild game, which is similar in comparison to the Mesolithic sites.

Finds of bear bones at the Pitted Ware culture sites of Ajvide on Gotland and Tråsättra in Uppland (Hallin 2008, 5; Magnell 2019) indicate that bear body parts were brought to regions with no bear population (Fig. 5). The island of Gotland, situated in the middle of the Baltic Sea, has never had any brown bear population (Lepiksaar 1986, 59). It can also be assumed that bears did not inhabit the island where the Tråsättra settlement was situated in the outer archipelago of the Mälarhavet. During the Neolithic, due to the Littorina transgression, the region of the Mälar valley in middle eastern Sweden had a shore-line that was about 40–20 m higher than today; it was thus a part of the Baltic Sea called Mälarhavet (Björck et al. 2019, 34–40).

The frequency of bears is slightly lower on sites in the northern parts of Sweden that date to the Early and Middle Neolithic in comparison with settlements from the southern parts (Fig. 4). Whether this means that the brown bear was hunted less often in the northern parts can be questioned. One explanation might be that most of the faunal remains from the northern parts of Sweden are burnt and fragmented, thus making the identification of bear bones more difficult in comparison with mainly unburned bones from southern regions.

The Bronze Age

The sites from the Late Neolithic and the Bronze Age show a lower occurrence of bears in comparison with the Middle Neolithic, with frequencies that are comparable to the sites from the Early Neolithic; this probably reflects a decreased importance of hunting for subsistence during this period. For the other predators (red fox, wolf, lynx) a decrease could also be noticed (cf. Fig. 3). Bear bones occur on only one site out of eleven Bronze Age ones from Scania, while bear remains occur at two settlements out of six sites from the region of Uppland (Fig. 5; Table 3). The lower occurrence of bears in the region of Scania, in comparison with Uppland, can be considered as an indication of the depletion of brown bears in this southernmost region of Sweden. The reconstruction of the vegetation in Scania, based on pollen analysis, shows deforestation; there was already a largely open cultural landscape during the Bronze Age in the agricultural plains of the western and southern areas of the region (Berglund 1991; Lagerås/Fredh 2019). This change in vegetation probably resulted in less suitable habitats for bears; together with an increasing human population and a higher hunting pressure this might have resulted in a depletion of brown bears in parts of Scania by the 2nd millennium BC. Due to the low number of sites with larger samples of faunal remains from the Bronze Age, the validity of this observation is yet uncertain and larger data sets are necessary to confirm these regional developments.

Sites dating to mainly the Late Neolithic and Bronze Ages in the northern parts of Sweden show an increased frequency of bear remains in relation to the Early and Middle Neolithic. This can possibly reflect an intensification of bear hunting in this period, but it may also have other explanations. Several of the Early and Middle Neolithic sites in northern Sweden are coastal settlements with a subsistence strategy largely based on the hunting of seals and, most likely, fishing, which means that terrestrial game such as bear generally occurs less frequently at these sites (Ekman/Iregren 1984, 20–26). As most of the Late Neolithic/Bronze Age settlements were situated inland, with a large focus on terrestrial game, such as elk and beaver, as a subsistence strategy (Ekman/Iregren 1984,

31–38), this probably means that encounters with, and the hunting of, bears were more likely in these areas.

In comparison with the southern parts of Sweden, the frequency of bears is higher at the Late Neolithic/Bronze Age sites in the northern parts, which probably reflects the differences in subsistence between the two regions (Fig. 4). In the northern parts hunting, gathering, and fishing were the basis of the economy during these periods, while subsistence in the southern parts was highly focused on agriculture and animal husbandry.

The Iron Age and Middle Ages

In the southern parts of Sweden, the occurrence of bears decreases even more during the Early Iron Age; bear bones occur at only 2 % of the sites (cf. Fig. 3). The only find from this period is from the settlement of Kyrsta in Uppland, where a claw (phalanx 3) was recovered in a well that has been dated to the 1st century BC (Sjöling/Bäckström 2006).

The decreasing occurrence of brown bears is possibly the result of an increased focus on the cultivation of crops and animal husbandry as a subsistence strategy during the Early Iron Age. A decrease in sites with bones of red fox, wolf, and lynx also indicates a general decrease in hunting in this period (cf. Fig. 3). Further, changes of the environment to a more open cultural landscape, and fewer suitable habitats for brown bears in certain agricultural core-areas in the southern parts of Sweden are thought to have resulted in a depletion of brown bears in certain regions.

The occurrence of bear remains increases again for Late Iron Age settlements and even more for sites from the Middle Ages (cf. Fig. 3). This does probably not reflect an increase of bear hunting in the southern parts of Sweden, but rather an increase in trading networks and in the circulation of bear skins. This is also indicated by the body part representation, which will be presented in the next section. Finds of bear claws occur occasionally in Iron Age burials from the region of Mälardalen and Gotland, especially from the Roman Iron Age to the early Vendel Period, which further indicates that bear skins were commodities of importance that were traded during this period (Iregren 1988; Lindholm/Ljungkvist 2016; cf. various papers, this volume).

The presence of bear bones at the trading sites of Bandelunda and Burge on the island of Gotland clearly shows that the zooarchaeological finds from the Late Iron Age and the Middle Ages represent the long-distance transportation of bear parts. Finds of bear remains from the Late Iron Age and the Middle Ages mainly occur in central places and towns, such as Uppåkra, Helgö, Birka, Sigtuna, and Lund. This most likely also reflects the importance of these places in trading and inter-regional contact. Finds of bear bones at Ekholmen, a castle situated in Västergötland, which also has a relatively high occurrence of bones from other wild game, indicate the presence of the brown bear in regions with woodland in southern parts of Sweden during the Middle Ages.

Historical sources from the Middle Ages, such as the laws (*landskapslagar*), also show the presence and hunting of brown bears in different regions of Sweden. In the *Upplandslagen* (the law of Uppland), the *Västmannalagen* (the law of Västmanland), the *Hälsingelagen* (the law of Hälsingland), as well as in the later law of King Magnus Eriksson, all from the 14th century, it is stated for everybody and everywhere that bear, wolf, and fox shall be killed with impunity. Also, the *Västgötalagen* (the law of Västergötland) states that the person who kills a bear owns it. The *Östgötalagen* (the law of Östergötland) dictates that everybody shall hunt bear, since he (the bear) is a superior force. "Superior force" here means that herders were not obliged to protect livestock from bear attacks as they were supposed to do for attacks by other predators, such as wolves. This law also prescribes that peasants should regularly perform communal drive hunts of bears and wolves. Worth mentioning is that the *Skånelagen*, the medieval law from Scania, does not mention rules regarding the hunting of bears. This can be interpreted as an indication that the animal was rare in the region. However, the *Skånelagen* mentions that it is punishable to keep and raise bears or wolves (Myrdal 2012, 180, 182–183).

The number of sites with larger faunal material from Norrland dating to the Iron Age and the Middle Ages is low, making interpretations of the occurrence of bears in the northern parts during these periods uncertain (cf. Fig. 5; Tables 4–5). However, the available data suggest an increase during the Iron Age, followed by a slight decrease in the Middle Ages (cf. Fig. 3). The peak occurrence of bear remains at northern Swedish sites during the Iron Age coincides with the period when bear claws (phalanx 3), usually interpreted to represent bear skins, occur relatively frequently in Roman Iron Age and early Vendel Period burials and with decreasing frequency until the Viking Period in the region of Mälardalen and Gotland (Lindholm/Ljungkvist 2016; cf. Lindholm/Ljungkvist, this volume). Possibly, the indications for an increase in bear hunting in the northern parts of Sweden reflect a high demand for bear skins to be exported to the southern parts during the Iron Age.

Utilisation of the brown bear: Body parts and butchering marks from the Mesolithic to the Middle Ages

The body part representation of brown bears from Mesolithic sites shows that usually all body parts were brought to the settlements (Figs. 6–7). Skulls and mandibles occur in rather low frequencies, but loose teeth, mainly molariform ones, are usually well represented and thus show that the heads were brought to the settlements (Fig. 6). Bones from the vertebral column are often present, mainly ribs and cervical vertebrae, as well as relatively high frequencies of long bones and bones from the paws, including phalanx 3 (claws). Further, this can be taken as an indication that the bear was often hunted in the vicinity of the settlements and few bones were left at the kill sites.

It is almost only at the Mesolithic sites that larger amounts of osteological bear remains have been found and where butchery marks could also be systematically analysed, which means that the butchering process can be studied in detail only for this period. The butchering is based on analyses of the sites of Ageröd I:HC, Tågerup, Ringsjöholm, Strandvägen, and Kanaljorden (Magnell 2006; Gummesson 2014; Gummesson et al. 2019). The frequency of butchery marks on bear bones is high (12–19 %); this is comparable to, or slightly higher than, the frequencies for commonly hunted animals, such as red deer and wild boar (Table 5). Cut marks from skinning have been found on the head (skull, mandible) as well as the metacarpals from the paws. Dismembering marks show that the bear carcasses had been cut up, the mandibles separated from the head, while the limbs had been sectioned in most joints from the shoulders and hips down to the phalanges of the paws. The filleting of meat occurred on different body regions, such as the mandibles, ribs, pelvis, and the long bones from anterior as well as posterior limbs (Fig. 8). All bear long bones seem to have been regularly fragmented and most likely marrow-fractured, as was the common practice for all other large wild game during the Mesolithic. Impact marks from marrow-fracturing have been found on the humerus and ulna.

Finds of tooth pendants – made of canines with drilled holes in the root apex – found at different sites such as Ringsjöholm, Skateholm, and Strandvägen, also show that the fangs of bears were worked and worn as ornaments and amulets (Larsson 1989, 372; Gummesson 2018, 62; Gummesson et al. 2019, 88). ZooMS (Zooarchaeology by Mass Spectrometer) and an analysis of leister points from Rönneholms Mosse have shown that long bones from bears were occasionally used for making bone points during the Mesolithic in Sweden (Jensen et al. 2020).

The body part frequency of the Neolithic sites differs from that of the Mesolithic ones by a larger proportion of bones from paws and a few from the trunk (Figs. 6–7). The lower occurrence of vertebrae and ribs might indicate hunting further away from the settlements during the Neolithic, with the trunk, after being stripped of meat, being left on the kill-sites more often in comparison with Mesolithic sites. The high proportion of bones from the paws may reflect transportation of this body part, probably as bear skins or amulets of bear claws, to the sites. Finds of carpal bones and claws from the sites of Tråsättra, which was situated on an island in the archipelago of eastern middle Sweden, and Ajvide on Gotland show the transportation of paws or skins over larger distances from the mainland of Sweden during the Middle Neolithic (HALLIN 2008, 5; MAGNELL 2019, 171).

The distribution of body parts from Bronze Age sites shows similarities to the Mesolithic ones with a rather equal distribution of bones from head, limb bones, and paws, but differs with its smaller amounts of bones from the trunk, as was the case for the Neolithic sites (cf. Figs. 6–7). This indicates that most parts of the bear were utilised, but the lower frequencies of bones from the trunk might reflect hunting at larger distances from the settlements, with certain body parts left on kill-sites. Most of the bear bones from the Bronze Age are from the site of Apalle in Uppland; because of this, body part frequencies of the period rather represent conditions at this site in particular and may not be representative for other settlements from the period.

The body part frequencies from the Iron Age and Middle Ages differ strikingly from the earlier periods in respect of bone finds from paws, mainly claws (phalanx 3) and a few finds of teeth, but there are no finds of bones from meaty parts (cf. Figs. 6–7). Most likely, the bones from paws represent bear skins. Either bones from the paws were left in raw skins brought to furriers in towns and settlements for further processing of the fur, or claws were still attached to bear skins. In most cases, these finds reflect the trading of bear skins, but in pre-Christian times they were probably also utilised in a symbolic sense, with skins used as wrappings or claws used as amulets.

Butchering marks on bear bones from the Viking Age cult site at Frösö church in Jämtland, with cut marks from skinning, dismembering of most joints and filleting of meat as well as chop marks from marrow fracturing and the extraction of canines, reveal that an intense utilisation of the bodies of bears also took place during the Late Iron Age in northern Sweden (MAGNELL/IREGREN 2010, 235–237; cf. contribution by MAGNELL on Frösö, this volume).

DEPOSITION AND HANDLING OF OSTEOLOGICAL BEAR REMAINS AT SETTLEMENTS

At most Mesolithic sites in Sweden, bear bones are found mainly unburned in cultural layers, scattered among the bones of other animals and lithics in seemingly random patterns, just like any other waste. However, at some sites there are indications of a special treatment of bear bones, reflecting a symbolic importance of the bear to people in Sweden during the Mesolithic.

A tooth pendant made of a bear canine found in burial 41 in the Mesolithic cemetery at the site of Skateholm I most likely had symbolic significance (Fig. 9). The burial is a double grave with an older man and a child aged up to about four years. The tooth pendant was placed along with four pieces of amber on the chest of the child (LARSSON 1989, 372–373). Tooth pendants from red deer and wild boar are rather common grave goods at Late Mesolithic burials in Sweden, while pendants of bear teeth are rare and probably had a specific symbolic meaning. Finds of bear figurines made of amber from the southern coastline of the Baltic Sea further indicate that brown bears had symbolic and ideological significance to the Stone Age cultures in southern Scandinavia (VANG PETERSEN 1998, 87–90; cf. GROSS/VANG PETERSEN, this volume).

The frequency of gnaw marks on bear bones from Mesolithic settlements differs between sites, indicating a variation in the handling of bear bones (Table 5). At the settlements of Ageröd I:HC and Ringsjöholm, situated in the area of Lake Ringsjön, 21 % of the bear bones exhibit gnaw marks by carnivores, most likely dogs, while only 12–13 % of the bones of wild boar are gnawed. At sites from other regions, such as Tågerup and Strandvägen, the pattern is the opposite with no gnaw marks on bear bones there (Table 5). At some sites, it seems like the bones of bear were deliberately fed to dogs, while at other sites bear bones were placed out of reach of dogs, indicating a possible ritual treatment of these bones.

At the Mesolithic site of Bökeberg III, a concentration of bear bones, together with bones of other predators and bones/antlers with ornaments, has been noted in refuse layers which have been interpreted as representing zones for ritual activities on the shore areas by the settlement (Karsten 2001). A parallel to this has been noted at the site of Strandvägen with its concentration of bear bones and remains of other predators, as well as human bones and depositions of artefacts in an area with underwater stone packings by the shore of the River Motala (Molin et al. 2014; Gummesson/Molin 2016; Gummesson et al. 2019, 86–88).

The most obvious evidence of a Mesolithic ritual treatment of bear bones comes from the site of Kanaljorden, situated in the region of Östergötland (cf. Fig. 1), best known for its depositions of human skulls placed on an underwater stone packing at the bottom of a small lake (Hallgren/Fornander 2016). Kanaljorden is the Mesolithic site with clearly the largest proportion of brown bear remains in relation to other mammalian taxa, and the bear is also the second most occurring animal after wild boar, which is not the case for any other Mesolithic settlement. A discovery of the first five cervical vertebrae found in their correct anatomical position, representing a deposition of a piece of meat from the neck of a bear, is also an indication of the special treatment of bears at this site (Fig. 10). Further, two complete mandibles indicate a special handling of the bones since mandibles of larger mammals were usually marrow-fractured during the Mesolithic. The spatial distribution of the osteological remains at Kanaljorden also indicates different ritual zones, with human skulls in the central part of the stone packing, wild boar in the eastern part, and brown bear in the southern part (Fig. 10; cf. Gummesson 2014; Gummesson et al. 2018, 81–86).

Bear bones at ritual sites and ritual zones at settlements indicate a symbolic significance of the bear during the Mesolithic. However, a similar ritual treatment of bones can also be noted for many other animals, such as wild boar, red deer, and other predators as well (Karsten 2001, 144–147; Magnell 2006, 86–89; Gummesson et al. 2018, 81–86). It can thus be questioned if there was a similar distinguishing treatment of bear bones within the Mesolithic cultures in southern Sweden as it was characteristic for bear ceremonialism among many later circumpolar groups of people (Hallowell 1926; Zachrisson/Iregren 1974; cf. various contributions, this volume).

There are only very few finds of bear bones from Early and Middle Neolithic sites in the southern parts of Sweden associated with the Funnelbeaker culture. At the site of Flädie in Scania, a humerus of a bear was found together with cattle bones in a pit. Cut marks show that meat has been cut from the bone, but, other than the cattle bones (and common practice during the Stone Age), the humerus has not been marrow-fractured, but has been placed complete and unbroken in the pit (Hellgren et al. 2020). This can possibly be one of few indications of ritual treatment of bear bones by the Funnelbeaker culture in Sweden.

At the Pitted Ware settlement of Äs, situated in the region of Västmanland (cf. Fig. 1), bear teeth from the maxilla were found inside a ceramic vessel (Lepiksaar 1974, 150). Whether the teeth represent parts of the fragmented skull of a bear once deposited in the vessel is a bit uncertain, but this find possibly reflects the ritual treatment of brown bear body parts. Tooth pendants made of bear canines also occur at Korsnäs, another Pitted Ware settlement situated in the region Södermanland (cf. Fig. 1; Sjöling 2000, 24). Most finds of bear remains from Pitted Ware settlements are claws, or other bones from paws, recovered from cultural layers, and probably represent bear skins. However, it is possible that these skins did not only function as garments or rugs or carpets, but also had a symbolic meaning and use in various ritual practices.

Finds of clay figurines and carved bones and stones that have been interpreted as depicting bear heads have also been found on sites associated with the Pitted Ware culture, which indicates that bears had a specific meaning to this Middle Neolithic group (Wyszomirska 1984, 106, 244–245; Björck et al. 2019, 141; cf. Lindström, this volume).

Bear bones from Bronze Age sites have mainly been found scattered in cultural layers, and it is difficult to associate the osteological bear remains with special treatment. However, the find of a bear canine from the site of Ryssgärdet indicates a symbolic meaning beyond a utilitarian function (Wigh 2008, 377, 386). A later parallel is represented by a find from the Late Iron Age settlement of Varla in Halland, where a bear canine has been found in a pit dated to the 6th–8th centuries AD (Johansson 1997; 120, 189). Tooth pendants made of bear canines have also been found in the rich chamber burial of a child at Björkå, Ångermanland, dating to the Migration Period, and the burial of an adult woman from Önsvala in Scania, dating to the 6th century, where the bear tooth was found in a container with other kinds of amulets (Fransson 2011, 101; Larsson 2013, 143–144). Tooth pendants made of bear canines have also been found in the town of Sigtuna, in layers from the 11th century (Wikell 2015, 6, 10). The impressive bear canines seem to have been used as amulets during different periods, probably with varying symbolic functions, such as apotropaic magic for protection and as signals of identities.

Most of the faunal bear remains from the Iron Age and the Middle Ages are, as mentioned earlier, either distal phalanges (claws) or bones from paws, which most likely represent former bear skins. The bear bones either originate from archaeological sites and contexts associated with trading or a higher social status. Besides this, bear skins were an exclusive commodity, and the bear hunt itself was a dangerous activity that gave the successful hunter prestige, which could be associated with heroic deeds and the ideology of the aristocracy of the Late Iron Age and the Middle Ages (Oehrl 2013).

Finds of bear remains originate from the town of Birka as well as the sites of Bandelunda and Burge on Gotland, which were trading sites (Larsson 1997; Karlsson 2001; Wigh 2001). The only find of a bear bone from a Late Iron Age settlement in Scania is a so far unpublished claw (phalanx 3) from the central place of Uppåkra, discovered in layers of a sequence of halls probably representing the seat of the rulers of this settlement. Bear bones have also been found at the castle of Ekholmen in Västergötland, a stronghold associated with royalty in the Middle Ages (Lepiksaar 1991). Claws found in pits and cultural layers dating to the 11–13th centuries of the town of Lund come from either the residence of a canon or buildings associated with the Dominican convent in this town. Besides these bear phalanges, bones from other fur animals were also found during the excavation, indicating crafts associated with pelts in this part of the town (Fig. 11; cf. Hellgren/Magnell 2020).

In the northern parts of Sweden, evidence for the special treatment of bear bones can be noted from the Neolithic until modern times. A deposition of bear bones from Aspnäset in the region of Ångermanland, dated to about 2500 BC, is the earliest evidence of osteological bear remains being used for ritual activities in northern Sweden (Mulk/Iregren 1995, 11; cf. Iregren, this volume). Depictions of bears have also been found on rock carvings at Nämforsen, Ångermanland, and rock paintings at the site of Flatruet, Härjedalen (Hallström 1960, 95). This further indicates that the bear has for at least five thousand years been considered a sacred animal in the northern parts of Sweden. Finds of unburned mandibles and bear teeth deposited on top of a cremation burial, along with the antlers of mainly elk and reindeer, at the site of Krankmårtenhögen in Härjedalen, dating to the Early Iron Age, are later examples of the ritual treatment of bear bones from the northern parts of Sweden (Ambrosiani et al. 1984). The Sami bear burials show that ritual ceremonies with bear remains continued until the 19th centuries (Zachrisson/Iregren 1974; cf. Iregren, this volume).

Conclusions

Finds of brown bear remains in the zooarchaeological record for settlements in Sweden reflect the importance of the bear in hunting, the use of its meat and skin, as well as the symbolic significance of the animal through time and in different geographic regions. The frequency of bears is high in periods when hunting played a significant part in subsistence, i.e. the Mesolithic and the Pitted Ware culture of the Middle Neolithic, and in northern Sweden until the Middle Ages. In periods with a larger focus on agriculture and animal husbandry, the frequency of bears is lower, and changes of the landscape and hunting pressure in certain agricultural core-areas in the southern parts of Sweden may have resulted in a depletion of the brown bear in certain regions already in the Late Bronze Age. The occurrence of bear bones at settlements and towns from the Iron Age and Middle Ages in the southern parts of Sweden rather reflects the trading of bear skins from hunting grounds in northern Sweden.

During the Mesolithic, the brown bear was frequently hunted and thus occurred at most sites. A slight decline over time from the Early to Late Mesolithic can possibly reflect a decrease in bear populations in areas around large settlements, due to hunting pressure and changes of vegetation and biotopes. The anatomical distribution with a representation of different body parts reflects hunting near to settlements, and butchering marks show the utilisation of skins, meat, and bone marrow, as well as of bones and canine teeth for the production of osseous tools and ornaments.

The few finds of bears at Early Neolithic settlements in the southern parts of Sweden indicate a decline in bear hunting as a reflection of the neolithisation and the decrease in the importance of hunting for subsistence during that period. The low frequency of bears at settlements from the Late Neolithic until the Middle Ages in the southern parts of Sweden is probably a reflection of economies based on agriculture and animal husbandry, but likely also of the depletion of bear populations in agricultural core areas of Southern Sweden. This depletion seems to have started during the Bronze Age in Scania and during the Iron Age in other regions. Bear finds from settlements and towns in the southern parts of Sweden from the Late Iron Age and the Middle Ages mainly consist of claws and other bones from paws, which most likely represent the trading of bear skins. In the northern parts of Sweden, an increase in the frequency of bears in the find material can be noted up until the Iron Age and the Middle Ages, which was probably due to an increasing demand for bear skins to export to other regions, as is reflected in the occurrence of bear claws in Iron Age burials on Gotland and in the Mälar valley.

Finds of bear bones in ritual contexts (excluding burials) in Sweden seem to be more closely associated with periods and cultures where hunting played a more significant role in subsistence and the way of life. The ritual treatment of bear bones can be viewed as a reflection of respect for certain predators by hunter-gatherers, who saw them as equal to humans. Further, ritual practices with bones also played a part in the social and symbolic interactions between humans, bears, and the spiritual world to ensure successful hunting.

In periods and cultures in which animal husbandry played a significant role in the economy, the examples of the ritual treatment of bear bones are few and thus reflect another relationship between humans and bears. Possibly, this is due to bears becoming a threat to livestock and a generally different relationship with wild animals among farming societies. The bones of bears from the settlements and towns of the Iron Age and the Middle Ages mainly represent bear skins, which may have had symbolic importance, but can mainly be interpreted as representing a commodity signalling wealth and inter-regional contacts.

The use of the impressive bear canines as tooth pendants and amulets from the Mesolithic until the Middle Ages reflects a symbolic importance of bears to humans in Sweden that has transcended through the ages. The exact symbolic meaning of the bear canines and their social signals likely varied between different periods, geographic areas, and social contexts, but can probably be related to an almost universal concept of strength and power associated with bears that prevails in the long-term relationship between humans and bears in Sweden, as in most parts of the world.

Bibliography

Aaris-Sørensen 1978: K. Aaris-Sørensen, Knoglematerialet fra den mellemneolitiska boplads ved Korsnäs. Rapport, Riksantikvarieämbetet och Statens historiska museer 8, 1978, 1–28.

Aaris-Sørensen 1980: K. Aaris-Sørensen, Depauperation of the mammalian fauna of the island of Zealand during the Atlantic period. Videnskablige Meddelelser fra Dansk Naturhistorisk Forening 142, 1980, 131–138.

Aaris-Sørensen 2009: K. Aaris-Sørensen, Diversity and dynamics of the mammalian fauna in Denmark during the last glacial-interstadial cycle, 115–0 kyr BP. Fossils and Strata Monograph Series 57 (Chichester 2009).

Ambrosiani et al. 1984: B. Ambrosiani/E. Iregren/P. Lahtiperä, Gravfält i fångstmarken. Undersökningar av gravfälten Smalnäset och Krankmårtenhögen, Härjedalen (Stockholm 1984).

Arnesson-Westerdahl/Ericson 1989: A. Arnesson-Westerdahl/P. Ericson, Den tidigmesolitiska faunan vid Hornborgasjön i Västergötland. Skaraborgs länsmuseum, unpubl. report (Skara 1989).

Berglund 1991: B. E. Berglund, The cultural landscape during 6000 years in southern Sweden: the Ystad Project. Ecological Bulletins 41 (Copenhagen 1991).

Berglund et al. 1992: B. E. Berglund/S. Håkansson/J. Lepiksaar, The cultural Late Weichselian polar bear (Ursus maritimus Phipps) in southern Sweden. Sveriges Geologiska Undersökningar Series Ca 81, 1992, 31–42.

Berlin 1930: H. Berlin, Förteckning över benfynd från stenåldersboplatsen vid Sjöholmen. Meddelanden från Lunds universitets historiska museum 1929–1930, 1930, 40–42.

Binford 1978: L. R. Binford, Nunamiut Ethnoarchaeology (New York 1978).

Binford 1981: L. R. Binford, Bones: Ancient Men and Modern Myths (New York 1981).

Björck et al. 1997: S. Björck/M. Rundgren/O. Ingolfsson/S. Funder, The Preboreal oscillation around the Nordic Seas: terrestrial and lacustrine responses. Journal of Quaternary Science 12(6), 1997, 455–465.

Björck et al. 2019: N. Björck/M. Artursson/K.-F. Lindberg, Tråsättra. Aspekter på säljägarnas vardag och symbolik. Arkeologisk undersökning, Stockholms län, Uppland, Österåkers kommun, Österåker socken, Tråsättra 1:14, Österåker RAÄ 553. Arkeologerna Rapport 2019:40 (Stockholm 2019).

Björklöf 2010: S. Björklöf, Björnen i markerna och kulturen (Möklinta 2010).

Boethius 2017: A. Boethius, Huseby klev and the quest for pioneer subsistence strategies: diversification of a maritime lifestyle. In: P. Persson/B. Skar/H. M. Breivik/F. Riede/L. Jonsson (eds.), The early settlement of Northern Europe – Climate, human ecology, and subsistence (Sheffield 2017) 99–128.

Boethius 2018: A. Boethius, Signals of sedentism: Faunal exploitation as evidence of a delayed-return economy at Norje Sunnansund, an Early Mesolithic site in southeastern Sweden. Quaternary Science Reviews 162, 2018, 145–168.

Boethius et al. 2020: A. Boethius/M. Kjällquist/O. Magnell/J. Apel, Human encroachment, climate change and the loss of our archaeological organic cultural heritage: Accelerated bone deterioration at Ageröd, a revisited Scandinavian Mesolithic key-site in despair. PLoS One 2020. https://doi.org/10.1371/journal.pone.0236105.

During 1986: E. During, The Fauna of Alvastra. An Osteological Analysis of Animal Bones from a Neolithic Pile Dwelling. OSSA 12, Supplement 1 (Stockholm 1968).

Ekman 1973: J. Ekman, Early Mediaeval Lund, the fauna and the landscape. An osteological investigation of bone remains from the Early Mediaeval settlement. Archaeologica Lundensia. Investigationes de Antiqvitatibus Urbae Lundae V (Lund 1973).

Ekman/Iregren 1984: J. Ekman/E. Iregren, Early Norrland 8. Archaeo-zoological Investigations in Northern Sweden. Kungliga vitterhets historie och antikvitets akademien (Stockholm 1984).

Ericson et al. 2003: P. Ericsson/A.-S. Kjellberg/A. Åkermark Kraft/B. Wigh, Osteologisk analys av djurbensmaterialet. I: Ullén, I. Bronsåldersboplatsen vid Apalle i Uppland. Arkeologi på väg, undersökningar för E18. Riksantikvarieämbetet UV Uppsala Rapport 1997:64 (Uppsala 2003).

Eriksson/Magnell 2001a: M. Eriksson/O. Magnell, Det djuriska Tågerup. Nya rön kring Kongemose- och Ertebøllekulturens jakt och fiske. In: P. Karsten/B. Knarrström (eds.), Tågerup. Specialstudier (Lund 2001) 156–237.

Eriksson/Magnell 2001b: M. Eriksson/O. Magnell, Jakt och slakt. In: P. Karsten (ed.), Dansarna från Bökeberg. Om jakt, ritualer och inlandsbosättning vid jägarstenålderns slut (Stockholm 2010) 49–77.

Fransson 2011: U. Fransson, Från tand till kammargrav. Om att vara barn, Att bli stor och att dö i en hög. In: F. Fahlander (ed.), Spåren av de små: Arkeologiska perspektiv på barn och barndom (Stockholm 2011) 97–122.

Gummesson 2009: S. Gummesson, Osteologisk rapport på material från Åloppe, Norrskog. Opublicerad rapport, Osteoarkeologiska forskningslaboratoriet, Stockholm universitet (Stockholm 2009).

Gummesson 2014: S. Gummesson, Osteologisk rapport av animala skelettfynd från Kanaljorden, Motala. Östergötland. Opublicerad rapport, Osteoarkeologiska forskningslaboratoriet, Stockholm universitet. OFL Rapport 2014:2 (Stockholm 2014).

Gummesson 2019: S. Gummesson, Rapport från osteologisk analys av benmaterial från Dagsmosse, Östergötland. Förundersökning av lokalen Jussberg, samt fyndplats Bårstad. Arkivrapport, Stiftelsen Kulturmiljövård, Västerås (Västerås 2019).

Gummesson/Molin 2016: S. Gummesson/F. Molin, The Mesolithic cemetery at Strandvägen, Motala, in eastern central Sweden. In: J. M. Grünberg/B. Gramsch/L. Larsson/J. Orschiedt/H. Meller (eds.), Mesolithic burials – rites, symbols and social organisation of early postglacial communities/Mesolithische Bestattungen – Riten, Symbole und soziale Organisation früher postglazialer Gemeinschaften. International Conference, Halle (Saale), Germany, 18th–21st September 2013. Tagungen des Landesmuseums für Vorgeschichte Halle 13/I (Halle/Saale 2016) 145–159.

Gummesson et al. 2018: S. Gummesson/F. Hallbren/A. Kjellström, Keep your head high: skulls on stakes and cranial trauma in Mesolithic Sweden. Antiquity 92, 2018, 74–90.

Gummesson et al. 2019: S. Gummesson/J. Karlsson/H. Kivikero/O. Magnell/J. Storå, Osteologisk analys av djurben från Strandvägen. Arkeologisk undersökning, Östergötlands län, Östergötland, Motala kommun, Motala socken, fastighet Strandvägen 1. RAÄ 290. Osteoarkeologiska forskningslaboratoriet: Stockholm universitet, OFL Rapport 2017:4 (Stockholm 2019).

Hallgren/Fornander 2016: F. Hallgren/E. Fornander, Skulls on stakes and skulls in water. Mesolithic mortuary rituals at Kanaljorden, Sweden 7000 BP. In: B. Grüneberg/B. Gramsch/L. Larsson/J. Orschiedt/H. Meller (eds.), Mesolithic burials – rites, symbols and social organisation of early postglacial communities/Mesolithische Bestattungen – Riten, Symbole und soziale Organisation früher postglazialer Gemeinschaften. International Conference, Halle (Saale), Germany, 18th–21st September 2013. Tagungen des Landesmuseums für Vorgeschichte Halle 13/I (Halle/Saale 2016) 161–174.

Hallin 2008: G. Hallin, Ajvideboplatsen. Rapport från arkeologisk undersökning av fornlämning nr. 171 på fastigheten Ajvide 2:1 i Eksta socken, Gotland. Avdelningen för arkeologi och osteologi, Högskolan på Gotland (Visby 2008).

Hallowell 1926: A. I. Hallowell, Bear ceremonialism in the Northern Hemisphere. American Anthropologist 28(1), 1926, 1–175.

Hallström 1960: G. Hallström, Monumental art of northern Sweden from the Stone Age: Nämforsen and other localities (Stockholm 1960).

Hansson 1992: A. Hansson, Rapport över arkeologisk undersökning av Hedningsgärdet, fornlämning nr. 169, Ljusnedal 1:1, Härjedalens kommun, Jämtlands län, Arkeologiska undersökningar utförda av Jämtlands läns museum (Östersund 1992).

Hellgren/Magnell 2020: F. Hellgren/O. Magnell, Osteologisk analys. In: S. Larsson (ed.), Kvarteret Sankt Mikael 16. Arkeologisk undersökning 2019. Kulturen Rapport 2020 (Lund 2020).

Hellgren et al. 2020: F. Hellgren/O. Magnell/L. Nilsson, Osteologisk analys. In: M. Artursson/N. Hyll (eds.), Fjelie 13 och 50, en tidigneolitisk långdös och boplatslämningar från brons- och järnålder. Skåne, Lomma kommun, Fjelie socken, fastighety 2:13 och 16:3, fornlämning Fjelie 13 och 50, Arkeologisk undersökning 2016 och 2017. Arkeologerna Rapport 2020:58 (Stockholm 2020).

Holm 1993: J. Holm, Settlements of the Hamburgian and Federmeser Culture at Slotseng, south Jutland. Journal of Danish Archaeology 10, 1991, 7–19.

Holmgren 1985: Å. Holmgren, Analys av benmaterialet från fornlämning 45 Ås sn, Jämtland. Appendix C. In: M. Olausson, Kyrklägdan i Ås. Arkeologisk undersökning av en boplats från folkvandringstid till medeltid (Östersund 1985).

Hårding 1990: B. Hårding, Vad benen berättar. In: S. Tesch (ed.), Nakt och människor i kungens Sigtuna (Sigtuna 1990).

Iregren 1988: E. Iregren, Finds of brown bear (*Ursus arctos*) in Southern Scandinavia – Indications of local hunting or trade? In: B. Hårdh/L. Larsson (eds.), Trade and Exchange in Prehistory. Studies in honour of Berta Stjernquist. Acta Archaeologica Lundensia Series 8° (Lund 1988) 295–308.

Iregren et al. 1990: E. Iregren/B. Ringberg/A.-M. Robertsson, The brown bear (*Ursus arctos* L) find from Ugglarp, Southernmost Sweden. The skeleton, its age and environment. Sveriges geologiska undersökning. Avhandlingar och uppsatser, Serie C:824 (Uppsala 1990).

Jensen et al. 2020: Z. T. T. Jensen/A. Sjöström/A. Fischer/E. Rosengren/L. T. Lanigan/O. Bennike/K. K. Richter/K. J. Gron/M. Mackie/M. F. Mortensen/L. Sørensen/D. Chivall/K. H. Iversen/A. J. Taurozzi/J. Olsen/H. Schroeder/N. Milner/M. Sørensen/M. J. Collins, An integrated analysis of Maglemosian bone points reframes the Early Mesolithic of Southern Scandinavia. Nature, Scientific Reports 10, 17244 (2020). https://doi.org/10.1038/s41598-020-74258-8.

Johansson 1997: F. Johansson, Osteologisk analys av djurbenen från Varla. In: L. Lundqvist/E. Schaller Åhrberg, Med kunglig utsikt. Varla under järnålder och tidig medeltid. UV Väst Rapport 26, 1997, 115–192 (Mölndal 1997).

Jonsson 1984: L. Jonsson, Djuren i staden, sett ur osteologisk synvinkel. In: O. Ehn/J. H. Gustafsson (eds.), Kransen. Ett medeltida kvarter i Uppsala (Uppsala 1984) 88–94.

Jonsson 1986: L. Jonsson, Animal bones from Bredasten. Preliminary results. In: M. Larsson (ed.), Bredasten, an Early Ertebölle site with a dwelling structure in South Scania. Meddelanden från Lunds Universitets Historiska Museum 1985/1986, 1986, 50–51.

Jonsson 1988: L. Jonsson, The vertebrate faunal remains from the Late Atlantic settlement at Skateholm in Scania, South Sweden. In L. Larsson (ed.), The Skateholm Project I: Man and Environment (Lund 1988) 56–88.

Jonsson 2005: L. Jonsson, Rapport over inledande osteologisk underökning. In: B. Nordqvist (ed.), Huseby Klev, en kustboplats med bevarat organiskt material från äldsta mesolitikum till järnålder, Bohuslän, Morlanda socken, Huseby 2:4 och 3:13. RAÄ 89, 485. UV-väst Rapport 2005:2 (Mölndal 2005).

Karlsson 2001: J. Karlsson, Djurbenen från Burge i Lummelunda Osteologiska aspekter av Gotländsk ekonomi och konsumtion mellan heden och kristen tid. Unpublished CD-uppsats, Högskolan på Gotland (Visby 2001).

Karlsson 2009: J. Karlsson, Ekonomi och konsumtion i den tidiga medeltidens Tälje: en osteologisk utvärdering. Fornvännen 104(3), 2009, 177–187.

Karsten 2001: P. Karsten, Dansarna från Bökeberg. Om jakt, ritualer och inlandsbosättning vid jägarstenålderns slut (Stockholm 2010).

KINDGREN 1995: H. KINDGREN, Hensbacka-Hogen-Hornborgasjön: Early Mesolithic coastal and inland settlements in western Sweden. In: A. Fischer (ed.), Man and Sea in the Mesolithic: Coastal Settlement Above and Below Present Sea Level. Oxbow Monograph 53 (Oxford 1995) 171–184.

LAGERÅS/FREDH 2019: P. LAGERÅS/E. D. FREDH, Long-term development of landscape openness and arable in an agricultural region of southern Sweden: the potential of REVEALS estimates using pollen records from wells. Vegetation History and Archaeobotany 19, 2019, 113–124.

LARSSON 1989: L. LARSSON, Late Mesolithic settlements and cemeteries at Skateholm, Southern Sweden. In: C. Bonsall (ed.), The Mesolithic in Europe. Papers presented at the third International Symposium Edinburgh 1985 (Edinburgh 1989) 367–378.

LARSSON 1992: L. LARSSON, Neolithic settlement in the Skateholm area, southern Scania. Meddelande från Lunds Universitets Historiska Museum 1991/1992, 1992, 5–44.

LARSSON 1994: L. LARSSON, The earliest settlement in Southern Sweden. Late Paleolithic Settlement remains at Finjasjön in the North of Scania. Current Swedish Archaeology 2, 1994, 159–177.

LARSSON 1997: N. LARSSON, Djurbenen berättar. Vikingatida djurhållning på Gotland utifrån en osteologisk analys av djurbensmaterialet från Bandlundeviken, Häffinds 1:9 Burs sn. Opublicerad uppsats, Stockholm universitet (Stockholm 1997).

LARSSON 2013: L. LARSSON, Rich women and poor men. Analyses of a cemetery at Önsvala in the hinterland of Uppåkra. In: B. Hårdh/L. Larsson (eds.), Folk, fä och fynd. Uppåkra studier 12 (Lund 2013) 133–161.

LEPIKSAAR 1974: J. LEPIKSAAR, Djurrester från den mellanneolitiska (gropkeramiska) boplatsen vid Äs, Romfartuna sn, Västmanland. In: L. Löfstrand (ed.), Yngre stenålderns kustboplatser. Undersökningarna vid Äs och studier i den gropkeramiska kulturens kronologi och ekologi (Uppsala 1974) 140–156.

LEPIKSAAR 1978: J. LEPIKSAAR, Bone Remains from the Mesolithic Ageröd I:B and I:D. In: L. Larsson, Ageröd I:B-I:D. A Study of Early Atlantic Settlement in Scania (Lund 1978) 234–244.

LEPIKSAAR 1982: J. LEPIKSAAR, Djurrester från den tidigatlantiska boplatsen vid Segebro nära Malmö. In: L. Larsson, Segebro. En tidigatlantisk boplats vid Sege ås mynning (Malmö 1982) 105–168.

LEPIKSAAR 1983: J. LEPIKSAAR, Animal remains from the Atlantic bog site at Ageröd V in Central Scania. In: L. Larsson, Ageröd V, an Atlantic Bog Site in Central Scania (Lund 1983) 159–168.

LEPIKSAAR 1986: J. LEPIKSAAR, The Holocene history of theriofauna in Fennoscandia and Baltic Countries. STRIAE 24, 1986, 51–70.

LEPIKSAAR 1991: J. LEPIKSAAR, Fynd av djurrester från senmedeltida borgen Ekholmen vid Trollhättan. In: C. Aarsrud (ed.), Skärvor och fragment. Kring medeltid i Älvsborgs län. Västgöta-Dal, en årsbok från Älvsborgs länsmuseum, 1989/1990 (Vänersborg 1991).

LINDHOLM/LJUNGKVIST 2016: K. -J. LINDHOLM/J. LJUNGKVIST, The bear in the grave. Exploitation of top predator and herbivore resources in 1st millennium Sweden, first trends from a long-term research project. European Journal of Archaeology 19(1), 2016, 3–27.

LYMAN 1994: R. L. LYMAN, Vertebrate Taphonomy (Cambridge 1994).

MAGNELL 2004: O. MAGNELL, Djuren och människan. In: M. Andersson (ed.), Kustslättens mötesplatser (Stockholm 2004) 51–86.

MAGNELL 2006: O. MAGNELL, Tracking Wild Boar and Hunters. Osteology of Wild Boar in Mesolithic South Scandinavia (Lund 2006).

MAGNELL 2011: O. MAGNELL, Djurben från Rönneholms mosse, osteologisk analys av material från utgrävningar 2010. Institutionen för Arkeologi och Antikens historia, Lunds universitet, Reports in Osteology 2011:7 (Lund 2011).

MAGNELL 2017: O. MAGNELL, Climate and wild game populations in South Scandinavia at the Holocene Thermal Maximum. In: G. G. Monks (ed.), Climate Change and Human Responses. A Zoorchaeological Perspective. Vertebrate Paleobiology and Paleoanthropology (Dordrecht 2017) 123–135.

MAGNELL 2019: O. MAGNELL, Benmaterialet. In: N. Björck/M. Artursson/K.-F. Lindberg, Tråsättra, aspekter på säljägarnas vardag och symbolik. Arkeologisk undersökning, Stockholms län, Uppland, Österåkers kommun, Österåker socken, Tråsättra 1:14, Österåker RAÄ 553. Arkeologerna Rapport 2019:40 (Stockholm 2019) 167–176.

MAGNELL/IREGREN 2010: O. MAGNELL/E. IREGREN, Veitstu hvé blóta skal. The Old Norse blót in the light of osteological remains from Frösö church, Jämtland, Sweden. Current Swedish Archaeology 18, 2010, 223–250.

MANNERMAA/VON MOSCINSKY 2001: K. MANNERMAA/K. VON MOSCINSKY, Nymölla vid Skräbeån i Skåne. Unpublished C-uppsats i Historisk Osteologi, Institutionen för arkeologi och antikens historia, Lunds universitet (Lund 2001).

MIINA et al. 2009: J. MIINA/P.-P. HOTANEN/S. KAUKO, Modelling the Abundance and Temporal Variation in the Production of Bilberry (*Vaccinium myrtillus* L.) in Finnish Mineral Soil Forest. Silva Fennica 43(4), 2009, 577–593.

MOLIN et al. 2014: F. MOLIN/G. GRUBER/L. HAGBERG, Motala – a North European Focal Point? In: F. Riede/M. Tallavaara (eds.), Lateglacial and Postglacial Pioneers in Northern Europe. BAR International Series 2599 (Oxford 2014) 91–102.

MULK/IREGREN 1995: I.-M. MULK/E. IREGREN, Björngraven i Karats. Duoddaris 9 (Jokkmokk 1995).

MYRDAL 2012: J. MYRDAL, Boskapsskötseln under medeltiden: en källpluralistisk studie. Nordiska museets handlingar 139 (Stockholm 2012).

NILSSON 2006: L. NILSSON, Djur och människor längs vägen. Malmöfynd nr 9. Malmö kulturmiljö (Malmö 2006).

NOE-NYGAARD 1995: N. NOE-NYGAARD, Ecological, Sedimentary and Geochemical Evolution of the Late-Glacial to Postglacial Åmose Lacustrine Basin, Denmark. Fossils and Strata 37 (Oslo, Copenhagen, Stockholm 1995).

NORDQVIST 1997: B. NORDQVIST, Västkusten. In: M. Larsson/C. Lindgren/B. Nordqvist (eds.), Regionalt och interregionalt: stenåldersundersökningar i Syd- och Mellansverige, Riksantikvarieämbetet (Stockholm 1997) 32–46.

Oehrl 2013: S. Oehrl, Bear hunting and its ideological context (as a background for the interpretation of bear claws and other remains of bears in Germanic graves of the 1st millennium AD). In: O. Grimm/U. Schmölcke (eds.), Hunting in northern Europe until 1500 AD – Old traditions and regional developments, continental sources and continental influences. Papers presented at a workshop organised by the Centre for Baltic and Scandinavian Archaeology, Schleswig, June 16th and 17th, 2011. Schriften des Archäologischen Landesmuseums, Ergänzungsreihe 7 (Neumünster 2013) 297–332.

Olson et al. 2011: C. Olson/N. Björck/J. Storå, Huts and Deposition of Refuse at Fräkenrönningen, a Neolithic Coastal Dwelling site in Eastern Middle Sweden. International Journal of Osteoarchaeology 21, 2011, 173–186.

Pales/Garcia 1981: L. Pales/M. Garcia, Atlas ostéologique pour servir à l'identification des Mammifères du Quaternaire 2 – Carnivores/Homme (Paris 1981).

Parducci et al. 2019: L. Paducci/I. G. Alsos/P. Unneberg/M. W. Pedersen/L. Han/Y. Lammers/J. S. Salonen/M. Väliranta/T. Slotte/B. Wohlfarth, Shotgun Environmental DNA, Pollen, and Macrofossil Analysis of Lateglacial Lake Sediments from Southern Sweden. Frontiers in Ecology and Evolution 7. https://doi.org/10.3389/fevo.2019.00189.

Petré 1980: B. Petré, Björnfällen i begravningsritualen, statusobjekt speglande regional skinnhandel. Fornvännen 75, 1980, 5–14.

Ricard/Messier 1996: J.-P. Ricard/C. Messier, Abundance, growth and allometry of red raspberry (*Rubus idaeus* L.) along a natural light gradient in a northern hardwood forest. Forest Ecology and Management 81, 1996, 153–160.

Sahlén et al. 2006: V. Sahlén/J. Swenson/S. Brunberg, Björnen i Sverige. En rapport från Skandinaviska Björnprojektet till den svenska Rovdjursutredningen. Skandinaviska Björnprojektet, Rapport 2006, nr. 4, 1–41 (Uppsala 2006).

Segerberg 1999: A. Segerberg, Bälinge mossar. Kustbor i Uppland under yngre stenåldern. Aun 26 (Uppsala 1999).

Servheen 1990: C. Servheen, The status and conservation of the bears of the world. International conference on bear research and management. Monographic series no. 2 (Victoria 1990).

Sjöling 2000: E. Sjöling, Benartefakterna från Korsnäs. En arkeoosteologisk analys av ben-, horn- och tandartefakter från en gropkeramisk lokal i Södermanland. Unpublished CD-uppsats i arkeoosteologi, Stockholms universitet (Stockholm 2000).

Sjöling 2013: E. Sjöling, Osteologisk analys. Boplatsmaterial. RAÄ 113, Österleden, område A, Vaksala 1:1, Vaksala socken, Uppland. UM 8202. In: M. Lucas/R. Lucas, Kring Vaksala prästgårds hage – från bronsålderliv till vendeltidsdöd i Vaksalas centralbygd. Arkeologisk förundersökning och särskild undersökning. Fornlämning 104, 113 & 322, Vaksala socken, Uppland. Upplandsmuseets rapporter 2013:31 (Uppsala 2013).

Sjöling/Bäckström 2006: E. Sjöling/Y. Bäckström, Osteologisk analys av benmaterialet från Kyrsta RAÄ 327, Kyrsta 39b. Bilaga 3. In: A. Onsten-Molander/J. Wikborg, Kyrsta. Del 2. Förhistoriska boplatslämningar. Undersökningar för E4, RAÄ 327 & RAÄ 330, Ärentuna socken, Uppland. SAU Skrifter 17 (Uppsala 2006).

Ukkonen/Mannermaa 2017: P. Ukkonen/K. Mannermaa, Jääkauden jälkeläiset. Suomen nisäkkäiden ja lintujen varhainen historia (Helsinki 2017).

Vang Petersen 1998: P. Vang Petersen, Rav, hjortetak og mesolitisk magi. Danefæ fra jægerstenalderen. Nationalmuseets Arbejdsmark 1998, 87–100.

Vretemark 1997: M. Vretemark, Från ben till boskap. Kosthåll och djurhushållning med utgångspunkt i medeltida benmaterial från Skara. Skara: Skrifter från Skaraborgs Länsmuseum 25 (Skara 1997).

Wigh 2001: B. Wigh, Animal Husbandry in the Viking Age Town of Birka and its Hinterland (Stockholm 2001).

Wigh 2008: B. Wigh, Benens vittnesbörd. Osteologisk analys av benmaterialet från Ryssgärdet. In: E. Hjärthner-Holdar/T. Eriksson/A. Östling (eds.), Mellan himmel och jord: Ryssgärdet, en guldskimrande bronsåldersmiljö i centrala Uppland. Arkeologi E4 Uppland, studier 5 (Uppsala 2008) 371–389.

Wikell 2015: R. Wikell, En björnkvinna eller bärsärk i Sigtuna? In: A. Söderberg/L. Tamm/R. Edberg/N. Källström/J. Runer (eds.), Situne Dei. Årsskrift för Sigtunaforskning och historisk arkeologi 2015 (Sigtuna 2015) 6–13.

Wyszomirska 1984: B. Wyszomirska, Figurplastik och gravskick hos nord- och nordösteuropas Neolitiska fångstkulturer. Acta Archaeologica Lundensia Series in 4°N °18 (Lund 1984).

Wyszomirska 1988: B. Wyszomirska, Ekonomisk stabilitet vid kusten. Nymölla III. En tidigneolitisk bosättning med fångstekonomi i nordöstra Skåne (Lund 1988).

Zachrisson/Iregren 1974: I. Zachrisson/E. Iregren, Lappish bear graves in Northern Sweden. Early Norrland 5 (Stockholm 1974).

Dr. Ola Magnell
Arkeologerna, National Historical Museums
Lund
Sweden
ola.magnell@arkeologerna.com

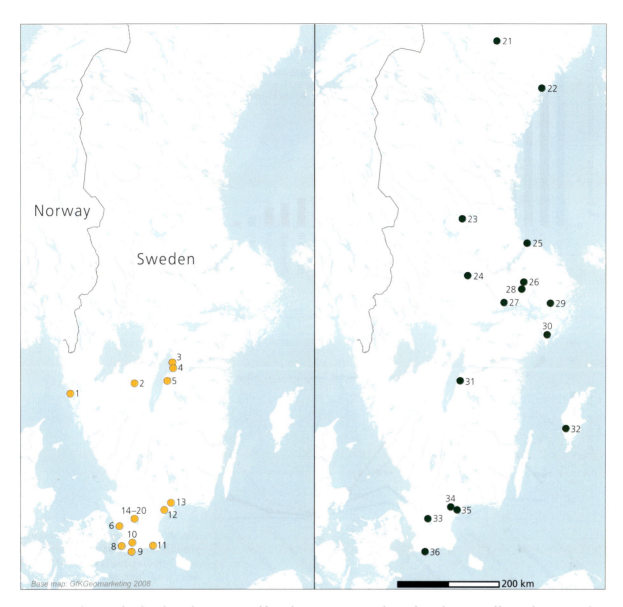

Fig. 1. Distribution of archaeological sites (NISP of faunal remains >100) with osteological remains of brown bear in Sweden. To the left – Mesolithic: 1: Huseby klev; 2: Almeö; 3: Kanaljorden; 4: Strandvägen; 5: Jussberg; 6: Tågerup I, II, III; 7: Segebro; 8: Arlöv I; 9: Skateholm I; 10: Bökeberg III; 11: Bredasten; 12: Nymölla III; 13: Sunnansund; 14: Ageröd I:HC; 15: Ageröd I:D; 16: Ageröd I:B; 17: Rönneholms Mosse; 18: Ringsjöholm; 19: Ageröd V; 20: Sjöholmen. To the right – Early and Middle Neolithic: 21: Åsele 1023–1024; 22: Bjästamon; 23: Ore 6; 24: Korsnäset/Grangärde 128; 25: Fräkenrönningen; 26: Sotmyra; 27: Äs; 28: Åloppe; 29: Tråsättra; 30: Korsnäs; 31: Alvastra; 32: Ajvide; 33: Sjöholmen; 34: Hunneberget; 35: Nymölla I; 36: Rävgrav. References and NISP of sites shown in Tables 1 and 2 (maps GIS department, ZBSA).

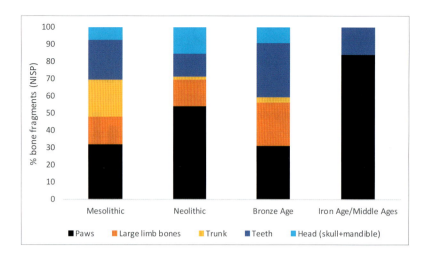

Fig. 6. Distribution of osteological remains of brown bear from five body parts found at settlements in southern parts of Sweden dating to different periods. The quantification is based on NISP and a compilation of several different sites from each period.

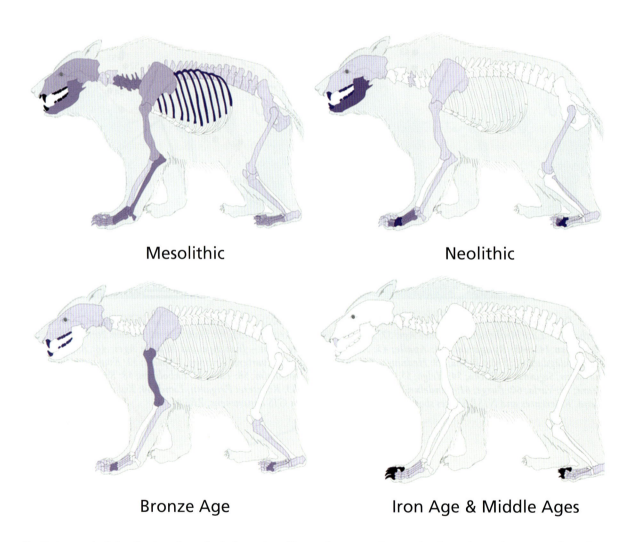

Fig. 7. Anatomical distribution of osteological remains of brown bear at settlement sites from the southern parts of Sweden, dating to different periods. Colouring indicates the presence of a bone element; the darker the colour, the higher the frequency of bone or teeth occurance (© 2003 ArcheoZoo.org; after PALES/GARCIA 1981, pl. 13).

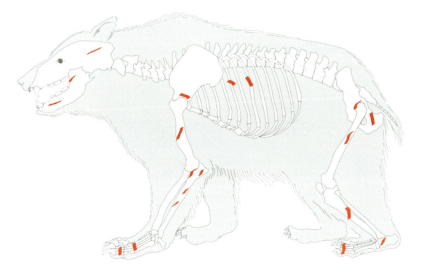

Fig. 8. Anatomical distribution of butchering marks on bear bones at settlement sites from the Mesolithic (© 2003 ArcheoZoo.org; after PALES/GARCIA 1981, pl. 13).

Fig. 9. Tooth pendants made of bear canines on the chest of a child from double burial 41 at the Late Mesolithic cemetery of Skateholm I (photo L. Larsson).

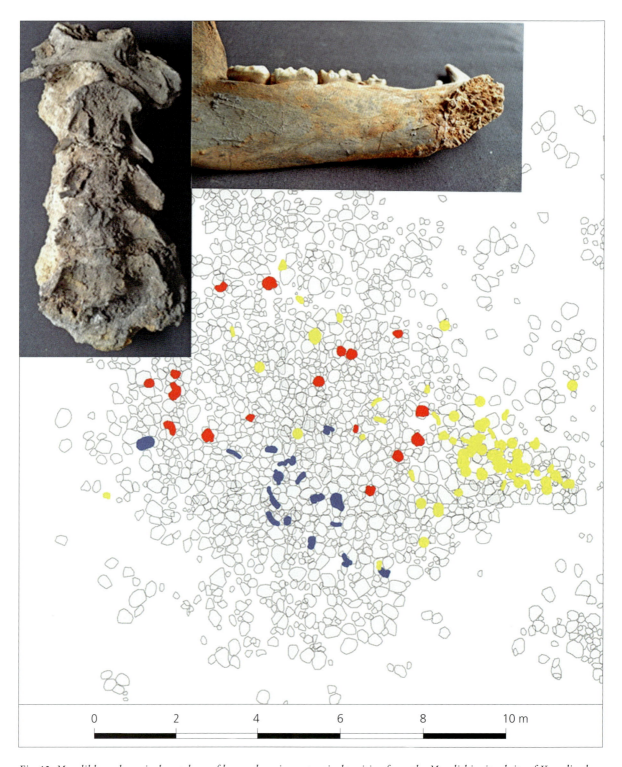

Fig. 10. Mandible and cervical vertebrae of brown bear in anatomical position from the Mesolithic ritual site of Kanaljorden. Spatial distribution of human bones (red), bones of brown bears (blue), and wild boar (yellow) on the stone packing (after Gummesson *2014, figs. 12; 16;* Gummesson *et al. 2018, fig. 9).*

Fig. 11. Phalanx 3 (claw) of a brown bear from the 11–13th centuries, found in the quarter Sankt Mikael in the town of Lund, in comparison with the skeleton of a recent bear (photo O. Magnell).

Table 1. Settlement sites with osteological remains of brown bear dating to the Mesolithic from Sweden, based on sites with total NISP (number of identified specimens) of mammals >100. EM = Early Mesolithic (9500–6400 BC); MM = Middle Mesolithic (6400–5400 BC); LM = Late Mesolithic (5400–3900 BC). Numbers of sites refer to numbers shown on maps of Fig. 1.

Site	County	Dating	NISP	Reference
1a: Huseby klev (Preboreal phase)	Bohuslän	EM	3	JONSSON 2005; BOETHIUS 2018
1b: Huseby klev (Boreal phase)	Bohuslän	EM	1	JONSSON 2005; BOETHIUS 2018
2: Almeö	Västergötland	EM	5	ARNESSON-WESTERDAHL/ERICSON 1989
3: Kanaljorden	Östergötland	MM	214	GUMMESSON 2014
4: Strandvägen	Östergötland	LM	26	GUMMESSON et al. 2019
5: Jussberg	Östergötland	EM	4	GUMMESSON 2019
6a: Tågerup phase I	Skåne	MM	23	ERIKSSON/MAGNELL 2001a
6b: Tågerup phase II	Skåne	MM	8	ERIKSSON/MAGNELL 2001a
6c: Tågerup phase III	Skåne	LM	2	ERIKSSON/MAGNELL 2001a
7: Segebro	Skåne	MM	17	LEPIKSAAR 1982
8: Arlöv I	Skåne	LM	2	JONSSON 1988
9: Skateholm I	Skåne	LM	2	JONSSON 1988
10: Bökeberg III	Skåne	LM	9	ERIKSSON/MAGNELL 2001b
11: Bredasten	Skåne	LM	10	JONSSON 1986; MAGNELL 2006
12: Nymölla III	Skåne	LM	1	WYSZOMIRSKA 1988
13: Sunnansund	Skåne	EM	19	BOETHIUS 2017
14: Ageröd I:HC	Skåne	EM	67	BOETHIUS et al. 2020
15: Ageröd I:D	Skåne	EM	2	LEPIKSAAR 1978
16: Ageröd I:B	Skåne	EM	3	LEPIKSAAR 1978
17: Rönneholms mosse	Skåne	EM	2	MAGNELL 2011
18: Ringsjöholm	Skåne	MM	21	MAGNELL 2006
19: Ageröd V	Skåne	MM	1	LEPIKSAAR 1983
20: Sjöholmen (lower layers)	Skåne	LM	12	BERLIN 1930; unpublished data

Table 2. Settlement sites with osteological remains of brown bear dating to the Early and Middle Neolithic in Sweden. Based on sites with total NISP (number of identified specimens) of mammals >100. EN = Early Neolithic (3900–3300 BC), MN = Middle Neolithic (3300–2300 BC), LN = Late Neolithic (2300–1800 BC); BA = Bronze Age (1800–500 BC); x = quantification of NISP unknown. Numbers of sites refer to numbers shown on maps of Fig. 1.

Site	County	Dating	NISP	Reference
21: Åsele 1023–1024	Ångermanland	MN (LN, BA)	3	Ekman/Iregren 1984
22: Bjästamon 23	Ångermanland	MN–LN	x	Ekman/Iregren 1984
23: Ore 6	Dalarna	MN	253	Ekman/Iregren 1984
24: Korsnäset/ Grangärde 128	Dalarna	MN	8	Ekman/Iregren 1984
25: Fräkenrönningen	Gästrikland	MN	2	Olson et al. 2011
26: Sotmyra	Uppland	MN	1	Segerberg 1999
27: Äs	Västmanland	MN–LN	12	Lepiksaar 1974
28: Åloppe	Uppland	MN	3	Gummesson 2009
29: Tråsättra	Uppland	MN	2	Magnell 2019
30: Korsnäs	Södermanland	MN	3	Aaris-Sørensen 1978
31: Alvastra	Östergötland	MN	3	During 1986
32: Ajvide	Gotland	MN	1	Hallin 2008
33: Sjöholmen (upper layer)	Skåne	MN	2	Berlin 1930; unpublished data
34: Hunneberget	Skåne	MN	1	Magnell 2004
35: Nymölla I	Skåne	MN	22	Mannermaa/Von Moscinsky 2001
36: Rävgrav	Skåne	EN–MN	x	Larsson 1992

Table 3. Settlement sites with osteological remains of brown bear dating to the Late Neolithic and Bronze Age in Sweden, based on sites with total NISP (number of identified specimens) of mammals >100. M = Mesolithic (9500–3900 BC); N = Neolithic (3900–1800 BC); LN = Late Neolithic (2300–1800 BC); BA = Bronze Age (1800–500 BC); EIA = Early Iron Age (500 BC–AD 400); IA = Iron Age (500 BC–AD 1050); x = quantification of NISP unknown. Numbers of sites refer to numbers shown on maps of Fig. 5.

Site	County	Dating	NISP	Reference
37: Arjeplog 532	Lappland	N, BA, EIA	3	Ekman/Iregren 1984
38: Arjeplog 522	Lappland	BA, EIA	7	Ekman/Iregren 1984
39: Tåsjö 154	Ångermanland	LN, BA, IA	2	Ekman/Iregren 1984
40: Bodum 26	Ångermanland	M, N, BA, IA	3	Ekman/Iregren 1984
41: Åsele 1028	Ångermanland	M, BA	2	Ekman/Iregren 1984
42: Rätan 145	Jämtland	LN–BA	5	Ekman/Iregren 1984
43: Ryssgärdet	Uppland	BA	2	Wigh 2008
44: Apalle	Uppland	BA	33	Ericson et al. 2003
45: Ängdala	Skåne	BA	x	Nilsson 2006

Table 4. Settlement sites with osteological remains of brown bear dating to the Iron Age and the Middle Ages in Sweden, based on sites with total NISP (number of identified specimens) of mammals >100. EIA = Early Iron Age (500 BC–AD 400); RIA = Roman Iron Age (0–AD 400; MP = Migration Period (AD 400–550); VP = Vendel Period (AD 550–800); VA = Viking Age (AD 800–1050); MA = Middle Ages (AD 1050–1500); x = quantification of NISP unknown. Numbers of sites refer to numbers shown on maps of Fig. 5.

Site	County	Dating	NISP	Reference
46: Arjeplog 508	Lappland	RIA	1	Ekman/Iregren 1984
47: Kyrklägdan	Jämtland	MA	2	Holmgren 1985
48: Frösö kyrka	Jämtland	VA	256	Magnell/Iregren 2010
49: Hedningsgärdet	Jämtland	MA	1	Hansson 1992
50: Kyrsta	Uppland	EIA	1	Sjöling/Bäckström 2006
51: Kv. Kransen, Uppsala	Uppland	MA	x	Jonsson 1984
52: Kv. Trädgårdsmästaren, Sigtuna	Uppland	VA–MA	3	Hårding 1990
53: Birka	Uppland	VA	9	Wigh 2001
54: Helgö	Uppland	RIA–VP	x	Kind information by B. Stolle, Osteoarkeologiska forskningslaboratoriet, Stockholms universitet
55: Helgeandsholmen, Stockholm	Uppland	MA	1	Vretemark 1997
56: Tälje	Södermanland	MA	1	Karlsson 2009
57: Ekholmen	Västergötland	MA	1	Lepiksaar 1991
58: Kv. Rådhuset, Skara	Västergötland	MA	1	Vretemark 1997
59: Varla	Halland	MP–VP	1	Johansson 1997
60: Burge	Gotland	MA	1	Karlsson 2001
61: Bandelundaviken	Gotland	VA	1	Larsson 1997
62: Uppåkra	Skåne	MP–VP	1	Unpublished data
63: Kv. Färgaren, Lund	Skåne	VA–MA	1	Ekman 1973
64: Kv. Sankt Mikael, Lund	Skåne	MA	2	Hellgren/Magnell 2020

Table 5. Frequency of butchering and gnawing marks on bones of brown bear and wild boar at four Mesolithic sites in % (cf. Magnell 2006; Gummesson et al. 2019).

	Butchering marks		Gnawing marks	
	Brown bear	Wild boar	Brown bear	Wild boar
Ageröd I:HC	19.0	9.7	20.7	12.9
Ringsjöholm	12.5	21.0	20.8	12.3
Tågerup	11.8	15.0	0.0	9.0
Strandvägen	15.4	3.3	0.0	3.7

Zooarchaeological brown bear (*Ursus arctos*) finds in eastern Fennoscandia

By Kristiina Mannermaa, Tuija Kirkinen and Suvi Viranta-Kovanen

Keywords: Fennoscandia, bears, human-bear relationships, osteology, animal hair identification

*Abstract: In Finland, the present-day brown bear (*Ursus arctos*) population is just over 2,000 animals, and the species has been classified as "near threatened" in the country. Today, bear tours are sold to tourists, during which the animals are fed with carcasses and then observed and photographed from special cabins. Simultaneously, the bear population is regulated by hunting, and bear meat is considered a delicacy (although rarely eaten). In ethnographic sources that describe the traditional Finno-Karelian bear hunting ritual in detail, the bear was treated as near-human with the potential to be an ancestor. The aim of these rituals was to secure the rebirth of the bear. In this paper, we present the zooarchaeological find material related to the bear, i.e. burnt and unburnt brown bear bone finds as well as keratinous bear pelt remains and hairs from sites in Finland and on the Karelian Isthmus, dating from the Mesolithic to medieval times (c. 10,000 cal BP to AD 1500). Based on these finds, we discuss the roles of the bear from its first immigration into the region to the Middle Ages.*

Introduction

The Finno-Karelian epic, ritual poetry and ethnographic sources are rich in references to the brown bear (*Ursus arctos*). At the core of this tradition is the slaying of a bear, a ritual drama which began with the killing of the bear in its winter den, culminated in the ritual wedding of the bear and a maiden, and ended by the returning of the bear's bones back to the circle of life. This was done by hanging the skull up on a tree and burying the bones at the foot of a sacred pine (Fig. 1; see Piludu, this volume; Holmberg [Harva] 1915; Haavio 1967; Honko 1993; Pentikäinen 2007; Krohn 2008; Siikala 2008; 2012).

The Finno-Karelian bear rite has its roots in circumpolar bear ceremonialism, which has been documented widely in northern America and Eurasia, e.g. among the Sámi and many Siberian peoples (e.g. Hallowell 1926; Äikäs et al. 2009, 118; Sarmela 2009, 80; see also Rydving, this volume; Iregren, this volume). The rite has been practiced by northern hemisphere hunting populations, hypothetically since the Palaeolithic (Zvelebil 1997; Sarmela 2009, 80; Conneller 2011, 365–366; Siikala 2012, 381). At the core of the Finno-Karelian tradition was the idea of the bear's divine origin and its relationship to humans (Itkonen 1948b, 364–366; Haavio 1967, 16–41; Krohn 2008, 146–164; Sarmela 2009, 79–94; Siikala 2012, 368–370; Witzel 2013; see also Russell 2012, 52–58, 168–170).

Archaeologist Knut Helskog has analysed these practices in the ethnohistoric record from the circumpolar arctic and examined the meanings that were associated with bears among prehistoric hunter-fisher-gatherer populations in northern Fennoscandia (HELSKOG 2012). However, his extensive analysis did not focus on Finland in particular, nor on bone and other perishable materials. Most notably, HELSKOG (2012) did not find a uniform pattern in the representation or contexts of bear finds; instead, he argues that clear variations exist in how bears were hunted, killed, ritualised and cosmologised. Accordingly, when Oliver Grimm studied bear phalanges and the remains of bear pelts found in Iron Age burials in Scandinavia and central Europe, he noticed that the material does not support a uniform explanation for the finds (GRIMM 2013).

In this paper, we combine the data from the zooarchaeological find material, i.e. burnt and unburnt brown bear bone finds as well as keratinous bear pelt remains and hairs from the Finnish sites dating from the Mesolithic to medieval times (*c.* 10,000 cal BP to AD 1500). We present the spatial and temporal distribution of archaeological brown bear bone finds across Finland and the Karelian Isthmus, and we discuss the data in terms of the immigration history of the brown bear in Finland as well as the utilisation of the species by prehistoric and early historical people during different cultural periods. The anatomical composition of individual bone assemblages was recorded to study possible differences between chronological periods, as well as between sites with different characters. Based on these results, we build a nuanced picture of the prehistoric and early historical ways of treating brown bear bones and skins in settlement sites, burials, and other ritual contexts.

Brown bear populations in Finland

General remarks

The brown bear is a part of the forest fauna in modern northern Eurasia. In Fennoscandia today, the bear population is regulated by hunting especially in the reindeer herding areas in Lapland. Hunting takes place in the autumn and is often performed with a special breed of dogs, the Karelian bear dog. The Karelian bear dog is an original Finnish breed, but these bold dogs are also used in other countries to hunt and scare bears away from human activities. The Karelian bear dog is probably descended from local dogs, and their traditional use for bear hunting may extend back to prehistory (POHJOISMÄKI et al. 2018).

Today, the brown bear is classified as "near threatened" (NT) in Finland by the IUCN Red List. It nearly faced extinction in all Nordic countries in the 19th century, and the populations have only recovered during the past 30 years (ZEDROSSER et al. 2001). The main reason for its recovery was the immigration of bears from Russia. The Finnish bear population continues to receive gene flow (migrating individuals) from Russia, but also, in the north, to a lesser extent from Scandinavian bears (KOPATZ et al. 2014). The gene flow from Finland to southern Scandinavia is very limited. The Finno-Russian bears remain largely isolated from Norwegian and Swedish bears.

This pattern is similar to that found in prehistory. During the re-colonisation – after the ice sheet had retreated in the early Holocene (*c.* 12,000 years ago onwards) – the Scandinavian bear population derived from the western lineage, whereas the Finnish bears originated from the eastern one. The two lineages spread from the Last Glacial Maximum (LGM) refugia in Europe. The western lineage had its refugia on the Iberian peninsula, as well as in Italy and the Balkans, whereas the eastern lineage had Carpathian origins (TABERLET/BOUVET 1994).

Bear remains are very common finds in postglacial subfossil materials in southern and central Europe. Among the order Carnivora, it is second only to the red fox in the frequency of occurrence (SOMMER/BENECKE 2005). The subfossil abundance is probably due to the fact that bears were widespread and common, but also that they hibernated in caves, where they occasionally died and their remains preserved well (e.g. Sabol 2005). Whereas most bear subfossils in southern and central Europe are unconnected with humans, this is not the case in Finland, where all (or most) of the material is human refuse or derives from bear burials or graves.

The earliest Finnish burnt bear bone find (from Malmio IA settlement in Savukoski, northern Finland) dates so early (*c.* 10,000 years; Nurminen 2020) that it must be very close to the actual arrival of the species.

Distribution history
The brown bear co-evolved in the Middle Pleistocene of Eurasia, and it co-occurred and probably co-evolved alongside Palaeolithic hunters. Both humans and bears became highly adapted to the Quaternary climatic fluctuations. It is most probable that humans have negatively affected the size of bear populations since the beginning of the Holocene (Albrecht et al. 2017).

It is likely that even during the LGM there were periods when fauna, including bears, were able to occupy territories beyond their refugia. The post-LGM bear populations show a consistent presence in the changing environments, e.g. in the northern Urals (Bachura/Kosintsev 2006) and the Alps (Döppes/Pacher 2014). Changes in seasonal temperatures may affect population sizes, but this is not shown in the fossil record (Albrecht et al. 2014). However, genetic evidence reveals that the lineage from the Carpathian refuge did not intermix with the southern or other lineages (Davison et al. 2011), which also indicates a distribution route for bears similar to the postglacial spread of humans (*c.* 12,000 cal BP onwards). The fossil (subfossil) record of bears is incomplete and does not provide a complete postglacial history. The earliest record in the Baltic area comes from Estonia and is dated to the Preboreal period (10,300 to 9000 years ago; Sommer/Benecke 2005).

Material and methods

Research area
The research area covers the Finnish mainland but excludes the Åland Islands (see Gustavsson/Ljungkvist, this volume), which most likely have never had a permanent brown bear population of their own. The Karelian Isthmus was formerly divided by the Finnish-Russian border until the border was moved following World War II; find material archived in Finland (before 1945) is included in the cases surveyed here (Fig. 2).

Zooarchaeological research material
The research material consists of zooarchaeological brown bear finds, i.e. hard and soft products and waste such as bones, teeth, claws, pelts and hair, excavated from archaeological contexts. A central factor affecting the preservation of organic materials in Finland is the acidity of soils, which causes the relatively fast decaying of unburnt bones and untanned skins (Arponen 2008; Hurcombe 2014, 93). Burning, however, improves the preservation of bones considerably (Ukkonen 2001).

Osteological material
The osteological material from the Finnish Stone Age consists almost exclusively of burnt bones. From the Iron Age on, both burnt and unburnt bones have been found, mostly depending on the character of the site, and burnt bones have been found especially in cremation burials and also, to a minor degree, in settlement site layers. The Stone Age assemblages consisting of burnt and fragmented bones present many challenges. Fragmentation affects the number of identifiable taxa (e.g. Ukkonen 2001; Seitsonen et al. 2017). The material can also be biased due to the morphology, size and histology (Vaneeckhout et al. 2013) of bones, which leads to uneven preservation (Iregren/

Jonsson 1973; Okkonen 1991; Lyman 1994, 386–390; Sigvallius 1994; Ukkonen 2001; Mannermaa 2008). For example, when a bone from a large animal breaks into smaller pieces, these pieces often have fewer diagnostic features compared to equivalent pieces of bones from smaller animals.

The archaeological data from the brown bear bone finds have been collected from Kirkinen (2017) and Ukkonen/Mannermaa (2017). These lists have been updated from the literature and the osteological reports from 2017–2020 archived in the Finnish Heritage Agency. Osteological analyses were conducted by several osteologists over a long period of time (Table 1). The treatment of the (burnt/unburnt) bones, anatomical elements, find context and geographical location of the finds were recorded for all assemblages.

Hair finds

Keratin animal fibres, such as wool and hair, are not as prone to acidity as e.g. bast fibres, and minuscule hair fragments have been found even in Stone Age contexts (Ahola et al. 2018; Kirkinen 2019a; see also Wilson 2008; Wilson/Tobin 2010; Tridico et al. 2014). Skin tissue, which is composed mostly of collagen, is not usually preserved in the acidic soils of Finland without tanning, except in special cases (see Arponen 2008). This means that pelts can be seen mostly as clumps of loose hair in archaeological assemblages. Hair, as an organic soft material, is prone to degradation. As hairs burn or char in a fire, they have been almost entirely preserved in inhumation burials. Hairs have been found especially in Late Iron Age and medieval inhumation burials. The inhumation burial tradition started in southwestern Finland during the 6th century AD, where it appears to be connected to at least some Scandinavian immigration (e.g. Raninen/Wessman 2015, 281–282). Inhumation traditions appear to be connected to the early spread of, probably vernacularised, Christianity, especially during the period AD 1000–1150; in eastern Finland and on the Karelian Isthmus inhumation appears during the 12th century, where it is also connected to Christianisation. The furnishing of graves with pelts and fur garments ended gradually in western Finland in the 13th century AD, during the period of the so-called Baltic Crusades when church-authorised Christianity was aggressively asserted on local cultures from the west (e.g. Ahola/Frog 2014, 42–43). In eastern Finland and on the Karelian Isthmus, the most recent furnished graves have been [14]C- and coin-dated to the 14th–16th centuries (Mikkola 2009, 184; 2012; Laakso 2014, 130). Regarding northern Finland, the animal hair materials from the 17th-century Forest-Sámi burial ground of Mukkala have been included in the study.

In burials, hairs have been found especially in contact with metal items, the toxic alloys of which prevent the activity of fungi and bacteria (cf. Edwards 1989; Solazzo et al. 2014). Also, microbial- and chemical-induced breakdown can create an anoxic state in the grave and, with the persistence of these conditions, increase the possibility of the preservation of hairs (Wilson et al. 2001, 215).

The data concerning hair finds have been collected from Schwindt (1893), Kirkinen (2015; 2017; 2019a) and Kirkinen et al. (2019; 2020a; b). The hairs have been identified by their morphological features; no DNA results were available because of the acidity of the soils in Finland. The context and geographical location of the finds were recorded for all assemblages.

Dating

Eight bones, i.e. just a fraction of the bear finds, have been radiocarbon-dated (see Table 2). Most of the research material was dated on the basis of the typology of the human-made finds, e.g. pottery or brooches from the sites or their context (see Table 3). The archaeological dating is, however, problematic. In Finland, the prehistoric settlement sites often lack stratigraphy, resulting in layers that may represent several occupation phases. Also, the Iron Age cremation cemeteries under level ground, in which the majority of burnt brown bear phalanges have been found, are collective by nature (Wessman 2010). For these reasons, many of the sites are categorised as multiperiodic. Only the Late Iron Age and early medieval inhumation burials as closed contexts can offer a relatively reliable archaeological dating for the bone and hair finds.

In this research, the material was divided into seven chronological periods after Ukkonen/Mannermaa (2017; cf. Fig. 2). Additionally, more detailed datings are presented in the text. All Stone Age periods 1–4 refer to populations practicing foraging economies. Farming and animal husbandry became a dominant subsistence only during the periods 5–6 (see Table 3).

Results

Osteological material

Brown bear bones have been identified at 77 archaeological sites. Both the number of sites containing brown bear bones and the number of identified bone fragments (431) is low, compared, for instance, to those for Eurasian elk (*Alces alces*; 4,050 fragments in 250 assemblages, situation in 2016). The spatial and temporal distribution of the sites is shown in Figure 2. Brown bear bones were recovered from all archaeological periods (Table 3) except Period 5 (Early Metal Period and Early Iron Age; 3750–1700 cal BP). However, some of the bones from mixed layers might derive from these periods, but this can only be confirmed by direct AMS-datings.

Eight brown bear fragments have been radiocarbon-dated (Table 2). The earliest finds come from the Early Mesolithic site of Malmio IA in Savukoski, the Late Mesolithic/Early Neolithic sites of Saamen Museo in Inari, Neitilä in Kemijärvi (northern Finland), and Käyrälampi in Kouvola (eastern Finland). Bones from the cremation cemetery of Rikalanmäki in Salo (southwestern Finland) date to the Late Iron Age.

All brown bear bones in Stone Age assemblages are burnt, and all historical bear finds are unburnt (Fig. 3). The majority of bear bones from the Iron Age are burnt. Unburnt bear bones are also found at some multiperiodic sites.

The distribution of bear finds by the character of their contexts is presented in Figure 4. Stone Age samples derive from occupation layers. An exception is Käyrälampi in Kouvola, a potential ritual deposition with several identified burnt bear bones from all skeletal parts and one burnt elk bone. Bear bones from Iron Age contexts derive from settlement sites, cemeteries and settlement-cemetery complexes. Historical bear finds were discovered at settlement sites (urban contexts in Tornio and Turku) and at other sites (the Sámi offering place of *sieidi* in Enontekiö and the Kuusistonlinna fortification at Kaarina).

The anatomical distribution of bear bones from different periods is shown in Figure 5. Phalanges form the majority of identified bear bones from all periods. Long bones are very rare. Some skull and mandible fragments or teeth have been identified in material from the Stone Age and historical times. Interestingly, practically all Iron Age bear finds are from phalanges; less than one percent of all finds derives from other elements than phalanges. The elements besides phalanges are two teeth used as pendants from Tursiannotko in Pirkkala, a complete canine from Ihananiemi in Sysmä, an ulna from Uusi-Ruskeala C in Hartola, and a molar and a metatarsus or metacarpus from Mulli in Raisio.

Most of the phalanges from the Iron Age and early medieval period, over 160 in total, are third phalanges, and they derive from cremation burials (Fig. 6). The find material consists of burned phalanges, or their proximal parts, as the fragile distal ends have often been fragmented by fire. The finds originate from over 20 cemeteries, the datings of which cover an almost 1000-year-long period of time from *c.* the 3rd to the 11th centuries. The majority of these sites (86 %) were so-called cremation cemeteries under level ground (400/600–1000 AD), which can be characterised as including both collective and individual cremations, with no visible structures above ground (for the definition see Wessman 2010, 19–24, 34).

Hair finds

Brown bear hairs (Fig. 7) have been identified in seven Late Iron Age and medieval inhumation burials at four cemeteries (Table 4). They have not been found in Stone Age or Bronze Age contexts. However, this might not be the real situation as soil samples from only two Stone Age burials have been analysed for fibres so far (Ahola et al. 2018; Kirkinen 2019b).

In Luistari cemetery (southwestern Finland), which is the largest Iron Age to historical period inhumation cemetery in Finland (Lehtosalo-Hilander 1982a; b; c; 2000), bear hairs have been found in three burials of the 23 graves analysed (Kirkinen 2015; 2017; Kirkinen et al. 2020b). Very interesting is the 11th-century female grave 56, in which minuscule bear hair fragments were found by microscope on top of the upper body of the deceased. These fragments were interpreted tentatively as the remains of a traditional *sieppuri* (Fig. 8), a garment made of a bear pelt by historical Sámi populations (Sirelius 1912, 50–51; Itkonen 1948a, 339). In the female burials 95 and 377, bear hairs have been hypothesised as the remains of grave furnishings (Kirkinen 2015).

Many of the burials with bear pelts are noteworthy due to their late date. In Ristimäki in Ravattula (southwestern Finland), the cemetery is located around the earliest known church foundations in Finland (12th–13th centuries), and the burials were Christian in character (Ruohonen 2017). Bear skin remains were identified from two female burials interpreted as grave furnishings (Kirkinen et al. 2020a). At Kekomäki in Kaukola (Karelian Isthmus), the richly furnished grave 1, with its remains of two female and two male individuals from the 13th century (Schwindt 1893, 16–32; Uino 1997, 233), contained bear hairs among other fur remains, such as unidentified furred animals, Cervidae, Phocidae, and mountain hare (Schwindt 1893; Kirkinen 2015).

At Mukkala in Savukoski (eastern Lapland), a former Forest-Sámi burial ground that has been coin-dated to the 17th century, several graves showed evidence of wrapping the deceased in animal pelts. In male grave III, the corpse was wrapped in a bear skin. Also, this burial falls into the Christianisation process of eastern Lapland, which is evidenced e.g. by the burial crosses and Christian motifs found at Mukkala (Leppäaho 1937; Kirkinen et al. 2019). As mentioned above, one grave in Mukkala also contained two bear tooth pendants.

Discussion

The chronology of zooarchaeological bear finds
Stone Age bear finds
Osteological assemblages that date to the Stone Age are almost entirely from settlement sites, where the burning of bones, e.g. in hearths, has enabled their preservation. The most numerous anatomical parts are phalanges, but metatarsals, metacarpals, tarsals, carpals and sesamoid bones have also been identified. From the head, cranial fragments and teeth are present, but not common. At first glimpse, the similar element distribution pattern looks as if it could indicate the use of pelts.

However, it is uncertain whether this anatomical composition reflects the real situation; the majority of the recorded fragments are dense bone elements, and it is this density that improves the survival of these elements when they are burned (Ukkonen 2001; Ukkonen/Mannermaa 2017). It is also this density that allows even small fragments of these anatomical elements to be identified. Teeth are rarely identified in burnt materials; burning destroys enamel, despite the fact that bone can survive better when burned or cremated.

Assuming that the bear element distribution and the small number of bear bones in Stone Age assemblages somehow reflect the real deposition situation, we can make some interpretations. The first, obvious interpretation would be that most Stone Age bear finds originate from skins. This picture would suggest that the importance of the uses and meanings associated with bear pelts and claws is already visible in the Stone Age remains. The lack of bone types other than the extremities at the settlement sites might indicate that a bear carcass was treated (ritually) in a different way than, for example, elk and reindeer (*Rangifer tarandus*) carcasses. The impression achieved, based on the burnt assemblages from the Finnish Stone Age, is that most of the bear bones were either deposited somewhere other than the occupation sites, or that they were deposited in the occupation sites but the bones were not thrown on to the fire.

The ritual treatment of certain animals or animal parts has been indicated in some Mesolithic burials, where osteological material has been very well preserved. For example, in Late Mesolithic Skateholm (Sweden) many red deer (*Cervus elaphus*) antlers have been deposited in human burials (LARSSON 2017). In graves at Late Mesolithic Yuzhniy Oleniy Ostrov in Karelia (northwestern Russia), it is mostly the incisors of elks and Eurasian beavers (*Castor fiber*) and the canines of brown bear that have been deposited (GURINA 1956; MANNERMAA et al. 2021).

To use the absence of bones as an indication of the special treatment of body parts is very risky in archaeology, and even more so when the material consists of burnt bone fragments. It is not a surprise that all bear bones from the Stone Age in Finland derive from occupation layers; bone material (human skeletons and possible animal bones and bone artefacts) has not been preserved in millennia-old inhumation graves in Finland. Even the preservation of enamel, which is the hardest part of bone, is extremely rare (AHOLA et al. 2018). Marja Ahola has summarised the data from the Stone Age inhumation burials, and none of them have yielded bear bones (AHOLA 2019). This does not necessarily represent the original burial situation.

Bear canine pendants are commonly found in northeast European Stone Age burials (e.g. GURINA 1956). We can use the Yuzhniy Oleniy Ostrov cemetery as an example of the treatment of bear remains. Typically, bear canines are commonly found in the graves, but no other bear body parts are present. This indicates that bear canines held special symbolic meaning and were thus preferred. How were the remaining bear bodies treated, and where were they deposited? Assemblages from Mesolithic occupation sites in the area do not contain many bear finds (SAVVATEEV/VERESHCHAGIN 1987; SEITSONEN et al. 2017). Again, this might be partly due to preservation circumstances; Karelia has similarly acidic soils as most of Finland (with a few exceptions, one of which is Yuzhniy Oleniy Ostrov). We cannot answer these questions here; at this stage, our knowledge of the Stone Age bone assemblages is too scarce. The material parallels Knut Helskog's research, in which he concludes that the lack of bones at the settlement sites might indicate the existence of site types that were preferred for the processing of bear bones (HELSKOG 2012, 217). Bear finds are also very rare at the Swedish Stone Age sites (see MAGNELL, on humans and bears in Sweden, this volume). Archaeological assemblages with bear canine pendants, such as Yuzhniy Oleniy Ostrov, and the absence of bear bones at occupation sites, support the results received from the Finnish Stone Age assemblages: Bear bones were already treated in a special way in the Stone Age in Finland (and northern Europe). The general paucity of bear bones at Stone Age sites indicates that bears were rarely killed and thus the meat might have been considered a special delicacy, or was a food consumed solely in ceremonial and ritual occasions.

The archaeological evidence of special bear (bone) burials is extremely scarce in northern Europe and derives mainly from historical time periods (see IREGREN, this volume). Thus far, the Stone Age assemblage from Käyrälampi in southeast Finland is a rare example of a potential ritual bear burial site. The results will be published in detail in a forthcoming publication (MANNERMAA et al. in prep.), but here we can conclude that there are no special reasons that oppose the idea of a ritual deposition, although the deposition did not exclusively contain bear bones.

Despite the problems in interpreting burnt bone assemblages, artistic depictions of bears suggest that in the Stone Age bears had a special status. This can be seen in the realistic-style stone artefacts depicting bears during the Neolithic and Early Metal Period (*c.* 4000–1000 cal BC; cf. CARPELAN 1975; MANNERMAA/NÚÑEZ forthcoming; Fig. 9a–c). Such bear-shaped artefacts are known from various parts of Finland. Unambiguous bears are present in the rock art in Norway and Sweden but, interestingly, they are not depicted in Stone Age and Early Metal Period rock paintings in Finland (Lahelma 2008, 26–27; Helskog 2012). Generally, elks (or other cervids) are the most common animals depicted in these areas. Also, on the Kola peninsula and in Karelia, elk form the majority of the rock art depictions, but several bear pictures are also present (Kolpakov 2008; Kolpakov et al. 2008). Among the 1,200 figures seen at Zalavruga (near today's Belomorsk on the White Sea), there are 20 bears, while elk and beluga whale (*Delphinapterus leucas*) are much more numerous. On the east side of Lake Onega, the most frequently depicted animal in rock carvings is the swan (*Cygnus* sp.), followed by the elk and then the bear (Kolpakov 2008; Kolpakov et al. 2008).

Bronze Age and Early Metal Period bear finds
Despite the bear-headed stone items dated to the Early Metal Period, no bear bones have been identified from the Bronze Age/Early Metal Period (see Table 1) osteological finds. Bone materials from this period are generally scarce and most assemblages derive from cairns. The results of osteological analyses from the Bronze Age and Early Metal Period (3750–1750 cal BP) were listed in an unpublished database prepared for a book (Ukkonen/Mannermaa 2017). The total number of bone specimens from this period is 5,875 (data collected until 2012), but only a fraction of these has been identified taxonomically. The identified species or animal groups are mainly seals, Eurasian beaver, Eurasian elk, and wild reindeer, with occasional finds of mountain hare, pine marten, red fox, and common otter. Also, some domestic animals were identified. Pirjo Lahtiperä has suggested that human bodies were wrapped in seal furs, and placed on seal bones in burial cairns (Lahtiperä 1970). However, cremated animal bones from cairns have not been properly investigated in Finland, nor have their dates been confirmed by AMS-dating.

Does the absence of bear bone finds mean a discontinuation of bear symbolism, or is the gap in the Early Metal Period artificial, due to the poor preservation of bone and biases in deposition and find contexts? In general, burnt bone materials from prehistoric sites in Finland do not allow a detailed interpretation of the uses of bears, because a strong taphonomic bias is always associated with burnt bone assemblages.

Iron Age bear finds
During the Iron Age, the anatomical composition of the bear bone finds resembles the trend already observed in the Stone Age material. But in contrast, bear bones are almost absent at the settlement sites. The high number of third phalanges detected from the Middle and Late Iron Age cremation cemeteries is evidence that bears were hunted.

Many researchers (e.g. Petré 1980; Schönfelder 1994; Mäntylä-Asplund/Storå 2010; Lindholm/Ljungkvist 2016; Ukkonen/Mannermaa 2017, 182; for a discussion, see Grimm 2013) have interpreted the phalanges found in the Iron Age cremation burials as the remains of burnt bear pelts. Evidence of the placing of a bear pelt at the pyre of the deceased has been recorded in the secondary burial in Högom mound 4, Sweden (Ramqvist 1992, 194–198; Grimm 2013, 285, 292). Also, the remains of bear pelts, sometimes with claws still attached, have been detected in some Iron Age inhumation burials in Scandinavia and central Europe (see Grimm 2013).

Also, in Finland and on the Karelian Isthmus, bear hairs have been identified in single Late Iron Age and medieval inhumations (Table 4). They are part of a tradition of furnishing the graves with the pelts of large mammals, most often of Eurasian elk and wild forest reindeer (*Rangifer tarandus fennicus*), but sometimes also with the skins of cattle and bear (Kirkinen 2015; 2019a). However, in some cases the amount of bear hair is so small that it is impossible to define its function, since the pelts have been used for fur garments as well (see Kirkinen et al. 2020a).

In some cases, a possible explanation for the presence of the bones of distal extremities is that bear paws could have been used for magic and healing, as was the case during historical times (e.g. Lehikoinen 2009, 39, 171–172). From a Late Iron Age context, bear paw remains have been found in a ritual cairn in Myllymäki in Hattula, southern Finland (Sarkamo 1970).

The almost total absence of bear tooth pendants in Iron Age – also Late Iron Age – graves is another interesting fact, especially when comparing Finnish material to the ones of the neighbouring countries (see Grimm 2013). Thus far, an impressive bear tooth pendant has been found only in Suotniemi cemetery (KM 2487:7; Fig. 10) on the Karelian Isthmus. Additionally, two bear tooth pendants have come to light at the Tursiannotko Late Iron Age/early medieval period settlement site in the Häme region (Moilanen 2017). The general absence of bear tooth pendants from Finland has been taken into consideration when examining the so-called bronze bear tooth pendants (Fig. 11). This is a relatively common artefact type, found especially in Late Iron Age female inhumations (*c.* AD 700–900). These pendants have been interpreted as being connected with female fertility, protective magic, and as being symbolic expressions of control over the wilderness (Asplund 2005; Riikonen 2005; Kivisalo 2008). However, Tõnno Jonuks has recently questioned previous explanations and focussed on the fact that these pendants do not display the identifiable features of real organic bear canines (Jonuks 2017). Instead, he presents other possible explanations, from them being replicas of canids' teeth to representing fantastic creatures, such as dragons, which all served as symbols of ruling families.

Historical era
Evidence for the use of furs and pelts in funerary contexts continues until relatively late in Finland. Bear pelts have been detected in graves that date to the phase of Christianisation. From the 15th century onwards, the church aimed to stop the tradition by collecting bear pelts from parishioners and by using them in front of the altar as carpets (Korhonen 1982a; b; see also Østergård 2009, 120–121). Animal skins were replaced by textiles, and especially by rugs in burials (pall-clothes), the tufts of which probably imitated fur (Pylkkänen 1974, 27–31).

The bear finds from northern Finland indicate the high status of the bear as late as during the 17th century. In eastern Lapland, two bear canines, turned into pendants by perforation, were found in the *noaidi* (burial V) grave at Mukkala in Savukoski, a Forest-Sámi burial ground (Leppäaho 1937). In this cemetery, a bear pelt was used as a wrapping in burial III (Kirkinen et al. 2019). In Enontekiö, western Lapland, four bear upper molars were found in the *sieidi* of Näkkälä, and they most likely represent a bear skull offering (Äikäs et al. 2009).

In Tornio, southwestern Lapland, nine bear phalanges were found in a cluster in a house foundation; they most likely represent a complete bear paw, which has been interpreted as a house offering (Puputti 2010, 60). Sonja Hukantaival has studied historical house offerings in detail and has listed examples in which bear skulls, teeth, claws and paws have been utilised (Hukantaival 2016).

What about the missing bones?
Hunter-gatherers
In the Stone Age burnt assemblages from Finland, taphonomy may have played a major role in affecting the body part and species distribution. Also, as many Stone Age bone assemblages are from mixed contexts, and these contexts are not properly investigated and dated, the interpretations of these rare finds must be kept on a relatively general level. Furthermore, it is not easy to identify the character of human behaviour (for example, rituals or waste management) at so-called occupation sites, so this has not been done systematically in Finland.

Nevertheless, bear bones are rare in Stone Age occupation contexts, and bear bones were rarely deposited in refuse fauna, i.e. waste from domestic activities. The treatment of bear bones is thus different from those of other big game animals such as elk and wild reindeer. Despite the small number of bear bones being a common phenomenon throughout the Stone Age, we must keep in mind that the Stone Age is a long period and contains many cultural phases and populations. There is no possibility to consider similarities or differences in attitudes towards the bear among prehistoric populations. However, a shared northern forest environment and the foraging economy can be considered as factors leading to potential similarities in attitudes toward bears and the place of the bear in culture, religion and mythology.

Farming populations

During the Iron Age, the present zooarchaeological material stresses that bear pelts were first placed on cremation pyres and in inhumations in Scandinavia and central Europe (MØHL 1978; PETRÉ 1980; SCHÖNFELDER 1994; GUSTAVSSON et al. 2014). The tradition spread to Finland from the west as part of a broader range of Germanic influence (KIRKINEN 2017). Later on, bear skins were placed in inhumation burials, which is evidenced by the hairs found in graves (KIRKINEN 2017; 2019; KIRKINEN et al. 2020b).

In Scandinavia, the discussion about burnt bear phalanges has concentrated mainly on the question of whether or not the claws were associated with the male elité and warriorship. For example, in Sweden, Karl-Johan Lindholm and John Ljungkvist have concluded that the bear phalanges were first associated with the elité and males, and later with the middle class, too (LINDHOLM/LJUNGKVIST 2016).

Also in Finland, the burial type in which the burnt bear phalanges were recorded has been interpreted to indicate the warrior cult and power (see SALMO 1938, 308–310; 1941; PIHLMAN 1990; SCHAUMAN-LÖNNQVIST 1996a; b; 1999; WICKHOLM/RANINEN 2006; RANINEN 2007; 2009; WESSMAN 2010, 62–66). Kirkinen's research on Finno-Karelian Kalevala-metric poetry (KIRKINEN 2017) yielded a group of poems and incantations that referred literally to the use of a burning skin for protection in war and in a mythical journey to the land of the dead. These verses have been connected to the same cultural sphere as the graves with phalanges (SIIKALA 1992, 294–296; 2002; 2014), which supports the hypothesis that these verses might be telling us about the cremation of bear pelts on pyres.

LINDHOLM/LJUNGKVIST (2016) interpret the lack of bear bones other than third phalanges as being connected with the trading of furs from the inland regions of Sweden. This is evident, e.g. on Gotland, which never had a local bear population (see JORDAHL et al., this volume). Correspondingly, fur trading is also a possible explanation in Finland. However, the ethnographic and historical sources for the Sámi and Finnic traditions of bear ceremonialism suggest an alternative explanation in that the bones of the bear were disposed of in a ritual manner, such as returning them to the forest, as was the case in later practices (e.g. HOLMBERG [HARVA] 1915, 47; ITKONEN 1948b, 366; HAAVIO 1967, 34; ÄIKÄS et al. 2009, 118; SIIKALA 2012, 388).

CONCLUSIONS

The interpretations of prehistoric and medieval human-bear relationships stress the importance of the bear to northern cultures. The characteristics of this relationship have been built on theories about the cosmologies of communities under study, extrapolating a central division between hunter-gatherer and farming groups. In the studies concentrating on hunter-gatherer populations, there is a tendency to see bears as ancestors, clan attributes, guardian spirits, and symbols of power (see e.g. HELSKOG 2012, with references; LOSEY et al. 2013), whereas concerning Iron Age farming groups bears and bear pelts have been seen as symbols of power, wealth and warriorhood (e.g. PETRÉ 1980; LINDHOLM/LJUNGKVIST 2016). This dualism is, of course, oversimplified. Based on the Finnish material, we highlight that the cultural context needs to be taken into account in explaining the roles of the bear for the community. On the basis of bear-related find material in Finland, it is possible to define geographical and/or chronological differences as well as the *longue durée* of a phenomenon in human-bear relationships.

The special powers of the bear are associated with skulls, canines, penis bones and third phalanges, as is supported by ethnographic examples from various parts in the circumpolar area (e.g. HALLOWELL 1926). Brown bear third phalanges are the most common finds in Stone Age occupation contexts, and it is tempting to hypothesise that most bear finds also from the Stone Age originate from skins. Yet, as mentioned earlier, taphonomic factors have likely biased the anatomical composition. The time span of the Stone Age in Finland covers approximately 7,000 years, and the amount of bear bones from this long period is only 102. It follows that this restricted assemblage is far from being representative. In the future, a detailed investigation of contexts and a systematic AMS-dating programme can shed more light on the Stone Age bear finds from Finland and help us to understand their meanings and functions.

As for the Iron Age find material, we follow KIRKINEN (2017) and suggest that it is possible to define at least two human-bear cultures, the Germanic one that cremated the pelts along with the deceased, and the traditional Finno-Ugric one in which the bear was appreciated as a divine forefather.

We suggest that both the placing of pelts, as well as phalanges or complete claws loosened from the skins, in the graves represents the qualities and strength of the bear. This is supported by ethnographic data according to which the strength of an animal – and sometimes of a human being – is located in its hair and nails (Fig. 12; e.g. PENTIKÄINEN 2007, 120–121, 139; LEHIKOINEN 2009, 171–172). Moreover, the bear was believed to have its origin in a curl of hair (PENTIKÄINEN 2007, 138–139). Therefore, we suggest that sometimes real pelts were cremated, sometimes maybe paws, and sometimes the claws were used instead – but they all served the same goal, the acquisition of the qualities of a bear to facilitate the change from the realm of the living to the one of death.

Before we can make any far-reaching conclusions about the bear throughout prehistoric times in Finland, we need a systematic re-analysis of bear bones from archaeological sites. At the moment, the data is in lists of osteological reports that have been issued by several osteologists. The contexts of many prehistoric bear finds are not clearly understood and direct AMS-datings should be conducted on multiperiod sites. Also, the inhumation burials from the Stone Age and the Iron Age should be studied systematically for animal hairs. Proper understanding of contexts would help in interpreting the bone finds and their chronologies.

ACKNOWLEDGEMENTS

We want to thank Pirkko Ukkonen for the initiation and inspiration to write this paper. Pirkko also provided unpublished materials to us. Several osteologists and archaeologists have helped us gather materials for this paper, a collective thank you to all of you! We express our deep gratitude to Frog, Oliver Grimm, and one anonymous reviewer for their valuable comments on the manuscript.

BIBLIOGRAPHY

AHOLA 2019: M. AHOLA, Death in the Stone Age. Making Sense of Mesolithic-Neolithic Mortuary Remains from Finland (ca. 6800–2300 cal BC) (Helsinki 2019).

AHOLA/FROG 2014: J. AHOLA/FROG, Approaching the Viking Age in Finland: An Introduction. In: J. Ahola/Frog/C. Tolley (eds.), Fibula, Fabula, Fact – The Viking Age in Finland. Studia Fennica Historica 18 (Helsinki 2014) 21–84.

AHOLA et al. 2018: M. AHOLA/T. KIRKINEN/K. VAJANTO/J. RUOKOLAINEN, On the scent of an animal skin: New evidence on Corded Ware mortuary practices and livelihoods in northern Europe. Antiquity 92, 2018, 118–131.

Äikäs et al. 2009: T. Äikäs/A.-K. Puputti/M. Núñez/J. Aspi/J. Okkonen, Sacred and Profane Livelihood: Animal Bones from Sieidi Sites in Northern Finland. Norwegian Archaeological Review 42(2), 2009, 109–122.

Albrecht et al. 2017: J. Albrecht/K. A. Barton/N. Selva/R. S. Sommer/J. E. Swenson/R. Bischof, Humans and climate change drove the Holocene decline of the brown bear. Scientific reports 7(1), 2017, 1–11.

Alenius et al. 2013: T. Alenius/T. Mökkönen/A. Lahelma, Early farming in the northern boreal zone: re-assessing the history of land use in southeastern Finland through high-resolution pollen analysis. Geoarchaeology 28(1), 2013, 1–24

Arponen 2008: A. Arponen, Arkeologisen materiaalin konservointi. In: P. Halinen/V. Immonen/M. Lavento/T. Mikkola/A. Siiriäinen/P. Uino (eds.), Johdatus arkeologiaan (Helsinki 2008) 225–235.

Asplund 2005: H. Asplund, The Bear and the Female. Bear-tooth pendants in Late Iron Age Finland. In: S. Mäntylä-Asplund (ed.), Rituals and Relations. Studies on Society and Material Culture of the Baltic Finns (Helsinki 2005) 13–30.

Bachura/Kosintsev 2007: O. Bachura/P. Kosintsev, Late Pleistocene and Holocene small-and large-mammal faunas from the Northern Urals. Quaternary International 160(1), 2007, 121–128.

Carpelan 1975: C. Carpelan, Älg- och björnhuvudföremål från Europas nordliga delar. Finskt Museum 82, 1975, 5–67.

Carpelan 1999: C. Carpelan, On the postglacial colonisation of Eastern Fennoscandia. In: M. Huurre/C. Carpelan/P. Halinen/T. Kirkinen/V. Laulumaa/M. Lavento/M. Lönnqvist (eds.), Dig it all. Papers dedicated to Ari Siiriäinen (Helsinki 1999) 151–172.

Conneller 2011: C. Conneller, The Mesolithic. In: T. Insoll (ed.), The Oxford Handbook of the Archaeology of Ritual and Religion (Oxford 2011) 358–370.

Davison et al. 2011: J. Davison/S. Y. Ho/S. C. Bray/M. Korsten/E. Tammeleht/M. Hindrikson/A. Cooper, Late-Quaternary biogeographic scenarios for the brown bear (*Ursus arctos*), a wild mammal model species. Quaternary Science Reviews 30(3/4), 2011, 418–430.

Döppes/Pacher 2014: D. Döppes/M. Pacher, 10,000 years of *Ursus arctos* in the Alps – A success story? Analyses of the Late Glacial and Early Holocene brown bear remains from alpine caves in Austria. Quaternary International 339, 2014, 266–274.

Edgren 1999: T. Edgren, Käännekohtia Suomen kivikaudessa. In: P. Fogelberg (ed.), Pohjan poluilla. Suomalaisten juuret nykytutkimuksen mukaan. Bidrag till kännedom av Finlands natur och folk 135, 1999, 281–293.

Edwards 1989: G. Edwards, Guidelines for dealing with material from sites where organic remains have been preserved by metal corrosion products. In: R. Janaway/B. Scott (eds.), Evidence Preserved in Corrosion Products: New Fields in Artifact Studies. United Kingdom Institute for Conservation of Historic and Artistic Works, Occasional Papers 8 (London 1989) 3–7.

Grimm 2013: O. Grimm, Bear-skins in northern European burials and some remarks on other bear-related furnishings in the north and middle of Europe in the 1st millennium AD. In: O. Grimm/U. Schmölcke (eds.), Hunting in northern Europe until 1500 AD. Old traditions and regional developments, continental sources and continental influences. Papers presented at a workshop organised by the Centre for Baltic and Scandinavian Archaeology, Schleswig, June 16th and 17th, 2011. Schriften des Archäologischen Landesmuseums, Ergänzungsreihe 7 (Neumünster 2013) 277–296.

Gurina 1956: Н. Н Гурина, Оленеостровский могильник. Izdatel'stvo Akademii Nauk SSSR 47. Materialy i Issledovaniia po Arkheologii SSSR (1956).

Gustavsson et al. 2014: R. Gustavsson/J.-E. Tomtlund/J. Kennebjörk/J. Storå, Identities in transition in Viking Age Åland. In: Frog/J. Ahola/J. Lucenius (eds.), The Viking Age in Åland. Insights into Identity and Remnants of Culture. Annales Academiae Scientiarum Fennicae. Humaniora 372 (Sastamala 2014) 159–186.

Haavio 1967: M. Haavio, Suomalainen mytologia (Porvoo 1967).

Hallowell 1926: I. Hallowell, Bear ceremonialism in the Northern Hemisphere. American Anthropologist 28(1), 1926, 1–175.

Helskog 2012: K. Helskog, Bears and Meanings among Hunter-fisher-gatherers in Northern Fennoscandia 9000–2500 BC. Cambridge Archaeological Journal 22(2), 2012, 209–236.

Holmberg (Harva) 1915: U. Holmberg (Harva), Lappalaisten uskonto. Suomen suvun uskonnot II (Porvoo 1915).

Honko 1993: L. Honko. Hunting: Introduction. In: L. Honko/S. Timonen/M. Branch (eds.), The Great Bear (Helsinki) 117–140.

Hukantaival 2016: S. Hukantaival, "For a Witch Cannot Cross Such a Threshold!" Building Concealment Traditions in Finland c. 1200–1950. Unpubl. PhD thesis, University of Turku (Turku 2016).

Hurcombe 2014: L. Hurcombe, Perishable Material Culture in Prehistory: Investigating the Missing Majority (London, New York 2014).

Iregren/Jonsson 1973: E. Iregren/R. Jonsson, Hur ben krymper vid kremering. Fornvännen 68, 1973, 97–100.

Itkonen 1948a: T. Itkonen, Suomen lappalaiset vuoteen 1945. 1. osa (Porvoo 1948).

Itkonen 1948b: T. Itkonen, Suomen lappalaiset vuoteen 1945. 2. osa (Porvoo 1948).

Jonuks 2017: T. Jonuks, Bronze Tooth Pendants from the Late Iron Age: Between Real and Fictional Zooarchaeology. Norwegian Archaeological Review 50(2), 2017, 135–148.

Kirkinen 2015: T. Kirkinen, The Role of Wild Animals in Death Rituals: Furs and Animal Skins in the Late Iron Age Inhumation Burials in Southeastern Fennoscandia. Fennoscandia Archaeologica XXXII, 2015, 101–120.

Kirkinen 2017: T. Kirkinen, "Burning pelts" – brown bear skins in the Iron Age and Early Medieval (0–1300 AD) burials in South-East Fennoscandia. Estonian Journal of Archaeology 21(1), 2017, 3–29.

Kirkinen 2019a: T. Kirkinen, Between skins. Animal skins in the Iron Age and historical burials in Eastern Fennoscandia. Unpubl. PhD thesis, University of Helsinki (Helsinki 2019).

Kirkinen 2019b: T. Kirkinen, Mikroskooppinen eläinkuitumateriaali kertoo vaatetuksesta ja turkisten käytöstä. Suomen Museo 126, 2019, 91–97.

Kirkinen et al. 2019: T. Kirkinen/A. Arponen/I. Van den Berghe, Globalization and tradition in Forest Sámi commemoration rituals. Textiles and animal skins in the 17th century burial ground in Mukkala, Eastern Lapland, Finland. Monographs of the Archaeological Society of Finland 7, 2019, 177–195.

Kirkinen et al. 2020a: T. Kirkinen/J. Riikonen/C. Dove/ J. Ruohonen, Identification and the Use of Fur and Feathers Excavated from the Late Iron Age and Early Medieval (12th–13th centuries) Ristimäki Cemetery in Kaarina, SW Finland. Fennoscandia Archaeologica XXXVII, 2020, 45–59.

Kirkinen et al. 2020b: T. Kirkinen/K. Vajanto/S. Björklund, Animal-hair evidence in an 11th century female grave in Luistari, Finland. Archaeological Textiles Review 62, 2020, 109–125.

Kivisalo 2008: N. Kivisalo, The Late Iron Age Bear-Tooth Pendants in Finland: Symbolic Mediators between Women, Bears, and Wilderness? Temenos 44, 2008, 263–291.

Koivisto et al. 2016: S. Koivisto/V. Laulumaa/K. Nurminen, Vanhoja luita ja uusia ajoituksia Kyrönmaan rautakauden röykkiöhaudoista. Monttu auki. Arkeologisia kenttätutkimuksia 1 (Helsinki 2016).

Kolpakov 2008: E. M. Kolpakov, Petroglyphs of Kanozero: typological analysis, in Kanozero Petroglyphs. The Kirovsk International Conference on Rock Art (Kirovsk 2008) 64–65.

Kolpakov et al. 2008: E. M. Kolpakov/A. I. Murashkin/ V. Y. Shumkin, The rock carvings of Kanozero. Fennoscandia Archaeologica 25, 2008, 86–96.

Kopatz et al. 2014: A. Kopatz/H. G. Eiken/J. Aspi/I. Kojola/C. Tobiassen/K. F. Tirronen/P. I. Danilov/S. B. Hagen, Admixture and gene flow from Russia in the recovering northern European brown bear (*Ursus arctos*). PLoS ONE 9(5), 2014. https://doi.org/10.1371/journal.pone.0097558.

Korhonen 1982a: T. Korhonen, Saaliseläimen talja kirkkouhrina. Suomen Museo 89, 1982, 45–68.

Korhonen 1982b: T. Korhonen, Saaliseläimen taljan ja kotieläimen vuodan käyttö uhreina. Suomen Antropologi 3, 1982, 96–121.

Krohn 2008: K. Krohn, Suomen suvun uskonnot. Suomalaisten runojen uskonto (Helsinki 2008 [originally published 1905]).

Laakso 2014: V. Laakso, Papinniemi in Uukuniemi and related archaeological sites of the eastern orthodox cultural area in Finland (Turku 2014).

Lahelma 2008: A. Lahelma, A Touch of Red: Archaeological and Ethnographic Approaches to Interpreting Finnish Rock Paintings. ISKOS 15, 2008.

Lahtiperä 1970: P. Lahtiperä, Luuaineiston analyysi. In: U. Salo/P. Lahtiperä (eds.), Metallikautinen asutus Kokemäenjoen suussa (Pori 1970) 199–219.

Larsson 2017: L. Larsson, Animals and animal depictions in the early Holocene of Northern Europe based on an antler adze with deer depictions. In: V. Brieske/A. Dickers/M. M. Rind (eds). Tiere und Tierdarstellungen in der Archäologie (Steinfurt 2017) 63–74.

Lehikoinen 2009: H. Lehikoinen, Ole siviä sikanen. Suomalaiset eläinuskomukset (Helsinki 2009).

Lehtosalo-Hilander 1982a: P.-L. Lehtosalo-Hilander, Luistari I. The Graves. Suomen Muinaismuistoyhdistyksen Aikakauskirja 82(1) (Helsinki 1982).

Lehtosalo-Hilander 1982b: P.-L. Lehtosalo-Hilander, Luistari II. The Artefacts. Suomen Muinaismuistoyhdistyksen Aikakauskirja 82(2) (Helsinki 1982).

Lehtosalo-Hilander 1982c: P.-L. Lehtosalo-Hilander, Luistari III. A burial ground reflecting the Finnish Viking Age society. Suomen Muinaismuistoyhdistyksen Aikakauskirja 82(3) (Helsinki 1982).

Lehtosalo-Hilander 2000: P.-L. Lehtosalo-Hilander, Luistari 4: a History of Weapons and Ornaments. Suomen Muinaismuistoyhdistyksen Aikakauskirja 107 (Helsinki 2000).

Leppäaho 1937: J. Leppäaho, Savukosken Mukkalan lappalaiskalmisto. Kotiseutu 3/4, 1937, 134–144.

Lindholm/Ljungkvist 2016: K.-J. Lindholm/J. Ljungkvist, The bear in the grave: exploitation of top predator and herbivore resources in first millennium Sweden – first trends from a long-term research project. European Journal of Archaeology 19(1), 2016, 3–27.

Losey et al. 2013: R. J. Losey/V. I. Bazaliiskii/A. R. Lieverse/A. Waters-Rist/A. Faccia/A. W. Weber, The bear-able likeness of being: ursine remains at the Shamanka II cemetery, Lake Baikal, Siberia. In: C. Watts (ed.), Relational archaeologies: humans, animals, things (London/New York 2013) 65–96.

Lyman 1994: L. Lyman, Vertebrate Taphonomy (Cambridge 1994).

Mannermaa 2008: K. Mannermaa, The Archaeology of Wings. Birds and People in the Baltic Sea Region during the Stone Age (Helsinki 2008).

Mannermaa/Núñez forthcoming: K. Mannermaa/ M. Núñez, Carving and shaping wood, stone and bone. In: J. Grünberg (ed.), Mesolithic Art – Abstraction, Decoration, Messages. International & Interdisciplinary Conference Halle (Saale), Germany, 19th–21st September 2019. Tagungen des Landesmuseums für Vorgeschichte Halle (Halle [Saale]).

Mannermaa et al. 2021: K. Mannermaa/R. Rainio/ E. Girya/D. Gerasimov, Let's groove: Attachment techniques of Eurasian elk (*Alces alces*) tooth pendants at the Late Mesolithic cemetery Yuzhniy Oleniy Ostrov (Lake Onega, Russia). Journal of Archaeological and Anthropological Sciences 13(3), 2021. https://doi.org/10.1007/s12520-020-01237-5.

Mäntylä-Asplund/Storå 2010: S. Mäntylä-Asplund/ J. Storå, On the archaeology and osteology of the Rikala cremations cemetery in Salo, SW Finland. Fennoscandia Archaeologica XXV, 2010, 53–68.

Miettinen 2012: T. Miettinen, Kymenlaakson esihistoriallinen kehitys. In: Y. Kaukiainen (ed.), Kymenlaakson historia. Osa I (Helsinki 2012) 32–79, 438.

Mikkola 2009: E. Mikkola, The Mikkeli Tuukkala cemetery – the 2009 excavations and new interpretations. Fennoscandia Archaeologica XXIV, 2009, 177–185.

Moilanen 2017: U. Moilanen, Riipuksena tai amulettina käytetyt karhunhampaat. Pirkanmaan maakuntamuseo (2017). https://tursiannotko.fi/2017/08/01/riipuksena-tai-amulettina-kaytetyt-karhunhampaat/

Møhl 1978: U. Møhl, Bjørnekløer og brandgrave. Dyreknogler fra germansk jernalder i Stilling. Kuml 1977, 119–129.

Nurminen 2020: K. Nurminen, Savukoski Sokli 2019 Malmio 1A KM 42446, Malmio 1B, KM 42447. Osteologinen analyysi kivikautisen asuinpaikan kaivauksen luista. In: J. Seppä/V. Laulumaa (eds.), Savukoski Malmio 1 ja 1B. Kivi- ja varhaismetallikautisen asuinpaikan kaivaus 5.8.–27.9.2019. Museoviraston arkeologisten kenttäpalveluiden tutkimusraportti/Unpublished excavation report (2020).

Okkonen 1991: J. Okkonen, Huomioita palaneen luun säilymisestä esihistoriallisen asuinpaikan kulttuurikerroksessa. Faravid 15, 1991, 157–161.

Pentikäinen 2007: J. Pentikäinen, Golden King of the Forest: The Lore of the Northern Bear (Helsinki 2007).

Petré 1980: B. Petré, Björnfallen i begravningsritualen – statusobjekt speglande regional skinnhandel? Fornvännen 75, 1980, 5–14.

Pihlman 1990: S. Pihlman, Kansainvaellus- ja varhaismerovinkiajan aseet Suomessa. Typologia, kronologia ja aseet ryhmästrategioissa. ISKOS 10, 1990.

Pohjoismäki et al. 2018: J. Pohjoismäki/S. Lampi/J. Donner/H. Anderson, Origins and wanderings of the Finnish hunting spitzes. PLoS ONE 13(6), 2018. https://doi.org/10.1371/journal.pone.0199992.

Puputti 2010: A.-K. Puputti, Bones in pits and ditches. A contextual approach to animal bone distribution in Early Modern Tornio. Journal of Nordic Archaeological Science 17, 2010, 53–64.

Pylkkänen 1974: R. Pylkkänen, The Use and Traditions of Medieval Rugs and Coverlets in Finland (Helsinki 1974).

Ramqvist 1992: P. Ramqvist, Högom. The excavations 1949–1984. Archaeology and environment 13, Högom part I (Umeå 1992).

Raninen 2007: S. Raninen, Kovia ja nimekkäitä miehiä. Persoonan konstituutio rautakaudella. Muinaistutkija 1, 2007, 18–28.

Raninen 2009: S. Raninen, Rautakautisia rajanylityksiä – sotaisa Suomi? Arkeologipäivät 2008, 45–51.

Raninen/Wessman 2015: S. Raninen/A. Wessman, Rautakausi. In: G. Haggrén/P. Halinen/M. Lavento/S. Raninen/A. Wessman (eds.), Muinaisuutemme jäljet. Suomen esi- ja varhaishistoria kivikaudelta keskiajalle (Helsinki 2015) 213–365.

Reimer et al. 2009: P. J. Reimer/M. G. Baillie/E. Bard/A. Bayliss/J. W. Beck/P. G. Blackwell/C. Bronk Ramsey/C. E. Buck/G. S. Burr/R. L. Edwards/M. Friedrich/P. M. Grootes/T. P. Guilderson/I. Hajdas/T. J. Heaton/A. G. Hogg/K. A. Hughen/K. F. Kaiser/B. Kromer/F. G. McCormac/S. W. Manning/R. W. Reimer/D. A. Richards/J. R. Southon/S. Talamo/C. S. M. Turney/J. van der Plicht/C. E. Weyhenmeyer, IntCal09 and Marine09 radiocarbon age calibration curves, 0–50,000 years cal BP. Radiocarbon 51, 2009, 1111–1150.

Riikonen 2005: J. Riikonen, Iron Age Aprons from Southwest Finland and Other Cloths and Pendants Worn on the Waist. In: S. Mäntylä-Asplund (ed.), Rituals and Relations. Studies on Society and Material Culture of the Baltic Finns (Helsinki 2005) 31–72.

Ruohonen 2017: J. Ruohonen, Ristimäki in Ravattula: on the remains of the oldest known church in Finland. In: J. Harjula/S. Hukantaival/V. Immonen/A. Randla/Tanja Ratilainen (eds.), New Visits to Old Churches: Sacred Monuments and Practices in the Baltic Sea Region (Cambridge 2017) 46–60.

Russell 2012: N. Russell, Social Zooarchaeology. Humans and Animals in Prehistory (Cambridge 2012).

Sabol 2005: M. Sabol, Bear assemblage from the Za Hájovnou Cave in Moravia (Czech Republic): sex ratio and age structure. Abhandlungen der Naturhistorischen Gesellschaft Nürnberg 45, 2005, 215–224.

Salmo 1938: H. Salmo, Die Waffen der Merowingerzcit in Finnland. Suomen Muinaismuistoyhdistyksen Aikakauskirja 42,1 (Helsinki 1938).

Salmo 1941: H. Salmo, Merovinkiaikaisen ratsusotilaan hautakalusto Euran pitäjän Pappilanmäestä. Suomen Museo 47, 1941, 11–39.

Sarkamo 1970: J. Sarkamo, Retulansaaren uhriröykkiö. Suomen Museo 77, 1970, 35–47.

Sarmela 2009: M. Sarmela, Finnish Folklore Atlas. Ethnic Culture in Finland 2. Suomalaisen Kirjallisuuden Seura 587 (Helsinki 2009; partially revised version in English). Published first in Finnish in 1994: http://www.kotikone.fi/matti.sarmela/folkloreatlas.pdf.

Savvateev/Vereshchagin 1978: Yu. A. Savvateev/N. K. Vereshchagin, Okhotnich'e-promyslovye zhivotnye I 1334 kamennyi inventar' naseleniya Karelii i yuzhnoy chasti Kol'skogo poluostrova epokh I 1335 neolita i rannego metalla. In: Yu. A. Savvateev (ed.), Mezoliticheskie pamyatniki Karelii (Petrozavodsk 1978) 181–1336.

Schauman-Lönnqvist 1996a: M. Schauman-Lönnqvist, The Vainionmäki society. In: P. Purhonen (ed.), Vainionmäki, a Merovingian Period Cemetery in Laitila, Finland (Helsinki 1996) 130–135.

Schauman-Lönnqvist 1996b: M. Schauman-Lönnqvist, Weapons. In: P. Purhonen (ed.) Vainionmäki, a Merovingian Period Cemetery in Laitila, Finland (Helsinki 1996) 53–62.

Schauman-Lönnqvist 1999: M. Schauman-Lönnqvist, The west Finnish warriors and the early Svea kingship (550–800 AD). In: T. M. Dickinson/D. Griffiths (eds.), The Making of Kingdoms. Papers from the 47th Sachsensymposium, York, September 1996. Anglo-Saxon Studies in Archaeology and History 10 (Oxford 1999) 65–70.

Schwindt 1893: T. Schwindt, Tietoja Karjalan rautakaudesta ja sitä seuraavilta ajoilta Käkisalmen kihlakunnan alalta saatujen löytöjen mukaan. Suomen Muinaismuistoyhdistyksen Aikakauskirja 13 (Helsinki 1893).

Schönfelder 1994: M. Schönfelder, Bear-claws in Germanic graves. Oxford Journal of Archaeology 13(2), 1994, 217–227.

Seitsonen et al. 2017: O. Seitsonen/S. Seitsonen/L. Broderick/D. Gerasimov, Burnt bones by Europe's largest lake: Zooarchaeology of the Stone Age and Early Metal period hunter-gatherers at Lake Ladoga, NW Russia. Journal of Archaeological Science: Reports 11, 2017, 131–146.

Seppä/Laulumaa 2020: J. Seppä/V. Laulumaa, Savukoski Malmio 1 ja 1B. Kivi- ja varhaismetallikautisen asuinpaikan kaivaus 5.8.–27.9.2019. Museoviraston arkeologisten kenttäpalveluiden tutkimusraportti/Unpublished excavation report (2020).

Sigvallius 1994: B. Sigvallius, Funeral Pyre. Iron Age Cremations in North Spånga. Theses and Papers in Osteology 1 (Stockholm 1994).

Siikala 1992: A.-L. Siikala, Suomalainen šamanismi. Mielikuvien historiaa. Suomalaisen Kirjallisuuden Seuran toimituksia 565 (Helsinki 1992; published in English: Siikala 2002).

Siikala 2002: A.-L. Siikala, Mythic Images and Shamanism. A perspective on Kalevala poetry. Folklore Fellows Communications 28 (Vammala 2002).

Siikala 2008: A.-L. Siikala, Myytit, riitit ja tietäjän toimet. Savo ja sen kansa. Savon historia Vol. VII. Suomalaisen Kirjallisuuden Seuran toimituksia 1192 (Helsinki 2008) 110–186.

Siikala 2012: A.-L. Siikala, Itämerensuomalaisten mytologia. Suomalaisen Kirjallisuuden Seuran toimituksia 1388 (Helsinki 2012).

Siikala 2014: A.-L. Siikala, Singing of incantations in Nordic tradition. Scripta Instituti Donneriani Aboensis 13, 2014, 191–205.

Sirelius 1912: U. T. Sirelius, Über einige Traggeräte und Umschlagetücher bei den finnisch-ugrischen Völkern. Suomen Muinaismuistoyhdistyksen Aikakauskirja XXVI, 1912, 29–58.

Solazzo et al. 2014: C. Solazzo/P. Walton Rogers/ L. Weber/H. Beaubien/J. Wilson/M. Collins, Species identification by peptide mass fingerprinting (PMF) in fibre products preserved by association with copper-alloy artefacts. Journal of Archaeological Science 49, 2014, 524–535.

Sommer/Benecke 2005: R. S. Sommer/N. Benecke, The recolonization of Europe by brown bears Ursus arctos Linnaeus, 1758 after the Last Glacial Maximum. Mammal Rev. 35, 2005, 156–164.

Stuiver/Reimer 1993: M. Stuiver/P. J. Reimer, Extended 14C data base and revised CALIB 3.0 14C age calibration program. Radiocarbon 35, 1993, 215–230.

Taberlet/Bouvet 1994: P. Taberlet/J. Bouvet, Mitochondrial DNA polymorphism, phylogeography and conservation genetics of the brown bear Ursus arctos in Europe. Proceedings of the Royal Society – B 255, 1994, 195–200.

Tridico et al. 2014: S. Tridico/S. Koch/A. Michaud/ G. Thomson/K. Kirkbride/M. Bunce, Interpreting biological degradative processes acting on mammalian hair in the living and the dead: which ones are taphonomic? Proceedings of the Royal Society – B 281, 2014. https://doi.org/10.1098/rspb.2014.1755.

Uino 1997: P. Uino, Ancient Karelia – Archaeological studies. Muinais-Karjala – Arkeologisia tutkimuksia. Suomen Muinaismuistoyhdistyksen Aikakauskirja 104 (Helsinki 1997).

Ukkonen 2001: P. Ukkonen, Shaped by the Ice Age: reconstructing the history of mammals in Finland during the Late Pleistocene and Early Holocene (Helsinki 2001).

Ukkonen/Mannermaa 2017: P. Ukkonen/K. Mannermaa, Jääkauden jälkeläiset. Suomen lintujen ja nisäkkäiden varhainen historia (Helsinki 2017).

Vaneeckhout et al. 2013: S. Vaneeckhout/A.-K. Salmi/ J.-A. Junno, Archaeological Refuse Fauna in Finland: Understanding the Role of Bone Combustion. Anthropozoologica 48(1), 2013, 125–134.

Wessman 2009: A. Wessman, Reclaiming the past: Using old artefacts as a means of remembering. Interarchaeologia 3, 2009, 71–88.

Wessman 2010: A. Wessman, Death, destruction and commemoration: tracing ritual activities in Finnish Late Iron Age cemeteries (AD 550–1150). ISKOS 18, 2010.

Wickholm/Raninen 2006: A. Wickholm/S. Raninen, The broken people: Deconstruction of personhood in Iron Age Finland. Estonian Journal of Archaeology 10(2), 2006, 150–166.

Wilson 2008: A. Wilson, The Decomposition of Hair in the Buried Body Environment. In: M. Tibbett/D. Carter (eds.), Soil Analysis in Forensic Taphonomy. Chemical and Biological Effects of Buried Human Remains (Boca Raton [Florida] 2008) 123–151.

Wilson/Tobin 2010: A. Wilson/D. Tobin, Hair after death. In: R. M. Trueb/D. J. Tobin (eds.), Aging Hair (New York 2010) 249–261.

Wilson et al. 2001: A. Wilson/R. Dixon/H. Dodson/ R. Janaway/A. M. Pollard/B. Stern/D. Tobin, Yesterday's hair – human hair in archaeology. Biologist 48(5), 2001, 213–217.

Witzel 2013: E. J. M. Witzel, The Origins of the World's Mythologies (Oxford) 375–420.

Zedrosser et al. 2001: A. Zedrosser/B. Dahle/J. E. Swenson/N. Gerstl, Status and management of the brown bear in Europe. Ursus 2001, 9–20.

Zvelebil 1997: M. Zvelebil, Hunter–gatherer ritual landscapes: spatial organisation, social structure and ideology among hunter–gatherers of northern Europe and western Siberia. Analecta Praehistoria Leidensia 29, 1997, 33–50.

Østergård 2009: E. Østergård, Woven into the Earth. Textiles from Norse Greenland (Aarhus ²2009).

Associate Prof. Kristiina Mannermaa
Department of Cultures
University of Helsinki
Helsinki
Finland
kristiina.mannermaa@helsinki.fi

Dr. Tuija Kirkinen
Department of Cultures
University of Helsinki
Helsinki
Finland
tuija.kirkinen@helsinki.fi

Docent Suvi Viranta-Kovanen
Department of Anatomy
University of Helsinki
Helsinki
Finland
suvi.viranta-kovanen@helsinki.fi

Fig. 1. A bear skull tree in Vuoskujärvi, Finnish Lapland, in 1914 (photo S. Paulaharju / Finnish Heritage Agency).

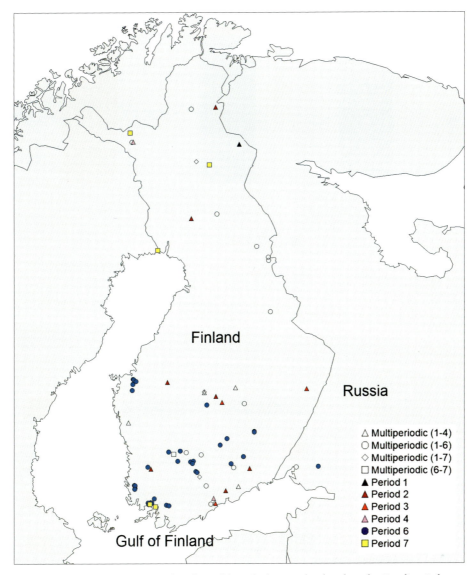

Fig. 2. The distribution of zooarchaeological bear finds in Finland and on the Karelian Isthmus. Period 1: Ancylus Lake stage, Mesolithic (11,200–8800 cal BP); 2: Litorina Sea stage, Mesolithic and Early Neolithic (8800–6000 cal BP); 3: Middle Neolithic (6000–5100 cal BP); 4: Late Neolithic (5100–3750 cal BP); 5: Early Metal Period (3750–1700 cal BP); 6: Late Iron Age and Early Medieval Period (AD 300–1500); 7: Urban Medieval and Historical Period (AD 1500 to present) (graphics T. Kirkinen).

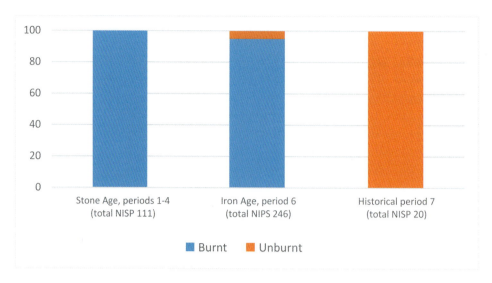

Fig. 3. The distribution of burnt/unburnt bear bone finds (in percentages). Sites with mixed Stone Age, Early Metal Period/Bronze Age, Iron Age and Historical period datings are excluded.

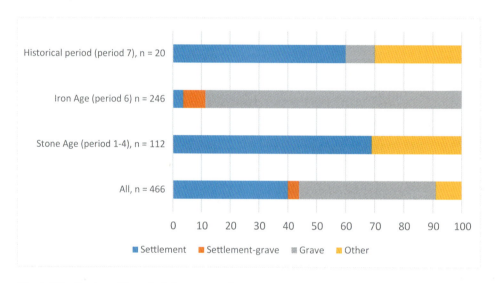

Fig. 4. Distribution of bear finds by contexts (in percentages).

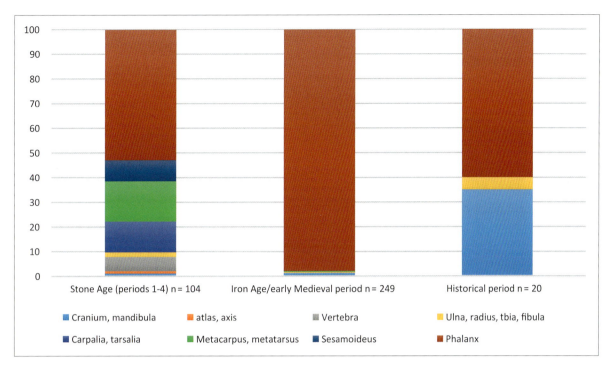

Fig. 5. Anatomical distribution of bear bones (in percentages). The following assemblages have been excluded from this graph: Stone Age – Rautalampi Mäntyranta (no information available about the anatomical element), Iron Age – Hämeenlinna Kalomäki 2, Laitila Rukoushuone, Raisio Pappilanmäki, Turku Ristimäki, Turku Kärsämäki, Uusikaupunki Kalmumäki (total number of identified bear bones unknown).

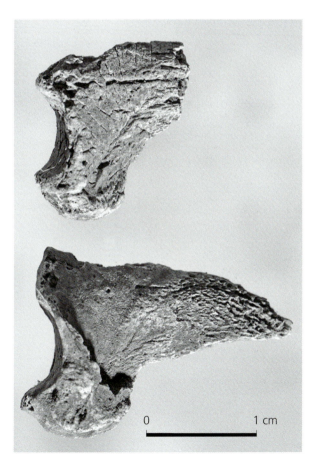

Fig. 6. Burnt third phalanges of a brown bear (photo T. Kirkinen).

Fig. 7. Bear hairs under light microscope (photo T. Kirkinen).

Fig. 8. A traditional sieppuri, *a short cape made from a brown bear pelt, from the 19th century. Note the bear's nose and eyeholes on the lowest part of the garment (SU4527:33; Finno-Ugric Collection at the National Museum of Finland; photo T. Kirkinen).*

Fig. 9. Bear figurines from Finland. a: Bear-headed stone axe from Paltamo, northern Finland; 191 x 41 x 43 mm (KM 13275:1; National Museum of Finland; photo Finnish Heritage Agency); b: Bear figurine made of flint from Taipalsaari, southeastern Finland. The item dates to the Neolithic; 55 x 24 x 7 mm (KM 31289:584; National Museum of Finland; photo R. Bäckman / Finnish Heritage Agency); c: Possible bear figurine made of amber, dated to the Neolithic. The item was found by divers in front of the famous Astuvansalmi rock painting site, eastern Finland; 34 x 18 x 11 mm (KM 27146:1; National Museum of Finland; photo Finnish Heritage Agency).

Fig. 10. Bear tooth pendant from the Iron Age Suotniemi inhumation cemetery, Karelian Isthmus, present-day Russia (KM 2487:7; National Museum of Finland; photo T. Kirkinen).

Fig. 11. So-called bear tooth pendants made of bronze from Kalanti, southwestern Finland. The items are dated to the Viking Age (KM 15131:3; National Museum of Finland; photo Finnish Heritage Agency).

Fig. 12. Bear's paw, from Evenks in Northern Asia, 19th– 20th century (Museum of the History of Religion, St. Peterburg, B-114-I). A bear's paw protected people from diseases and misfortunes. Following this belief, Evenks usually hung it above the child's cradle. Sometimes a paw was suspended on a reindeer's neck to guard it from predators (photo Museum of the History of Religion, St Petersburg [Russia]).

Table 1. Archaeological bear bone finds in Finland and on the Karelian Isthmus. See periods in Fig. 2. KM = National Museum of Finland; TYA = Archaeological collections at the University of Turku; TMM = Turku Provincial Museum.

Site	Collection	Cat. no.	Excavator	Excavation year	Osteologist	Burnt/ unburnt	Character of site	Period	NISP	U. arctos	H. sapiens
Savukoski Malmio 1A	KM	42446	Seppä & Laulumaa	2019	Katariina Nurminen	burnt	occupation	period 1	1	1	
Alajärvi Heikinkangas and Rasi	KM	11895	Luho	1948	Ann Forstén	burnt	occupation	Period 2	18	18	
Askola Pappila Perunamaa Saunapelto	KM	13068	Luho	1952	Jukka Jernvall	burnt	occupation	Period 2	3	1	
Inari Nellimjoen suu S	KM	24376	Sohlström	1988	Stella From	burnt	occupation	Period 2	251	1	
Kokemäki Kraviojankangas	KM	20584	Heikkurinen	1979	Mikael Fortelius	burnt	occupation	Period 2	1,135	1	
Rovaniemi Jokkavaara	KM	26610	Karjalainen	1991	Pirkko Ukkonen	burnt	occupation	Period 2	309	2	
Rovaniemi Jokkavaara	KM	21834	Torvinen	1982	Pirkko Ukkonen	burnt	occupation	Period 2	576	1	
Äänekoski Kapeenkoski	KM	25301	Kankkunen, Halinen	1989	Pirkko Ukkonen	burnt	occupation	Period 2	1	1	
Laukaa Hartikka	KM	23697	Taskinen	1987	Pirkko Ukkonen	burnt	occupation	Period 3	24	1	
Rääkkylä Vihi 1	KM	30460	Pesonen	1997	Pirkko Ukkonen	burnt	occupation	Period 3	384	1	
Vantaa Stenkulla	KM	29954	Katiskoski	1996	Pirkko Ukkonen	burnt	occupation	Period 3	1,691	1	13
Vantaa Maarinkunnas	KM	30464	Leskinen	1997	Niklas Söderholm, Pirkko Ukkonen	burnt	occupation	Period 3	1358	1	25
Enontekiö NE Tunturipolku	KM	28751	Halinen	1993	Pirkko Ukkonen	burnt	occupation	Period 4	233	34	39
Tuusula Kirkkotie	KM	26697	Ruonavaara	1991	Sirpa Nummela	burnt	occupation	Period 4	10	1	
Hartola Uusi-Ruskeala c	KM	37985	Pesonen	2009	Kristiina Mannermaa	unburnt	occupation	Period 6	807	1	0
Hämeenlinna Varikonniemi	KM	23703	Schulz	1987	Jukka Jernvall	unburnt	occupation	Period 6	652	1	3
Mikkeli Valkola	KM	14074	Turunen	1956	Pirkko Ukkonen	unburnt	occupation	Period 6	82	2	0

Cont. Tab. 1.

Cont. Table 1.

Site	Collection	Cat. no.	Excavator	Excavation year	Osteologist	Burnt/ unburnt	Character of site	Period	NISP	U. arctos	H. sapiens
Raisio Mullin edus-pelto	TYA	619, 631, 642, 667	Pietikäinen	1994–1997	Ulla Tupala	unburnt	occupation	Period 6	4,183	3	1
Sysmä Ihananiemi	KM	32291	Poutiainen	2000	Kristiina Mannermaa	unburnt	occupation, cremation?	Period 6	7,32	1	0
Kokemäki Käräjämäki	KM	32705	Taivainen	2001	Kristiina Mannermaa	burnt	cremation	Period 6	104	7	70
Kouvola (Jaala) Pukkisaari	KM	19915, 29097, 30871	Miettinen, Mertanen	1994, 1995, 1996	Kati Salo	burnt	cremation	Period 6	726	13	703
Laihia Mujanvainio (cairn 1b)	KM	10621:18–35	Meinander		Tarja Formisto, Katariina Nurminen	burnt	cremation	Period 6	16	4	523
Laihia Mujanvainio (cairn 3)	KM	10856:12d,18	Meinander		Katariina Nurminen	burnt	cremation	Period 6	2	1	443
Laitila Vainionmäki A		27777			Tarja Formisto	burnt	cremation	Period 6	?	2	
Laitila Vainionmäki A		24389			Tarja Formisto	burnt	cremation	Period 6	?	9	
Laitila Vainionmäki B	KM	34726	Luoto	2004	Kati Salo	burnt	cremation	Period 6	316	1	461
Laitila Kylämäki		?				burnt	cremation	Period 6	?	5	
Laitila Rukoushuone-Kansakoulunmäki		?			Pirjo Lahtiperä	burnt	cremation	Period 6	?	?	
Lempäälä Päivääniemi	KM	23749	Katiskoski	1987	Tarja Formisto	burnt	cremation	Period 6	1	3	1
Lieto Merola		?			Pirjo Lahtiperä	burnt	cremation	Period 6	?	?	
Mikkeli Latokallio	KM	11070	Sarasmo	1939	Pirkko Ukkonen	burnt	cremation	Period 6	32	2	28
Nastola Skinnari	KM	31607			Tarja Formisto	burnt	cremation	Period 6	?	2	
Käkisalmi Suotniemi	KM	2487:7	Schvindt	1885			inhumation	Period 6		1	
Pirkkala Tursiannotko	KM	39258	Schvindt	2012	Auli Bläuer	unburnt	occupation	Period 6		2–3	

Cont. Tab. 1. ▲

Cont. Table 1.

Site	Collection	Cat. no.	Excavator	Excavation year	Osteologist	Burnt/ unburnt	Character of site	Period	NISP	U. arctos	H. sapiens
Raisio Siiri 1	TYA	413, 454, 494, 554	Pitkänen, Fagerström, Fagerström & Lehtonen	1987–1990	Anne-Mari Liira	burnt	cremation	Period 6	2,737	60	2,121
Salo (Halikko) Isoriihenmäki	KM	18837	Sarvas	1972	Mikael Fortelius	burnt	cremation	Period 6	44	3	13
Salo (Halikko) Rikalanmäki	TYA	89,105,106	Seppänen	1976–1978	Jan Storå	burnt	cremation	Period 6	368	20	10
Turku Kärsämäki		?				burnt	cremation	Period 6	?	5	
Turku Ristimäki		?			Pirjo Lahtiperä	burnt	cremation	Period 6	?	?	
Uusikaupunki Kalmumäki					Pirjo Lahtiperä	burnt	cremation	Period 6	?	?	
Vaasa (Vähäkyrö) Kaavontönkkä	KM	9520	Hackman	1932	Katariina Nurminen	burnt	cremation	Period 6	11	6	48
Valkeakoski Koirakivi	KM	41247	Moilanen	2017	Anne-Mari Liira	burnt	cremation	Period 6	215	1	15
Vöyri Lågpeltkangas	KM	11295	Tegengren	1941	Tarja Formisto	burnt	cremation	Period 6	?	1	
Vöyri Soldatstomparen	KM	11132	Tegengren	1939	Tarja Formisto	burnt	cremation	Period 6	?	2	
Jämsä Hiidenmäki	KM	33293	Vanhatalo	2002	Niklas Söderholm	burnt	cremation	Period 6	?	14	
Raisio Pappilanmäki		?			Pirjo Lahtiperä	burnt	cremation	Period 6	?	1	
Tampere Vilusenharju	KM	18556			Pirjo Lahtiperä	burnt	cremation, inhumation	Period 6	?	54	
Kaarina Ravattula Ristimäki	TYA	914	Ruohonen	2015	Anne-Mari Liira	unburnt	inhumation	Period 6	4	1	
Vöyri Pörnullbacken	KM	31395	Viklund, Löeffler, Risla	1998	Barbro Hårding	burnt	cemetery-cremation-occupation	Period 6	412	7	340
Vöyri Pörnullbacken	KM	31395	Viklund, Löeffler, Risla	1999	Barbro Hårding	burnt	cemetery-cremation-occupation	Period 6	684	4	542

Cont. Tab. 1. ▲

Cont. Table 1.

Site	Collection	Cat. no.	Excavator	Excavation year	Osteologist	Burnt/ unburnt	Character of site	Period	NISP	U. arctos	H. sapiens
Hämeenlinna Kalomäki 2	KM	?				burnt	cremation-inhumation-occupation	Period 6	?	?	
Hattula Myllymäki	KM	17291		1967	Björn Kurtén	burnt	cremation-occupation	Period 6	?	5	
Hämeenlinna Riihimäki	KM	30304	Seppälä	1997	Niklas Söderholm	burnt	cremation-occupation	Period 6	1,361	2	1,308
Savukoski Mukkala	SU	5187	Leppäaho	1934		unburnt	inhumation	Period 7	?	2	
Tornio keskusta	KM	2002081	Herva	2002	Anna-Kaisa Puputti	unburnt	urban	Period 7	1	10	
Turku Åbo Akademin tontti (Y504)	TMM	21816	Pukkila	1998	Auli Tourunen	unburnt	urban	Period 7	14,935	2	1
Enontekiö Näkkälä	KM	37851	Äikäs	2008	Anna-Kaisa Puputti	unburnt	other	Period 7	50	4	
Kaarina Kuusiston Piispanlinna			Suna	?	Inari Kylänen	unburnt	other	Period 7	?	2	
Kouvola (Valkeala) Käyrälampi	KM	24604	Miettinen	1985	Kristiina Mannermaa	burnt	other	Period 2	73	34	
Kristiinankaupunki Räväsen	KM	29610	Laulumaa	1996	Pirkko Ukkonen	burnt	occupation	Multiperiodic (1–4)	264	1	
Loviisa (Liljendal) Kvarnbacken	KM	9273, 18900, 19152	Äyräpää, Pohjakallio, Pohjakallio	1930, 1972, 1973	Mikael Fortelius	burnt	occupation	Multiperiodic (1–4)	1	2	
Rautalampi Mäntyranta	KM	29442	stray finds		Pirkko Ukkonen	burnt	occupation	Multiperiodic (1–4)	35	1	
Saarijärvi (Summassaari) Moilanen	KM	12234	Luho	1949	Ann Forstén	burnt	occupation	Multiperiodic (1–4)	1	8	
Vantaa Asola	KM	20164	Väkeväinen	1977	Kati Salo	burnt	occupation	Multiperiodic (1–4)	187	1	
Viipuri Häyrynmäki	KM	5620:543	Ailio	1909, 1910	Mikael Fortelius	burnt	occupation	Multiperiodic (1–6)	194	1	
Viipuri Häyrynmäki	KM	5428:611	Soikkeli	1912	Mikael Fortelius	burnt	occupation	Multiperiodic (1–6)	577	1	

Cont. Tab. 1.

Cont. Table 1.

Site	Collection	Cat. no.	Excavator	Excavation year	Osteologist	Burnt/unburnt	Character of site	Period	NISP	U. arctos	H. sapiens
Enontekiö Valkea-järvi E	KM	26760	Halinen	1991	Pirkko Ukkonen	burnt	occupation	Multiperiodic (1–6)	105	1	
Inari Saamen museo	KM	27808	Seppälä	1993	Pirkko Ukkonen	burnt	occupation	Multiperiodic (1–6)	776	9	
Kangasala Pohti-olampi and Tiilitehdas	KM	32000	Schulz	1999	Kristiina Mannermaa	burnt	occupation	Multiperiodic (1–6)	426	3	
Kangasala Pohtio-lampi ja Tiilitehdas	KM	32554	Schulz	2000	Kristiina Mannermaa	burnt	occupation	Multiperiodic (1–6)	1,51	2	
Kemijärvi Neitilä 4	KM	15671	Sarvas	1962	Pirkko Ukkonen	burnt	occupation	Multiperiodic (1–6)	146	2	
Kuhmo Järvelä I	KM	27295	Kontio	1992	Pirkko Ukkonen	burnt	occupation	Multiperiodic (1–6)	168	2	
Lahti (Nastola) Kilpisaari 1	KM	32180	Poutiainen	2000	Kristiina Mannermaa	burnt	occupation	Multiperiodic (1–6)	637	3	
Pieksämäki Naarajärvi	KM	22019	Jussila	1983	Mikael Fortelius	burnt	occupation	Multiperiodic (1–6)	128	1	
Pälkäne (Luopioinen) Hietaniemi	KM	16822	Miettinen	1965	Anna Pirkkalainen	burnt	occupation	Multiperiodic (1–6)	202	4	
Pälkäne (Luopioinen) Hietaniemi	KM	17374	Miettinen	1967	Anna Pirkkalainen	burnt	occupation	Multiperiodic (1–6)	40	4	
Pälkäne (Luopioinen) Hietaniemi	KM		Miettinen		Leif Blomqvist	burnt	occupation	Multiperiodic (1–6)	1	3	
Pälkäne (Luopioinen) Hietaniemi	KM	17131	Miettinen	1966	Anna Pirkkalainen	burnt	occupation	Multiperiodic (1–6)	80	3	
Riihimäki Sinivuokkoniemi	KM	30884	Matiskainen	1998	Pirkko Ukkonen	burnt	occupation	Multiperiodic (1–6)	298	1	
Suomussalmi Kellolaisten tuli	KM	14831	Huurre	1959	Mikael Fortelius	burnt	occupation	Multiperiodic (1–6)	179	1	
Suomussalmi Särkkä	KM	18322	Huurre	1970	Mikael Fortelius	burnt	occupation	Multiperiodic (1–6)	145	1	

Cont. Tab. 1.

Cont. Table 1.

Site	Collection	Cat. no.	Excavator	Excavation year	Osteologist	Burnt/ unburnt	Character of site	Period	NISP	U. arctos	H. sapiens
Taivalkoski Tervaniemi	KM	28128	Saukkonen	1993	Pirkko Ukkonen	burnt	occupation	Multiperiodic (1–6)	106		18
Taivalkoski Tervaniemi I	KM	28687	Raike	1994	Pirkko Ukkonen	burnt	occupation	Multiperiodic (1–6)	8,129	1	
Vantaa Erikas	KM	19430	Sarkki	1974	Kati Salo	burnt	occupation	Multiperiodic (1–6)	54	1	
Saarijärvi Karjalais-pirti/Rusavierto	KM	32195	Leskinen	2000	Pirkko Ukkonen	burnt	occupation	Multiperiodic (1–7)	3,875		19
Saarijärvi Karjalais-pirtti/Rusavierto	KM	29406	Schulz	1995	Pirkko Ukkonen	burnt	occupation	Multiperiodic (1–7)	683	1	
Saarijärvi Karjalais-pirti/Rusavierto	KM	31616	Leskinen	1999	Kristiina Mannermaa	burnt	occupation	Multiperiodic (1–7)	2,685	1	
Saarijärvi Karjalais-pirti/Rusavierto	KM	31616	Leskinen	1999	Nina Peltonen	burnt	occupation	Multiperiodic (1–7)	296	1	
Sodankylä Autiokenttä 1	KM	20585	Honkanen	1979	Pirkko Ukkonen	unburnt	occupation	Multiperiodic (1–7)	504	1	
Janakkala Taurula	KM	24745, 26065	Schulz	1989, 1990	Sirpa Nummela	unburnt/burnt	occupation	Multiperiodic (1–7)	1,304	1	
Pirkkala Tursiannotko	KM	39258	Raninen	2012	Auli Bläuer	unburnt	occupation	Multiperiodic (6–7)	1,424	2	

Table 2. Radiocarbon dates from archaeological brown bear (Ursus arctos) finds in Finland. [1]UKKONEN/MANNERMAA 2017; [2]MIETTINEN 2012; [3]KIRKINEN 2017; [4]MÄNTYLÄ-ASPLUND/STORÅ 2010; [5]KOIVISTO et al. 2016; [6]SEPPÄ/LAULUMAA 2020. KM = National Museum of Finland, Helsinki; TYA = Archaeological collections at the University of Turku. *Calibration: Calib 6.0.1 (STUIVER/REIMER 1993); calibration data IntCal09 (REIMER et al. 2009).

Site	Collection no.	Date BP	Lab. no.	Date cal BP
Inari Saamen Museo	KM 27808:341	7154 ± 44[1]	Hela-3000	8007–7947
Kouvola Valkeala Käyrämpi	KM 24604	7130 ± 45[2]	Hela-2826	8001–7933
Kemijärvi Neitilä	KM 19671:1084	6963 ± 41[1]	Hela-3121	7843–7732
Mikkeli Tuukkala Valkola	KM 14074		Hela-3587	720–605
Salo Rikala	TYA 105:322	1515 ± 35[4]	Ua-36963	1483–1347
Vaasa Vähäkyrö	KM 9520	1190 ± 30[5]	BETA-358440	1220–1030
Kaavontönkkä			Ua-66392	
Savukoski Sokli Malmio	KM 42446:83	8820 ± 61[6]	Hela-1885	10,167–9667
Enontekiö Näkkälä	KM 3785	830 ± 21[7]		785–691

Table 3. Chronological periods. CW = Comb Ware; Jäkärlä = Jäkärlä Ware; EAW = Early Asbestos Ware; Sär 1 = Säräisniemi 1 Ware; Kierikki = Kierikki Ware; Pyheensilta = Pyheensilta Ware; Pöljä = Pöljä Ware; Jysmä = Jysmä Ware; CorW = Corded Ware; Kiukainen = Kiukainen Ware; Palajguba = Palajguba Ware; ST = Sarsa-Tomitsa Ware; Vardøy = Vardøy Ware; Lovozero = Lovozero Ware; Sär 2 = Säräisniemi 2 Ware; Kjelmøy = Kjelmøy Ware; Luukonsaari = Luukonsaari Ware; Sirnihta = Sirnihta Ware; Anttila = Anttila Ware; Paimio = Paimio Ware; Morby = Morby Ware. Note that in Finnish archaeology the Neolithic Period does not refer to farming but to the appearance of pottery. Agriculture was introduced into Finland and the Baltic area very slowly and mainly towards the end of the Neolithic (e.g. CARPELAN 1999; EDGREN 1999; ALENIUS et al. 2013).

Period	Date	Pottery type	Subsistence
7 – Urban Medieval and Historical Period	AD 1500–present	Imported Wares	Agriculture, hunting-fishing-fowling-gathering
6 – Late Iron Age and Early Medieval Period	AD 300–1500	Iron Age Wares; Karelian Ware, Medieval Imported Wares	Agriculture, hunting-fishing-fowling-gathering
5 – Early Metal Period	3750–1700 cal BP	ST, IT (Vardøy), Lovozero, Sär 2 (Kjelmøy, Luukonsaari, Sirnihta, Anttila, Kainuu), Paimio, Morby	Hunting-fishing-fowling-gathering, agriculture
4 – Late Neolithic	5100–3750 cal BP	Pyheensilta, Pöljä, Jysmä, CorW, Kiukainen, Palajguba II, etc.	Hunting-fishing-fowling-gathering, early agriculture
3 – Middle Neolithic	6000–5100 cal BP	CW 2, CW 3, Kierikki	Hunting-fishing-fowling-gathering, (early agriculture)
2 – Litorina Sea stage, Mesolithic & Early Neolithic	8800–6000 cal BP	CW 1:1, CW 1:2, Jäkärlä, EAW, Sär 1	Hunting-fishing-fowling-gathering
1 – Ancylus Lake stage, Mesolithic	11,200–8800 cal BP	No pottery	Hunting-fishing-fowling-gathering

Table 4. Archaeological bear hair finds in Finland and on the Karelian Isthmus (SCHWINDT 1893; KIRKINEN 2015; 2017; 2019a; KIRKINEN et al. 2019; KIRKINEN et al. 2020a; b). KM = National Museum of Finland; TYA = Archaeological collections at the University of Turku; SU = Finno-Ugric Collection at the National Museum of Finland.

Site	Burial	Date	Collection no.	Function
Eura Luistari	56 (female)	11th century	KM 18000, several subnumbers	garment?
Eura Luistari	95 (female)	9th century	KM 18000: 2075	grave furnishing
Eura Luistari	377 (female)	11th century	KM 18000: 4272, 4273	grave furnishing
Ristimäki Ravattula	4/2014 (female)	13th–14th centuries	TYA 912:523b, 523f	grave furnishing
Ristimäki Ravattula	18/2016 (female)	13th–14th centuries	TYA 993:173	grave furnishing
Kaukola Kekomäki	1 (male and female)	13th century	KM 2489: 4, 6, 14?	grave furnishing
Savukoski Mukkala	III (male)	17th century	SU 5187: 14, 15	grave furnishing

The history of the brown bear (*Ursus arctos* L.) in the northern German lowlands

By Ulrich Schmölcke

Keywords: Archaeozoology, zooarchaeology, human-animal relationship, history of hunting, mammal extinction

Abstract: Brown bears had been widely distributed in the northern German lowlands since the final centuries of the last Ice Age 15,000 years ago and remained part of the fauna until their disappearance over the last two millennia. They became extinct in the northwestern region by the Stone Age, but survived for longer in parts of eastern Germany, where the last individual was killed in the 18th century. It is evident that the meaning and relevance of bears changed a number of times and in fundamental ways. The rarity of bear remains in Mesolithic and Neolithic bone assemblages indicates a lack of human interest in bear hunting, for either spiritual or profane reasons. Later, in the pre-Roman and Roman Iron Age, bears played a role in funeral practices. In those days, bear claws were quite regularly used in burial contexts, and this new role not only had a cultural and religious meaning, but must have created a demand for bear parts and an increased pressure on hunting. In medieval times, bears became extinct in nearly all regions west of the river Elbe, but in the east, which was settled by Slavs c. 1000–1200 CE, bears still occurred. In these regions, bears were a constant, if not common, hunting prey in medieval times. However, also in the regions east of the Elbe, the growing human population density, large-scale habitat loss, massive direct persecution, and finally, the capture of bears for public entertainment, led to the extirpation of the last populations.

Introduction

The brown bear (*Ursus arctos* L.) is a common part of the natural fauna of central Europe, though this is hard to believe for many people today as the species disappeared from most of its regions hundreds of years ago. The history of the European bear, which was intensively analysed during the first decade of the 21st century (summarised by Davison et al. 2011), began during the last glacial period, when bears lived, at least regionally, in subarctic environments north of the Mediterranean (Sommer/ Benecke 2005, fig. 1; Edwards et al. 2014). The species' northern central European history began during the final stages of the last Ice Age between 15,000 BCE and 9600 BCE, when brown bears occurred in central Europe in both the colder and warmer phases of the Late Glacial (Sommer/ Benecke 2005, fig. 2). The bear's presence in environments that ranged from subarctic to densely wooded areas demonstrates its high level of adaptability to very different ecological conditions; it just needs enough food, such as berries, honey, roots, carrion, fish, or small ruminants (Jakubiec 1993). The hibernation behaviour of bears depends on the availability of food, but females spend the winter in self-made dens, where they give birth to two to three cubs, which stay with their mother for two years. Bears are not sexually mature before the age of four to five years.

Less is known about the ecology of the northern German lowland populations of brown bears. Only SCHMIDT (1856, 15–16) provides some information about the food and behaviour of the last Pomeranian bears. He reports that bears sometimes tried to catch a single prey animal by jumping on its back and biting its neck, but their favourite food were berries followed by pears, grapes, ants, and honey. According to Schmidt, they lived in forests with large, hollow oak trees. This corresponds with modern observations in eastern Europe, where the brown bear also prefers extensive woods with old trees (JÜRGENSON 1974). However, recent bear populations in Europe are completely restricted to remote areas, whereas archaeological data show them in a great variety of landscape contexts (cf. KUNST/PACHER 2019).

The hunting of bears is widely known to be a highly dangerous and risky undertaking, even if SCHMIDT (1856, 16), reporting about Pomerania in early modern times, tells us that an attacked bear acts *"tapfer, aber einsichtslos"*, meaning "brave but without insight" (cf. OEHRL 2013 for a more general view). As the author explains, their defensive behaviour, their presentation of their own body to the hunter, and their attempts to crush humans instead of biting them, makes it easy and quite safe to shoot bears. In any case, the risk involved in encountering bears in pre-firearm eras is mirrored in their name in the Germanic languages, since the German *Bär*, the English *bear*, the Scandinavian *bjørn* and other similar names have their root in *bher-, originally meaning just "brown" (KLUGE/VON SEEBOLD 2012). Using such an indirect name is common when the actual name of something or somebody is a taboo. By calling bears just "the brown one" people avoided calling up the potentially dangerous animal (KLUGE/VON SEEBOLD 2012). In (prehistoric) people's imagination, a bear hearing humans talking about "the brown" would not be interested, whereas it might be by people naming it directly (SCHMÖLCKE et al. 2017; see NEDOMA and UDOLPH, this volume).

The main task of the present study is to create an overview of the former appearance of bears in the German lowlands and to investigate changes in the human-bear relationship in the area under consideration through time. Changes and developments in the human perception of the bear are always related to cultural customs and practices, so it makes sense to use a chronological timeline based on the main cultural periods for the diachronic structure. It is clear that such a structure simplifies archaeological insights and units, but it is also suitable in the context of the questions to be answered here. Thus, the postglacial period will be divided into the following stages, each of which will be analysed and discussed separately: Mesolithic (9600–4000 BCE), Neolithic (4000–1800 BCE), Bronze Age (1800–500 BCE), Pre-Roman Iron Age (500–1 BCE), Roman Iron Age (0–500 CE), Early Medieval period (500–1050 CE), Late Medieval period (1000–1500 CE), and early modern times (1500–1750 CE). Even if the cultural development was relatively uniform and synchronous in the investigated area, sometimes there are strong discrepancies (e.g. the Neolithic began much earlier in the Brandenburg area than in Schleswig-Holstein). In such cases, the cultural stage has priority over the chronology.

DATA BACKGROUND AND FIRST AMAZING OUTCOMES

The archaeozoological data for this study are gathered from the huge data collection "The Holocene History of the European Vertebrate Fauna", which was built up under the leadership of Angela von den Driesch (Ludwig Maximilians University, Munich), Norbert Benecke (German Archaeological Institute, Berlin), and Dirk Heinrich (Christian Albrechts University, Kiel) in the 1990s (HEINRICH et al. 2016). With respect to northern Germany – defined here as part of the central European plain north of 52°N and between the modern borders of the Netherlands and Poland (Fig. 1) – the data collection lists 429 archaeological sites with altogether more than one million animal remains (= NISP =

Number of Identified Specimens) dating from the Late Glacial to early modern times. Seventy sites comprise remains of brown bear; the total number of bear remains is 557, these are 0.7 % of the remains of wild animals – and 0.06 % including remains from domesticated ones. Excluding sites with unclear or period-overlapping dating (such as "from 10th to 18th centuries"), reduces the number of irrelevant sites and records, and the remaining number is both statistically significant and meaningful (Table 1). This remains true even if the find list is incomplete, because it only sporadically comprises the results of excavations published after 2000.

The hunting of wild mammals did not always have the same degree of importance in the past. It is possible to measure the relevance of a specific hunted species by comparing its relative proportions in an archaeozoological assemblage with the relative proportions of other hunted species. If the proportion of a species is high, it is likely that the hunting of it played a considerable role for the local human group. Naturally, during the Late Glacial and the Mesolithic, when human meat demand was satisfied nearly exclusively by the exploitation of wild animal resources – the dog was only sporadically eaten (EWERSEN/SCHMÖLCKE 2013) – the proportion of wild species' remains in the excavated bone assemblages reaches maximum values of near 100 % (Table 1). Hunting was often still important during the Neolithic, though with a successively declining tendency. Later on, the exploitation of wild animals lost its significance, and, between the Bronze Age and early modern times, only 1–6 % of archaeological animal remains derive from wild species.

As we will see, the persecution of bears also did not always have the same relevance in the past. A first overview of the data shows that, from a quantitative point of view, bears were never important. In most historical periods, their remains constitute around 0.5 % of hunted wild animals, with the exception of the Roman Iron Age, during which they constitute about 3 % (Table 1). Red deer, roe deer, and wild boar were quantitatively always much more relevant. There is, however, one extraordinary divergence from this, when during the pre-Roman Iron Age nearly half of all identified remains from wild mammals derive from brown bears! In the paragraph about the Iron Age, some pages below, this astonishing phenomenon will be discussed and explained. At this point, it should just be remembered that comparisons between proportions of different animal species at different sites are no simple undertaking at all. Among many other factors, social and ecological circumstances as well as site-functional aspects must be considered, and sometimes it requires background information, ranging from species behaviour to excavation technique (cf. KUNST 2014).

Next to species proportions, constancy is a valuable measurement used in this paper. Constancy means the percentage presence of e.g. brown bear remains in a number of sites or periods (SCHMÖLCKE 2003; 2013). The absolute number of individual specimens from bears at one of the sites (or periods) is not important, and herein constancy has some similarity to the biogeographical concept of nestedness (ULRICH et al. 2009). Constancy just compares the presence or absence of a species. Comparisons of the constancy of bear records at all excavated archaeological sites with animal remains in the area of interest show that bears are found at between 9 and 35 % of postglacial sites, depending on their dating (Table 1). The medium constancy is 17 %, which means that statistically, in northern Germany, six archaeological sites need to be excavated in order to find at least one single part of a bear! It is remarkable that this value does not differ very much from the Mesolithic to the Early Medieval period. Only the Bronze Age with its lower values and the Iron Age and Late Medieval eras with their higher values show greater differences. It will be one of the tasks of the present paper to find and discuss possible reasons for this distinctive picture.

The database used for this study comprises all kinds of archaeological features, from settlements to single animal skeletons to human burials. Even if it makes less sense, when addressing many of the research questions, to compare results from large medieval settlement excavations with those from single Iron Age grave pits, I have decided to do so in the present paper – at least in the tables. The main reason for this is that a differentiation between temporal, spatial, or functional categories that are too different reduces both the statistical testimony and the clarity of main developments concerning the human-bear relationship. In the present text, differentiations with respect to different functional feature groups will be made and discussed if necessary.

Prior to historic times, there are no written sources, and it is difficult to reconstruct the human perception of the largest central European predator and the general relationship between bear and human. Working with archaeozoological data means always to take into account that the approach to reconstructing the former distribution of a species will necessarily be related to, and influenced by, former human settlement patterns and archaeological research activities. Thus, the additional consideration of data from other scientific disciplines is important. Besides archaeological remains, it is also necessary to include other historical sources such as written testimonies, pictures, and even memorials. Only with the existence of such sources, alongside the archaeological ones, will our knowledge about the deep-rooted nature of human-bear relationships find a stable basis. In the area of interest, these kinds of sources are available only from *c.* 1000 CE onward, i.e. with the beginning of the Late Medieval period. A scientific study of bears that systematically investigates and analyses non-archaeological sources has, up to now, been published only for the area previously known as the German Democratic Republic (BUTZECK et al. 1988; a related approach is used by OEHRL 2013). As it includes some regions that are also relevant for the present article, this study is an unprecedented and indispensable addition to the archaeozoological data.

Brown bears in the German lowlands during the Mesolithic

At the end of the last glacial period, when the climate began to warm up and the first forests began to cover the region we are studying, the brown bear became a widely occurring animal (SOMMER/BENECKE 2005). Due to sharing the whole area of their habitat with Mesolithic hunter-fisher-gatherer communities, bears were a potential prey of human hunters. It is remarkable that, during the only postglacial period when human life was based exclusively on the exploitation of natural resources, the relative NISP of bear remains reaches, with less than 0.3 % of the wild mammal remains, its minimum value (Table 1). To make it clearer: Arithmetically, you need to identify more than 300 Mesolithic animal bones to find a bear! Thus, bear bones are a rarity in Mesolithic faunal assemblages – not only in our area of interest but e.g. also in Austria, where they are completely lacking during the Mesolithic (KUNST/PACHER 2019). Even at the archaeological sites of Tribsees, Hohen Viecheln, and Friesack, which have yielded the highest number of bear remains (for locations see Fig. 1), they reach only between 0.3 and 0.7 % of the wild animal NISP, and often sites with numerous identified animal remains yield only few, if any, bear bones (Table 2). Late Mesolithic occupation sites such as Rüde in Schleswig, Rosenhof in Holstein, Timmendorf-Nordmole I in Mecklenburg, and Lietzow-Buddelin in western Pomerania (all dated to the 5th millennium BCE) are examples for this observation. They also indicate that bears were less frequent, if not absent, in coastal areas, probably because the landscape there, structured by the sea with bights, bays, and bogs, did not allow for the existence of extensive woods, the bear's optimal habitat (cf. AARIS-SØRENSEN 2009 for Denmark).

The data indicate that bears were equally distributed throughout the Mesolithic "natural landscape"– the small groups of Mesolithic hunter-gatherers influenced ecological processes only locally and did not cause permanent environmental changes (GROSS et al. 2019) –, but show regional preferences due to ecological, namely vegetation conditions. Bearing in mind that many excavations have been carried out in coastal areas with their excellent preservation conditions, the constancy of bear records at Mesolithic sites seems to be relatively high: 14 %, which is not far away from the values of later epochs. This figure does not mirror an economic importance – which did not exist at all –, but shows a wide range of the species outside the coastal region.

Since the remains, not only of bears but also of other Carnivoria, are found only rarely at northern central European Mesolithic sites and with only few bones per site (Charles 1997; Schmölcke/Nikulina 2022), such species were obviously not a regular hunting prey. Even the fact that artificially perforated bear canines have been repeatedly found, is not contradictory. They were just part of the general and common practice of using the teeth of large animals as clothing accessories (Gramsch 2012). It is possible that, during the Mesolithic, humans and bears avoided direct contact with each other, which will have been relatively easy due to the small population densities of both. Comparisons with modern hunter-gatherers show that it is likely that spiritual reasons might also have forbidden the persecution of bears. Many native peoples of northern Eurasia and America have admired the brown bear as a kind of mediator between animals and humans up to modern times, and for that reason they normally did not kill it (Hallowell 1926). Often, the killing of a bear was only allowed involving the performance of strict and complex rituals before and after the hunt (Duerr 2010, 76–79). Such bear ceremonialism is typical in nearly all wooded parts of the northern hemisphere, and bears were always considered to be "kings of the forest", a holy animal, or, as said before, a kind of mediator between animals and humans (Hallowell 1926; Pastoureau 2007; Pentikäinen 2007).

An archaeozoological analysis of carcass treatment is not possible due to the low number of bear bones per site. Friesack 4 (NISP 9) and the Polish site of Dąbki (NISP 15) indicate that if bears were hunted during the Mesolithic, then, at least sometimes, only selected parts of the animals were transported back to the settlement. At Friesack 4, only metapodials, phalanges, and fragments of bear pelvis have been recorded (Schmölcke 2019), whereas at Dąbki, apart from a shoulder blade and a rib, all other bear finds are remains from the paw (Schmölcke/Nikulina 2015). Such representation patterns indicate that two kinds of bones were taken to the settlements – those that remained in the pelt after skinning (phalanges and metapodials), and those representing parts of the body with the best meat quality. Obviously, the Mesolithic foragers consumed most parts of the bear at the kill site and took only the pelt and selected parts of the bear's body back to their base camp. Such a special treatment of the animal characterises the bear as a very particular hunting object (Schmölcke et al. 2017).

Brown bears in the German lowlands during the Neolithic

Especially during the Early Neolithic, animal remains from archaeological sites are proof of a mixed economy (Hartz/Schmölcke 2013; Schmölcke in press). Although the subsistence of the people was now based mainly on domesticated animals, hunting was still an important part of the subsistence economy. Summarising the whole Neolithic era in the area of interest, the remains of wild mammals comprise more than 40 % of all animal bones (Table 1). In Neolithic assemblages, bear remains occur only sporadically; within the remains of wild animal species, they reach a proportion of less than 0.6 %, and, arithmetically, only every seventh excavated site yields at least one bear part (constancy 16 %). The maximum values for bears among the wild species were never higher than 1.9 % (Waren-Stinthorst) and 1.7 % (Hüde I); it has more often been the case that some single finds yielded amounts of less than 1 % (Table 3). The game-rich assemblage from Hüde I, which dates to about 2800 BCE, provides the largest number of bear remains out of all the sites considered (NISP 145) and can be of high importance for our knowledge of Neolithic bear hunting methods, carcass treatment, and bone modifications. Hüde I is the only site that offers the possibility of significant analyses concerning age structure, skeletal part representation, and cut marks. Unfortunately, the published analyses are not, as yet, very detailed (Hübner et al. 1988, 54–55). They show that all bear remains derive from relatively large adult animals and that, as seen at some Mesolithic sites, two parts of the skeleton dominate the material by far, namely the paws with metapodials and phalanges (43 % of the finds in Hüde I) and the head with cranium, mandibles and teeth (30 %). This is a remarkable difference to all the other animal species at Hüde I. An explanation for this unequal distribution of the different skeletal parts is that often only parts of the carcass reached the human settlement, in particular those parts correlated to pelts or those suitable for the fabrication of tooth pendants. The authors of the Hüde I study mention furthermore the high degree of fragmentation of postcranial bear remains but provide no details (Hübner et al. 1988, 41). At another site, Wolkenwehe-Heidmoor, which was repeatedly occupied in the 4th millennium BCE, Ewersen (2007, 86) found clear cut marks on all eight limb bones and vertebrae, and interprets them as being the result of decapitation, skinning, systematised disarticulation, and the consumption of bear meat. In one case, chop marks show the intention of removing the claw from the rest of the paw. Together, Heidmoor and Hüde I demonstrate the intensive exploitation of bear carcasses.

The quantity and constancy of bear remains from the Neolithic are low, but the proportion of bear within the hunted animals increases compared to the Mesolithic. Certainly, the high number of finds in Hüde I influence the values positively, but an increasing number of records and amounts can also be found for other large predator species (Schmölcke/Nikulina 2022). In the northern German lowlands, wolves, lynxes, wild cats, and bears seem then to have become a slightly more common object of hunting than before, and in the Netherlands this development is even more evident (Kuijper et al. 2016, tab. 2). Certainly, the number of records is still low, but the data might indicate a change in the overall tradition of, and motivation for, hunting. Thanks to the presence and breeding of livestock, the exploitation of wild animals was no longer the only human meat resource, and it is therefore likely that new, alternative motives and reasons for hunting developed. One new key reason for hunting was surely to guard livestock from predators and the crops from herbivores. Even if there is no archaeological indication of whether the meaning of the bear changed for the people, it cannot be ruled out that bears also achieved a new status in religious, symbolic, and ideological systems (Boyle 2006; Hartz/Schmölcke 2013).

Even if the killing of bears was potentially slightly more common during the Neolithic than during the Mesolithic, in the archaeozoological data there is no evidence that the Neolithic farmers influenced the general occurrence and distribution of the bear (Sommer/Benecke 2005). In addition, there is also no indication of a purposeful reduction of bear populations. Only from Denmark had brown bears obviously already disappeared, at the latest during the Neolithic (Aaris-Sørensen 2009). Since there is no evidence at all in Danish archaeozoological records for an intensive persecution previous to the species' disappearance, the reasons it vanished are likely to have been the special spatial preconditions: Due to the worldwide sea-level rise, Denmark was divided into islands and the Jutland Peninsula during the Mesolithic between 7000 and 6000 BCE, and this might have affected local bear populations negatively by reducing suitable habitats. Certainly, the vulnerability of the bear population was increased by the fragmentation of its habitat.

Apart from Denmark, the range of bears in northern central Europe was, at the end of the Neolithic, generally similar to that of Mesolithic times. But this is probably no longer the case when we go more into regional details. It cannot be excluded that the Mid-Holocene warming period, with its maximum between 5000 to 3000 BCE, reduced the reproductive rate of the species and caused population declines. Bears consume more energy resources in warm winters, because they wake up more often and go in search of food. As a result, less energy is available for caring for new-borns, which reduces the number of offspring that survive to maturity (Albrecht et al. 2017). This might be particularly the case in regions or periods where the warming was greater in winter than in summer. A long series of mild or even very mild winters without any phases of frost characterises, for example, the climate in northern Germany between 3800 and 3300 BCE (Feeser/Dörfler 2015; Czymzik et al. 2016). One consequence of such a series of warm winters was potentially a decreasing reproduction success in bears, but the mild winter climate also caused a large rise in human population density (Hinz et al. 2012) – again with negative consequences for the bear by habitat limitation and by the rising need for hunting in order to protect livestock. Therefore, it can be expected that, at least in the favourite regions of Neolithic farmers, a decline in the bear populations took place. The archaeozoological data, however, do not allow at present to reconstruct such range dynamics of bears on a local or regional level.

For the first time in the area under consideration, depositions of burned distal phalanges (claws) of bear are recorded in some human graves during the Neolithic. One complex of cremation burials at Westerhammrich (East Frisia), dated to about 2800 BCE, is known; here in four different graves between one and five bear claws with traces of burning have been found (Bärenfänger 2009). We can understand them as the remains of bear skins with third phalanges still attached, which were burnt together with the deceased during cremation ceremonies, or as *pars pro toto* for bear skins. At present it remains open as to whether this was only a local tradition.

Impressive and significant examples of pendants made of bear teeth came to light in the flat grave cemetery of Ostorf (Mecklenburg), dating to around 3200 BCE, where necklaces have been found combining hundreds of single bear teeth, and also the teeth of many other animal species, in several human graves (Lehmkuhl 2007). These finds prove again the practice of bear hunting and the use of bear parts as clothing accessories, but they might also indicate that bears started to play a role in funeral traditions. At present, however, it is impossible to assess if this was a common practice. Perforated canine teeth from Neolithic settlement contexts were excavated at the site of Oldenburg LA 232, which was occupied around 3000 BCE (Fig. 2; Brozio et al. 2018), and Wolkenwehe-Heidmoor (Ewersen 2007, fig. 47). At Heidmoor, a perforated second metatarsal was found (Lehmkuhl 1986).

Brown bears in the German lowlands during the Bronze Age

The Bronze Age in the northern German lowlands shows phases of economic boom and bust, and it is the same for human population density and size. Generally, the exploitation of nature by humans and the anthropogenic changes to the landscape were much more intensive than in the previous Neolithic era (Kneisel et al. 2019). However, since settlements were now often established in areas unfavourable for the preservation of bones (e.g. on sandy hills), archaeozoological statements about the distribution and meaning of the bear in the Bronze Age are extremely limited. It is not only the number of available archaeological sites that decreases drastically compared to the Neolithic – from 80 to 11 sites containing wild species –, but the number of wild animal remains drops by 87 % to a NISP of only 284 (Table 1). Only 5 % of all excavated animal remains now derive from wild species; obviously hunting no longer had any economic, and probably no social, relevance.

Under these circumstances, it is not surprising that Bronze Age bear remains are nearly completely lacking. There is just a single grave from the Middle Bronze Age (1500–1250 BCE in the area of concern) excavated at Nützen, where, next to other unburned grave goods, a burned bear distal phalanx has been found (Schmid 1981). Up to now, this seems to be the only record of bears in the area of interest. It remains speculative whether the local, quite intensive anthropogenic deforestation during the Bronze Age at least in some regions caused a reduction or even the extinction of bear populations. But even if this was the case, the brown bear is an adaptable species, and if the habitat destruction was only temporary and not combined with active and intensive persecution, bears should have been able to re-colonise lost ranges. It is likely that only in centres of human population the bear population was already significantly lower compared to that of the Mesolithic.

From the pre-Roman Iron Age to the Early Medieval period, altogether 20 cremation burials in the area of investigation include single or several bear claws. As we have seen, this tradition was already rooted in the Neolithic, but as a common concept it started about 200 BCE (cf. the discussion in Schönfelder 1994, 220) and was widely distributed some generations later in the period between 100 BCE to 50 CE (Table 4; cf. Beermann 2016). In the area of interest, 110 bear claws have been excavated from 14 Iron Age graves at nine sites; the number per burial ranges from one to 15 (Table 4). The geographical distribution of the sites shows them spreading from Lower Saxony to Brandenburg, and in the north to Holstein, but there is also a striking cluster in the Lower Elbe area, surrounded by the burial grounds of Alt Mölln, Ehestorf, Ehestorf-Vahrendorf, Harsefeld, and Putensen (cf. Fig. 1, nos. 1, 18, 31, 52).

Whether it was the burning of bear skins or the alternative use of single claws as *pars pro toto*, as a common tradition the ritual of using bear claws seems to represent a complete change in the human-bear-reception. It is likely that the area of interest was at least influenced by the various circumpolar bear cult ceremonies distributed throughout all the wooded regions in the northern hemisphere. Despite all the differences between regions, times, cultures, and peoples, the general concept of these ceremonies is the perceived role of the bear as a kind of special mediator between nature and humans (Hallowell 1926). In contrast to those eras in which the killing of the "king of the forest" was strictly ritualised, in the Iron Age it must have become a more frequent undertaking. As opposed to earlier periods, a demand now developed for ursine burial equipment. It seems to be realistic but speculative that this went hand in hand with new initiation rituals, in which a young man had to kill a boar or a bear in order to be admitted to the adult community (Oehrl 2013, with examples). It is, in any case, an indication of a complete change in human-bear relationships that the brown bear is, in the northern German Iron Age, represented exclusively through canines and distal phalanges in human burial contexts. This means there is a total lack of bear remains in profane occupation waste deposits at settlements (Table 5) – as is also known to be the case in the northern French Iron Age (Méniel 2001). It is, however, hard to say if the total lack of bear remains at Iron Age settlements may be the result of a spiritual taboo, which decreed that bear carcasses should not be taken into inhabited areas but only to burial grounds. However, the reasons can also have been more technical: Hunting generally played a minor role in Iron Age subsistence and therefore there are, at best, small quantities of wild mammal remains at every excavated site, with wild mammal remains even missing sometimes (Table 5). In total, the proportion of wild animal remains at Iron Age archaeological sites is only 1.2 %, reaching its all-time minimum (cf. Table 1). Under such circumstances, it is far more likely that species with both economic value and high population density would be recorded, such as deer species or wild boar, than species with limited economical value and low population density, such as bear or carnivores. Therefore, it is, from a statistical point of view, rather optimistic to expect to find bear remains in occupation waste deposits of sites with wild mammal remains that total only 1 % or less.

Coming back to funeral practices, Iron Age burials containing bear remains could be female or male, rich or poor. The number of claws differs and ranges from one to fifteen, but is of course dependent on the excavation method used and the local preservation conditions. This is one reason why it is not always certain whether the distal phalanges were originally part of a skin. In any case, the fact that the burial context creates an association with the spiritual or magical qualities bestowed by Europe's strongest animal is obvious, and a potential decorative function of the claws is also undeniable (Schmölcke et al. 2017, 904–905). Context studies demonstrate that during the Iron Age, bear claws as burial goods were part of a widespread pattern of beliefs (Schönfelder 1994; Kivisalo 2008; Grimm 2013; Oehrl 2013; Beermann 2016; Lindholm/Ljungkvist 2016; Kirkinen 2017).

The variety of social positions might indicate that bear remains as grave goods were connected more to personal virtue or prestige of the deceased than to her or his role or function in society (Schönfelder 1994). Since bears were seen as a Germanic prestige item, we can assume that they were not hunted in the Iron Age as a regular prey but because of their value and significance, and after they had been killed they appeared as a meaningful grave offering (Schönfelder 1994).

Due to all these circumstances it is impossible to obtain information about potential changes in the bear's distribution during the Iron Age. We can suspect that a human-driven landscape change by an agriculture more intensive than ever before (cf. Behre 2008), together with a still-increasing deforestation, reduced suitable habitats and caused an ongoing fragmentation of the population. However, the exact degree of habitat fragmentation and the question of how much population density and the number of individual animals decreased remain open.

Brown bears in the German lowlands during the Roman Iron Age

Burials furnished with bear claws as part of the remains of a skin or as single grave goods date back to the pre-Roman part of the Iron Age, but, at least in Scandinavia, this tradition reached its heyday with hundreds of examples in the period from the late Roman Iron Age to the beginning of the Early Medieval period, i.e. the time frame between the 3rd to the 6th centuries CE (Grimm 2013; Lindholm/Ljungkvist 2016). Detailed studies from Sweden have shown that the human exploitation of bears in the Iron Age resulted in a considerable decrease in certain populations and even local extinctions (Lindholm/Ljungkvist 2016). There is a long, intensive, ongoing discussion about the remains of bears, in particular claws, in northern Germanic graves. They might be understood as the remains of a fur as a noble bedding or wrapping for the deceased, as spiritual items, luxury goods, objects of prestige, insignia of a hero or a berserk, or hunting trophies (cf. Oehrl 2013; Kirkinen 2017, for references and discussion). In the area discussed in the present paper, however, such features are scarce during this period and are known from only three locations (Table 6). The finds from Süderbrarup still belong to southern Scandinavia, and two other sites with bear phalanges in burial contexts further south in Mecklenburg (Parum) and in the Börde district in Saxony-Anhalt (Haldensleben) point to a larger, but not common, distribution of this custom in the central European lowlands.

We can surmise that the hunting of wild animals was of minor interest in the daily life of humans during the Roman Iron Age, even though the state of research might be insufficient to prove this, since in northern central Europe only relatively few Roman Period sites have yielded animal bones (Schönfelder 1994, tab. 2; Schmölcke/Breede 2011). Altogether, the proportion of wild species reaches only 1.3 % of the excavated animal remains from this period; this is the same proportion as in the previous pre-Roman Iron Age (cf. Table 1). Red deer (*Cervus elaphus*) was the only species of economic relevance, but it is remarkable that, after roe deer (*Capreolus capreolus*) and wild boar (*Sus scrofa*), the brown bear is relatively common, reaching about 3 % of all wild mammal remains. This is its all-time maximum, which is based, however, upon just 34 remains from funeral contexts. When claws from the burials are excluded, the proportion decreases to less than 1 %. The latter value is much more realistic, since the constancy of bear records is only 13 %, a very low rating, which shows that bear hunting was definitively not common. This applies especially to the western half of the investigated area, whereas bear records are more common east of the river Elbe (Schmölcke/Breede 2011). The bear remains excavated in the eastern region indicate that quite probably the carcasses of bears were taken to the settlements (Fig. 3). It remains open as to whether the relative frequency of the species in archaeological settlement assemblages is connected with initiation rituals (Ranke 1976; cf. Oehrl 2013) or with the protection of livestock (Teichert 1973).

As for the pre-Roman Iron Age, conclusions about the range of bears in the central European lowlands during the Roman Iron Age are difficult. It is, however, remarkable that all seven records from the period in question originate east of a line from Süderbrarup – Hildesheim (Tables 6–7, plus the site in Magdeburg-Cracau). Does such a distribution of records mean that bears were already extinct in most regions west of the river Elbe during the Roman Iron Age? As we will see below, the answer is "probably yes".

Brown bears in the German lowlands during medieval times

The medieval period on the northern German plain must be divided into two main stages: first, the time from the end of the Roman Iron Age to about 1000 CE; second, the period between about 1000 and 1500 CE. Generally speaking, the first period of time, labelled here Early Medieval, is characterised in the area of interest by a neighborhood of different peoples with clearly distinct traditions and cultures. Most relevant in the present context are the Franks and Saxons living more or less to the west of the river Elbe, and the Slavs to the east of it. In the second period, the Late Medieval, culture and lifestyle are much more uniform in the area of investigation.

The Early Medieval period

Concerning the distribution of bear remains during the Early Medieval we find a clear distinction in the northern German plain between the areas of different cultures (MÜLLER 2013). Whereas in the area of the Danish people in the north and the area of the Franks and Saxons in the west, only one single site (Haithabu) has yielded bear bones, there are 13 records in the area of the Slavs east of the Limes Saxoniae and the river Elbe (Fig. 4). The constancy differs between 2 % in the west (total number of excavated sites n = 41) and 30 % in the east (n = 43 sites). Either brown bears had already disappeared from the whole western part of our area of investigation, or the hunting of bears had a special meaning for the Slavs. Certainly, both can be true. A constancy of 33 % is the highest value found in the present study of settlements of a special period or region, and this value alone indicates a special social context of the bear or a special function of bear hunting. The causes of such regular bear hunting in a society can range from livestock protection to different kinds of rituals. Until today, it has not been possible to illuminate the backgrounds of the phenomenon but, since the bear remains derive exclusively from settlement contexts, there is at least no evidence for the importance of the bear in funeral practices.

Apart from differences between cultural groups, it appears more pronounced in the Early Medieval era than in previous periods that high bear proportions among wild species are not linked to high values for game in general, but correlate to the settlement's function. It is highly remarkable that large excavations of Early Medieval trading and market sites often yield tens or even hundreds of thousands of animal remains, but not a single record of a bear. The archaeozoological assemblages of such sites regularly yield, almost exclusively, the remains of domestic animals and only an extraordinarily small amount of wild species. Examples of trading sites with huge amounts of bones but less game are Menzlin with 0.2 % remains of wild animals (BENECKE 1988), Groß Strömkendorf with 0.4 % (SCHMÖLCKE 2004), and Elisenhof with 0.9 % (REICHSTEIN 1994). Commonly, red deer, roe deer, and hare (*Lepus europaeus*) are found, but bear remains are lacking. In the dataset used, there are only eleven Early Medieval excavations that yield a minimum of 200 remains of wild species and, at these sites, bears reach maximum proportions at Groß Raden (1.7 %), Starigard/Oldenburg (2.2 %), and Mecklenburg (2.6 %). These sites were among the most important Slavonic fortifications in those days, which at the same time served as noble's seats, and this applies also for most of the other sites mentioned in Table 8.

Within the time span stretching from the late 5th to the 7th centuries, a new tradition of human burials with bear-related furnishings reached parts of central Europe, but now they were different in their meaning compared to earlier periods (BEERMANN 2016; GRIMM 2013, and references therein). However, this new tradition affected only western and southern central Germany (GRIMM 2013, 290), and in the area of interest there is only the burial site at Liebenau, which yielded records of bear distal phalanges in five cases (Table 9). As is characteristic for the Merovingian region, these bear remains were deposited exclusively in female burials. Potentially, their purpose served a wide range of uses from decoration to hunting skills, from magic to healing (summarised by KIVISALO 2008, 273–274).

Late Medieval and early modern times
At the beginning of the Late Medieval era, bears had survived only in the former Slavonic area, which was subsequently conquered from the west, though the constancy of bear records reaches its second maximum value during the Late Medieval period! While in archaeological bone assemblages from Early Medieval times, bear remains are found at only 18 % of the sites in the German lowlands, this value increased to 23 % during the Late Medieval period (cf. Table 1). The fact that bear remains have been recorded at every fourth Late Medieval site is especially remarkable, when we take into account that in those days the original range of the brown bear in central Europe had already been massively reduced. This means that, during the Late Medieval period, for the first time the location of archaeological bear records does not correspond with the areas inhabited by wild bears. Probably the high constancy reflects two very different matters at the same time; namely, a high persecution level of the last remaining bears east of the Elbe river, as well as the presentation of captured bears as attractions in towns. It is, of course, not apparent in every single case whether an archaeological bear bone from an urban context is from such a "dancing bear", but the number of urban bear records alone makes this conclusion likely. In total, more than half of all Late Medieval records (10 of 19) derive from a clearly urban context.

At the same time, the number and frequency of bear remains are low at nearly all archaeological sites, a fact illustrated by the proportion of bear remains among finds of wild animals, which reaches a mean value of only 0.7 %. At Late Medieval sites with statistically significant amounts of wild mammal remains bears reach maximum proportions at Hitzacker-Weinberg (1.2–1.6 %), whereas the proportions are less than 1 % at all other relevant sites (Table 10).

In northern Germany, bears became generally rare and then extinct much earlier in densely settled and farmed lowland areas than in mountain regions (BUTZECK et al. 1988); a correlation that can also be found in Austria and Switzerland (KUNST/PACHER 2019). But BUTZECK et al. (1988) also mention another reason for the rapid decline of the bear in Late Medieval/early modern times – the development of hunting weapons and techniques, in particular the distribution of firearms.

As mentioned above, the quantity of bear remains in Late Medieval archaeozoological assemblages is low, but at the same time the high constancy reflects clearly that "hunting pressure" meant in those days that people began trying to kill every single bear they encountered, whether the bear was searching for food near settlements, discovered during regular hunts, or found accidentally (LINDNER 1940; for examples cf. SCHMIDT 1856). At least in parts of central Europe, the authorities allowed the hunter to keep the dead bear after registration (LINDNER 1940). The changing attitude towards the bear as an enemy of man is caused both by the massively increasing human population, with its need of large herds to feed it, and, after 1000 CE, the change in religion also in the east. Christianisation transformed the image of bears in a drastic way from powerful and respected "kings of the forest" to redoubtable "beasts" with a close connection to the devil (MOLSDORF 1926, 133). It was for the purpose of taming and conquering the devil that the practice of dancing bears spread over northern Germany and many other parts of Europe in medieval times (BRUNNER 2010, 139–146).

During Late Medieval times and later, bears were already being kept as exotic attractions in special bear-pits (Hauck 1963; Butzeck et al. 1988). In order to provide such entertainment, it was necessary to catch living bears, and for this purpose special kinds of traps came into use. Their distribution can be used as indicator for the last refugia of bears, but they are scarce in our area of interest and more widely distributed in the mountains of Thuringia and Saxony (see below; cf. Butzeck et al. 1988, 39–42). So-called *Bärengärten* (bear gardens) as a part of the garden culture of the elite were already becoming popular at the beginning of the 17th century, but these, too, were distributed mostly further south. The only *Bärengarten* in our study region was located in Oranienburg, where, at least in 1630 and 1732, bears from eastern populations arrived (Butzeck et al. 1988, 45).

The last centuries of the brown bear's presence in northern Germany and adjacent regions

Brown bears were part of the fauna of northern central Europe from the Late Glacial to modern times; their disappearance was a successive process that took more than 1,000 years and shows a temporal gradient from the west – with the earliest extinctions there – to the east (Fig. 5). The reasons why bears became extinct differ from region to region.

In the Netherlands, where brown bears were also present during almost the entire postglacial period, they disappeared during the Early Medieval period (Verhagen 1989). There is no evidence that hunting could have played a central role in the disappearance, but it is likely that the deforestation of large parts of the country destroyed the bears' habitat (Kuijper et al. 2016). A very late and extraordinary record from the Netherlands is a perfectly preserved left front leg including a paw found at Noordwijk, which has been directly radiocarbon dated to 880–970 cal. CE (1140 ±30 BP, GrA-66477; Kuijper et al. 2016) – probably the remains of one of the very last wild bears of this country.

Also in the western part of the study area, the North Rhine-Westphalian and Lower Saxony regions from the Netherlands to about 9°E, bears were already absent in Early Medieval times; they probably became very rare or even extinct here before the beginning of the common era (Figs. 4–5). Even given that hunting did not play a major social and economic role, the scarcity of bear remains in this region is striking. It can be assumed that the presence of the grave goods at the Liebenau burial field, dating to before the 7th century CE, indicate that hunters must have killed some of the last remaining animals living there. The single later record that refers to bears in this part of the investigated area stems from the site at Osnabrück's *Domplatz* in the High Medieval city centre, and probably concerns a dancing bear. Written sources mention two more Late Medieval records from this border region between the lowlands and the low mountain range – in 1445 a bear was killed near the town of Soest, and another one was killed one year later near Münster. It can therefore be supposed that these three animals belong to the same relict population living in the Münsterland or its surroundings until up to the end of the Late Medieval period (Rehage 2020).

In Schleswig-Holstein, according to Waldemar's *Erdbuch* from 1231, the last bear was killed in the early 13th century near the mouth of the Schlei Bay on the Baltic Sea (Lindner 1940). It is likely that the bones found in Schleswig (Schild site) originate from animals of the same population. However, also tamed bears occurred in Late Medieval towns, and it cannot be ruled out that the archaeological bear remains from Schleswig (Schild site), and also from Lübeck (Alfstraße 36/38; Hundestraße 13–17 sites) derive from such animals. The latest archaeological finds of bear remains in this area are most likely those from Bischofswarder and Scharstorf, dating to the 9th century. It is maybe no accident that both sites were located in the border region between Saxons and Slavs – this forested region called Limes Saxoniae was much less settled than all other parts of Schleswig-Holstein. The bear remains from Haithabu might be of the same age but, since Haithabu was a pre-urban central marketplace, these finds were possibly obtained through exchange. They do not necessarily indicate the presence of bears in the area concerned, particularly if it is taken into account that the bear had already disappeared from neighbouring Denmark in Neolithic times (AARIS-SØRENSEN 2009). All later Danish bear remains found all over the country in Iron Age to Early Medieval graves are tooth pendants and claws, thought to be imports from Sweden (AARIS-SØRENSEN 2009).

In the central and eastern areas of the Lower Saxon parts of the lowlands and the northern half of Saxony-Anhalt, there are two High Medieval bear remains from towns – Hildesheim (Domhügel site) and Hitzacker (Weinberg site) – as well as from the castle of Burg Bodenteich. If these remains are from dancing bears, the extinction of bears in central Lower Saxony also occurred during the Early Medieval period. At Bardowick (Kirchenhügel site) and Hämerten there are two records from this period, and two more come from the Roman Iron Age (Hildesheim-Bavenstedt and Magdeburg-Cracau), which indicate a presence of the species in this region until at least the last quarter of the 1st millennium CE. The Harz Mountains were for a longer time a refuge for bears, since a local authority captured a bear in Werningerode at the end of the 15th century and, a few decades later in 1573, another bear was captured nearby (BUTZECK et al. 1988). In 1686 the last bear was shot in the Harz Mountains; a memorial was later erected at the place where it was killed (BUTZECK et al. 1988, fig. 14). Towards the east, in the northern half of the state Brandenburg, including Berlin, there are quite a lot of medieval settlements containing bear remains. Since most of them have been found in town centres, such as Berlin-Köpenick (Schloßinsel site), it is hard to say how long wild bears lived in this region; they probably occurred in some wooded regions up until the Late Medieval era.

This is definitely the case for the Mecklenburgian lakeland further north. Most of the medieval records deriving from this area date to the time between the 11th to 13th centuries, including those from Wustrow (Fischerinsel), Groß Nemerow (Krickow-Hanfwerder), Zirzow, Vipperow, Teschendorf, and Teterow. They indicate that bears were widely distributed in the lakeland in those days and possibly not very scarce. Later, in early modern times, the population crashed, but it took until the 18th century to extirpate the species in the investigated area completely (BUTZECK et al. 1988).

It was in the Oder region and the Polish eastern part of Pomerania that the last bear population occurred (SCHMIDT 1856). Two aspects are remarkable about this region and period: First, three finds and records of the killings of complete bear families or single very young bears in 1727, 1730, and 1750 show the presence of a still-existing, but very small permanent Pomeranian bear population. The very last bear was killed in 1750, as a result of continuous hunting pressure. Second, the real economic damage caused by bears was small, and, although the Pomeranian authorities tried to extirpate the species by paying rewards for each bear, cases of bears being killed had become extraordinary events (JAKUBIEC/BUCHALCZYK 1993). In the 26 years between 1724 and 1750, in an area of about 300 x 150 km, people observed only 29 bears, including cubs, and these bears were held responsible for only 18 (sic!) dead cattle in the whole of Pomerania (calculated after SCHMIDT 1856). Attacks on humans were completely unknown in Pomerania in those days.

In the Polish lowlands and East Prussia, the last bears also vanished in the middle of the 18th century (JAKUBIEC/BUCHALCZYK 1993). The last refuge of bears in lowland Poland was Białowieża forest, on the border with Belarus, but here the last individuals were killed between 1873 and 1878 (KARPIŃSKI 1949). Today, only very occasionally, single individuals migrate from the east to Białowieża forest (SELVA 2019). Later, bears remained only in the south, where the last stable and slowly growing bear population of central Europe continues to exist today, in the Polish and particularly the Slovakian Carpathian Mountains (JAKUBIEC/BUCHALCZYK 1993; FERNÁNDEZ et al. 2012).

Conclusion

The bear was, over the years and by many cultures, seen by humans as a very respectable being, not just as a an ordinary animal but a kind of "king of nature" (Schmölcke et al. 2017; Schmölcke/ Nikulina 2022). In hunter-gatherer societies in present-day northern Germany, bears were not hunting trophies, nor they were considered as food competitors. This observation is in some contrast to the value of brown bear meat, which contains more calories per kilogram than that of all other terrestrial mammals available to humans (U.S. Department of Agriculture 2016). Taking this into account, it is not surprising that, from Late Medieval and early modern times, several recipes for bear meat are known (Pasda 2003, 52–53), and Gessner/Forer (1563, 33–35) list many possibilities for the use of a bear carcass for culinary or medical purposes. It is likely that humans and bears, for long time in history, avoided direct contact with each other. From the human perspective, hunting bears was very risky in relation to any potential benefits, and often there may also have been spiritual reasons for forbidding bear hunting. In addition, bears can tolerate living in much smaller population densities than e.g. herbivores, such as deer. Only in the Iron Age and the Roman Period did the killing of bears become more common in northern Germany, but this was possibly only for particular funeral purposes. Especially, the use of bear skins or single bear claws in burial rituals in Fennoscandia endured for nearly 1,000 years (see different contributions, this volume). The northern German plain was less affected by this tradition, but in several cases a connection to either death or funeral practices, or to noble hunting performed by the social elite, can be recorded.

As we have seen, the number of bear remains at archaeological sites as an indicator for the degree of human persecution remained low for a long time. We can assume that, prior to early modern times, hunting pressure was not strong enough to be the only reason for the bear's disappearance from large parts of its original range. It was foremost the alteration of its habitat that caused decreasing population densities and made the contact between adjoining populations more and more difficult. Extinctions of bears ultimately led to small and fragmented relict populations, first on local, later on regional levels. Relict populations without contact with their neighbours are vulnerable to stochastic events and to the loss of key individuals (Linnell et al. 2005). This applies in particular to predator species with their low reproduction rate. Often it needs only an increase in, or just the appearance of, negative factors to move the complete population into a situation dangerous for its survival.

In many parts of the investigated area, it was the medieval cultural landscape with high, and often still increasing human densities that reduced even the last remote areas of bears to critical points. Today, when conservation challenges include both the bear's large area requirements and its predatory behaviour, some stable bear populations can exist in countries with mean human population densities of 80 people/km^2 (Linnell et al. 2001), but, historically, bear populations started to decline at human population densities of 4 people/km^2 (Woodroffe 2000). The reasons for these differences are complex (Zedrosser et al. 2011), but to illustrate the limitation of suitable habitats for bears some calculations should be made: At the end of the Late Medieval period, about 9 million people lived in Germany (Pfister 1994); given an area of 540,000 km^2 and a regular human distribution (which is, of course, a massive simplification), a theoretical population density of 17 people/km^2 can be calculated. This theoretical value increased in *c.* 1800 to a population density of 40 people/km^2 (Pfister 1994). Even if these values are of a more theoretical nature, they underline the habitat limitation of bears. Based on historical written sources (Butzeck et al. 1988, 32–37) the effects of a growing human population and forest exploitation on bears can be studied in detail for the northeastern part of our study area.

In addition to anthropogenic landscape changes, the growth of the human population, hunting with firearms, and shifts in spiritual or religious customs influenced the presence of brown bears in northern Germany in a very negative way. In the course of Christianisation, but especially during Late Medieval and early modern times, bears were demonised. Similar to wolves, official orders signified killing them as a universal responsibility. It was finally the result of such intensive persecution that these large animals disappeared from wide parts of the European continent (Huber/Swenson 2013). Recently, in several European regions a comeback can be observed, partly promoted and stimulated in the context of wildlife conservation or restoration programs. These are the result of a slowly changing attitude toward this species in central Europe (as reflected in Kalb 2007 and especially obvious in Bürglin 2015).

Bibliography

Aaris-Sørensen 2009: K. Aaris-Sørensen, Dynamics of the mammalian fauna in Denmark. Fossils and Strata 57 (Copenhagen 2009).

Albrecht et al. 2017: J. Albrecht/K. A. Bartoń/N. Selva/R. S. Sommer/J. E. Swenson/R. Bischof, Humans and climate change drove the Holocene decline of the brown bear. Scientific Reports 7(1):10399, 2017, 1–10.

Bärenfänger 2009: R. Bärenfänger, Die bislang ältesten Brandbestattungen Niedersachsens. Die neolithischen Gräber in Westhammrich bei Leer. Archäologie in Niedersachsen 12, 2009, 33–35.

Beermann 2016: S. Beermann, Bärenkrallen und Bärenfelle in Brand- und Körpergräbern der vorrömischen Eisenzeit bis Völkerwanderungszeit in Mittel- und Nordeuropa. Universitätsforschungen zur Prähistorischen Archäologie 279 (Bonn 2016).

Behre 2008: K.-E. Behre, Landschaftsgeschichte Norddeutschlands: Umwelt und Siedlung von der Steinzeit bis zur Gegenwart (Neumünster 2008).

Benecke 1988: N. Benecke, Archäozoologische Untersuchungen an den Tierknochen aus der frühmittelalterlichen Siedlung von Menzlin. Materialhefte zur Ur- und Frühgeschichte Mecklenburgs 3 (Schwerin 1988).

Benecke 1994: N. Benecke, Archäozoologische Studien zur Entwicklung der Haustierhaltung in Mitteleuropa und Südskandinavien von den Anfängen bis zum ausgehenden Mittelalter (Berlin 1994).

Boyle 2006: K. Boyle, Neolithic Wild Game Animals in Western Europe: the Question of Hunting. In: D. Serjeantson/D. Field (eds.), Animals in the Neolithic of Britain and Europe (Oxford 2006) 10–23.

Brozio et al. 2018: J. P. Brozio/D. Filipovic/U. Schmölcke/W. Kirleis/J. Müller, Mittel- bis jungneolithische Siedlungshinterlassenschaften zwischen 3300–2600 v. Chr. – Der Fundplatz Oldenburg LA 232 im Oldenburger Graben, Ostholstein. Praehistorische Zeitschrift 93, 2018, 185–224.

Brunner 2010: B. Brunner, Bär und Mensch. Die Geschichte einer Beziehung (Darmstadt 2010).

Bürglin 2015: R. Bürglin, Sie sind wieder da. Bär, Luchs und Wolf erleben (Stuttgart 2015).

Butzeck et al. 1988: S. Butzeck/M. Stubbe/R. Piechocki, Beiträge zur Geschichte der Säugetierfauna der DDR, Teil 1: Der Braunbär *Ursus arctos* Linné 1758. Hercynia 25, 1988, 27–59.

Charles 1997: R. Charles, The Exploitation of Carnivores and Other Fur-bearing Mammals during the Northwestern European Late and Upper Paleolithic and Mesolithic. Oxford Journal of Archaeology 16, 1997, 253–277.

Czymzik et al. 2016: M. Czymzik/S. Dreibrodt/I. Feeser/F. Adolphi/A. Brauer, Mid-Holocene humid periods reconstructed from calcite varves of the Lake Woserin sediment record (north-eastern Germany). The Holocene 26, 2016, 935–946.

Davison et al. 2011: J. Davison/S. Y. W. Ho/S. C. Bray/M. Korsten/E. Tammeleht/M. Hindrikson/K. Østbye/E. Østbye/S.-E. Lauritzen/J. Austin/A. Cooper/U. Saarma, Late-Quaternary biogeographic scenarios for the brown bear (*Ursus arctos*), a wild mammal model species. Quaternary Science Review 30, 2011, 418–430.

Duerr 2010: J. Duerr, Von Tierhütern und Tiertötern. Mythos und Ethik der Jagd im kulturhistorischen Vergleich (Bonn 2010).

Edwards et al. 2014: C. J. Edwards/S. Y. W. Ho/R. Barnett/P. Coxon/D. G. Bradley/T. C. Lord/T. O'Connor, Continuity of brown bear maternal lineages in northern England through the Last-glacial period. Quaternary Science Reviews 96, 2014, 131–139.

Ewersen 2007: J. Ewersen, Die Tierknochenfunde aus der neolithischen Siedlung Heidmoor, Kr. Segeberg. Untersuchungen und Materialien zur Steinzeit in Schleswig-Holstein 4 (Neumünster 2007).

Ewersen/Schmölcke 2013: J. Ewersen/U. Schmölcke, Untersuchungen zur Haltung und Nutzung von Haushunden auf meso- und neolithischen Fundplätzen im nördlichen Deutschland. In: B. Ramminger (ed.), Studien zur Jungsteinzeit in Norddeutschland. Universitätsforschungen zur prähistorischen Archäologie 240 (Bonn 2013) 267–299.

Feeser/Dörfler 2015: I. Feeser/W. Dörfler, The early Neolithic in pollen diagrams from eastern Schleswig-Holstein and western Mecklenburg—evidence for a 1000 year cultural adaptive cycle? In: J. Kabaciński/S. Hartz/D. C. M. Raemaekers/T. Terberger (eds.), The Dąbki Site in Pomerania and the Neolithisation of the North European Lowlands (c. 5000–3000 calBC) (Rahden/Westf. 2015) 291–306.

Fernández et al. 2012: N. Fernández/N. Selva/C. Yuste/H. Okarma/Z. Jakubiec, Brown bears at the edge: modeling habitat constraints at the periphery of the

Carpathian population. Biological Conservation 153, 2012, 134–142.

Gessner/Forer 1563: C. Gessner/C. Forer, Thierbuch, das ist ein kurtze beschreybung aller vierfüssigen so wohl zahmer als wilder Thieren [etc.] (Zürych 1563).

Glykou 2016: A. Glykou, Neustadt LA 156. Ein submariner Fundplatz des späten Mesolithikums und des frühesten Neolithikums in Schleswig-Holstein. Untersuchungen zur Subsistenzstrategie der letzten Jäger, Sammler und Fischer an der norddeutschen Ostseeküste. Untersuchungen und Materialien zur Steinzeit in Schleswig-Holstein und im Ostseeraum 7 (Kiel, Hamburg 2016).

Gramsch 2012: B. Gramsch, Die Schmuckfunde von Friesack, Fundplatz 4, Lkr. Havelland. Veröffentlichungen zur brandenburgischen Landesarchäologie 46, 2012 (2014), 7–26.

Grimm 2013: O. Grimm, Bear-skins in northern European burials and some remarks on other bear-related furnishings in the north and middle of Europe in the 1st millennium AD. In: O. Grimm/U. Schmölcke (eds.), Hunting in northern Europe until 1500 AD. Old traditions and regional developments, continental sources and continental influences. Papers presented at a workshop organized by the Centre for Baltic and Scandinavian Archaeology, ZBSA. Schriften des Archäologischen Landesmuseums, Ergänzungsreihe 7 (Neumünster 2013) 277–296.

Gross et al. 2019: D. Gross/E. Corradini/U. Schmölcke/M. Zanon/W. Dörfler/S. Dreibrodt/I. Feeser/S. Krüger/H. Lübke/D. Panning/D. Wilken/H. Piezonka, Adaptations and transformations of hunter-gatherers in forest environments: New archaeological and anthropological insights. The Holocene 29, 2019, 1531–1544.

Hallowell 1926: A. I. Hallowell, Bear ceremonialism in the northern hemisphere. American Anthropologist 28, 1926, 1–175.

Hartz/Schmölcke 2013: S. Hartz/U. Schmölcke, From the Mesolithic to the Neolithic – Stone Age hunting strategies in the southwestern Baltic Sea area. In: O. Grimm/ U. Schmölcke, Hunting in northern Europe until 1500 AD. Old traditions and regional developments, continental sources and continental influences. Papers presented at a workshop organized by the Centre for Baltic and Scandinavian Archaeology, ZBSA. Schriften des Archäologischen Landesmuseums, Ergänzungsreihe 7 (Neumünster 2013) 21–40.

Hartz et al. 2011: S. Hartz/H. Jöns/H. Lübke/U. Schmölcke/C. von Carnap-Bornheim/D. Heinrich/S. Klooss/F. Lüth/S. Wolters, Prehistoric Settlements in the south-western Baltic Sea area and Development of the Regional Stone Age Economy. Final report of the SINCOS-II-subproject 4. Bericht der Römisch-Germanischen Kommission 92, 2011 (2014), 77–210.

Hauck 1963: K. Hauck, Tiergärten im Pfalzbereich. In: A. Gauert (ed.), Deutsche Königspfalzen 1 (Göttingen 1963) 30–74.

Heinrich et al. 2016: D. Heinrich/A. von den Driesch/N. Benecke, Holozängeschichte der Tierwelt Europas. Datacollection ed. by IANUS 2016: http://dx.doi.org/10.13149/001.mcus7z-2.

Hinz et al. 2012: M. Hinz/I. Feeser/K. G. Sjögren/J. Müller, Demography and the intensity of cultural activities: an evaluation of Funnel Beaker Societies (4200–2800 cal BC). Journal of Archaeological Science 39, 2012, 3331–3340.

Huber/Swenson 2013: D. Huber/J. Swenson, Brown bear, *Ursus arctos*. In: S. Deinet/C. Ieronymidou/L. McRae/I. J. Burfield/R. P. Foppen/B. Collen/M. Böhm (eds.), Wildlife comeback in Europe: the recovery of selected mammal and bird species (London 2013) 140–149.

Hübner et al. 1988: K.-D. Hübner/R. Saur/H. Reichstein, Die Säugetierknochen der neolithischen Seeufersiedlung Hüde I. Göttinger Schriften zur Vor- und Frühgeschichte 23, 1988, 35–142.

Jakubiec 1993: Z. Jakubiec, *Ursus arctos* Linnaeus, 1758. In: J. Niethammer/F. Krapp (eds.), Handbuch der Säugetiere Europas 5 (Wiesbaden 1993) 254–300.

Jakubiec/Buchalczyk 1993: Z. Jakubiec/T. Buchalczyk, The brown bear in Poland: its history and present numbers. Acta Theriologica 32, 1993, 289–306.

Jürgenson 1974: P. B. Jürgenson, Braunbär, *Ursus (Ursus) arctos* Linnaeus 1758. In: V. G. Heptner/N. P. Naumov/P. B. Jürgenson/A. A. Sludski/A. F. Cirkova/A. G. Bannikov, Die Säugetiere der Sowjetunion 2 (Jena 1974) 396–446.

Kalb 2007: R. Kalb, Bär, Luchs, Wolf – verfolgt, ausgerottet, zurückgekehrt (Graz, Stuttgart 2007).

Karpiński 1949: J. J. Karpiński, La régénération de l'ours dans la Forêt de Białowieża. Bijdragen tot de Dierkunde 28, 1949, 218–236.

Kirkinen 2017: T. Kirkinen, "Burning pelts" – brown bear skins in the Iron Age and early medieval (1–1300 AD) burials in south-eastern Fennoscandia. Estonian Journal of Archaeology 21, 2017, 3–29.

Kivisalo 2008: N. Kivisalo, The Late Iron Age Bear-Tooth Pendants in Finland: Symbolic Mediators between Women, Bears, and Wilderness? Temenos 44, 2008, 263–291.

Kluge/Von Seebold 2012: F. Kluge/E. Von Seebold, Etymologisches Wörterbuch der deutschen Sprache (Berlin 2012).

Kneisel et al. 2019: J. Kneisel/W. Dörfler/S. Dreibrodt/S. Schaefer-Di Maida/I. Feeser, Cultural change and population dynamics during the Bronze Age: Integrating archaeological and palaeoenvironmental evidence for Schleswig-Holstein, Northern Germany. The Holocene 29, 2019, 1607–1621.

Kuijper et al. 2016: W. J. Kuijper/I. K. A. Verheijen/A. Ramcharan/H. van der Plicht/T. van Kolfschoten, One of the last wild brown bears (*Ursus arctos*) in the Netherlands (Noordwijk). Lutra 59, 2016, 49–64.

Kunst 2014: G. K. Kunst, Wildtierreste als umwelthistorische Quellen in der Frühgeschichte: Eine Reflexion anhand römerzeitlicher Beispiele aus Ostösterreich. In: H. Friesinger/A. Stuppner (eds.), Mensch und Umwelt – Ökoarchäologische Probleme in der Frühgeschichte. Mitteilungen der Prähistorischen Kommission 84, 2014, 45–56.

Kunst/Pacher 2019: G. K. Kunst/M. Pacher, Brown bear remains in prehistoric and early historic societies: case studies from Austria. Berichte der Geologischen Bundesanstalt 132, 2019, 89–121.

Lehmkuhl 1986: U. Lehmkuhl, Archäozoologische und typologische Untersuchungen an Metapodia-Anhängern des Neolithikums und der Bronze- bis frühen Eisenzeit in der DDR. Jahrbuch für Bodendenkmalpflege Mecklenburg 1986 (1987), 19–38.

Lehmkuhl 2007: U. Lehmkuhl, Die Tierbeigaben aus den Gräbern von Ostorf (Mecklenburg) unter besonderer Berücksichtigung der durchbohrten Tierzähne. Mit einem Beitrag zu den Tierbeigaben aus den Gräbern von Tangermünde bei Stendal. Bericht der Römisch-Germanischen Kommission 88, 2007 (2010), 385–414.

Lindholm/Ljungkvist 2016: K.-J. Lindholm/J. Ljungkvist, The Bear in the Grave: Exploitation of Top Predator and Herbivore Resources in First Millennium Sweden – First Trends from a Long-Term Research Project, European Journal of Archaeology 19, 2016, 3–27.

Lindner 1940: K. Lindner, Geschichte des deutschen Weidwerks 2: Die Jagd im frühen Mittelalter (Berlin 1940).

Linnell et al. 2001: J. D. C. Linnell/J. E. Swenson/R. Andersen, Predators and people: conservation of large carnivores is possible at high human densities if management policy is favourable. Animal Conservation 4, 2001, 345–349.

Linnell et al. 2005: J. D. C. Linnell/C. Promberger/L. Boitani/J. E. Swenson/U. Breitenmoser/R. Andersen, The linkage between conservation strategies for large carnivores and biodiversity: the view from the "half-full" forests of Europe. In: J. C. Ray/K. H. Redford/R. S. Steneck/J. Berger (eds.), Large Carnviores and the Conservation of Biodiversity (Washington [DC] 2005) 381–399.

Méniel 2001: P. Méniel, Les Gaulois et les animaux. Élevage, repas et sacrifices (Paris 2001).

Molsdorf 1926: W. Molsdorf, Christliche Symbolik der mittelalterlichen Kunst (Leipzig 1926).

Müller 2013: U. Müller, Grenzen, Grenzgänger, Grenzregionen. In: M. Krieger/F. Lubowitz/B. S. Frandsen (eds.), 1200 Jahre deutsch-dänische Grenze. Aspekte einer Nachbarschaft (Neumünster 2013) 47–69.

Oehrl 2013: S. Oehrl, Bear hunting and its ideological context (as a background for the interpretation of bear claws and other remains of bears in Germanic graves of the 1st millennium AD). In: O. Grimm/U. Schmölcke (eds.), Hunting in northern Europe until 1500 AD. Old traditions and regional developments, continental sources and continental influences. Papers presented at a workshop organized by the Centre for Baltic and Scandinavian Archaeology, ZBSA. Schriften des Archäologischen Landesmuseums, Ergänzungsreihe 7 (Neumünster 2013) 297–332.

Pasda 2003: K. Pasda, Tierknochen als Spiegel sozialer Verhältnisse im 8.–15. Jahrhundert in Bayern. Praehistorika Monographien 1 (Erlangen 2004).

Pastoureau 2007: M. Pastoureau, L'Ours. Histoire d'un roi déchu (Montrouge 2007).

Pentikäinen 2007: J. Pentikäinen, Golden King of the Forest. The Lore of the Northern Bear (Helsinki 2007).

Pfister 1994: C. Pfister, Bevölkerungsgeschichte und historische Demographie 1500–1800 (München 1994).

Ranke 1976: K. Ranke, Bär. In: H. Beck/H. Jankuhn/K. Ranke/R. Wenskus (eds.), Reallexikon der Germanischen Altertumskunde 2 (Berlin, New York ²1976) 46–48.

Rehage 2020: H.-O. Rehage, Braunbär *Ursus arctos*. In: AG Säugetierkunde NRW – Online-Atlas der Säugetiere Nordrhein-Westfalens (2020): www.saeugeratlas-nrw.lwl.org.

Reichstein 1994: H. Reichstein, Die Säugetiere und Vögel aus der frühgeschichtlichen Wurt Elisenhof. Studien zur Küstenarchäologie Schleswig-Holsteins, Serie A: Elisenhof 6 (Frankfurt am Main 1994).

Schmid 1981: E. Schmid, Bestimmung von verbrannten Tierknochen aus Leichenbrand Grab II, Hügel 13 von Nützen, Kreis Segeberg. In: I. Kühl, Eine Leichenbrandbestattung mit Bärenkralle aus der mittleren Bronzezeit, Gemeinde Nützen, Kreis Segeberg. Die Heimat 88, 1981, 225–227.

Schmidt 1856: T. Schmidt, Zur naturgeschichtlichen Statistik der in Pommern ausgerotteten Säugethiere. Jubelschrift zur vierhundertjährigen Stiftungsfeier der Universität Greifswald (Stettin 1856).

Schmölcke 2003: U. Schmölcke, Die Stetigkeit als archäozoologische Bewertungsmethode: Beispiele aus Paläoichthyologie (frühmittelalterlicher Seehandelsplatz Groß Strömkendorf) und Paläoökologie (Neolithikum Schleswig-Holsteins). Beiträge zur Archäozoologie und Prähistorischen Anthropologie 4, 2003, 195–203.

Schmölcke 2004: U. Schmölcke, Nutztierhaltung, Jagd und Fischfang: Zur Nahrungsmittelwirtschaft des frühgeschichtlichen Handelsplatzes von Groß Strömkendorf, Landkreis Nordwestmecklenburg (Lübstorf 2003).

Schmölcke 2013: U. Schmölcke, Once more: sample size estimation in archaeozoology. Offa 69, 2013, 239–247.

Schmölcke 2016: U. Schmölcke, Die Säugetierfunde vom präboreal- und borealzeitlichen Fundplatz Friesack 4 in Brandenburg. In: N. Benecke/B. Gramsch/S. Jahns (eds.), Subsistenz und Umwelt der Feuchtbodenstation Friesack 4 im Havelland. Ergebnisse der naturwissenschaftlichen Untersuchungen (Wünsdorf 2016) 45–116.

Schmölcke 2019: U. Schmölcke, Early Mesolithic hunting strategies on red deer, roe deer and wild boar in Friesack 4, a three-stage Preboreal and Boreal site in northern Germany. In: H. Lübke/D. Groß/J. Meadows/D. Jantzen (eds.), Working at the sharp end: from bone and antler to Early Mesolithic life in Northern Europe. Untersuchungen und Materialien zur Steinzeit in Schleswig-Holstein und im Ostseeraum 10 (Kiel, Hamburg 2019) 239–254: doi 10.23797/9783529018619-8.

Schmölcke in press: U. Schmölcke, Forschungsstand Tiernutzung, Tierhaltung und Jagd. In: W. Dörfler/W. Kirleis (eds.), Landwirtschaft, Landnutzung und Landschaft der ersten Ackerbauern in Norddeutschland und angrenzenden Gebieten zur Zeit der Trichterbecher- und Einzelgrabkultur. Differenzierung von Landwirtschaft und Umwelt als Grundlage früher Monumentalität im Neolithikum Norddeutschlands (Kiel in press).

Schmölcke/Breede 2011: U. Schmölcke/M. Breede, Neues zur Fauna der Germania libera: Knochenfunde von Wildtieren aus Völschow, Mecklenburg-Vorpommern. Beiträge zur Archäozoologie und Prähistorischen Anthropologie 7, 2011, 53–63.

Schmölcke/Nikulina 2015: U. Schmölcke/E. Nikulina, Mesolithic beaver hunting station or base camp of supra-regional Stone Age fur trade? New archaeozoological and archaeogenetic results from Dąbki 9. In: J. Kabaciński/S. Hartz/D. C. M. Raemaekers/T. Terberger (eds.), The Dąbki Site in Pomerania and the Neolithisation of the North European Lowlands (c. 5000–3000 calBC) (Rahden/Westf. 2015) 65–86.

Schmölcke/Nikulina 2022: U. Schmölcke/E. A. Nikulina, Humans and their relationship to large carnivores in Central Europe from foragers to modern times: A survey. In: K. F. Hillgruber/T. Terberger (eds.), The *Homotherium* finds from Schöningen and big cats of the Ice Age. Forschungen zur Urgeschichte aus dem Tagebau in Schöningen 4 (Mainz 2022) 129–150.

Schmölcke et al. 2017: U. Schmölcke/D. Groß/E. A. Nikulina, Bears and beavers. 'The Browns' in daily life and spiritual world. In: B. V. Eriksen/A. Abegg-Wigg/R. Bleile/U. Ickerodt (eds.), Interaction without borders. Exemplary archaeological research at the beginning of the 21st century (Schleswig 2017) 901–916.

Schönfelder 1994: M. Schönfelder, Bear-claws in Germanic graves. Oxford Journal of Archaeology 13, 1994, 217–227.

Selva 2019: N. Selva, New Hope 140 Years after Extirpation of Brown Bears in Białowieża Forest, Poland–Belarus. International Bear News 28, 2019, 15–16.

Sommer/Benecke 2005: R. S. Sommer/N. Benecke, The recolonization of Europe by brown bears *Ursus arctos* Linnaeus, 1758 after the Last Glacial Maximum. Mammal Review 35, 2005, 156–164.

Swenson 1999: J. E. Swenson, Does hunting affect the behavior of brown bears in Eurasia? Ursus 11, 1999, 157–162.

Teichert 1973: M. Teichert, Haustierhaltung, Jagd und Fischfang in einigen germanischen Siedlungen zur La-Tène-Zeit und römischen Kaiserzeit. In: J. Matolcsi (ed.), Domestikationsforschung und Geschichte der Haustiere (Budapest 1973) 263–274.

Ulrich et al. 2009: W. Ulrich/M. Almeida-Neto/N. J. Gotelli, A consumer's guide to nestedness analysis. Oikos 118, 2009, 3–17.

U.S. Department of Agriculture 2016: U.S. Department of Agriculture, National Nutrient Database: http://ndb.nal.usda.gov/.

Verhagen 1989: M. Verhagen, De beer in de Nederlandse pre- en protohistorie. Cranium 6, 1989, 65–71.

Woodroffe 2000: R. Woodroffe, Predators and people: using human densities to interpret declines of large carnivores. Animal Conservation 3, 2000, 165–173.

Zedrosser et al. 2011: A. Zedrosser/S. M. J. G. Steyaert/H. Gossow/J. E. Swenson, Brown bear conservation and the ghost of persecution past. Biological Conservation 144, 2011, 2163–2170.

PD Dr. habil. Ulrich Schmölcke
Centre for Baltic and Scandinavian Archaeology (ZBSA)
Stiftung Schleswig-Holsteinische Landesmuseen
Schloss Gottorf
Schleswig
Germany
ulrich.schmölcke@zbsa.eu

Fig. 1. Topographic regions and archaeological sites mentioned in the text. 1: Alt Mölln; 2: Baabe; 3: Bacherswall; 4: Bardowick; 5: Basedow; 6: Bentumersiel; 7: Berlin-Blankenburg; 8: Berlin-Köpenick; 9: Bistoft; 10: Bondebrück; 11: Boomborg/Hatzum; 12: Bosau-Bischofswarder; 13: Brandenburg; 14: Burg Bodenteich; 15: Dąbki; 16: Deutsch Wusterhausen; 17: Döhren; 18: Ehestorf; 19: Eilsleben; 20: Elisenhof; 21: Feddersen Wierde; 22: Friesack 4; 23: Gielde; 24: Gommern; 25: Groß Nemerow; 26: Groß Raden; 27: Groß Strömkendorf; 28: Haithabu; 29: Haldensleben; 30: Hämerten; 31: Harsefeld; 32: Hildesheim; 33: Hitzacker; 34: Hohen Viecheln; 35: Hüde I; 36: Husby; 37: Kablow; 38: Leese; 39: Liebenau; 40: Lietzow-Buddelin; 41: Lübeck; 42: Magdeburg; 43: Mecklenburg; 44: Menzlin; 45: Münster; 46: Neu-Plötzin; 47: Neustadt/Holstein; 48: Nützen; 49: Ostorf; 50: Parchim-Löddigsee; 51: Parum; 52: Putensen; 53: Ralswiek-Augustenhof; 54: Rosenfelde; 55: Rosenhof; 56: Rüde; 57: Runstedt-Elzrandsiedlung; 58: Scharstorf; 59: Schleswig; 60: Seedorf; 61: Seinstedt-Erbbrink; 62: Soest; 63: Starigard/Oldenburg; 64: Süderbrarup; 65: Teschendorf; 66: Teterow; 67: Timmendorf-Nordmole; 68: Tribsees; 69: Vipperow; 70: Völschow; 71: Waltersdorf; 72: Wangels; 73: Waren-Stinthorst; 74: Werningerode; 75: Westerhammrich; 76: Wiesenau; 77: Wolkenwehe; 78: Wüste Kunersdorf; 79: Wustrow; 80: Zeestow; 81: Zirzow; 82: Rerik (map based on TUBS, wikipedia).

Fig. 2. Perforated canine teeth from late Neolithic (3200–3000 BCE) brown bears have been found not only in the cemetery of Ostorf (Mecklenburg-Western Pomerania), but also – depicted here – in settlement contexts such as Oldenburg LA 232, Schleswig-Holstein (photo J. Schüller, Landesmuseen Schleswig-Holstein).

Fig. 3. Brown bear (Ursus arctos) *mandible including fourth premolar from the Roman Iron Age site of Völschow, Mecklenburg-Western Pomerania (photo M. Breede, after SCHMÖLCKE/BREEDE 2011, fig. 2).*

Fig. 4. Early Medieval records of brown bears (Ursus arctos) in northern Germany. The distribution indicates a reduced range that was limited to regions northeast of the river Elbe (map based on TUBS, wikipedia).

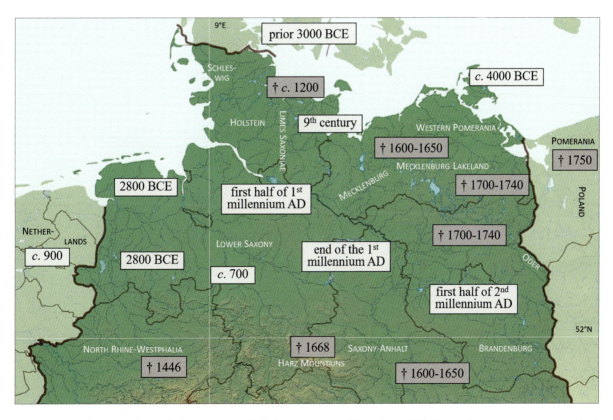

Fig. 5. Last archaeological records (light grey) and the last mentions of bears being killed according to written sources (dark grey; cf. text and BUTZECK et al. 1988) in northern Germany and adjacent areas (map based on TUBS, wikipedia).

285

Table 1. Data basis. NISP: Number of Identified Specimens; NISP (%): proportion of bear remains within NISP of wild mammals. *All these remains are distal phalanges deriving from human burials. **When excluding the comprehensive assemblages from two so-called hunting stations (Hüde I, Parchim-Löddigsee) the value is reduced to 42 %. The complete original data are available online (cf. HEINRICH et al. 2016).

Period	Sites (n)	Wild mammals (n)	Wild mammals (%)	Bear NISP	Bear Sites	Bear Constancy (%)	Bear NISP (%)
Late Glacial	18	4,430	100	0	0	0	0
Mesolithic	43	10,918	97.2	29	6	14.0	0.27
Neolithic	80	32,534	64.7**	179	13	16.3	0.55
Bronze Age	11	284	5.3	1	1	9.1	0.35
Iron Age	26	227	1.2	110*	9	34.6	48.5
Roman Iron Age	54	1,164	1.3	40	7	13.0	3.44
Early Medieval	83	12,468	4.5	83	15	18.1	0.67
Late Medieval	82	16,349	5.4	115	19	23.2	0.67
Early Modern	17	408	5.1	0	0	0	0
Total	**414**	**79,449**		**557**	**70**		

Table 2. Number and proportion of brown bear remains at Mesolithic archaeological sites in northern Germany. Listed are only sites with a minimum of 200 identified wild mammal remains. NISP: Number of Identified Specimens of wild mammals. References: Friesack 4 – SCHMÖLCKE 2016; Rosenhof, Timmendorf-Nordmole I, Lietzow-Buddelin, and Rosenfelde – HARTZ et al. 2011. All further references and the complete original data are available online (cf. HEINRICH et al. 2016) or in BENECKE 1994. For the location of the sites, see Fig. 1.

Site	Dating (BCE)	Type	NISP all	NISP bear	% bear
Friesack 4	9th millennium	Riverside	4,986	15	0.3
Hohen Viecheln	9th millennium	Lakeside	1,498	9	0.6
Rosenhof	5th millennium	Coastal	686	0	0
Timmendorf-Nordmole I	5th millennium	Coastal	551	0	0
Lietzow-Buddelin	5th millennium	Coastal	516	0	0
Rosenfelde	5th millennium	Coastal	485	0	0
Rüde	5th millennium	Lakeside	330	0	0
Bondebrück	6th millennium	Lakeside	302	1	0.3
Tribsees	8th millennium	Lakeside	277	2	0.7
Ralswiek-Augustenhof	5th millennium	Coastal	256	0	0
Seedorf (pit C)	5th millennium	Lakeside	200	0	0

Table 3. Number and proportion of brown bear remains at Neolithic archaeological sites in northern Germany. Listed are only sites with a minimum of 200 identified wild mammal remains. NISP: Number of Identified Specimens of wild mammals. References: Wolkenwehe-Heidmoor – EWERSEN 2007; Neustadt/Holstein – GLYKOU 2016; Timmendorf-Nordmole III – HARTZ et al. 2011. All further references and the complete original data are available online (cf. HEINRICH et al. 2016) or in BENECKE 1994.

Site	NISP wild	NISP bear	% bear
Hüde I	8,679	145	1.7
Parchim-Löddigsee	6,712	4	0.1
Wolkenwehe	4,776	0	0
Wolkenwehe-Heidmoor	4,511	10	0.2
Neustadt/Holstein	3,617	1	<0.1
Basedow	562	2	0.4
Baabe	549	0	0
Eilsleben	490	4	0.8
Bistoft	373	0	0
Wangels MN V	513	0	0
Waren-Stinthorst	269	5	1.9
Timmendorf-Nordmole III	227	0	0

Table 4. Northern German burials from the pre-Roman Iron Age with distal phalanges of brown bear (after SCHÖNFELDER 1994, tab. 1 and BEERMANN 2016, with references therein).

Site, grave number	Dating	Sex	Social position	Bear claws
Döhren, f. 55	450–300 BCE	female	leading	4
Leese, gr. 1995	400–150 BCE	female?	unknown	14
Ehestorf-Vahrendorf, gr. 513	350–120 BCE	unknown	poor	15
Ehestorf-Vahrendorf, gr. 1060	350–120 BCE?	unknown	high	10
Alt Mölln LA 13, gr. 107	250–150 BCE	unknown	normal	6
Husby, gr. 1033	120–50 BCE	male	leading	2
Ehestorf, gr. 1060	1st ct. BCE	male	high	10
Harsefeld, gr. 136	1st ct. BCE	unknown	high	3
Harsefeld, gr. 141	1st ct. BCE	male	high	3
Putensen, gr. 175	1–50 BCE	male?	poor	14
Putensen, gr. 215	1–50 BCE	male?	poor	14
Putensen, gr. 392	1–50 BCE	male?	poor	1
Putensen, gr. 394 or 397	1–50 BCE	male?	normal	12
Neu-Plötzin, 1934/2520	end of 1st century BCE	female?	leading	2

Table 5. Archaeological sites from the pre-Roman Iron Age with a minimum of about 300 identified mammal remains. NISP: Number of Identified Specimens of mammals. References and the complete original data are available online (cf. HEINRICH et al. 2016) or in BENECKE 1994.

Site	NISP all	NISP wild	% wild	NISP bear	% bear
Boomborg/Hatzum	12,244	23	0.2	0	0
Wüste Kunersdorf	5,058	42	0.8	0	0
Gommern	398	10	2.5	0	0
Runstedt-Elzrandsiedlung	307	0	0	0	0
Zeestow	298	0	0	0	0

Table 6. Northern German burials from the Roman Iron Age and Migration Period with distal phalanges of brown bear (after SCHÖNFELDER 1994, tab. 1 and BEERMANN 2016, with references therein).

Site, grave number	Dating	Sex	Social position	Bear claws
Parum, gr. 103 or 107	AD 1–50	female + male	high	12
Haldensleben (Südhafen), gr. 6	1st–2nd century	female	leading	2
Haldensleben (Südhafen), gr. 7	1st–2nd century	unknown	poor	1
Haldensleben (Südhafen), gr. 8	1st–2nd century	unknown	poor	1
Süderbrarup, gr. 402	5th century AD	unknown	leading	12
Süderbrarup, gr. 934	4th century AD	male?	leading	6

Table 7. Number and proportion of brown bear remains at Roman Iron Age settlement excavations in northern Germany. Listed are sites with a minimum of 1,000 identified mammal remains, which includes all sites with a minimum of 200 identified wild mammal remains. NISP: Number of Identified Specimens of mammals. References and the complete original data are available online (cf. HEINRICH et al. 2016) or in BENECKE 1994.

Site	NISP all	NISP wild	% wild	NISP bear	% bear
Feddersen Wierde	49,557	237	0.5	0	0
Waltersdorf	6,691	261	3.9	4	1.5
Bentumersiel	4,943	13	0.3	0	0
Hildesheim-Bavenstedt	4,701	52	1.1	1	1.9
Völschow	3,945	49	1.2	1	2.0
Gielde-Am Kaiserstein	3,475	243	7.0	0	0
Deutsch Wusterhausen	1,465	36	2.5	0	0
Seinstedt-Erbbrink	1,284	7	0.5	0	0
Kablow	1,210	87	7.2	0	0

Table 8. Number and proportion of brown bear remains at Early Medieval archaeological sites in northern Germany. Listed are only sites with a minimum of 200 identified wild mammal remains. NISP: Number of Identified Specimens of wild mammals. References and the complete original data are available online (cf. HEINRICH et al. 2016) or in BENECKE 1994.

Site	NISP all	NISP wild	% wild	NISP bear	% bear
Haithabu	255,829	5330	2.0	16	0.3
Wiesenau	3,037	1230	32.0	0	0
Scharstorf	16,074	1102	6.9	2	0.2
Bischofswarder	5,235	774	14.8	10	1.3
Hitzacker-Weinberg (7th–10th century)	15,935	460	2.9	3	0.7
Groß Raden	19,236	525	2.7	9	1.7
Brandenburg (Dominsel)	18,949	500	2.6	4	0.8
Starigard/Oldenburg (horizon 2–5)	25,899	494	1.9	11	2.2
Berlin-Blankenburg	655	382	58.3	4	1.0
Mecklenburg	12,445	265	2.2	7	2.6
Bacherswall	2,626	212	8.1	2	0.9

Table 9. Northern German Early Medieval burials with bear claws (BEERMANN 2016 lists Liebenau as the only known Early Medieval location). References and the complete original data are available online (cf. HEINRICH et al. 2016) or in BENECKE 1994.

Site, grave number	Dating	Sex	Social position	Bear claws
Liebenau, R13/B5	5st–7th century	female	unknown	2
Liebenau, 47 from R13/B5	5st–7th century	female	unknown	2
Liebenau, R14/B3	5st–7th century	female	high	3
Liebenau, R14/B4	5st–7th century	female	normal	1
Liebenau, S13/B1	5st–7th century	female?	high	5

Table 10. Number and proportion of brown bear remains at Late Medieval archaeological sites in northern Germany. Listed are only sites with a minimum of 200 identified wild mammal remains. NISP: Number of Identified Specimens of wild mammals. References and the complete original data are available online (cf. HEINRICH et al. 2016) or in BENECKE 1994.

Site	NISP wild	NISP bear	% bear
Hanfwerder	4,640	17	0.4
Hitzacker-Weinberg (12th–16th century)	2,509	44	1.8
Hitzacker-Weinberg (11th–12th century)	2,449	29	1.2
Berlin-Köpenick (Schlossinsel)	3,020	3	0.1
Schleswig (Schild)	1,349	3	0.2
Fischerinsel near Wustrow	729	1	0.1
Zirzow	552	1	0.2
Neu-Nieköhr	267	0	0
Kietzwerder near Prillwitz	239	0	0

In the company of bears: The role and significance of the bear from the perspective of the Holocene hunter-gatherer-fishers of the East European Plain forest zone (10th–3rd millennium BC)

By Ekaterina A. Kashina and Anastasia A. Khramtsova

Keywords: Brown bear, hunter-gatherer-fishers, East European Plain forest zone, portable art, burial rites, settlement layers, rock art

Abstract: The bear was the constant neighbour of prehistoric hunter-gatherer-fishers of the East European Plain forest zone (10th–3rd millennium BC) and also a part of their hunting prey. Nevertheless, scholars usually emphasise its special spiritual role, as it was quite different from the roles of other species of the boreal forest animal realm for both ethnographically-known Siberian indigenes and Holocene hunters. Here, we have made an attempt to put together and analyse all groups of the material culture sources which can give us some hints about the status, significance and symbolic meaning of the brown bear in the Holocene East European Plain forest zone – portable art, rock art, and osseous bear remains in settlement and burial contexts. These data show the significant presence of bear bones in kitchen waste and among bone tools, the sporadic presence of bear images in petroglyphs and cemetery materials, and the complete absence of them in Mesolithic/Neolithic portable art up until the start of the Final Stone/Bronze Age, around 3000 cal BC, when its presence increases.

Introduction

The "bear image topic" has been among the most studied ones of north Eurasian ethnography and archaeology for at least one hundred years. The enormous bulk of literature is represented mostly by the ethnographical studies of the Siberian aborigines but also by archaeological publications. These discuss a wide range of themes and can be grouped in the following way: 1) case studies that discuss the peculiarities of the bear image in both material culture and folklore, and 2) analytical studies that discuss the perception of the bear by particular Siberian aboriginal peoples, or by prehistoric societies. The most popular topics for ethnographical study are the bear festivals conducted in the 19th–20th centuries (see BAKELS/BOER, this volume), and the role of the brown bear as an ancestor in aboriginal spiritual belief. For archaeological studies, the most prominent topic is the role of the cave bear in Palaeolithic spirituality, based on bear skull finds in European caves (BAHN 2011).

The topic we are going to present in this paper lies chronologically between the above-mentioned Palaeolithic archaeology and Siberian aboriginal ethnography, and it deals with the mid-Holocene forest hunters of the East European Plain. We have decided that there is no need to refer to the numerous ethnographical pieces of evidence concerning the place of the bear in the worldview of the indigenous peoples of the taiga, as these are mostly collected at the Exchange for Local Observation and Knowledge of the Arctic (ELOKA).[1] There has been, from time to time, a temptation for scholars to extrapolate the ethnographical data from the reconstruction of the prehistoric hunters' spiritual sphere, but in this study we will try to escape this methodology and will analyse as deeply as possible the purely archaeological evidence of bear representations that were created by mid-Holocene hunter-gatherer-fishers – primarily those of the Central Russian regions (the Volga-Oka interfluve), while at the same time pointing out the analogies from Scandinavia and the eastern Baltic states.

The main questions are: What was the meaning of the bear for these people? Which peculiarities can be revealed in the process of analysing the bear image in portable and rock art (petroglyphs), and what can be discovered after the analysis of bear remains in the settlement layers/debris and burial inventories? One starting point for us was a paper by Knut Helskog, discussing the role of the bear based mainly on prehistoric Scandinavian rock art, in which he regrets that bear effigies and bear bones are very scarce finds in Scandinavia (see HELSKOG 2012, 217). Russian materials yield far more potential for conducting the same kind of study, namely portable art representing the bear as well as northwest Russian petroglyphs depicting bears, and the burial and settlement materials, i.e. brown bear osseous remains. On the other side, we sincerely hope that this paper can be a successful addition to the already-existing collection of papers in English, devoted both to northwest Russian petroglyphs and to the East European forest zone portable art (KOLPAKOV/SHUMKIN 2012; KASHINA/ZHULNIKOV 2015; POIKALAINEN/ERNITS 2019; MANTERE/KASHINA 2020; KASHINA/EMELYANOV 2020).

AIMS OF THE PAPER

Focusing initially on the portable art items of the East European forest zone, we aim to achieve the most objective conclusions, and for this reason, we will also discuss several additional issues, such as:
- the number and the peculiarities of petroglyphic bear images at Lake Kanozero, Lake Onega and the White Sea;
- the role of the bear as prey, based on random materials in osteological collections from Central Russian Neolithic sites;
- the use of bear remains in rituals, based on burial inventories, from a wide time span (generally dating to around 6200–3000 cal BC), as well as on the so-called "ritual hoards" of Central Russia (dating to around 3600–2700 BC).

The conclusion arrived at by HELSKOG (2012, 232) in the course of his bear image study of Scandinavian rock art, supplemented by other data sources, is quite ambiguous, postulating that the variation in bear image representation reflects local identities within a general system of beliefs. If we recognise the generally common climate conditions, fauna, vegetation, and the hunter-fisher-gatherer way of life, which continued for much longer in the East European forest zone than in most parts of western Europe, the perception of the bear image in Scandinavia and the northwestern Russian taiga forests in the mid-Holocene could and should have been very similar. The observed change of this perception over the course of time did not necessarily occur with the advent of agriculture and herding, but probably during more ancient periods, namely the Mesolithic, the Neolithic, and the Bronze Ages, and there were still hunter-gatherer-fishers in the Bronze Age taiga zone.

In the course of this study, based on the analysis of all the represented data which, in general, constitutes the archaeological materials dated to 6200–2000 BC, we will make an attempt to reveal the

---

1 ELOKA – an Arctic research data management program that combines local traditional knowledge and local observations data from Indigenous Arctic residents utilising effective and appropriate western methods to properly share Arctic Indigenous Knowledge. Cf. https://eloka-arctic.org/bears.

role and the meaning of the bear for the mid-Holocene hunter-gatherer-fishers of the East European Plain forest zone.

The background of the study

During the time period in question, the preferable living places were the lake lowlands, and especially those areas near the river mouths that were connected to a lake. These lowlands contained rich food resources, such as fish and waterfowl, so in many cases the most comfortable spots were settled for several millennia, and there are large multi-period sites (Burov 2011). The territory of the East European Plain forest zone is so huge that it still contains a lot of blind spots that have not been studied by archaeological excavations or surveys.

The beginning of the Neolithic period, according to Russian nomenclature, coincides with the moment the first ceramics appeared, which had a different chronology in the southern and northern areas of the forest zone. In any case, the time of 6000–5000 BC was generally the transitional period in most parts of the forest zone, which saw not only the appearance and development of the earliest types of ceramic ware (Kashina/Zhulnikov 2015) but also the global change in flint-knapping technologies – the change to the use of a flake as the main preform, instead of a blade (Lozovski 2003). The information about house building in the period 6200–3700 cal BC is quite scarce. At some Central Russian settlements, remains of Early Neolithic round huts with hearths were discovered, which could be reconstructed as light constructions above ground, covered by hides. The remains of such constructions were detected for the whole Mesolithic period as well, but they are usually quite ambiguous (Leonova 2004). The multi-period character of many settlements introduces further confusion into the discussion of the history of house building.

Nevertheless, in the 1970s the general Neolithic settlement system was reconstructed, based on survey data, by Vladimir Sidorov for the Moscow region and partly for the adjacent ones. The settlements and campsites of 5000–3700 cal BC of the so-called Pitted-Comb Ware culture were mapped, based on a wide range of excavations, test pits and hand-collected materials representing this highly-recognisable ceramic type, which demonstrated the dense network of inhabited spots, not absolutely contemporaneous of course, but dated to a wide time range of more than a millennium. These settlement clusters were usually surrounded by "empty zones" of around 30 km in diameter, recognised as the hunting/gathering grounds (Sidorov 1975). Probably, this mode of settling can be extrapolated in general to the whole East European forest zone, including the later period of 3700–2700 BC, but excluding the upland areas that do not contain the proper water bodies (for example, the Smolensk Upland). The building of semi-subterraneous large dwellings of more than 100 m² started around 3700 BC, simultaneously in Finland, northwestern and Central Russia. This is of great interest, as it marks not only the general increase of sedentism (Herva et al. 2014) but also the known changes in the social structure. Together with this, a crucial change in the ceramic technology took place: In the region of the Pit-Comb Ware and Comb-Pit Ware cultures, organic admixture started to prevail over the mineral one, and thus the Late Comb Ware appeared in the circum-Baltic, while in the central and in part of the northern Russian zones the ceramic type cardinally changed from Pit-Comb Ware to the Volosovo Ware, which features shells and bird down as the main admixtures. The core reasons for these movements/changes are generally unknown. Later, around 2900 BC, new inhabitants came into the forest zone, marking the beginning of the Bronze Age. One branch of the Corded Ware cultures that spread over the Russian northwest and centre is known as the Fatyanovo culture (dated to around 2900–2000 cal BC). This culture was represented by semi-sedentary or sedentary herders, who settled the river valleys and obtained the first imported bronze items. Recent genetic data suggest that there was almost no mixing between hunter-gatherers and herders (Kraynov 1972; Nordqvist/

HEYD 2020, 17). Another branch, known as the Battle Axe culture, together with influences from the territories of Sweden and Finland, also affected the Karelian and northern Russian territories. The relationship between the newcomers and the forest hunter-gatherer-fishers still remains poorly studied. They obviously contacted each other, despite inhabiting different eco-niches, as both of them were widely practicing hunting, gathering and fishing. The last hunter-gatherer-fishers remained at the centre of the East European Plain forest zone until at least the Early Iron Age (the end of the 1st millennium BC), and in some taiga areas for much longer.

THE BEAR IMAGE IN PORTABLE ART

The portable art of the East European hunter-gatherers is an outstanding phenomenon, which reveals the wide and multi-faceted use (mostly in everyday life) of sculptural and graphical effigies representing humans and animals of different species. The astonishing fact is that the bear image is almost absent in this. The portable art of the Final Mesolithic and the Early Neolithic (around 6200–4000 BC) is represented by a total of only about a dozen carved sculptures and pendants of bone/antler with the images of humans, snakes, elk, and birds (see for example JONUKS 2016), possible sledge-runners with elk heads, and the elk-head staffs made of antler (MANTERE/KASHINA 2020), but absolutely no bear images.

The next period (around 4000–3000 BC) was the time when portable art blossomed. The total number of finds is around 650 to date, and several groups can be distinguished – ceramic sculptures featuring human, bird, snake, beaver, otter, and elk images (KASHINA 2007), sculptural and graphic images on special (ritual?) ceramic vessels depicting waterfowl and humans (PESONEN 1996; KASHINA 2006; KASHINA/ZHULNIKOV 2015), carved sculptures of a different kind, and, finally, flint sculptures (KASHINA 2002; 2005). Carved pendants were widely distributed over the East European Plain forest zone during this period and consisted of numerous bird and snake images but also, rarely, those of beaver, elk, fish, and even a seal and a wisent. The other categories of carved sculptural items are represented by tools, which were probably used occasionally during rituals. These are differently-shaped bone lancets and spatulae with bird silhouettes or bird heads, wooden ladles and spoons with waterfowl heads, a wooden oar with bird heads, a bone pin with an elk head, etc. (KASHINA 2005). Again, the antler elk-head staffs were common for this time range. And finally, several finds of carved bear images were detected (Fig. 1).

Two wooden ladle handle tips, shaped into distinct bear heads, were found in the settlement layers at Usvyaty IV (Pskov region, Russia) and Sārnate (Latvia) (Fig. 2.1–2). Obviously, these items copy the well-known and widely disseminated (according to peat-bog site materials) ladles with waterfowl-shaped handle tips (KASHINA/CHAIRKINA 2011), but they look very unnatural – a swan has a long neck, but not a bear. This makes us assume that the waterfowl handles came first, while the bear handles appeared later. If we compare the absolute dates, obtained in the context of both sites, we see that Sārnate was occupied in the Final Stone Age, at the turn from the 4th to the 3rd millennium BC, and Usvyaty IV is believed to have been inhabited some centuries later – around the mid-3rd millennium BC, which was the period of the Early Bronze Age, according to the local chronology (BERG-HANSEN et al. 2019; BRONZEZEIT: EUROPA OHNE GRENZEN 2013, 346). The next find is a bone spatula from the settlement layer of Usvyaty IV, dated to the second half of the 3rd millennium BC (BRONZEZEIT: EUROPA OHNE GRENZEN 2013, 346; see Fig. 2.3). Taking into account the presence of a wide series of bone spatulae and lancet finds decorated with bird figures or heads and dated to around 3600–3300 cal BC in Latvia, Estonia, and adjacent regions of Russia (BĒRZIŅŠ et al. 2014), we can presume the same process: The bird image that was depicted on these special tools preceded that of the bear. A bone item (9 cm in length) with a probable bear head at the top of the rod was found in the settlement layer of the Abora I site, Latvia, which is dated to between the Final Stone Age and the beginning of the Bronze Age (Fig. 2.4; Loze 1983). The rod is decorated with groups of short incisions, but the general purpose of this object is unknown. The single find of a fragment of a burnt wooden item (4.3 cm in length) resembling a bear head was found in the same area at Lake Sennitsa (Pskov region) and looks much larger than the bear heads from Sārnate and Usvyaty IV, so it is hard to attribute it correctly (Bronzezeit: Europa ohne Grenzen 2013, 360). A bone spoon with a bear head was found in a dwelling at the Imerka VIII site (Republic of Mordovia) and most probably dates to the turn from the 4th to the 3rd millennium BC or a slightly later period. It should be mentioned that material remains of primitive bronze smelting were detected at this settlement (Korolev/Stavitskiy 2006; see Fig. 2.5). A bear metatarsal with a sculpted bear head on the epiphysis was found in the multi-period settlement of Chornaya Gora (Ryazan region) and, according to the stratigraphy, may belong to the Volosovo or Bronze Age materials, as may the above-mentioned Imerka VIII item (Tsvetkova 1969, 34–36; Fig. 2.6). Functionally, it could have been a sinew-thread-treating tool, as it has several grooves around the bone diaphysis. Two carved pendants made of bone and of amber, respectively, which could possibly have represented a bear, are known from the settlement of Tamula I (around 3900–2600 BC; bone pendant: Fig. 2.7) and the Tamula burial ground (burial XII, 3000–2600 cal BC) in Estonia (Jaanits 1957; Tõrv 2016, 186). The precise date of the bone figurine is unknown, but the modelling of its muzzle is reminiscent of the Sārnate wooden ladle mentioned earlier. The amber figurine is very schematic and should be attributed generally only as a mammal. Two wooden ladles or spoons, each with a handle tip carved into the shape of a mammal (interpreted as a bear head) were found in peatbogs in Laukaa (central Finland) and in Humppila (south Finland). Stylistically, these heads look more or less similar to the Sārnate and Usvyaty ladles but the actual date of the Finnish ones is unknown (Immonen 2002). Summing up the overview of the ritually carved bear representations, it seems quite clear that they represent a small group that leans geographically more or less to the eastern Baltic area, and all date to a slightly later time in comparison with most of the figurative portable art of the East European Plain forests, which leads up to the chronological period attributed in the local chronologies as the Early Bronze Age.

Around 3300 cal BC, and probably up until 2000 cal BC, another phenomenon, flint sculptures, appeared and disseminated throughout a large part of the East European forest zone, mainly Central Russia. Most of the finds come from multi-period settlement layers, thus their precise chronology is unknown. The common use of bifacial technology and pressure flaking gave way to the production of flattened flint sculptures of humans (Kashina 2002) and animals. Their total number is currently around 180 pieces. Remarkably, this tradition did not reach the territories of the modern Baltic states and Finland, as there were almost no flint quarries and the good flint came to these areas from elsewhere. Flint knapping technology does not allow the making of a detailed figure, so, in some cases, the interpretation of an image can be ambiguous. Series of sculptures have similar silhouettes, which helps in distinguishing several morphological groups such as a human (a simple one, a female, a supernatural creature), a bird, a snake, a fish, an unidentified mammal, an elk, a beaver (an otter?), a wild-boar, a white whale, and, finally, a bear.

Only three flint sculptures can be unequivocally identified as bears. All of them were single finds. The first one is from Zimnyaya Zolotitsa (Arkhangelsk region, the White Sea shore). It is of an outstanding size – 9.5 cm long (Fig. 3.1). According to S. Zamyatnin, it could represent a polar bear, as it has quite an elongated muzzle (Zamyatnin 1948). Two other flint sculptures – from the Bagon Otmel' (5.9 cm high) and Sinyaya Gora (5.2 cm high) sites – were found in the Tver region. The distance between these sites is only 100 km. The objects have very similar silhouettes with their bent hind limbs (Fig. 3.2–3), which are slightly reminiscent of the human profile representations on the northwest Russian petroglyphs and also of two unique, human profile flint sculptures from the north Russian areas – the Olskiy Mys (Arkhangelsk region) and Vis II (Komi Republic) sites.

By silhouette and size, the flint sculptures strongly resemble the pendants carved from bone, and there is no doubt that they were partly contemporaneous. The presence of notches on some flint bird, snake, mammal, and fish sculptures suggests that all of them were pendants, including even those without special notches. As for the flint sculptures depicting humans, the presence of arms, legs or limbs easily allowed the sculptures to be attached to clothes or skin/leather equipment (?). Sometimes, the worn surface is clearly visible, so probably these items had been in use for a long time. The hypothesis of these pendants' totemic essence was proposed earlier (KASHINA/EMELYANOV 2020) but, among the whole assemblage of flint sculptures, the bear image is clearly underrepresented. Only beaver/otter, human profile and white whale images are rarer: human (en face; n = 93), bird (n = 28), snake (n = 15), mammal, unidentified (n = 14), fish (n = 13), elk (n = 7), bear (n = 3), beaver/otter (n = 2), human (in profile; n = 2), white whale (n = 1). As the study of carved pendants has already shown, only two probable bear pendants are known from Estonia, but in Central and North Russia they are completely absent.

The rarity of the known carved and flint-knapped bear sculptures provokes the question about the real affiliation of the bear to the group of totemic animals. The prominent Russian ethnographer and archaeologist Mikhail Kosarev has argued that the totemic essence of the bear image in Siberian mythologies was greatly exaggerated by the generations of ethnographers who took part in the process of studying the bear festival. According to the field materials collected from different Siberian peoples, the bear was associated with the underworld but its kinship with humans has never been revealed (KOSAREV 2003, 101–104). According to M. Kosarev, the same misunderstanding (i.e. postulation of totemic essence) has taken place in ethnographic and archaeological literature concerning the perception of the elk image in the Siberian aboriginal world-view. In fact, the elk image was generally connected with positive, universal, and solar symbolism, but not with human ancestry. The prehistoric worshipping of mainly bird species as totemic ancestors by hunter-gatherer-fishers appears much more real because of the deep connection between the bird image and the notion of the human soul, and, according to the archaeological finds of bird pendants, it has been practised since the Upper Palaeolithic period (for example in Siberia, around 20,000 years ago). This hypothesis fits well with the map of the East European forest zone portable art distribution in the Final Stone Age and at the beginning of the Bronze Age (around 3500–2500 cal BC). The elk-head staffs, for example, which existed continuously from at least the Final Mesolithic, were definitely the largest items among the portable art collection, reaching more than 40 cm in length (MANTERE/KASHINA 2020), which recalls the unique spiritual role of this animal. The bird pendants prevail among the whole collection of portable art, and are believed to be the images of totemic ancestors (KASHINA/EMELYANOV 2020), while pendants in the form of elk and bear are almost absent. Also, we can assume that all known items with a carved bear image appear very late, around 3000 cal BC, or even after this time.

THE BEAR IMAGE IN WEAPONRY

If we presume that the elk and the bear have never been totem animals, but embodied far more universal notions and probably were even recognised as opposed entities, the time has come to analyse the hammer-axes with zoomorphic heads. These items appeared initially in antiquarians' collections of the 19th century but they still remain the most mysterious finds, as none of them have been found in a particular settlement or burial context until now. Their total number, including fragments, is 55, 27 of which were found in northwest and northern Russia. Typologically they are represented by two main groups – the mace (the animal head with the handle hole in it) and the axe (the animal head at one end of an item). They represent elk and bear heads but, in some rare cases, unidentified species occur, which were interpreted as a catfish, a seal, or an otter, as well as a phallus or an unclear image.

These stone items were found in the large territory that includes Sweden, Finland, Karelia, the Komi Republic and the Vologda, Tver and Yaroslavl regions (Carpelan 1977; Zhulnikov 2012), and all of them are occasional finds (Fig. 4.1–3). Researchers C. Carpelan and A. Zhulnikov recognise their ritual purpose, as these axes frequently have an abnormally small or narrow shaft hole, or even lack this, and a blunted blade. Also, some items were unfinished, unpolished and demonstrated traces of picketage technique, which provides evidence of their probably local production (Zhulnikov 2002). According to the observations of A. Zhulnikov, the animal-head axes had different prototypes among the "ordinary" axes, which allow a differentiated dating. Some of them, having a rectangular cross-section, can be dated to between the Final Stone Age and the beginning of the Bronze Age, around 3300–2500 BC (the time of Asbestos Ware in Karelia, and Porous Ware with shell admixture in other northern Russian regions). The others, with a ridge and a cylindrical hole and "casting seams", recall the axes of the Battle Axe culture and can be dated to 2800–2300 BC. In the same period, the Globular Amphorae culture (Fatyanovo) influenced the East European Plain forest zone, which is evident by the changes that occurred in the ceramics named "Fatyanoid Ware" by some researchers (Volkova 2019). It should be noted that almost all animal-head axes were found outside the territories of the Corded Ware, Globular Amphorae (Fatyanovo), and Battle Axe cultures and, despite the vague dates, most probably belonged to the taiga societies, which generally kept to the hunter-gatherer-fisher way of life. A. Zhulnikov also argues that the increased meaning of the bear image in spiritual life was caused by the intensification of violent clashes in the forest zone: The bear, which, judging by these stone axes, was generally much more "popular" than the elk, had become a symbol of "Man the warrior". The item itself could have been a scepter, a symbol of a warrior's power. Alternatively, it may have been used as a weapon for sacrifices and was ultimately intended to be hidden or buried in the vicinity of a settlement forever, we can only speculate, e.g. after the owner's death or, for example, during a peace conclusion ritual (e.g. "to bury the tomahawk"; Zhulnikov 2012, 70–72). The number of objects is considerable, bearing in mind the fact that all these axes are single finds, hidden or discarded outside the settlements. Thus, we may propose that maybe each adult male possessed one. In this sense, the axes with sculptural heads could have been more or less synonymous in their symbolic meaning to the elk-head staffs (Mantere/Kashina 2020).

In the territory of present-day Russia, the majority (13 out of 27) of sculpted axes feature a bear head. All pieces are quite different in size, detail and rock type and, though some of them show a resemblance to each other, it is quite clear that the bear image carved on the axes was not strictly canonised. Also, the axe forms are quite different, which points to differences in their dating. Finally, it should be mentioned that bear-axes found in the territory of modern Sweden, Finland, and Russia are nearly double the number of elk-axes, which makes them the most frequently encountered pieces of portable art of the forest zone from the 3rd millennium BC. This probably means that the roles of the two most mighty and notable animal images in the hunter-gatherers' spiritual life underwent a cardinal change. The borrowing of the bear image by the Finnish/northern Russian, most probably still non-herding, societies and its outstanding popularity in the context of ritual weaponry may bear witness to the deep connections between herders and hunter-gatherer-fishers in the taiga area (Mantere/Kashina 2022).

The bear image in petroglyphs

As K. Helskog mentioned in his paper (Helskog 2012), the bear image occurred quite rarely on Scandinavian petroglyphs and paintings, except in Alta, Norway. One of the most intriguing scenes found in Alta is the bear hunt in the winter season. The features of natural rock were used to represent a landscape and a story, which includes episodes of chasing the prey using skis, the den as a natural cavity, the moment of killing, etc. Remarkably, the same compositions, sometimes without the particular landscape details, are known on northwestern Russian petroglyphs at Lake Kanozero (the Kola Peninsula, two compositions), Zalavruga/Vyg (Belomorsk, at least five compositions), and Lake Onega (South Karelia, three compositions; see Zhulnikov 2006, 115–117, 152–153; Kolpakov/Shumkin 2012, 300). The similarities between Scandinavian and northwest Russian petroglyphs were discussed many times, and both K. Helskog and E. Kolpakov concluded that large-scale regional differences prevail over the similarities of the image sets or some multi-image compositions. Even if the mythological features were common, the sites developed independently (Kolpakov/Shumkin 2012, 230; Kolpakov 2019, 29). It should be mentioned that the age of different petroglyphic groups varies and there is sometimes a wide chronological range: Alta – 4800–2700 BC, Onega – 5000–2000 BC, Belomorsk – 4000–2000 BC, Kanozero – 4000–1000 BC (Gjerde 2018; Zhulnikov 2010; Kolpakov/Shumkin 2012; Blyshko/Zhulnikov 2020).

The petroglyphic representations show that bears were hunted with spears, most likely in winter/early spring during hibernation. On the Belomorsk petroglyphs, the bow and arrow was also shown being used for the bear hunt. Zhulnikov (2006, 152–153) interprets the presence/absence of bear trails as the evidence of wintertime/summertime hunting (referring to bear trails left on the snow). The only scene of "bear the hunter" exists at Lake Onega – a bear chasing an elk. A quite similar one, in Alta, with a bear and a reindeer, is referred to by Helskog (2012, fig. 13). In addition, A. Zhulnikov stresses that petroglyphs represent mythological images and scenes as a kind of manual for young community members, and their creation acted as a repetition of mythological events in the present (Zhulnikov 2006, 115; Zhulnikov/Kashina 2010). Helskog (2012, 219) assumes "the use of bear figures in narratives or stories and possibly rituals", and "some of these illustrate stories that we know bears could not have been engaged in, except in legends and myths", but at the same time he admits that there were also bear scenes that were drawn from life. In summing up the analysis of bear images on petroglyphs, we assume that the bear was in general depicted rarely, though in some areas numbers of images were notably higher.

Osseous bear remains and tools in settlement layers

The only available source that allows a rough reconstruction of the scale of bear hunting is from the period 3500–2700 BC, and it takes the form of several, large osteological collections of the Volosovo culture settlements at the Volga-Oka interfluve. At these sites, excavated mostly in Soviet times, bone and antler had sometimes been rather well preserved in the dark, humified sandy sediments of the large semi-subterraneous dwellings or near the pit dwellings. There were not only artefacts and their debris, but also the far more numerous kitchen waste items consisting of mammal, fish, and bird bones. These unique collections of thousands of bone units are still poorly studied. Nevertheless, some available archive data from the 1970s' osteological studies concerning the River Oka sites of Chornaya Gora and Vladychino (Ryazan region) contain important information about bear hunting.

At the Chornaya Gora settlement, excavated during several seasons at the beginning of the 1970s (Andreeva 1974a; see Fig. 1), 197 bear bones were found, which represented no less than 15 individuals, while the total amount of mammal bones was 2,783. In this group, the bear bones were less numerous than those of elk and marten. Bear skeletal parts such as a skull, a mandibula, a scapula and pelvic and long bones were detected, of which most had been broken and which, according to E. Andreeva, bear witness to the consumption of bear meat and bone marrow. The bone measurements point to the highly variable age of the hunted individuals – from cubs to large animals. Twenty-four bear bones had been modified and were typologically defined as tools and personal ornaments – eight fibulae and two parts of a radius were turned into awls and have a pointed tip; three radius bones were used as chisels, one as a hoe; two ulna bones were used as daggers and a dented point, respectively, and, finally, pendants with a drilled hole were made from six canines and four phalanx bones.

At the Vladychino settlement (data are available only for the year 1974; see Fig. 1), which was situated close to the previous site on the opposite shore of the River Pra, 55 bear bones out of altogether 12,600 bones (including mammal, bird, and fish bones) were obtained (ANDREEVA 1974b). Here, the bear bones were less numerous than the elk, marten and wild boar bones. In this brief osteological report, E. Andreeva mentions the similarity between both sites in the composition of mammal osseous remains, but points out the small number of broken bear bones. Thus, according to the presented osteological data, though it is quite old, random, and covers only a part of each site, the bear was definitely hunted. It was consumed and its bones were used for the making of a wide range of tools and ornaments.

A reasonable question can be asked: Was the elk the most desired hunting prey for the Mesolithic forest zone hunters, being able to supply household requests as regards food, hides and the raw material for tools? The answer comes by way of an example from Inuit ethnography: It turns out that brown bear fur and skin were believed to be extremely precious due to their warmth (bedding) and durability (boat coverings, skin ropes, skins with fat for boat construction), respectively. The bear guts used for parkas were able to repel water, and for this reason they were appreciated more than seal guts (FIENUP-RIORDAN 2007). Thus, many products from the killed bear had unique qualities that were superior to those provided by elk and other hunted mammals. It seems that these particular features influenced in many ways the complex of rites, beliefs, and restrictions connected with the brown bear, which are widely known ethnographically.

Some categories of tools made from bear bones have already been mentioned above. The studies of bone implements from the East European Plain forest zone Mesolithic/Neolithic collections were rarely aimed at distinguishing which particular mammal species provided the raw material for one tool or another. Let us hope that soon studies will appear that analyse these bone tools in a more detailed way. At present, only random examples can be observed from the neighbouring settlements of Shagara II and Velikodvorye I (Ryazan and Moscow regions; see Fig. 1), situated not far from the two earlier-described sites with osseous collections within the same lake system and containing even richer osseous collections than those described. The third settlement of Volodary (Nizhniy Novgorod region; see Fig. 1) is situated apart from the other settlements in the Volga river basin. All these settlements belong to the same chronological phase of the Volosovo culture, dated approximately to 3500–2700 BC. In all given cases, we have no further information about the exact settlement context of the bear bone finds, thus we can only speculate that these were items that had probably been lost, or maybe deliberately hidden in a dwelling. They are published here for the very first time.

One of the most frequent ways to make a bear canine pendant was by drilling a hole. At the same time, fastenings that consisted of carved, interlocking notches existed (Fig. 5.1–2). Another puzzling case is the canine with its root deliberately sawn off for an unknown reason (Fig. 5.3). And finally, a unique example is represented by an awl, found at Volodary, made of a horizontally-split bear canine with its tip sharpened and bearing visible traces of polishing (Fig. 5.4). Considering the widely-shared notion about the ritual role of bear canines as a part of garment decoration with deep spiritual value and meaning for the forest Mesolithic/Neolithic hunters, which seems to be generally difficult to decipher (see e.g. LARSSON 2006, 281), the last example points at the quite profane function of a bear canine, obviously chosen because of its potential to be fashioned into an awl.

The presence of bear baculum bones in the settlement layers of Shagara II and Velikodvorye I is evidenced by three items (Fig. 5.5–7). Two of them are unmodified (Fig. 5.5–6), but the last one has a wide, drilled hole in the bottom part and a broken tip that bears traces of heavy polishing (Fig. 5.7). Probably, the latter was used as an awl, and the hole was made for fastening. The only similar piece known so far comes from the Early Neolithic (burial 3 in Kubenino, Arkhangelsk region, around 4900 cal BC), where it was a part of a necklace-like set of five items that included bone tools and a sculpted pendant (KASHINA et al. 2021, 81). The modified baculum from Shagara II (Fig. 5.7) can be recognised in parallel with the canine-awl item from Volodary (Fig. 5.4) as a profane household tool made of the suitable raw bone that was close at hand at that moment.

As we finish this section and start with the bear bone burial finds, we argue that, judging by the given examples, the meaning of the bear as an extremely special and sacred animal should not be exaggerated. Despite all the difficulties and risks accompanying the bear hunt, reasons of economy and convenience might have played a role in the use of bear bones, and the bear was definitely not a taboo animal.

Bear remains in burials

Let us now turn to the more "traditional", ritual perspective, from which the role of a bear has been more frequently observed in the world of prehistoric hunter-fisher-gatherers; in particular, to the position of the bear and its faunal remains in the Early to Middle Holocene hunter-fisher-gatherer burial practices. Before proceeding to examine the particular cases and local developments of the use of bear remains in mortuary rituals, it is important to give a brief overview of the East European Plain forest zone burial practices as such.

The Early to Middle Holocene burials in the East European Plain forest zone are characterised mostly by solitary inhumations; however, partial and full cremations also took place (STOKOLOS 1984; KAZAKOV 2011, 29). Moreover, based on the lack of burial remains in some large-scale areas, such as the Middle Volga, we may assume the practice of archaeologically invisible burial forms and/or very moderate archaeological surveys and excavations in the area. The bodies were buried with and without post mortem manipulation of the corpses. As for the body positioning, stretched supine and crouched on one side are among the most common positions of the inhumations; however, the diversity of funeral rites depended on the chronological and geographical frame and some other variables. Therefore, such positions as stretched prone, sitting or standing as well as burials of separate body parts could also have been a choice of the Early to Middle Holocene hunter-gatherer-fishers. The deceased were buried without burial goods or, contrarily, equipped with tools, personal ornaments, weaponry, objects of art or faunal remains. Burials containing osteological remains of bears were also detected.

In regard to the collection of data, the present study employs published and unpublished (archive) sources as well as material from selected archaeological collections (Sakhtysh, Volodary) to analyse the role and specificity of the use of bear remains throughout the time span in question. The majority of the unmodified bear remains were not taken into collections after the excavations (besides the bear metacarpals from Sakhtysh II), therefore their description is based on published and archive sources. Difficulties in the records of bear remains and their association with graves emerge due to the location of a large number of the burial sites in sandy soils, which in many cases hinders the detection of grave borders. Another noteworthy aspect is the position of many burial sites within the multi-period settlements (e.g. Berendeevo I, Yazykovo I, Maslovo Boloto V, etc.), which can have obstructed the chronological and cultural identification of objects and led to post-depositional alterations, such as random artefacts that have fallen into a grave from the settlement layer. Hence, our database attentively takes into account the spatial and stratigraphical distribution of the artefacts and their association with the graves.

The first question aimed to detect the difference between the use of bear osteological remains at settlements and burial sites. A comparison revealed that bear faunal remains observed within the burial contexts differ in most cases in their concentration and representation from those found within the settlements. In particular, among the remains found within the Early to Middle Holocene burial sites, bear teeth, skulls, metapodiae, phalanges, and metacarpal parts are the most frequent. Other skeletal parts were rarely used.

The second set of questions aimed to identify and attempt to explain possible variations of burial contexts in which bear osteological remains were involved. Our dataset demonstrated that some remains were placed within the graves, but other bear bone finds, such as solitary bear fangs, bear jaws, phalanges and, rarely, skulls, occur outside the burial pits, not as stray debris but as a part of other burial ground objects, represented for example by the clusters of various bear bones, or even bear skulls that are accompanied by artefacts or deliberately placed over them (Kostyleva/Utkin 2010, 30). The known differentiation in these burial contexts might indicate the various actions and specific behaviour within their background (Figs. 6–7).

As shown in Figure 7, both modified and unmodified bear remains can be detected *inside the graves*. Let us now consider these cases more closely. The unmodified bear remains found inside the graves derive from 13 burials at the following sites: Chornaya Gora, Ksizovo 6, Sakhtysh I, II, Vasilyevskiy Kordon 17, Volodary, Yazykovo I, and Yuzhniy Oleniy Ostrov (Figs. 6–7). Skulls, jaws, incisors, metacarpi, metapodia, and phalanges were placed in different ways – at the feet, close to the head, slightly above the corpse, or were found somewhere else among the bones of the deceased. An interesting observation is that the last pattern is traced in collective burials, where the human bodies had usually undergone a great deal of manipulation (Chornaya Gora, grave 50; Volodary, grave 7).

Thus far, the following peculiarities of location and choice of unmodified bear bones can be identified:
1) All graves that contain bear bones belong to the Final Neolithic Volosovo culture, with the exception of Yuzhniy Oleniy Ostrov (Final Mesolithic) grave 116, which contained a bear jaw inside the grave, and grave 132, which contained some bear bones;
2) sometimes various selected skeletal parts were found near each other (e.g. jaws/metacarpi or phalanges/metapodia), in other cases only homogenous skeletal parts were placed in a grave, for example, bear skulls;
3) in all cases, jaws and metacarpi were the most frequent finds.

Bear teeth without any fastening elements that had been placed inside graves prompt the question of how the teeth were attached to clothes or other items. Maybe some sort of adhesive of animal, fish, or plant origin was used. For instance, at the Yazykovo I site (grave 2) the unmodified bear teeth were arranged in a row along both thigh bones (Kostyleva/Utkin 2010, 158). Cases of animal tooth pendants placed in a row together with unmodified teeth, found in the Mesolithic-Neolithic Zvejnieki burial ground, were also analysed by Larsson (2006).

Regarding the modified bear remains found inside 74 graves, these are represented by bear incisors, fangs, and a phalange (as in the unique case at Yuzhniy Oleniy Ostrov, grave 125), placed in a particular order, most frequently on the body or in its close vicinity. Modified bear remains were detected at the sites of Chornaya Gora, Minnyarovo, Sakhtysh II, IIa, Shagara I, Yazykovo I, and Yuzhniy Oleniy Ostrov (Fig. 7).

Our observations found that when represented in clusters or arrangements, the ornaments were concentrated either along the lower limbs or in the pelvic area; in the arrangement, the modified bear teeth were combined with teeth of other species or represented exclusively. Different techniques were applied to modify the fastening at the proximal part – grooving around, grooving partially, grooving together with a hole drilling, drilling only, modifying from both sides (making the proximal part either slightly or considerably thinner) and then drilling, and cutting off the proximal tip (Fig. 8a). Among these, the tendency of a more frequent use of the grooving technique in the Late Mesolithic materials and the almost complete replacement of it by the drilling technique in the Late Neolithic were detected.

We now move on to observe the osteological remains of bears found *outside the graves*. Analysis shows that there is only one case, Minino 2, that represents objects with modified bear remains outside the graves (teeth pendants placed in the pits; see Sorokin/Hamakawa 2014, 71). Another 11 cases of bear remains found within the cemetery but outside the graves derive from Berendeevo I, Karavaikha, Maslovo Boloto V, Popovo, Sakhtysh II, and Volodary. These include unmodified remains, such as skulls, jaws, metacarpi, pelvic bones, sternum, and other post-cranial parts, which occur singly as well as accompanying a cluster of artefacts or other faunal remains (e.g. Oshibkina 2016, 798–799).

Some of the most remarkable features with osteological bear remains found outside the graves come from the Sakhtysh II site. In one case, the cranial and postcranial parts of two bears were located in the vicinity of grave 4; in two other cases, bear front limb bones – long bones in the so-called "ritual hoard" 1 and metacarpi in "ritual hoard" 10 – were placed in anatomical order over a cluster of artefacts (Fig. 8b). Finally, the bones of a bear paw were found in a cluster, among burnt faunal remains and artefacts ("ritual hoard" 12; cf. Kraynov 1988, 38–44). The term "ritual hoard" refers to a cluster of artefacts and faunal remains, as found within the Volosovo burial sites (3500–2700 cal BC); often they were deliberately placed in a pit, sometimes they were covered with ochre, were broken, and had undergone thermal treatment.

To sum up, within the Early to Middle Holocene burial sites in the East European Plain forest zone bear remains were represented in both modified and unmodified form inside as well as outside the graves. There is always the methodological uncertainty that items can have accidentally ended up in a grave or cemetery sediment. However, it has been established that bear remains are relatively rare in Mesolithic/Early Neolithic burials and are usually represented by teeth, thus they could have been used as a part of burial garments, adornment sets, or other items required to furnish the grave, such as blankets, shrouds, bags, quivers, or other accessories. A large increase in bear bone remains in burials and grave-side ritual objects is recorded in the Volosovo period (around 3500–2700 cal BC), when there was a marked rise not only in the number of objects but also in the variety of skeletal parts of bears involved in the rituals and the ways they were represented across the cemeteries.

Discussion

After having analysed the bulk of sources concerning the bear image in portable and rock art, as well as bear remains in settlement and burial contexts, we can conclude that there is an imbalance among the analysed groups of archaeological finds. Being strikingly rare, in fact almost invisible, in the portable art from the Mesolithic/Neolithic, the bear image was increasingly used in different spots of the forest zone close to the start of the Bronze Age, revealing itself in wood (ladles), bone (pendants, spatulae, spoon), and flint sculpture as well as later, in the 3rd and 2nd millennia BC, as a sculpted decoration of ritual stone axes, which were widespread in Scandinavia and the Russian North. The bear image is completely absent among the sets of the Neolithic ceramic sculptures of the Comb and Comb-Pitted Ware cultures, while such species as waterfowl, beaver, elk, and snake are represented there (Kashina 2007). These images have constantly occurred in the other groups of portable art at least since the Final Mesolithic, but the brown bear has not. In fact, we could suspect that a bear-sculpture representation was a kind of taboo for a long time.

The most interesting fact is that the bear image then appears on items that already existed in the 4th millennium BC (wooden ladles, bone spatula, and carved pendants/flint sculptures), but which had previously depicted other images, such as waterfowl (ladles/spatulae) and human/waterfowl/snake/elk/beaver (pendants). The figural axes normally featured not only bear but also elk heads, the peculiarities of which visibly echoed the iconography of the antler elk-head staffs, known since the end of the 7th millennium BC.

As for the petroglyphs, which unfortunately are not dated as precisely as the objects of portable art, the bear image is very rare in most of the rock art groups. This feature, as well as the intention to portray bear traces and dens instead, can be interpreted as an attempt to elude the bear image itself while dropping hints of its presence. Quite the same conclusion can be derived from the Mesolithic/Neolithic burial and grave-side contexts with their bear tooth pendants, skulls, and especially paw bones. At the same time, it contradicts the universal recognition of the bear as a much desired and honourable hunting prey, and relevant scenes of bear hunting do exist on petroglyphs as well. The settlement cases that we cited here, together with the Inuit example, stress the economical need for the bear (which was probably even essential for survival) in the boreal prehistoric and indigenous communities.

Osseous remains of bears in burials represent an ambiguous data set, because the organic materials survive quite randomly in the East European forest zone territory. The situation of cemeteries at the same spots as the multi-period settlements has also to be considered. Nevertheless, the known abundance of bear bone finds at the cemeteries of the Central Russian Volosovo sites, together with some rare portable art pieces representing the bear in flint and bone, deserves further attention to study the supposed shifts in Final Neolithic/Bronze Age spirituality.

Conclusion

In the territory of the East European forest zone Mesolithic/Neolithic, the general development of bear depictions over the millennia was as follows: Hunter-gatherer-fishers rarely depicted the bear (perhaps it was a kind of a taboo) but hunted it, consumed its meat and used its bones in households and rituals. The bear image at last gradually appeared in portable art on the turn from the Stone to the Bronze Age. The Bronze Age of the huge territories of the northern latitudes of Russia is understudied, and the main differences from the more southern territories are the very moderate amount of metal finds and the seeming absence of animal husbandry and agriculture. Nevertheless, the migration of the Corded Ware derivates definitely influenced the ceramic traditions of the local hunting communities all over the East European Plain forest zone and was probably the initial cause of the spread of the bear image, at least in portable art, and, later, the spread of figural axes, of which those that are known feature the bear head but were found outside the frames of the Corded Ware culture areas. Recent analyses of genetic data suggest that there was almost no admixture between hunter-gatherers and herders on the East European Plain. The clearly observed increase in the bear image in portable art provides evidence of interactions between hunter-gatherers and herders, probably since the herders' first appearance in the forest zone. Their presence lasted for about a millennium and could have caused some ideological changes that were also reflected in the material culture of the hunters.

The brown bear definitely played one of the key roles in the mid-Holocene forest zone societies' spirituality and livelihood, but its representations are rarely met until the turn from the 4th to the 3rd millennium BC, the time period of the turn from the Stone to the Bronze Age. Also, the bear was constantly part of the prey, not only in prehistory but during following epochs as well, and definitely had a recognised place in pagan mythology and folklore.

Acknowledgements

We are grateful to Dr. Pavel Kosintsev for the zooarchaeological consultation, to Dr. Roman Smolyaninov and Dr. Elena Kostyleva for the permission to use the burial data obtained in the Lipetsk and Ivanovo regions in our study, to Ville Mantere and Dr. Aleksandr Zhulnikov for the consultation on the figural axes topic, to Aija Macāne, to Dr. Arkadiy Korolev, to Dr. Mark Shakhnovich, and to Normunds Grasis for the permission to publish particular photos, to Dr. Oliver Grimm for his kind invitation to participate in this volume, and finally we want to express our gratitude to the reviewers.

Bibliography

Andreeva 1974a: E. G. Andreeva, Fauna drevney stoyanki Chornaya Gora. Archiv otdela arkheologicheskikh pamyatnikov sektora A Gosudarstvennogo Istoricheskogo Muzeya, 1974, folder 78, file 103.

Andreeva 1974b: E. G. Andreeva, Nauchniy otchot o zhivotnykh Vladychenskoy stoyanki po kostnym ostatkam iz arkheologicheskikh raskopok 1974 g. Archiv otdela arkheologicheskikh pamyatnikov sektora A Gosudarstvennogo Istoricheskogo Muzeya, 1974, folder 79, file 108.

Bahn 2011: P. G. Bahn, Religion and Ritual in the Upper Palaeolithic. In: T. Insoll (ed.), The Oxford Handbook of the Archaeology of Ritual and Religion (Oxford 2011) 344–357.

Berg-Hansen et al. 2019: I. M. Berg-Hansen/H. Damlien/M. Kalniņš/I. Zagorska/A. Schülke/V. Bērziņš, Long term variation in lithic technological traditions and social interaction: the Stone Age of the Eastern Baltic (Latvia), 10500–2900 cal BC. Fennoscandia Archaeologica XXXVI, 2019, 6–32.

Bērziņš et al. 2014: V. Bērziņš/U. Brinker/C. Klein/H. Lübke/J. Meadows/M. Rudzīte/U. Schmölcke/H. Stümpel/I. Zagorska, New research at Riņņukalns, a Neolithic freshwater shell midden in northern Latvia. Antiquity 88, 2014, 715–732.

Blyshko/Zhulnikov 2020: D. V. Blyshko/A. M. Zhulnikov, Rekonstruktsiya petroglificheskogo svyatilischa na myse Peri Nos VI. Uchonye zapiski Petrozavodskogo gosudarstvennogo universiteta 42(6), 2020, 8–14.

Bronzezeit. Europa ohne Grenzen 2013: Bronzezeit. Europa ohne Grenzen. 4.–1. Jahrtausend v. Chr. Ausstellungskatalog/Exhibition catalogue (St. Petersburg 2013).

Burov 2011: G. M. Burov, Rybnaya lovlya v epokhu mezolita na Yevropeyskom Severe Rossii. Rossiyskaya Arkheologiya 2, 2011, 5–15.

Carpelan 1977: C. Carpelan, Älg- och björnhuvudföremal från Europas nordliga delar. Finskt Museum 1975, 1977, 5–67.

Fienup-Riordan 2007: A. Fienup-Riordan, Yuungnaqpiallerput. The Way We Genuinely Live. Masterworks of Yup'ik Science and Survival (Seattle 2007).

Gjerde 2018: J. M. Gjerde, An overview of Stone Age rock art in northernmost Europe – what, where and when? Rock Art of the White Sea. I. South-Korea: Ulsan Petroglyph Museum 2018, 204–225.

Gurina 1956: N. N. Gurina, Oleneostrovskiy mogil'nik. Materialy i issledovaniya po arkheologii SSSR 47 (Moskva, Leningrad 1956).

Helskog 2012: K. Helskog, Bears and meanings among hunter-fisher-gatherers in Northern Fennoscandia 9000–2500 BC. Cambridge Archaeological Journal 22(2), 2012, 209–236.

Herva et al. 2014: V.-P. Herva/K. Nordqvist/A. Lahelma/J. Ikäheimo, Cultivation of Perception and the Emergence of the Neolithic World. Norwegian Archaeological Review 47(2), 2014, 141–160.

Immonen 2002: V. Immonen, Functional ladles or ceremonial cutlery? A cultural biography of prehistoric wooden spoons from Finland. Acta Borealia 19, 2002, 27–47.

Jaanits 1957: L. Jaanits, Neue Gräberfunde auf dem spätneolithischen Wohnplatz Tamula in Estland. In: Studia Neolithica in honorem Aarne Äyräpää. SMYA 58 (Helsinki 1957) 81–100.

Jonuks 2016: T. Jonuks, A Mesolithic human figurine from River Pärnu, South-West Estonia: a century-old puzzle of idols, goddesses and ancestral symbols. Estonian Journal of Archaeology 20(2), 2016, 111–127.

Kashina 2002: E. Kashina, Anthropomorphic flint sculpture of the European Russian forest zone. Anthropologica at Praehistorica. Bulletin de la Société royale belge d'Anthropologie et de Préhistoire 113, 2002, 51–60.

Kashina 2005: E. Kashina, Iskusstvo malykh form neolita-eneolita lesnoy zony Vostochnoy Yevropy. PhD thesis, Moscow State University (Moskva 2005).

Kashina 2006: E. A. Kashina, Izobrazheniya cheloveka na neo-eneoliticheskoy keramike lesnoy zony Vostochnoy Yevropy. In: A. Y. Martynov (ed.), Pervobytnaya i srednevekovaya istoriya i kultura Yevropeyskogo Severa: problemy izucheniya i nauchnoy rekonstruktsii. (Arkhangelsk 2006) 92–97.

Kashina 2007: E. A. Kashina, Nabory lepnykh zoomorfnykh skulptur v neolite-eneolite lesnoy zony Vostochnoy Yevropy i Finlyandii. In: A. Y. Trufanov (ed.), Mif, obryad i ritual'niy predmet v drevnosti (Ekaterinburg, Surgut 2007) 125–135.

Kashina/Chairkina 2011: E. A. Kashina/N. M. Chairkina, Derevyannaya posuda s navershiyami v vide golov vodoplavayushchikh ptits na territorii Zauralya, lesnoy zony Vostochnoy i Severnoy Yevropy. Vestnik Novosibirskogo Gosudarstvennogo Universiteta, Series History/Philology 10(7), 2011, 157–169.

Kashina/Emelyanov 2020: E. Kashina/A. Emelyanov, In bird we trust: bone bird pendants made by forest hunter-gatherer-fishers in the central part of the East European Plain (3500–2700 BC). In: O. Grimm (ed.), Raptor on the fist – falconry, its imagery and similar motifs throughout the millennia on a global scale. Advanced studies on the archaeology and history of hunting, edited by the ZBSA 2.1 = Advanced studies in ancient iconography II (Kiel, Hamburg 2020) 403–416.

Kashina/Zhulnikov 2015: E. Kashina/A. Zhulnikov, Vessel guardians: sculpture and graphics related to the ceramics of North-Eastern European hunter-gatherers. Documenta Praehistorica XLII, 2015, 289–295.

Kashina et al. 2021: E. Kashina/M. Ahola/K. Mannermaa, Ninety years after: new analyses and interpretations of Kubenino hunter-gatherer burials, north-western Russia (c. 5000 cal BC). Quaternary International 574, 2021, 78–90.

Kazakov 2011: E. P. Kazakov, Pamyatniki epokhi kamnya v Zakamye. Arheologicheskiy ocherk 1 (Kazan 2011).

Kolpakov 2019: E. M. Kolpakov, Petroglify Kol`skogo poluostrova i Severnoy Fennoscandii. PhD thesis summary, Institute for the History of Material Culture (St. Petersburg 2019).

Kolpakov/Shumkin 2012: E. M. Kolpakov/V. Y. Shumkin, Rock Carvings of Kanozero (St. Petersburg 2012).

Korolev/Stavitskiy 2006: A. I. Korolev/V. V. Stavitskiy, Primokshanye v epokhu rannego metalla (Penza 2006).

Kosarev 2003: M. F. Kosarev, Sibirskiy totemizm: illuziya i deystvitelnost`. In: L. A. Chindina (ed.), Arheologo-etnograficheskiye isssledovaniya v yuzhno-tayozhnoy zone Zapadnoy Sibiri (Tomsk 2003) 100–105.

Kostyleva/Utkin 2010: E. L. Kostyleva/A. V. Utkin, Neo-eneoloticheskiye mogil'niki Verkhnego Povolzhya i Volgo-Okskogo mezhdurechya. Planigraficheskiye i khronologicheskiye struktury (Moskva 2010).

Kraynov 1972: D. A. Kraynov, Drevneyshaya istoriya Volgo-Okskogo mezhdurechya. Fatyanovskaya kultura (Moskva 1972).

Kraynov 1988: D. A. Kraynov, O religioznykh predstavleniyah plemyon volosovskoy kul'tury. In: B. A. Timoschuk (ed.), Drevnosti slavyan i Rusi (Moskva 1988) 38–44.

Larsson 2006: L. Larsson, Tooth for a tooth for a grave. Tooth ornaments from the graves at the cemetery of Zvejnieki. In: L. Larsson/I. Zagorska (eds.), Back to the Origin. New research in the Mesolithic-Neolithic Zvejnieki cemetery and environment, northern Latvia (Lund 2006) 253–288.

Leonova 2004: E. V. Leonova, Mezoliticheskiye zhilischa Volgo-Okskogo mezhdurechya (k probleme interpretatsii istochnika). In: H. A. Amirkhanov (ed.), Problemy kamennogo veka Russkoy ravniny (Moskva 2004) 49–68.

Loze 1983: I. A. Loze, Akmens laikmeta māksla Austrumbaltijā (Riga 1983).

Lozovski 2003: V. V. Lozovski, Perekhod ot lesnogo mezolita k lesnomu neolitu v Volgo-Okskom mezhdurechye (po materialam stoyanki Zamostye 2). In: V. I. Timofeev (ed.), Neolit-eneolit Yuga i neolit Severa Vostochnoy Yevropy (novye materialy, issledovaniya, problemy neolitizatsii regionov) (St. Petersburg 2003) 219–239.

Mantere/Kashina 2020: V. N. Mantere/E. A. Kashina, Elk-head staffs in Prehistoric North-Eastern Europe and North-Western Russia: signs of power and prestige? Oxford Journal of Archaeology 39(1), 2020, 2–18.

Mantere/Kashina 2022: V. N. Mantere/E. A. Kashina, Zoomorphic stone maces and axes in the forest zone of north-eastern Europe: Manifestations of interaction between hunter-gatherers and cattle-herding groups in the 3rd millennium BC. In: A.-K. Salmi (ed.), Oodeja Mikalle – Odes to Mika – Оды Мике: Festschrift for Professor Mika Lavento on the occasion of his 60th birthday (Helsinki 2022) 35–44.

Nordqvist/Heyd 2020: K. Nordqvist/V. Heyd, The Forgotten Child of the Wider Corded Ware Family: Russian Fatyanovo Culture in Context. Proceedings of the Prehistoric Society 86, 2020, 65–93.

Oshibkina 2016: S. V. Oshibkina, Funeral rituals of the population of the Eastern Lake Onega region. In: J. M. Grünberg/B. Gramsch/L. Larsson/J. Orschiedt/H. Meller (eds.), Mesolithic burials – rites, symbols, and social organization of early postglacial societies (Halle/Saale 2016) 793–808.

Pesonen 1996: P. Pesonen, Rääkkylan joutsenet ja muita kampakeramiikan linnunkuvia. In: H. Ranta (ed.), Kentältä poimittua 3. Kirjotelmia arkeologian alalta. Museoviraston arkeologian osaston julkaisuja 6 (Helsinki 1996) 5–14.

Poikalainen/Ernits 2019: V. Poikalainen/E. Ernits, Rock carvings of Lake Onega II. The Besov Nos Region. Karetski and Peri localities (Tartu 2019).

Sidorov 1975: V. V. Sidorov, Geografiya neolita Podmoskovya. In: P. M. Kozhin/L. V. Koltsov/M. P. Zimina (eds.), Pamyatniki drevneyshey istorii Yevrazii (Moskva 1975) 94–101.

Sorokin/Hamakawa 2014: A. N. Sorokin/M. Hamakawa, Geoarkheologicheskiye obyekty Zabolotskogo torfyanika na territorii Yevropeyskoy Rossii. Izvestiya Irkutskogo gosudarstvennogo universiteta 10, 2014, 50–93.

Stokolos 1984: V. S. Stokolos, Staro-Nagayevskiy mogil'nik. In: N. Y. Merpert (ed.), Epokha medi Yugo-Vostochnoy Yevropy (Kuybyshev 1984) 22–42.

Tõrv 2016: M. Tõrv, Persistent Practices. A Multi-Disciplinary Study of Hunter-Gatherer Mortuary Remains from c. 6500–2600 cal BC, Estonia. Dissertationes Archaeologiae Universitatis Tartuensis 5 (Tartu 2016).

Tsvetkova 1969: I. K. Tsvetkova, Urkasheniya i skulptura iz neoliticheskogo poseleniya Chornaya Gora. In: V. P. Levashova (ed.), Ekspeditsii Gosudarstvennogo Istoricheskogo muzeya (Moskva 1969) 25–38.

Volkova 2019: E. V. Volkova, Otrazheniye v goncharnykh traditsiyah kontaktov fatyanovsko-balanovskogo i pozdnevolosovskogo naseleniya. Samarskiy Nauchniy Vestnik 8, 2(27), 2019, 129–136.

Zamyatnin 1948: S. N. Zamyatnin, Miniatyurnye kremnyoviye skulptury v neolite Severo-Vostochnoy Yevropy. Sovetskaya Arkheologiya X, 1948, 85–123.

Zhulnikov 2002: A. M. Zhulnikov, O nakhodke v Pribelomorye figurnogo topora-molota s obukhom v vide golovy medvedya. Tverskoy Arkheologicheskiy Sbornik 5, 2002, 439–441.

Zhulnikov 2006: A. M. Zhulnikov, Petroglify Karelii. Obraz mira i miry obrazov (Petrozavodsk 2006).

Zhulnikov 2010: A. M. Zhulnikov, O khronologii naskalnykh izobrazheniy Belogo morya. Uralskiy Istoricheskiy Vestnik 1(26), 2010, 62–69.

Zhulnikov 2012: A. M. Zhulnikov, K voprosu o khronologii i funktsii skulpturnykh navershiy is kamnya. In: A. V. Suvorov (ed.), Istoriya i arkheologiya Russkogo Severa: sbornik materialov nauchnoy konferentsii, posvyaschennoy 60-letiyu so dnya rozhdeniya N. V. Guslistova (Vologda 2012) 69–72.

Zhulnikov/Kashina 2010: A. M. Zhulnikov/E. A. Kashina, Obraz ptitsy v iskusstve neolita-eneolita lesnoy zony Vostochnoy Yevropy. Rossiyskaya arkheologiya 2, 2010, 5–17.

Dr. Ekaterina A. Kashina
Department of Archaeology
State Historical Museum
Moscow
Russia
eakashina@mail.ru (corresponding author)

PhD student Anastasia A. Khramtsova
ROOTS Young Academy
Graduate School "Human Development in Landscapes"
Christian-Albrechts-University
Kiel
Germany

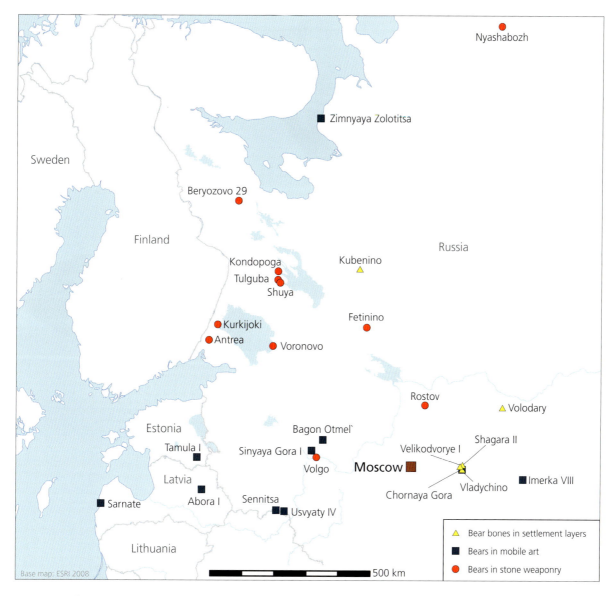

Fig. 1. Distribution of sites mentioned in the text with different categories of items connected with the bear (map GIS department, ZBSA).

Fig. 2. Carved bear images in portable art. 1; 3: Usvyaty IV, Pskov region; 2: Sārnate, Latvia; 4: Abora I, Latvia; 5: Imerka I, Republic of Mordovia; 6: Chornaya Gora, Ryazan region; 7: Tamula I, Estonia (photos E. A. Kashina [1–4; 7], A. Korolev [5], A. Macāne [6]). Objects 1 and 3 © The State Hermitage Museum, St. Petersburg (Russia).

Fig. 3. Flint bear images in portable art. 1: Zimnyaya Zolotitsa, Arkhangelsk region; 2: Bagon Otmel', Tver region; 3: Sinyaya Gora 1, Tver region (photos I. Seden'kov, State Historical Museum, Moscow [1–2], E. A. Kashina [3]).

Fig. 4. Axes with bear heads. 1: Nyashabozh, Komi Republic; 2: Volgo, Tver region; 3: Beryozovo 29, Republic of Karelia (photos I. Seden'kov, State Historical Museum, Moscow [1–2], M. Shakhnovich [3]).

Fig. 5. Bear canines and baculum bones found in settlement layers. 1, 4: Volodary, Nizhniy Novgorod region; 2–3, 5: Velikodvorye I, Moscow region; 6–7: Shagara II, Ryazan region (photos I. Seden'kov, State Historical Museum, Moscow).

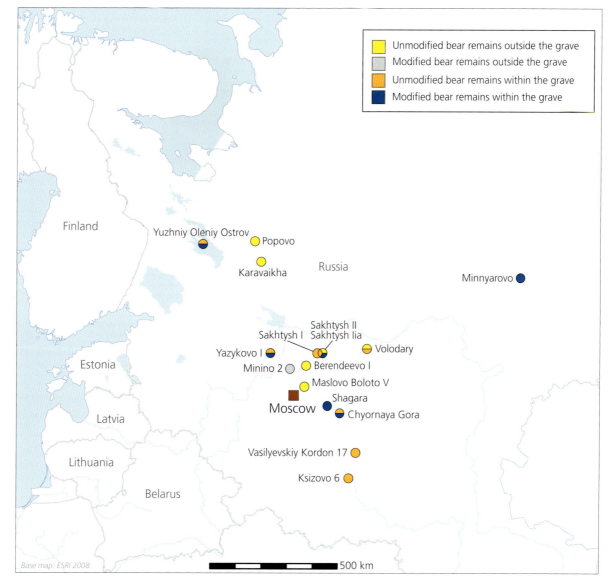

Fig. 6. Early to Middle Holocene burial sites with bear remains within (map GIS department, ZBSA).

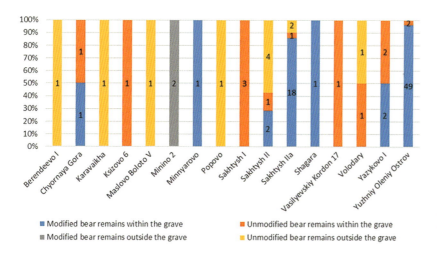

Fig. 7. Distribution of bear remains within the cemeteries (number of items).

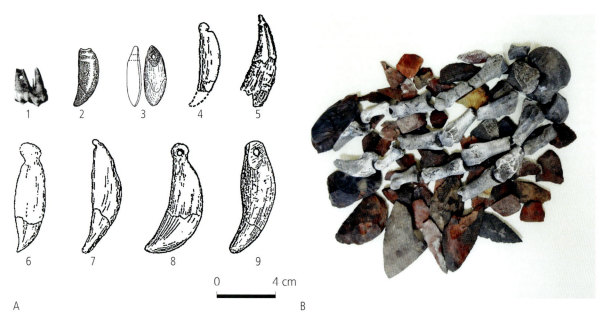

Fig. 8. a: Various methods of bear tooth pendant fastening (after Gurina 1956; Kazakov 2011; Kostyleva/Utkin 2010) – 1: Sakhtysh IIa, grave 7b; 2: Minnyarovo; 3: Sakhtysh IIa, grave 64; 4: Yuzhniy Oleniy Ostrov (YOO), grave 9; 5: YOO, grave 46; 6: YOO, grave 4; 7: YOO, grave 151; 8: YOO, grave 114; 9: YOO, grave 100; b: The "ritual hoard" 10, Sakhtysh II, Ivanovo region, in form of the installation exhibited at the Archaeological Museum of the Ivanovo State University (photo A. A. Khramtsova).

Chapter 6

Bears in archaeo(zoo)logical, focused analysis (northern Europe)

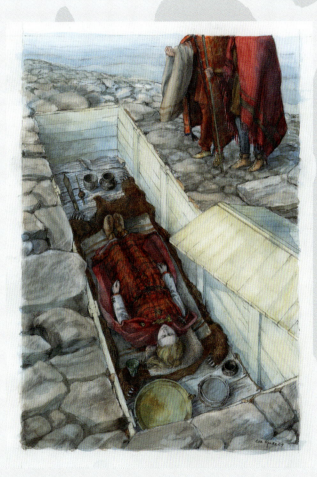

The burial of the Krosshaug petty queen, with the deceased placed on a bear skin, from the Migration Period of southwestern Norwegian Rogaland (see Grimm, Bear skin burials, this volume; image © E. Gjerde, The Museum of Archaeology, University of Stavanger, Norway).

The White One: How to frame the narrative of the world's oldest intact polar bear skeleton, specimen S10673 from Finnøy, southwestern Norway, in a museum display

By Kristin Armstrong Oma and Elna Siv Kristoffersen

Keywords: Polar bear, animal remains, museum display, museum studies, Norway

Abstract: Once upon a time, in the white wilderness, a white bear lumbered along close to the coastline. Twelve thousand four hundred years later he was found under the floor of a laundry room. His remains were excavated and brought to the nearest museum, at that time called The Museum of Archaeology in Stavanger. This article discusses the various narratives that surround the polar bear, ranging from what can be glimpsed of his life to his role in the scientific narrative, the story of how he was discovered and excavated, and how the public has engaged with him. Finally, the article discusses the ethics of exhibiting human and animal remains and how museums can be arenas for reflective practices. The role of a museum is to disseminate the multi-layered story of this polar bear; a story that starts with a once living, breathing creature that roamed the icy landscape at the decline of the last glacial period, and the ensuing biography of his skeletal remains – as well as their scientific value. Here, we unpack these stories and examine them in the light of museum practices, both regarding dissemination strategies and collection strategies. How are historically-embedded, ontological rationales active in steering the slant of the story as regards how museums organise their collections and make the collections come alive by creating narratives for them? And how do the different stories, related to the polar bear itself, combine into a charismatic story?

Introduction – three polar bear stories

It could begin like this: Once upon a time, in the white wilderness, a white bear lumbered along. He was large and looming, but the vastness of the snowy landscape left him barely discernible. The wind whistled and howled, and the icy whiteness on the ground was so white it glimmered as though bright blue. The desolate landscape did not hold many beings. Yet. The white bear felt a change was in the air; there was an unknown smell of an ancient fear – fire. But also something even stranger; far away, a speck of yellow, movement. His hackles up, he walked backwards and turned towards the edge of the ice and slipped silently into the big blue.

Or it could begin like this: In the summer of 1976, Sverre and Reidun Asheim started digging out the foundation for their new basement on their property on the island of Finnøy, in the county of Rogaland in southwestern Norway (Fig. 1). Strange, large bones emerged, yellow with age, larger than human and even cattle bones (although they were deemed initially to be bovine remains). The bones were kept in a box for some years until, as chance would have it, a boy visited a Stone Age excavation hosted by the Museum of Archaeology and suggested to the archaeologist that these bones might be of interest to the museum. Geologist Hanne Thomsen, one of the excavators and a researcher who was to be involved with the project, was alerted to the find and was eventually able to carefully excavate the remaining bones and bring them all to the museum (KRISTOFFERSEN 2017).

Most often, however, the story begins like this: *Ursus maritimus* Phipps is a subspecies of the taxon *Ursus* and is a large, white bear, singularly adapted to life in the Polar regions. It is currently classified as a vulnerable species, due to climate change (MOLNÁR et al. 2020). The oldest known almost complete, well-preserved specimen was found in southwestern Norway in 1976 and was a male, specimen S10673 (BLYSTAD et al. 1983). He has come to contribute to the understanding of the development of the polar bear and is mentioned in various articles on the development of the species (e.g. INGÓLFFSSON/WIIG 2008; BACHMANN et al. 2013; WIIG et al. 2015). An almost complete skeletal polar bear is an exceptionally rare find, since polar bears live and die on pack ice and sink to the bottom of the sea when they die, from whence their bones are hardly ever retrieved (BACHMANN et al. 2013). The skeleton of this bear is thus an extraordinary find and a very important missing link in polar bear research. The find site is, however, far from what would, today, be a natural habitat for a polar bear. Situated in southwestern Norway, Finnøy is one of the outer islands in the Boknafjord in Ryfylke and is currently in a temperate zone where temperatures seldom drop beneath 0° Celsius in winter. The island is part of Stavanger municipality, with a size of 104 km². Currently, it has some 3,000 residents, with Judaberg its only town.

All of these narratives are equally relevant and reveal aspects of the near-complete Late Weichselian polar bear (*Ursus maritimus* Phipps) skeleton, which has been dated to 12,400 BP (KRISTOFFERSEN 2017, 15–17); each story is singularly accurate and relevant within its context. In chronological order, they present particular points of time that are embedded in the remains of this polar bear. From the elemental beginnings of the polar bear immersed in the ice and sea of the Weichselian period to the elaborate scientific description, they present a brief history with a huge time-depth. For the polar bear, at species level there is a development from a perfectly suitable environment with plentiful space, sea, ice, food, to being driven far north, pushed to the last frontiers of ice, threatened by extinction and in dire straits. Also, they reflect the presence of human beings in a specific place but over a long period of time, with their wide diversity in human experience, knowledge of the world and the language to describe it; humans who range from a few human pioneers following the retreating edge of the ice to a well-developed civilisation, represented by the couple who wanted to expand their living quarters and the scientists who studied the remains of the bear.

The role of a museum is to disseminate a fundamentally multi-layered story. This story encompasses different layers of material studies, and includes the bear remains and their geological, geographical and archaeological contexts. The story originated with the bear himself – a once living, breathing creature that roamed the icy landscape at the decline of the last glacial period – and the ensuing biography of his skeletal remains – as well as their scientific value. In this article, we will unpack these stories and critically examine them in light of the museum practice of the Museum of Archaeology at the University of Stavanger. Our study includes dissemination strategies and collection strategies. The main question we wish to raise is: How are historically-embedded, ontological rationales active in steering the slant of the story as regards how museums organise their collections and bring them to life by creating narratives for them? And how do the different stories, and the polar bear itself, combine into a charismatic story?

The science story I. Species specificities: A highly specialised seal hunter on sea ice

Living in a polar environment requires the development of specialised behaviours and biological traits. Polar bears are well protected from cold temperatures thanks to their dense, multi-layered coat of which every hair is translucent, with no pigment. Their dark skin and thick layer of fat, the fur on their feet, their small, round ears and the short tail are engineered to retain body heat, and the long, narrow skull warms the air as the bear breathes. They are excellent hunters because they have enormous strength, white fur that blends seamlessly with their surroundings, a long muzzle that gives them a superb sense of smell, long claws that give them a good grip on their prey, broad feet that allow them to walk on thin ice, and they are fast, excellent swimmers (Massimi 2012, 30–39).

Polar bears are the most important predators in the arctic ecosystem. They are amongst the largest carnivorous mammals in the world and can survive by going without eating for long periods if they must. In comparison, the brown bear only fasts when it goes into hibernation in winter. Polar bears have four teats, unlike the brown bear's six, and the females have smaller litters. They have been regarded as the friendliest of all bears – sociable, curious, and endlessly patient (Massimi 2012, 30–39).

The science story II. A polar bear skeleton from Late Weichselian marine sediments

An almost complete, well-preserved polar bear skeleton, *Ursus maritimus* Phipps, was found in 1983 in Late Weichselian marine sediments on Finnøy, southwestern Norway (Blystad et al. 1983). The bones belonged to a large, 26–30-year-old male, ^{14}C-dated to 10,660 ± 80 years BP. In 2017, the date was collated using OxCal 4.3, and the more precise date is now 12,050–12,600,[1] i.e. *c.* 12,400 BP. The find has unequivocally shown that polar bears were present in the high arctic marine environment in southwestern Norway at the very end of the Ice Age.

Only bones from the fore limbs, a few vertebrae and most of the toes and fingers were missing. The polar bear was identified as a male through a completely preserved penis bone. Judging by the size and development of the bones, he was large and fairly old. A preliminary study of the structure of one of the canine teeth indicated an age between 26 and 30 years, which is quite high for a polar bear.

Fossil finds of polar bears in former glaciated areas are extremely rare. This can, at least partly, be attributed to the fact that polar bears for the most part live and die on pack ice, where upon death their remains slip into the ocean and become sedimented in the sea bed at great depths and in inaccessible Arctic areas, which makes their preservation in terrestrial sediments exceptional (Ingólfsson/Wiig 2008; Bachmann et al. 2013). When polar bear fossils are located, often only single bones or bone fragments have been found. The nearly complete and well-preserved skeleton from Finnøy must, therefore, be regarded as sensational.

The find was first published in a note in *Norsk Geologisk Tidsskrift* (Blystad et al. 1983), later in two museum catalogues (Thomsen 1985; Kristoffersen 2017) and, due to how rare and important it is to the understanding of the development of the polar bear, it is mentioned in various articles on this topic (e.g. Ingolffsson/Wiig 2008; Bachmann et al. 2013; Wiig et al. 2015). The bear has thus become a very important specimen for research on the evolution of polar bears.

---

[1] Thanks to Professor Jan Mangerud, Department of Earth Science & Bjerknes Centre for Climate Research, for carrying out the new correlation together with Asbjørn Simonsen.

The science story III. The role of the Finnøy bear in understanding the evolution of bears

The ancestors of bears were carnivorous predators, but at some point they transitioned to an omnivorous diet. This change can be read from the evolution of their dental characteristics; rounded cheek teeth are found in omnivores and herbivores rather than the sharp cheek teeth found in carnivores. The extinct cave bears mostly ate plants and had rounded cheek teeth. In contrast, polar bears are characterised by sharp cheek teeth due to their specialised adaptation to a mostly carnivorous diet, although they have also been observed feeding on seaweed in times of need (Balto 2020). But rather than having the dental characteristics of predators, polar bears have teeth shaped like those of other bears, only with sharper edges. Such morphological changes can happen relatively rapidly, and fossil specimens from transitional phases in periods when such radical changes happen are rarely found. The polar bear from Finnøy has turned out to be just such a transitional specimen. His cheek teeth are more primitive, rounded and gnarly compared to those of present-day polar bears. This demonstrates how the polar bear has changed noticeably over the last 13,000 years.

The evolutionary histories of the brown bear and the polar bear are interlinked, and they successfully produce fertile offspring (Ingólfsson/Wiig 2008, 455). Dental characteristics such as rounded or sharp cheek teeth indicate whether they are one or the other. However, in times of upheaval and climate change their interbreeding has assured the survival of the species and also led to morphological changes through time, as several changes in climate and thus living conditions drove brown bears and polar bears either together or apart.

The Finnøy bear lent his mandible to the research on the Late Pleistocene polar bear fossil found on the Poolepynten on Svalbard, the oldest remains of a polar bear ever found (130,000–115,000 BP; e.g. Ingolfsson/Wiig 2008; Bachmann et al. 2013; Wiig et al. 2015). Measurements of the Finnøy mandible were, together with a Danish find from Asdal, in Vendsyssel on the northernmost part of Jutland (Aaris-Sørensen/Petersen 1984), compared with the one from Poolepynten and were crucial to its sex determination as a male. Having its penis bone intact, there is no doubt about the sex of the Finnøy bear. The comparative study of him as well as other available subfossil polar bear remains also revealed that there was no significant change in the size of polar bears during the Late Quaternary (Ingólfsson/Wiig 2008, 455, 458, 461).

The polar bear has become a symbol of the debates concerning the effect of climate change in the Arctic, and these rare, ancient finds play a significant role in this regard. The Poolepynten bear, in particular, has contributed to the understanding of the consequences of the survival of this species, by comparative analyses with other polar bears and brown bears. New molecular methods have demonstrated a close kinship between these subspecies of bears as well as a large genetic variation within brown bears (Miller et al. 2012). The so-called ABC bears that have lived in relative isolation in Alaska (a genetically isolated population of brown bears from the Admiralty, Baranof, and Chichagof islands in Alaska's Alexander Archipelago) are particularly similar to polar bears. In fact, the Poolepynten bear is equally similar to the ABC bears as regards current polar bears, even though his appearance and jaw were similar to that of a polar bear. These studies have provided fresh insights into the development of species that live in extreme habitats and their adaptation to changing climates and can therefore be used for future projections of the fate of polar bears by modelling different climate change scenarios (e.g. Molnár et al. 2020).

Analyses of ancient DNA, inherited through both patrilinear and matrilinear descent, have indicated the age of divergence of the polar bear as a species, with an initial split around 4–5 million years ago, and "since their divergence from brown bears, polar bears have embarked on their own evolutionary pathway and developed their own unique genomic signatures" (Miller et al. 2012). Having said that, there is evidence for a successful admixture between species at several points in time, at both 600,000 and 3–400,000 years ago at the least. Only successful breeding between a polar bear female and a male brown bear can explain these divergent splits between the species. Therefore, it seems that a hybridisation between the polar bear and the brown bear has happened on several occasions during the Pleistocene, and the latest such significant event probably took place around 150,000 years ago (Edwards et al. 2011). Large-scale climatic changes during the lifespan of polar bears have led to shared habitats and thus the potential for interbreeding and hybridisation in periods with warmer climatic conditions. This is also happening in our time. The polar bear has evolved to singularly cope with an extreme ecosystem and, as such, its survival as a species is not a given unless it is capable of adaptations in the future. The study of how a species has adapted to climatic changes in the past can go some way towards predicting to what extent it is suited to adapt to future climate changes.

In conclusion, recent research demonstrates that, as a species, the polar bear is significantly older than previously thought and that subsequent, frequent hybridisations between the species, including genetic exchange, have happened in several periods when the planet has undergone dramatic climate change, up until recently. Possibly, these hybridisations have ensured the survival of the polar bear as a species during previous climatic changes. The challenge for present-day polar bears might be the rapid pace of the current climate change.

The find story. How the polar bear emerged from under the laundry room

The most important finds often come to light by coincidences. Several random circumstances led to the discovery of the Finnøy polar bear skeleton. The find came to the attention of the scientific staff at the then Museum of Archaeology in Stavanger when an astute and curious boy came to visit a Stone Age excavation in the summer of 1982 in Sandvika by the Gandsfjord in Sandnes municipality. The boy told the excavation staff that "we have some old bones in the basement". During a stressful excavation, most archaeologists would have responded with a polite "Oh, how interesting!" but left it at that. However, one of the field workers was geologist Hanne Thomsen, and her response was: "Do you want me to come and have a look?". And fortunately, she did.

The find story had already started in the summer of 1976, when Sverre and Reidun Asheim dug out the foundation for their house at Judaberg on the island of Finnøy, about 100 m behind the old dairy. When the basement floor had been levelled, they dug a wire into the floor and saw several large bones in the clay. The well-preserved bones were parts of the lower jaw and of the left upper jaw, ribs, two vertebrae and bones from the fore and hind limbs of a large animal. Uncertain of which kind of animal the bones could stem from, they held on to them and put them in a box originally used for margarine. The box with the bones was subsequently gifted to the boy's father, Arvid Magnussen, a family friend with an interest in such things. Eventually the box was placed in the basement, where the boy had his room. When Hanne Thomsen saw them, she immediately realised their potential scientific value and that the bones were worth a closer study. The boy's father brought them to the museum, where it was quickly established that the bones were parts of the skeleton of a polar bear – and that it would have to be very old. After all, polar bears had not lived in the region since the ice retreated, 12,000 years ago.

Two Quaternary geologists and one botanist hurried to Finnøy for a closer inspection of the find site. The owners of the house informed them that there were more bones, but they were currently embedded in the clay under the floor of the laundry room in the basement. The owners permitted the floor to be broken open and donated all bones that were found to the museum.

In the autumn of 1982, an excavation team went to Judaberg to try to retrieve the rest of the polar bear. The team was made up of botanist Asbjørn Simonsen, geologists Per Blystad and Hanne Thomsen from the Museum of Archaeology in Stavanger, as well as zoologist Rolf Lie from the Museum of Zoology in Bergen. Arvid Magnussen contributed as a photographer and local contact.

Cuddly polar bear teddies wearing a shirt with "Finn" printed on the chest are also on sale in the museum shop (Fig. 7). Our impressions in the following are based on reactions and comments from the public that have been passed on from daily encounters between our guides[3] and the public.

Our intention with this exhibition has been that the Finnøy bear should stand out as a symbol of the effect of climate change and therefore not simply reveal the region's pre-human past; his remains speak to the fragile present circumstances of polar bears and the wider environment, as well as their precarious future and imminent threat of extinction. According to our guides, he succeeds in his mission. It makes an impression on the visitors that he struggled to survive in the same way as the polar bear does today, thousands of years later. He also seems to manage to communicate the changes the landscape went through after the Ice Age, by explaining why the sea has turned into dry land. Large-scale climatic and landscape changes may be better understood through the authenticity of an individual you actually can see and who lived through them. In the same way, as an individual, he seems to contribute to a better understanding of deep time. However, the public meet him in various ways and recognise different aspects of him and his story. Children are perceptive and often notice things differently. A little schoolboy wondered why several bones had writing on them (the museum's numbers), and several children associate the spine of the skeleton with marshmallows. Another child found him "a bit scary".

It is our impression that the bear(s)[4] has been able to reach out to people in his skeletal form in the same way as the Viste boy. Both exhibits share an old age, and they are both authentic individuals. These are qualities that seem to enable them to reach out towards people in various age groups – from small children to grown-ups and often both together as families. It is of importance that the bear was already able to do so from his old place in the café area. Even though he was among people at this location, and sort of partook in activities such as concerts and lectures, the narrow showcase was randomly and not especially favourably situated. This was also the case with the Viste boy, in his old and dusty showcase and its random placement in the middle of the old exhibition. These are facts that might suggest that it is the skeletons in themselves, animal or human, with an agency of their own, that reach out to people and not so much the way they are exhibited. However, the beautiful new exhibit room with the polar bear skeleton on white ice sheet and blue sea bathed in northern light, certainly adds to the visitors' experience and gives them space for reflection. Even though the Finnøy bear is now more isolated, he has a whole room to himself, and we regard his new environment as a more respectful way to exhibit him.

The bear and his story are also well known locally on Finnøy Island where he was discovered. Apart from being old and rare, the history of how he was found has fascinated many, as well as the many coincidences that saved him for the future. People from the island often come to the museum to see him and, in connection with the opening of the new exhibition, the history of the discovery was repeated in the local papers in interviews with the owners of the house where he was found. The bear even entered into the Atlas Obscura[5] with the headline "Now on Display: An Ice-Age Polar Bear Skeleton Found in a Laundry Room. The startling story behind a new exhibit at a Norwegian archaeology museum". Finally, he has become a comfort to tourists who expect to see polar bears in the streets of Norway (an expectation that some still have).

---

3   The museum guides Ellen Hagen and Heidi Wevle have generously shared their experience from working in the exhibition and meeting the audience for many years and have followed the Finnøy bear from his old showcase to his new exhibit room.
4   Even though we refer to the Finnøy bear as one individual, it is important to keep in mind that the exhibit is in fact a composite of two polar bears.
5   https://www.atlasobscura.com/articles/ice-age-polar-bear-laundry-room-norway.

The charisma story. How a prehistoric bear touches us today

Charismatic and looming in popular culture, polar bears seem to hold endless fascination for humans, and have become a "flagship species" since the beginning of the 21st century – a popular, charismatic species that is currently a rallying point that serves to promote conservation and activism (Flinterud 2013, 204; Lorimer 2007, 919; 2015, 139). The popularity of the polar bear as a charismatic species in recent decades testifies to its agency as well as its ability to evoke something larger than itself, and it becomes entangled in relationships with the scientific community as well as with the public. A prehistoric animal that has existed in various contexts, in the past as a living animal and in the present as a skeleton – albeit as a skeleton that retains the connection to the living animal it once was – the Finnøy bear has, after his death, become entangled in the lives of those who found him under their basement floor, and those who thereafter excavated his remains and brought him to the museum. When found, he entered history anew and re-engaged in social relations of other ages; his biography attained a new layer. Through reincarnations and recontextualisation, an object accumulates an extended biography – "beyond different systems of understanding" (Joy 2009, 541, with references to Gillings/Pollard 1999; MacGregor 1999; Moreland 1999), and the same applies to a prehistoric animal skeleton.

Further, the Finnøy bear engaged the scientists who worked on cataloguing his skeletal elements, who dated him and thus discovered his rarity and subsequently special status in the scientific community. Following this, he worked on the minds and imaginations of those that planned and executed the exhibitions where he is the main subject on display. His legacy continues to weave him into the stories and involvements of the audience who experience him in the exhibition, as a mixture of aesthetics, species specificities, find story and the plight of polar bears today contribute to bring to mind issues such as the freezing Arctic and climate change, the sixth extinction, and the cuteness of polar bear cubs. All of this goes to show that this polar bear has an uncanny ability to enter new relationships and stories, again and again.

Like monuments, animals accumulate biographies from the changes of the world around them and can accumulate lengthy biographies generated by passing through hands and changing spatial contexts (Gillings/Pollard 1999, 179–180). Gavin MacGregor states that prehistoric artefacts "may include a number of resurrections relating to their movement between different ages or different systems of understanding" and stresses the importance of a sensory approach to the study of extended biographies (MacGregor 1999, 258).

Myths, stories and the naming of polar bears in circumpolar cultures testify to their ability to evoke wonder in humans and to the longevity of this fascination. The Inuits' beliefs were based on animism, i.e. that the world and its components, as well as objects, are endowed with spiritual properties and a soul, and the polar bear, named Nanuq, was a powerful animal (Dijkhuizen 2020).

As well as being a "flagship species", the polar bear is currently also perceived as "a signature Arctic species" (Miller et al. 2012, 14295) and has become embroiled in a particular narrative trajectory; that of the threat to biodiversity from climate change. "Because of the uncertain long-term status of the polar bear and its charismatic nature, this species has become a focal point for discussions concerning the impact of global climate change on biodiversity" (Miller et al. 2012, 14296). Dying polar bears have become synonymous with global warming, and polar bears have shifted from iconic symbols of extreme environmental adaptation prior to 1990, to clarion calls that blend the sixth extinction (Kolbert 2014) with climate change (Bachmann et al. 2013; Miller et al. 2012), and function as rallying points for conservationists. For example, the World Wildlife Fund (WWF) frequently uses breathtaking photos of majestic and iconic polar bears to market their campaigns. This has led to critique along the lines of (ab)using charismatic and beautiful animals such as polar bears and tigers to trigger emotional responses, but also to the detrimental effect of other, less charismatic animals, equally in need of protection and conservation (e.g. Lamb 2020).

At the Museum of Archaeology, University of Stavanger, human remains are, as previously described, exhibited in a separate room entitled "Meet the Humans" (Kristoffersen/Armstrong Oma 2016). Rather than exhibiting the remains as scientific data, they are presented as once-living humans, and discussions on ethics leading up to the exhibitions served as a means to safeguard a respectful frame for the human remains. In several cases, this included an exact – as far as possible – representation of their find context, alongside the required micro-environments to safeguard preservation within the exhibition cases, which made a noticeable dent in the budget of the exhibition. The room is furnished not as a tomb but with a wall-paper that looks like a soil profile (stratigraphy) to give a feel for being underground, inside the archaeological context from whence the bones have been retrieved. The choice of exhibiting the remains in a respectful manner was made in tandem with an explicit goal of providing the audience with a frame of reference for considerations of an existential nature, such as what does it mean to be human now, compared to in the past? How do different cultures deal with death and loss?

The polar bear is currently exhibited in a separate space and the environment he lived in has been abstractly recreated by hanging his bones in his true size on top of broken glass, which looks like ice, with white walls illuminated by blueish-green lights and the sound of the whistling wind over the icy tundra. We safeguard the presentation of his individual personality and presence – his agency – by referring to him as "he" and by exhibiting him as an individual rather than as purely scientific data, although the exhibition texts utilise both of these aspects of the bones. We may not be able to claim that we exhibit him as a subject rather than an object, but the locals have given him the nickname Finn, since he was discovered on the island of Finnøy.

As such, the polar bear is, in our slightly rewritten words of Howard Williams, "afforded prominent individual personality and presence in the museum" (Williams 2019). His original life-world is given import and his biography is presented as something that matters, not just as an educational tool but as an all-pervasive part of a fundamentally foreign past, when this part of Norway was on the cusp between the last glaciation and the first Stone Age pioneers.

However, we are culpable of the omission of the fact that, as previously noted, because the reconstructed polar bear skeleton was not fully complete when excavated, bones from another polar bear from Bergen Museum have been added to supplement missing bones. Thus, he is in reality a composite of different individuals, a fact that is currently not relayed to the public.

Further, the polar bear is not the only bear on display in the museum. In the "Meet the Humans" exhibition of human remains, cremation graves from the Late Roman Period (CE 200–400) and Migration Period (CE 400–550) that include bear claws are exhibited. These bear remains have not been afforded the same individual personality and presence in the exhibition set-up, rather they are displayed as grave goods and a result of a human funerary tradition. However, their agency and species-specific bear-ness is disseminated by highlighting the bear's potential cosmological role as animals that hibernate and emerge with young cubs in the spring, thus becoming a symbol of birth and regrowth following death – or in the case of bears, rebirth as reemergence following hibernation.

As part of the dissemination of the polar bear, it was suggested that the museum would purchase a taxidermied polar bear to display in another part of the museum away from the exhibition hall, near the café and the reception area. The rationale was that a "cuddly" stuffed bear would appeal to the younger audience and create more interest in the skeletal bear in the exhibition. Further, such a bear would be educational and help the audience imagine what the skeletal bear would have looked like. The purchase never happened because the museum leadership argued that it is unethical to partake in the trade of taxidermied animals threatened by extinction. This is mentioned to demonstrate how the polar bear evoked multi-faceted ethical discussions that engaged museum workers in charge of making the exhibition, those in charge of marketing and public communication, and the museum leaders.

Beyond these observations, the polar bear elicits a response by his very otherness, as the remnant of a once-living creature he serves the same purpose as that of Yorick's skull in Hamlet – a *memento mori*. Polar bear Iorek Byrnison in Philip Pullman's beloved Northern Lights trilogy, His Dark Materials, might be inspired by the following line in Shakespeare's Hamlet, in which Hamlet reminisces about Yorick: "Alas, poor Yorick! I knew him, Horatio […] he hath borne me on his back a thousand times" – just like Iorek the *panserbjørn* carries Lyra Silvertongue on his back. These and other associations might play out in the visitor's inner eye and remind them of how all things must pass. Given the stark outlook for polar bears as a species, this is what they might all become, skeletons in a museum. Stories matter, and a story is also shaped by the way we create displays of past objects and afford them space to be experienced as past beings. The construction of stories by means of a museum display must also play a part in the ethical considerations that are so vital when creating exhibitions.

Exhibition as a reflective practice – creating an arena to contemplate the big questions

In conclusion, we return to Williams (2019) and his blog post dedicated to the discussion of whether or not it is ethical to exhibit animals: "Neither animal nor human remains should be displayed by default. Neither animal nor human remains should be removed from display as a blanket policy. The ethics lie not in their presence or absence, but in the character and efficacy of their deployment in relation to envisioned goals of particular exhibits." In our experience, the ethics rest in the reflection by the museum workers when creating exhibition visions and strategies. By bringing ethics into the process, the goals for the exhibitions are challenged and, as a result, the exhibitions become more grounded in both the agency, context and species specificities of those that are on display, whether they be human or animal.

In the case of the Finnøy bear, we have concluded that, by exhibiting him, we are allowing his remains to continue to engage with the public. Repatriation is a moot point, since the bear is exhibited no more than 30 km from where he perished. By his display, the museum can disseminate scientific stories, the find story, and narrative stories as well as charisma stories. These stories affect all of us – and they will shape our future. Similar to Howard Williams' visit to Weston Park Museum in Sheffield, England, UK, where the polar bear is "a named animal (that) becomes an animal-ancestor for the museum – known by generations of visitors and familiar to them" (Williams 2019), the Finnøy bear is somewhat of a mascot for the Museum of Archaeology. Ultimately, the exhibition becomes a site within the museum to reflect on themes such as the majestic stature of male polar bears, their remarkable adaptation to an extreme environment, their resilience to global changes through thousands of years, and how polar bears might become extinct forever in the near future (Molnár et al. 2020). By affording him a prominent place that unfolds his presence and individual personality, we also open a space for the affection of the public and for the desire to return and visit him again and again, across the generations. He can thus become a stepping stone into experiencing other individuals in the museum, and our job as curators and custodians is to make sure that the agency afforded to the Finnøy bear is extended to the remains of other animals.

Acknowledgements

We are grateful to the editors of this volume for their remarks, and to Howard Williams for his thoughtful and inspiring comments.

Bibliography

Aaris-Sørensen/Petersen 1984: K. Aaris-Sørensen/K. S. Petersen, A Late Weichselian find of a polar bear (*Ursus maritimus*) from Denmark and reflections on the paleoenvironment. Boreas 13, 1984, 29–33.

Armstrong Oma 2018a: K. Armstrong Oma, The sheep people. The ontology of making lives, building homes and forging herds in Early Bronze Age Norway (Sheffield 2018).

Armstrong Oma 2018b: K. Armstrong Oma, Transformative theft of past and present: The human-horse bond reflected in the Viking Period Gausel bridle. In: M. Vedeler/I. M. Røstad/E. S. Kristoffersen/Z. T. Glørstad (eds.), Charismatic objects – from Roman times to the Middle Ages (Oslo 2018) 125–146. https://doi.org/10.23865/noasp.51.

Bachmann et al. 2013: L. Bachmann/J. Aars/Ø. Wiig, Hvor gammel er Isbjørnen som art? Naturen 3, 2013, 115–120.

Balto 2020: A. K. Balto, The polar bear's diet (2020). www.npolar.no/en/newsarticle/the-polar-bears-diet/.

Bang-Andersen 2017: S. Bang-Andersen, The first one thousand years. Human colonization and differentiated landscape use in South-Western Norway, 10,000-9000 BP. In: H. P. Blankholm (ed.), Early economy and settlement in Northern Europe. Pioneering, research use, coping with change (Sheffield 2017) 275-310.

Blystad et al. 1983: P. Blystad/H. Thomsen/A. Simonsen/R. W. Lie, Find of a nearly complete Late Weichselian polar bear skeleton, *Ursus Maritimus* Phipps, at Finnøy, southwestern Norway: a preliminarily report. Norsk Geologisk Tidsskrift 63, 1983, 193–197.

Dijkhuizen 2020: B. Dijkhuizen, The mythical legends of the Inuit (2020). https://vocal.media/wander/the-mythical-legends-of-the-inuit.

Edwards et al. 2011: C. J. Edwards/M. A. Suchard/P. Lemey/J. J. Welch/I. Barnes/T. L. Fulton/R. Barnett/T. C. O'Connell/P. Coxon/N. Monaghan/C. E. Valdiosera/E. D. Lorenzen/E. Willerslev/G. F. Baryshnikov/A. Rambaut/M. G. Thomas/D. G. Bradley/B. Shapiro, Ancient Hybridization and an Irish Origin for the Modern Polar Bear Matriline. Current Biology 21, 2011, 1–8.

Flinterud 2013: G. Flinterud, Polar bear Knut and his blog. In: L. E. Thorsen/K. A. Rader/A. Dodd (eds.), Animals on display. The Creaturely in Museums, Zoos and Natural History (State College 2013) 192–213.

Gillings/Pollard 1999: M. Gillings/J. Pollard, Non-portable stone artefacts and contexts of meaning. The tale of Grey Wether (www.museums.ncl.ac.uk/Avebury/stone4.htm). World Archaeology 31(2), 1999, 179–193.

Ingólfsson/Wiig 2008: Ó. Ingólfsson/Ø. Wiig, Late Pleistocene fossil find in Svalbard: the oldest remains of a polar bear (*Ursus maritimus* Phipps, 1744) ever discovered. Polar Research 28, 2008, 455–462.

Joy 2009: J. Joy, Reinvigorating object biography. Reproducing the drama of object lives. World Archaeology 41(4), 2009, 540–556.

Kolbert 2014: E. Kolbert, The Sixth Extinction. An Unnatural History (London 2014).

Kristoffersen 2017: E. S. Kristoffersen (ed.), Isbjørnen fra Finnøy. Rogaland ved istidens slutt. AmS-Småtrykk 88 (Stavanger 2017).

Kristoffersen 2018: E. S. Kristoffersen, The charisma of extended biographies and aesthetics. Migration period cases. In: M. Vedeler/I. M. Røstad/E. S. Kristoffersen/Z. T. Glørstad (eds.), Charismatic objects – from Roman times to the Middle Ages (Oslo 2018) 57–71. https://doi.org/10.23865/noasp.51.

Lamb 2020: G. Lamb, What are charismatic species and why do they matter for conservation? (2020). https://medium.com/age-of-awareness/the-connection-people-looking-at-animals-looking-at-people-2eb306515854.

Lorimer 2007: J. Lorimer, Nonhuman charisma. Environment and planning D. Society and Space 25:5, 2007, 911–932.

Lorimer 2015: J. Lorimer, Wildlife in the Anthropocene. Conservation after Nature (Minneapolis 2015).

MacGregor 1999: G. MacGregor, Making sense of the past in the present. A sensory analysis of carved stone balls. World Archaeology 31(2), 1999, 258–271.

Massimi 2012: E. M. Massimi (ed.), Dyr i fare (Oslo 2012).

Miller et al. 2012: W. Miller/S. Schuster/A. J. Welch/A. Ratan/O. C. Bedoya-Reina/F. Zhao/H. L. Kim/R. C. Burhans/D. I. Drautz/N. E. Wittekind/L. P. Tomsho/E. Ibarra-Laclette/L. Herrera-Estrella/E. Peacock/S. Farley/G. K. Sage/K. Rode/M. Obbard/R. Montiel/L. Bachmann/Ó. Ingólfsson/J. Aars/T. Mailund/Ø. Wiig/S. L. Talbot/C. Lindqvist, Polar and brown bear genomes reveal ancient admixture and demographic footprints of past climate change. PNAS 109(36), 2012, 14295–14296. www.pnas.org/cgi/doi/10.1073/pnas.1210506109.

Molnár et al. 2020: P. K. Molnár/C. M. Bitz/M. M. Holland/J. E. Kay/S. R. Penk/S. C. Amstrup, Fasting season length sets temporal limits for global polar bear persistence. Nat. Clim. Chang. 10, 2020, 732–738. https://doi.org/10.1038/s41558-020-0818-9.

Mooallem 2014: J. Mooallem, Wild ones: A sometimes dismaying, weirdly reassuring story about looking at people looking at animals in America (New York 2014).

Moreland 1999: J. Moreland, The world(s) of the cross. World Archaeology 31(2), 1999, 194–213.

Thomsen 1985: H. Thomsen, Isbjørnen fra Finnøy. AmS-Småtrykk 15 (Stavanger 1985).

Wiig et al. 2015: Ø. Wiig/J. Aars/L. Bachmann/Ó. Ingólfsson, Verdens eldste isbjørnfossil og hva det har lært oss. Ottar 308(5), 2015, 49–53.

Williams 2019: H. Williams, Is it Ethical to Display Animals in Museums? Blogpost (2019). https://howardwilliamsblog.wordpress.com/2019/01/13/is-it-ethical-to-display-animals-in-museums/.

OTHER SOURCES

KRISTOFFERSEN/ARMSTRONG OMA 2016: E. S. Kristoffersen/K. Armstrong Oma, Møt menneske. Exhibition at the Museum of Archaeology, University of Stavanger.

BRYNDIS SNÆBJØRNSDOTTIR/WILSON: Bryndis Snæbjörnsdottir/M. Wilson, "nanoq: flat out and bluesom". Art project, 2001–2006. Series of photographs from Snæbjörnsdóttir/Wilson's survey of taxidermied polar bears in the British Isles. Collections of Reykjavík Art Museum, Iceland and Nevada Museum of Art, USA. https://snaebjornsdottirwilson.com/_dsf9635/.

Prof. Kristin Armstrong Oma
Museum of Archaeology
University of Stavanger
Stavanger
Norway
kristin.a.oma@uis.no

Prof. emer. Elna Siv Kristoffersen
Museum of Archaeology
University of Stavanger
Stavanger
Norway
siv.kristoffersen@uis.no

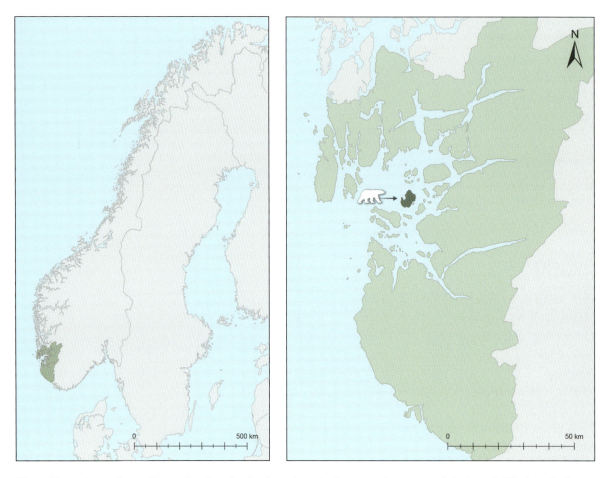

Fig. 1. Finnøy in southwest Norwegian Rogaland, where the polar bear remains came to light (map I. Tinglum Bøckman; based on © Kartverket).

Fig. 2. The Finnøy polar bear under excavation. Spine and rib bone (photo H. Thomsen, Museum of Archaeology, University of Stavanger).

Fig. 3. The Finnøy polar bear under excavation. Femur bone with remains of the stomach, with dissolved bones from seal, visible as a brownish layer above the bone (photo H. Thomsen, Museum of Archaeology, University of Stavanger).

Fig. 4. The Finnøy polar bear in his old place in the showcase to the left in the café area (photo Museum of Archaeology, University of Stavanger).

Fig. 5. The Finnøy polar bear in the new exhibition at the Museum of Archaeology, front view (photo A. G. Øvrelid, Muesum of Archaeology, University of Stavanger).

Fig. 6. The Finnøy polar bear in the exhibition at the Museum of Archaeology, side view (photo A. G. Øvrelid, Museum of Archaeology, University of Stavanger).

Fig. 7. The Finnøy polar bear as a cuddly toy with a T-shirt with his name, Finn (photo Museum of Archaeology, University of Stavanger).

The bear minimum. Reconsidering ursine remains and depictions at Pitted Ware culture (c. 3200–2300 BC) sites in Sweden[1]

By Tobias Lindström

Keywords: Sweden, Middle Neolithic, 4th millennium, 3rd millennium, predator, human-animal relations

Abstract: This paper discusses the remains and possible depictions of bears found at Middle Neolithic Pitted Ware culture (PWC) sites in Sweden in relation to ethnographic accounts of northern Eurasian bear ceremonies. The bear has been defined by some as a prominent animal in PWC cosmology, but bear remains occur at only eleven PWC sites, most commonly in the form of teeth or limb bones. A couple of zoomorphic figurines have tentatively been interpreted as depicting bears, but most of these have few distinguishing features and can best be described as generic. Even though the bear remains are comparatively rare inclusions at these sites, some of them seem to have been treated and deposited in "special" ways such as inside a clay pot or adjacent to hearths and cooking pits. Due to the varying treatment of bear remains in the investigated area, it is suggested that any special status of bears in the PWC cosmology must be argued for on a site-to-site basis rather than applied wholesale to the entire PWC complex.

Introduction

The hunting of bears and the ritual disposal of their bones seem to be practices of considerable antiquity. It has been suggested that the ritual treatment of bears was already being carried out in the Upper Palaeolithic, as is exemplified by finds of red ochre-stained bones from cave bears (*Ursus spelaeus*) found in Belgian caves (Germonpré/Hämäläinen 2007). A possible ritual use of bear remains has also been archaeologically identified at Bronze and Iron Age sites in different parts of Eurasia (e.g. Magnell/Iregren 2010; Kosintsev et al. 2018). In many northern Eurasian hunter-gatherer communities the bear is considered a particularly evocative and potent figure and is often associated with deities and spirits (e.g. Zolotarev 1937; Kimura 1999; Pedersen 2001; Rydving 2010; Westman Kuhmunen 2015). The bears' size and appearance, their ability to stand and walk on their hind legs (much like humans), and the danger they pose to those who cross their path are some of the most commonly cited reasons why bears are almost universally feared and admired. Sámi and Finnish noa names for the bear include "golden king of the forest", "keeper of the forest" or simply "forest" (see Piludu, this volume; Pentikäinen 2005, 9; Herva/Lahelma 2019, 80–81).

---

[1] This paper is dedicated to the memory of my grandmother, Vivi-Anne Dahlberg, who passed away on October 16, 2019.

Inquiries into bear remains found in prehistoric archaeological contexts frequently draw upon the spectacular bear ceremonies attested in ethnographic studies of northern Eurasian hunter-gatherer communities (e.g. Zvelebil 2003; Germonpré/Hämäläinen 2007; Losey et al. 2013; Gasilin/Gorbunov 2019). The ethnographic record has often proved to be a useful source for hypotheses and interpretations that are not readily apparent to archaeologists, although caution may be advised against uncritically applying recent practices to the archaeological record. Some archaeologists view the bear ceremonies documented in northern Eurasia as relics from the Stone Age, essentially providing direct insight into bear-related ritual activities that took place in the distant past (e.g. Gasilin/Gorbunov 2019, 73). The idea is not a new one; the bear ceremonies conducted by different groups in northern Eurasia were regarded by past anthropologists as representatives of a single primordial bear ritual (e.g. Hallowell 1926). Modern historians of religion are, however, sceptical about such claims (Rydving 2010). In his comparison between the bear ceremonies of the Khanty in Siberia and the Sámi in northern Fennoscandia, Håkan Rydving points out that important differences are found in the focus and priorities of the peoples conducting the ceremonies; the Khanty consider the festivities and the entertainment as the most important aspects of the ceremony, while the Sámi put more emphasis on the feast and the subsequent burial of the bear's remains (Rydving 2010, 42; see also Rydving, this volume). Also, most of the similarities that can be observed in the bear ceremonies of the Khanty and the Sámi are found in other types of rituals as well and are thus not specific to the treatment of bears (Rydving 2010, 43).

In this paper, I will concern myself with bear-related archaeological finds attributed to the Swedish Pitted Ware culture (PWC) of the Middle Neolithic period (*c.* 3200–2300 BC). "PWC" is the conventional name for a hunter-gatherer complex found primarily in coastal areas of southern Scandinavia, with a concentration of "typical" PWC sites on the eastern Swedish mainland and the Swedish islands of Gotland and Öland as well as the Finnish Åland islands (Fig. 1). These typical sites often yield large amounts of pottery sherds emanating from the characteristic pointed-based and pit-decorated clay pots that have given the culture its name. Seasonal sites are also found in northeastern Denmark, while the presence of the PWC in western Sweden and southern Norway has been a matter of debate due to the comparatively small amounts of pottery in these areas (see e.g. Larsson 2009, 56).

Large amounts of bones from seal and fish at the PWC sites strongly suggest a marine subsistence, even if large terrestrial animals such as wild boar and elk as well as smaller animals and birds are also present in the faunal assemblages. Stable isotope analyses on human remains confirm that the diet was based on marine protein (Eriksson et al. 2008; Fornander et al. 2008). Small zoomorphic and anthropomorphic clay figurines are also a somewhat regular occurrence on PWC sites in eastern Sweden, albeit often in small numbers (but see Björck et al. 2020). It has been suggested that some of these depict bears (Runeson/Kihlstedt 2018, 62; Björck et al. 2020, 138–141), and this will be further discussed later on.

The ample finds of animal bones in cultural layers, graves and deposits, coupled with the making of zoomorphic clay figurines, have led scholars to propose that the PWC was characterised by a cosmology and/or a cultural identity that was deeply entwined with wild animals and the practice of hunting (Wyszomirska 1984, 114–116; Björck 2003; 2011; Gill 2003, 150–152; Stenbäck 2003, 204; Fornander et al. 2008). On the basis of supposed bear depictions among the clay figurines the bear has been mentioned as an important figure in an animistic cosmology, rooted in the Mesolithic, that the PWC supposedly shared with other hunter-gatherers to the north and northeast in Finland, the Baltic states, and Russia during the Neolithic (Wyszomirska 1984, 209; Björck 2003; 2011). Some archaeologists have also drawn parallels between the PWC and recent hunter-gatherers in northern Eurasia with regards to subsistence strategies, social structure, ideology, and ritual practice (e.g. Zvelebil 1997; Mannermaa 2008).

A potentially illuminating approach might be to compare the bear-related finds at PWC sites to the treatment of bears among ethnographically studied northern Eurasian hunting-gathering societies conventionally described as "animistic". The term "animism" carries considerable baggage that stems from its Eurocentric origin as a way of explaining the alleged ontological shortcomings of non-European "primitives" (Tylor 1871; Harvey 2006, 5–9). A more current anthropological scholarship has sought to expel some of the primitivist connotations of the term and to reclaim it as a valuable analytical tool for discussing communities in which humans are but one kind of person among a myriad of persons (e.g. Bird-David 1992; 1999; Harvey 2006, xi).

In the following section, a short overview of the ethnographic accounts of bear ceremonies conducted by recent hunter-gatherers in northern Eurasia will be presented, and the methodological utility of these accounts will be discussed with reference to the material residues produced during the aforementioned ceremonies. Most of the examples presented in the following paper revolve around subspecies of the brown bear (*Ursus arctos*), so the vernacular term "bear" refers specifically to brown bears, while other species of bear are mentioned by their common and scientific names when they occur.

Bear ceremonies and their archaeological implications

Various bear ceremonies have been practiced by many groups of hunter-gatherers in northern Eurasia throughout the years. These groups include, but are not limited to, the Sámi in northern Fennoscandia, the Khanty in Siberia, and the Yukaghir, Ulch, Orochon, and Ainu in eastern Russia and northern Japan. In his seminal work on bear ceremonialism in the northern hemisphere, the American anthropologist Alfred Irving Hallowell highlighted what he perceived to be striking similarities in the bear-human relationships of several societies in northern Eurasia (Hallowell 1926). An overall similarity between these societies is that bears are considered to be intimately connected to the spirit world as well as able to engage in reciprocal social relations with humans (Ingold 2000, 92–94; Nadasdy 2007; Willerslev/Pedersen 2010).

The reciprocal relations between bears and humans require humans to show the bear respect, which is why the process of hunting, killing and butchering of bears is circumscribed by a host of rules and regulations. The bear is often talked about as a guest that provides the humans with gifts (its meat, pelt, and gall) and must therefore be given prayers, gifts, and a splendid feast in return (Irimoto 1996, 297). This ethos of reciprocity, gift giving and hospitality is further highlighted in the terminology employed in connection with bears in many places. Words and phrases such as "getting" the bear or "sending it away" are used rather than "hunting" or "killing", implying an ontology in which the bear gives itself willingly to the hunters and its spirit must be guided back to the spirit realm (Zolotarev 1937; Irimoto 1996; Westman Kuhmunen 2015, 81). Among the Tsaatang in northern Mongolia this ontology manifests itself in a fascinating way. There, custom dictates that when a hunter tracks a bear and the bear climbs up into a Siberian larch, the hunter must end his pursuit and walk away. Similarly, if an unarmed member of the Tsaatang should climb up into a Siberian larch when pursued by a bear, the bear is expected to honour this reciprocal relation by walking away (Pedersen 2001, 415).

Two main types of bear ceremonies among the hunter-gatherers in northern Eurasia can be identified: they involve the hunting and killing of an adult bear, and the capturing and rearing of a bear cub in the settlement (Zolotarev 1937; Irimoto 1996; Kwon 1999; Yamada 2018). The most well-known and well-documented of these bear ceremonies is perhaps the Ainu *Iyomante*, the sending-off of a bear reared in the settlement, which concludes with the bear being paraded around and shot with blunt ceremonial arrows, before being shot with a sharp, fatal arrow in the heart and subsequently strangled between two wooden poles (KIMURA 1999, 95). The bear is then given gifts and prayers, the skin and head are removed, and the meat boiled and eaten at a communal feast. Several days of feasting and honouring the bear ensue, and the ceremony is not over until all the meat has been eaten (KITAGAWA 1961, 149). A hole is also made in the bear's cranium in order to let the spirit out: For female bears, the hole is made on the right side, and for males on the left side (MASUDA et al. 2006). The skull is then wedged between the prongs of a forked wooden pole at a designated altar in or near the settlement and decorated with wooden shavings (KITAGAWA 1961, 148; Fig. 2). Archaeological finds do, however, indicate that bear skulls were sometimes deposited in other places as well. DNA-analyses of bear skulls found in a rock shelter in central Hokkaido show that the skulls emanate from different geographical areas, and that they might be the remnants of bear ceremonies conducted by various groups (MASUDA et al. 2006).

Sámi bear ceremonies are mostly known from the clergyman Pehr Fjellström's account of the southern Sámi bear hunt (FJELLSTRÖM 1981, originally published in 1755). The account tells us that when a bear is killed in its den, it is left in the woods while the hunters return home singing a song to let the settlement know they have killed a bear. Upon returning, they are met by women who look at them through brass rings and spit chewed alder bark at them and their hunting dogs (FJELLSTRÖM 1981, 20; WESTMAN KUHMUNEN 2015, 82). The hunters are then subjected to ritual seclusion in a designated tent overnight, before they can return to collect the bear carcass the following day. The bear is skinned and the meat carved off in a certain order, and thereafter it is cooked and served by the men in separate vessels for men and women, who also eat in different parts of the camp (FJELLSTRÖM 1981, 26). After all the meat has been eaten, the bones are either directly buried in anatomical order (FJELLSTRÖM 1981, 30–33), or placed on a platform to prevent scavenging before subsequent burial (WESTMAN KUHMUNEN 2015, 84). Archaeologically and osteologically studied Sámi bear graves in Sweden provide a perspective that is not visible in the ethnohistorical accounts. While the ethnohistorical material stresses that all bones must be intact (see FJELLSTRÖM 1981, 31), the known bear graves often feature intact skulls and shoulder blades, while the other bones are fragmented or cleaved in a manner consistent with the extraction of marrow (see IREGREN; SOMMERSETH, this volume; ZACHRISSON/IREGREN 1974, 83–84; JENNBERT 2003).

For the Khanty in Siberia, the discovery of a bear or an active bear den is taken to mean that the bear wants to visit the settlement. The bear is then killed and taken to the settlement, where it is presented with gifts and then skinned, leaving the paws and skull in the pelt (JORDAN 2003, 115–120). There are also examples of the pelt having been already removed in the forest, with the rest of the carcass left behind (RYDVING 2010, 38–39). The bear's cranium is cracked open and the brain eaten by the hunters while the other bones are carefully disarticulated, with some meat left around the joints, and then deposited in bodies of water, quite unlike the bones of other animals (JORDAN 2003, 123). The bear skulls are placed on the roofs of storehouses, with their jaws wired shut with branches in order to prevent revenge if the bear remembers its killer in its next life (JORDAN 2003, 123). A bear's skull is, however, only kept by the Khanty while the hunter who killed it is alive, after which the skull is deposited in water or in the forest. The practice of wiring the bear's jaws shut with branches to prevent vengeful attacks is also found among the Yukaghir, who then place the skull with the other defleshed bones on a wooden platform or in a tree (HALLOWELL 1926, 143–144; WILLERSLEV 2013, 51).

It is important to keep in mind that substantial parts of the ethnographically studied bear ceremonies would leave little to no traces in the archaeological record. These parts include, for example, non-material practices such as songs, stories and games as well as perishable paraphernalia used during the ceremonies. Still, while many aspects of the bear ceremonies do not leave tangible material traces that would survive for millennia, the skeletal remains are often treated in a special manner. Skulls are evidently the most important bone element in the ethnographic record (see also GERMONPRÉ/ HÄMÄLÄINEN 2007). Crania or whole skulls feature prominently, from the Sámi prescribing that the cranium must be complete at burial, to the caching and display of bear crania seen among the Ainu. The jaws are in some cases wired shut, but it would take exceptional conditions for young, flexible branches to be preserved for thousands of years. In some cases, holes are made in the bear's cranium either unilaterally left or right or in the back. Other skeletal parts might also be important, since for example the Khanty leave both the skull and the paws in the pelt and consider the soul to reside in these (JORDAN 2003, 115–120). Similarly, the Navajo tribe in the southwestern USA have used the paws of bears (both grizzly bear, *Ursus arctos horribilis*, and American black bear, *Ursus americanus*) for making medicine bags (PAVLIK 1997). Regarding the special treatment of certain bone elements, it is also worthwhile to mention the Neolithic cemetery of Shamanka II in southern Siberia. There, bacula (penis bones) from bears have been found in numerous graves, and in two of the graves bear crania were deposited face down in the filling above the human skeletons (LOSEY et al. 2013). Bacula from bears are attested from the Palaeolithic and have been interpreted as symbols of the strength and power of the bear, and the use of them as tools has been suggested to be a means of transmitting the "bear power" to humans (GIEMSCH 2017).

BEAR-RELATED FINDS ON PWC SITES IN SWEDEN

Looking at the long-term trends, there is a noticeable decline in the amount of recovered bear remains following the shift from the Mesolithic to the Early Neolithic (after *c.* 4000 BC), with very few finds until the latter part of the Middle Neolithic (*c.* 2900–2300 BC), where an increase attributable to the hunting subsistence of the PWC can be noted (see MAGNELL, this volume). Bear remains are, however, comparatively few even in the PWC area, occurring at eleven sites in total, while bear figurines or depictions are explicitly suggested at three sites (Fig. 3; Table 1). Finds from the Middle Neolithic horizons at Sjöholmen and Nymölla are included in the table, but it should be noted that these are multi-period sites with Mesolithic and Early Neolithic settlement layers as well, and a certain degree of intermingling can therefore not be completely ruled out (see for example FORSSANDER 1930; WYSZOMIRSKA 1979).

As is evident in Table 1, most of the bear remains recovered from PWC sites in Sweden have been found in cultural layers, where they occur among the remains of more common animals such as seal, fish, pig, elk, beaver, and waterfowl. There are, however, some exceptions to this, as there are three Swedish PWC sites with documented instances of bear remains having been found in contexts that suggest deliberate and selective deposition. In contrast to the ethnographic examples, where skulls are generally singled out, the possible deposits of bear remains at the Swedish PWC sites include teeth and bones from the extremities.

Perhaps the most intriguing example of the selective treatment of bear remains comes from Äs in Västmanland. Bones from pig, elk, and seal dominated the overall assemblage at the site, but two maxillary teeth from a bear were reportedly found placed inside a clay pot (LEPIKSAAR 1974, 150–152). At the newly excavated site of Tråsättra in Uppland, a distal phalanx and a carpal bone, possibly emanating from the same bear, were found next to a cooking pit in what has been interpreted as a ritual deposit (BJÖRCK et al. 2020, 174; MAGNELL 2020). The only other possible animal bone deposit at the site consisted of an otter mandible found in the remnants of a hut construction (BJÖRCK et al. 2020, 174).

It has been suggested that a bear ulna, which was recovered from the enigmatic Middle Neolithic pile dwelling site in Alvastra, Östergötland, may be the remnant of some sort of ritual activity following a successful hunt (DURING 1986, 138). The ulna was found together with two flint arrowheads close to one of the numerous hearths at the site. The absence of cut marks on the ulna possibly indicates that selective handling might have taken place, since most of the long bones from other species found in the pile dwelling were broken or otherwise damaged. Based on the size of the ulna, the Alvastra individual is estimated to have been of greater stature than most brown bears. The large number of hearths at the Alvastra pile dwelling, the signs of use by both the Funnel Beaker culture and the PWC around 3000 BC, as well as finds of dispersed human bones and a cut-marked human cranium, have led to the site being interpreted as an aggregation site where people met to socialise, trade, and conduct ceremonies (Frödin/Fürst 1919; Browall 1986, 65; Malmer 2002, 103–107).

With the exception of the Sjöholmen site, skeletal elements from the torso are lacking in the PWC material, which might indicate that the finds derive mainly from bear pelts rather than from complete carcasses (see Lepiksaar 1974, 150–152). As has previously been mentioned, this is also attested among the Siberian Khanty, who remove the pelt from the killed bear with the paws and the head still attached (Jordan 2003, 115–120). In Iron Age grave settings, finds of bear phalanges are sometimes interpreted as the remnants of a pelt onto which the deceased was laid when buried (Lindholm/Ljungkvist 2016; Kirkinen 2017), although it has been pointed out that finds of bear claws need not always imply the presence of a pelt (Grimm 2013). As of yet, no bear remains have been recovered from clear PWC burial contexts, but a distal phalanx was, however, recovered during excavations in 2007 at the large PWC site of Ajvide on Gotland, famous for its many burials (currently 85, see e.g. Wallin 2015). The phalanx was unearthed from a deep, dark cultural layer containing bones from common prey animals such as seal and pig, as well as human cranial fragments (Norderäng 2008, 5). The bear phalanx is likely unrelated to these fragments, as human remains are often found scattered in the cultural layers at Ajvide as well as at other PWC sites (Wallin 2015; Götherström et al. 2002). Additional bear remains, in the form of distal phalanges and teeth, have supposedly also been found at Ajvide, but the precise nature of these finds is unclear as they are mostly mentioned to showcase the far-reaching contacts of the Middle Neolithic trading networks (e.g. Österholm 1989, 187; 1991, 17; Burenhult 1997, 18; Molnar Appelblom 1997, 86; Wallin 2017, 123). At the present time, the finds from Ajvide are not fully systematised and are thus largely inaccessible to researchers.

The presence of at least one bear phalanx on Gotland is noticeable given the fact that the island has never had a resident bear population. Remains of other non-local species are present at PWC sites on Gotland, found in the form of worked beaver teeth, antler pieces from elk or deer as well as tooth beads made of elk teeth (Janzon 1974, 78; Lindqvist/Possnert 1997, 69; Burenhult 2002). Radiocarbon dating and DNA-evidence also suggest that the hedgehog was introduced to Gotland by the PWC people (Lindqvist/Possnert 1997; Fraser et al. 2012). These people were not strangers to the concept of acquiring and transporting animals, or parts thereof, from the mainland to the settlements on Gotland. This evidently included bear remains to a limited extent.

Even though the Danish PWC sites fall outside the scope of this paper, a cursory comparison between the Swedish sites with bear remains and an extraordinary deposit at the Danish PWC site of Kainsbakke, Djursland, raises questions (see Klassen/Gregersen, this volume). There, the remains of ten bears in the form of largely intact crania and phalanges have been found alongside pottery, flint tools and debitage, and the remains of other animals, such as fish bones, mollusc shells, horn cores from aurochs, mandibles from wild boar, elk and red deer, as well as some disarticulated human bones, in a large pit (Wincentz 2020, 48–52). Several of the bears were of advanced age and some of them have, through Strontium isotope analysis, been determined to be of non-local origin (Makarewicz/Pleuger 2020, 301–303; Klassen et al. 2020a). Crania and phalanges are not typically indicative of subsistence practices, but might instead indicate that pelts from bears killed in other parts of Scandinavia were brought to the site. The finds from Kainsbakke are more comparable with the northern Eurasian bear ceremonies, particularly since the skulls feature prominently, and have been interpreted as the remnants of "shamanistic rituals" adopted from the Swedish PWC (Klassen et al. 2020a; b). Comparisons with the east Swedish evidence for possible bear rituals fall short, since the Kainsbakke pit is more elaborate than anything found at the Swedish PWC sites.

Moving on to bear-related artefacts, a hitherto unique find is a bear ulna from Åloppe that has been fashioned into some sort of tool (Fig. 4). The artefact has been interpreted by some as an elk sculpture (INDREKO 1955; WYSZOMIRSKA 1984, 51), while others have questioned whether the schematic elk-like appearance really can be considered deliberate (CARPELAN 1977, 36–37). The only other example of a PWC artefact made of bear remains comes from Korsnäs and consists of a bead made of a perforated bear tooth (OLSSON et al. 1994; SJÖLING 2000).

As previously mentioned, there are a number of zoomorphic artefacts that have been singled out as possible bear depictions. Despite the existence of many zoomorphic clay figurines, exceedingly few can confidently be assigned to specific natural species, due to the fragmentation and abstraction of most of these artefacts. It has been suggested that one of two quadruped clay figurines found at Åby in Östergötland depicts a bear, although this is again presented with some caution due to the vague characteristics and fragmented state of the figurine (RUNESON/KIHLSTEDT 2018, 62; Fig. 5a). Six clay figurines from the aforementioned site of Tråsättra have been suggested to depict bears (BJÖRCK et al. 2020, 138–141; Fig. 5c–d). These clay figurines are, however, very dissimilar and at least one of them (Fig. 5d) might be considered anthropomorphic, since its facial markings are similar to those found on anthropomorphic figurines (see CEDERHVARF 1912; NUÑEZ 1986). The differences that are visible among these suggested bear depictions might indicate that either no standardised conventions for how to make bear figurines did exist, i.e. bear depictions could vary considerably from figurine maker to figurine maker, or that these figurines are not all meant to resemble bears.

Another artefact that has been suggested as a bear depiction is a peculiar greenstone club found at Stora Vika, Södermanland (FORNMINNESAVDELNINGEN 1966; Fig. 5b). Its exact find context is unknown since it was found by chance during landscaping, but finds of Pitted Ware pottery sherds at the same locality makes attribution to the PWC probable, as do the proximity to several confirmed PWC sites in the surrounding area (BAUDOU 1966; ISAKSSON et al. 2003). The stone club from Stora Vika is highly schematic and thus difficult to positively identify as a clear bear depiction. Conversely, several very naturalistic bear heads are seen on stone clubs that have been recovered as stray finds in northern Sweden and Karelia, mainly outside of the PWC area and of possible Late Neolithic date (see ALMGREN 1906; AILIO 1912; CARPELAN 1977).

Assessing scales: The bear minimum of the PWC

In comparison with the ethnographic accounts presented, the archaeological evidence from the Swedish PWC sites is variable in nature and seemingly less focused on particular skeletal elements. The prominent position of the cranium in the ethnographic sources is not reflected in the Swedish PWC material. The finds of bear remains at the Swedish PWC sites are sparse, and the treatment of them is obviously understated in comparison with the ethnographic accounts of bear ceremonialism among northern Eurasian hunter-gatherers. The selective handling and deposition of the bear's skull visible in the ethnography is markedly contrasted with the absence of skull bones from bears at the Swedish PWC sites, where teeth and paws are the most common elements found. This, in connection with the relative paucity of bear remains on these sites, does not seem to readily invite comparison with the exuberant bear ceremonies, which involve a large part of the community and specialised handling of entire bear carcasses, among recent hunter-gatherers in northern Eurasia. That is not to say that the comparisons are unhelpful, merely that they suggest that the handling of bear remains in the PWC was less formalised.

Knut Helskog suggests that the absence of bear bones, other than distal phalanges and teeth, in a large part of Fennoscandia might in itself be an indication of the special treatment afforded to bear remains from the Mesolithic and onwards (HELSKOG 2012). Of course, there need not be a direct correlation between the importance of bears in a society and the identifiable presence of bears in the material/archaeological record of that society. This is, for example, seen among Andean groups, where the spectacled bear (*Tremarctos ornatus*) cannot be readily identified in the iconography but is still an important cosmological figure (Paisley/Saunders 2010). A similar case has been proposed for the curious absence of bears in Old Norse iconography, when they are obviously significant in other cultural media (Hedeager 2004, 249; but see Oehrl 2013; and Oehrl, this volume, for a different view on the matter). The case does, however, get more complicated when all we have is the material record of a society, and while I agree with Helskog in principle, i.e. that absence might point to selective deposition, this is of course hard to prove. The discussion then boils down to how we can make sense of an unfamiliar archaeological phenomenon when our oft-used ethnographic models are seemingly inapplicable.

The location of many PWC sites on small islets in an archipelagic landscape is likely a sound explanation for there not being many bear bones present, since bears are known to roam larger stretches of, preferably forested, land (Magnell 2020). The bear ceremonies of various peoples in northern Eurasia were conducted in forested areas with breeding bear populations, where people conceivably witnessed and interacted with bears on a fairly regular basis. This is perhaps not readily comparable with the geographical distribution of bear remains at the Swedish PWC sites. Here I side with Nurit Bird-David regarding the "scale-blindness" of some archaeological research on animism, where focus has been placed upon a general, supra-regional notion of animistic collectives and of "all animals as persons" that might not have been relevant at a local level (Bird-David 2018). There has been a concept of "Animism with a capital A", when there are in fact plural animisms that should be considered (Willerslev 2013, 49). Animism is not an abstract, overarching worldview, but rather a relational and situational engagement, according to Bird-David (2018, 307), who takes the example of how an elephant that passes through a hamlet without trampling the huts is considered a person by her Indian Nayaka informants due to its respectful behaviour towards people. Conversely, an elephant that does not display such considerate behaviour is not regarded as a person.

I do agree with the fundamental importance of animals for the PWC and the similarities between the PWC and the Mesolithic and Neolithic hunter-gatherers of northeastern Europe, but to posit the bear as a recurring and important icon in an animistic cosmology attributable to the PWC is to sidestep the problems associated with the identification of the species depicted among the zoomorphic artefacts. If the zoomorphic figurines presented here (cf. Fig. 5) are all meant to resemble bears, it would suggest that no overarching conventions dictating the "proper" way of depicting bears existed. If the bear was indeed an important cosmological figure for the PWC, one would perhaps expect a higher degree of conformity in the depictions. It is also problematic, given the small amount of bear remains at the PWC sites. If we wish to understand prehistoric human-animal relations, our inquiries might benefit from the realisation that, at the centre, were the interactions between humans, animals, and plants, whose lives intersected on a regular basis (see for example Ingold 2013 on anthropology beyond humanity; see also Pilaar Birch 2018).

If we adjust the scale, from the grand animistic complex to what has happened at each site, we can focus on the indications of intentional "special" deposition, such as the bear teeth in the clay pot at Äs and the remains found near to hearths or cooking pits at other sites. The bear remains do not seem to have been treated in a manner that diverged all that much from how other animal remains were treated at the PWC sites, apart from the fact that bear remains were transported to Gotland not as artefacts, but in an apparently unmodified state. The animal bones included in graves and depositions at PWC sites suggest that the remains of certain animals such as, for example, seal, porpoise, dog, pig, and hedgehog were regularly employed in depositional acts, and it is reasonable to assume that the species of the animal and the anatomical element used were not inconsequential (see for example Lindström 2020). Viewed through this lens, the use of animal remains might have been a way to employ or direct the animal agencies within (e.g. CONNELLER 2004; OVERTON 2016). Bear remains could have constituted a particularly potent animal substance, as several finds seem to indicate special deposits. Perhaps the bear remains still contained a certain "bear-ness" that could be harnessed. Instead of grand rituals directed towards spirits or a spirit collective, as in the cited ethnographies, we might envision these deposits as small-scale acts of depositional "magic" carried out by the PWC hunter-gatherers in order to affect the world, bring about change, and ameliorate certain aspects of their day-to-day lives (see BRÜCK 1999; HOFMANN 2020).

Concluding remarks

No uniform way of relating to bears can be proposed from the osteological material found at the Swedish PWC sites, and the zoomorphic artefacts are too schematic to positively identify as bear depictions. The supposed status of the bear as a particularly prominent figure in an overarching PWC cosmology is thus difficult to establish. This should, however, not be misconstrued as an assertion that bears were unimportant, since bear remains were obviously important enough to transport from mainland areas with breeding populations of bears out onto the archipelago and even further away, across the sea to Gotland. The ethnographic accounts of the northern Eurasian bear ceremonies have, despite not being entirely applicable, aided the discussion of the bear-related finds on PWC sites. It is perhaps not surprising that bear remains might have been treated differently in areas with less dense (or non-existent) bear populations. Furthermore, even though the bear remains are scarce in comparison to the remains of many other animals such as seal, pig, elk, or fish, several finds point to intentional deposition near hearths, cooking pits, and, in one instance, inside a clay pot.

Bibliography

AARIS-SØRENSEN 1978: K. AARIS-SØRENSEN, Knoglematerialet fra den mellemneolitiske boplads ved Korsnäs, Grödinge socken, Södermanland. En zoolgisk [sic] gennemgang. Statens historiska museer Rapport 1978:8 (Stockholm 1978).

AILIO 1912: J. AILIO, Zwei Tierskulpturen. Finska fornminnesföreningens tidskrift XXVI, 1912, 257–282.

BAUDOU 1966: E. BAUDOU, Korrelationsberäkningar och lagerföljd på en gropkeramisk boplats vid Stora Vika. Fornvännen 61, 1966, 129–139.

BERLIN 1941: H. BERLIN, Benfynden från stenåldersboplatsen i Gualöv. Meddelanden från Lunds universitets historiska museum 1940–1941, 1941, 151–152.

BIRD-DAVID 1992: N. BIRD-DAVID, The Giving Environment: Another Perspective on the Economic System of Gatherer-Hunters. Current Anthropology 31(2), 1992, 189–196.

BIRD-DAVID 1999: N. BIRD-DAVID, "Animism" Revisited: Personhood, Environment, and Relational Epistemology. Current Anthropology 40(1), 1999, 67–79.

BIRD-DAVID 2018: N. BIRD-DAVID, Size Matters! The Scalability of Modern Hunter-Gatherer Animism. Quaternary International 464, 2018, 305–314.

BJÖRCK 1998: N. BJÖRCK, Fräkenrönningen: En "by" för 5000 år sedan. En gropkeramisk boplats. Arkeologisk undersökning, E4:an, RAÄ 399, Valbo socken, Gästrikland (Gävle 1998).

BJÖRCK 2003: N. BJÖRCK, The Neolithic Coastal Settlements: Cosmology and Ideology in a Baltic Sea Perspective. In: C. Samuelsson/N. Ytterberg (eds.), Uniting Sea: Stone Age Societies in the Baltic Sea Region (Uppsala 2003) 20–42.

BJÖRCK 2011: N. BJÖRCK, De bortglömda: Berättelsen om östra Sveriges neolitiska jägare och samlare. In: P. Bratt/R. Grönwall (eds.), Gropkeramikerna: Rapport från ett seminarium 2011 (Stockholm 2011) 12–28.

BJÖRCK et al. 2020: N. BJÖRCK/M. ARTURSSON/K.-F. LINDBERG (eds.), Tråsättra: Aspekter på säljägarnas vardag och symbolik. Arkeologisk undersökning. Stockholms län, Uppland, Österåkers kommun, Österåker socken, Tråsättra 1:14, Österåker 553 (Stockholm 2020).

BROWALL 1986: H. BROWALL, Alvastra pålbyggnad: Social och ekonomisk bas (Stockholm 1986).

BRÜCK 1999: J. BRÜCK, Ritual and Rationality: Some Problems of Interpretation in European Archaeology. European Journal of Archaeology 2(3), 1999, 313–344.

BRÄNNBORN et al. 2007: P. BRÄNNBORN/J. JARDFELT/A. KARLSSON, Fågel, fisk och mitt i mellan: Osteologisk analys av materialet från mesolitiska-neolitiska boplatsen Sjöholmen. Unpubl. Bachelor's thesis in historical osteology, Lund University (Lund 2007). http://lup.lub.lu.se/student-papers/record/1320699.

Burenhult 1997: G. Burenhult, Säljägare och svinherdar på Ajvide. In: G. Burenhult (ed.), Ajvide och den moderna arkeologin (Stockholm 1997) 15–21.

Burenhult 2002: G. Burenhult, The Grave-Field at Ajvide. In: G. Burenhult (ed.), Remote Sensing 2 (Stockholm 2002) 31–167.

Carpelan 1977: C. Carpelan, Älg- och björnhuvudföremål från Europas nordliga delar. Finskt Museum 82 (1975), 1977, 5–67.

Cederhvarf 1912: B. Cederhvarf, Neolitiska lerfigurer från Åland. Finska Fornminnesföreningens Tidskrift 26, 1912, 307–323.

Conneller 2004: C. Conneller, Beoming Deer: Corporeal Transformations at Star Carr. Archaeological Dialogues 11(1), 2004, 37–56.

During 1986: E. During, The Fauna of Alvastra: An Osteological Analysis of Animal Bones from a Neolithic Pile Dwelling (Stockholm 1986).

Ericson 1980: P. Ericson, Korsnäset: En gropkeramisk inlandsboplats i Dalarna. Unpubl. Bachelor's thesis in archaeology, Stockholm University (Stockholm 1980).

Eriksson et al. 2008: G. Eriksson/A. Linderholm/E. Fornander/M. Kanstrup/P. Schoultz/H. Olofsson/K. Lidén, Same Island, Different Diet: Cultural Evolution of Food Practice on Öland, Sweden, from the Mesolithic to the Roman Period. Journal of Anthropological Archaeology 27, 2008, 520–543.

Fjellström 1981: P. Fjellström, Kort berättelse om lapparnas björna-fänge, samt deras der wid brukade widskeppelser (Umeå 1981).

Fornander et al. 2008: E. Fornander/G. Eriksson/K. Lidén, Wild at Heart: Approaching Pitted Ware Identity, Economy and Cosmology through Stable Isotopes in Skeletal Material from the Neolithic Site Korsnäs in Eastern Central Sweden. Journal of Anthropological Archaeology 27, 2008, 281–297.

Fornminnesavdelningen 1966: Fornminnesavdelningen, Aktuellt: Från fornminnesavdelningens verksamhetsområde. Fornvännen 61, 1966, 190–191.

Forssander 1930: J. E. Forssander, En märklig stenåldersboplats vid Sjöholmen. Meddelanden från Lunds universitets historiska museum IV, 1930, 25–42.

Fraser et al. 2012: M. Fraser/S. Sten/A. Götherström, Neolithic Hedgehogs (Erinaceus europaeus) from the Island of Gotland Show Early Contacts with the Swedish Mainland. Journal of Archaeological Science 39(2), 2012, 229–233.

Frödin/Fürst 1919: O. Frödin/C. M. Fürst, Har skalpering förekommit i Norden under stenåldern? RIG 2, 1919, 193–204.

Gasilin/Gorbunov 2019: V. V. Gasilin/S. V. Gorbunov, Bear and Dog Remains from an Agnevo River Sacred Site (Central Sakhalin). Anthropology and Archaeology of Eurasia 58(1/2), 2019, 72–94.

Germonpré/Hämäläinen 2007: M. Germonpré/R. Hämäläinen, Fossil Bear Bones in the Belgian Upper Paleolithic: The Possibility of a Proto Bear-Ceremonialism. Arctic Anthropology 44(2), 2007, 1–30.

Giemsch 2017: L. Giemsch, Bear Necessities? On Potential Uses of the Ursine Baculum (Os penis) in Archaeological and Ethnological Contexts. In: P. Fasold/L. Giemsch/K. Ottendorf/D. Winger (eds.), Forschungen in Franconofurd. Festschrift für Egon Wamers zum 65. Geburtstag (Regensburg 2017) 41–53.

Gill 2003: A. Gill, Stenålder i Mälardalen (Stockholm 2003).

Grimm 2013: O. Grimm, Bear-Skins in Northern European Burials and Some Remarks on Other Bear-Related Furnishings in the North and Middle of Europe in the 1st millennium AD. In: O. Grimm/U. Schmölcke (eds.), Hunting in Northern Europe until 1500 AD. Old Traditions and Regional Developments, Continental Sources and Continental Influences. Papers presented at a workshop organised by the Centre for Baltic and Scandinavian Archaeology, Schleswig, June 16th and 17th, 2011. Schriften des Archäologischen Landesmuseums, Ergänzungsreihe 7 (Neumünster 2013) 277–296.

Gummesson 2008: S. Gummesson, Som man bäddar ligger benen. Åloppe, Norrskog. Rummet på en mellanneolitisk kustlokal. Unpubl. Master's thesis in osteoarchaeology, Stockholm University (Stockholm 2008).

Götherström et al. 2002: A. Götherström/N. Stenbäck/J. Storå, The Jettböle Middle Neolithic Site on the Åland Islands: Human Remains, Ancient DNA and Pottery. European Journal of Archaeology 5(1), 2002, 42–69.

Hallowell 1926: A. I. Hallowell, Bear Ceremonialism in the Northern Hemisphere. American Anthropologist 28(1), 1926, 1–175.

Harvey 2006: G. Harvey, Animism: Respecting the Living World (New York 2006).

Hedeager 2004: L. Hedeager, Dyr og andre mennesker: Mennesker og andre dyr i dyreornamentikkens transcendentale realitet. In: A. Andrén/K. Jennbert/C. Raudvere (eds.), Ordning mot kaos: Studier av nordisk förkristen kosmologi (Lund 2004) 219–287.

Hedell 1905: L. Hedell, Up, Nysätra sn, Åloppe. Undersökning 1902, inv.nr 11730. Bestämningar av däggdjursben. Rapport 1905 (Stockholm 1905).

Helskog 2012: K. Helskog, Bears and Meanings among Hunter-Fisher-Gatherers in northern Fennoscandia 9000–2500 BC. Cambridge Archaeological Journal 22(2), 2012, 209–236.

Herva/Lahelma 2019: V. P. Herva/A. Lahelma, Northern Archaeology and Cosmology: A Relational View (Oxon 2019).

Hofmann 2020: D. Hofmann, LBK Structured Deposits as Magical Practices. In: D. Hofmann (ed.), Magical, Mundane or Marginal? Deposition Practices in the Early Neolithic Linearbandkeramik Culture (Leiden 2020) 113–147.

Indreko 1955: R. Indreko, Eine Knochenfigur aus Åloppe. Annales Societatis Litterarum Estonicae In Svecia II (1950–1954), 1955, 47–56.

Ingold 2000: T. Ingold, The Perception of the Environment: Essays on Livelihood, Dwelling and Skill (London 2000).

Ingold 2013: T. Ingold, Anthropology Beyond Humanity. Suomen Antropologi 38(3), 2013, 5–23.

Irimoto 1996: T. Irimoto, Ainu Worldview and Bear Hunting Strategies. In: J. Pentikäinen (ed.), Shamanism and Northern Ecology (Berlin 1996) 293–303.

Isaksson et al. 2003: M. Isaksson/M. Ahlbeck/T. Fors, Marsta 3:5: Avgränsning av en boplats från yngre stenålder. Arkeologisk förundersökning RAÄ 710, Sorunda sn, Södermanland. Rapporter från Arkeologikonsult 2003:1090 (Upplands Väsby 2003).

Janzon 1974: G. O. Janzon, Gotlands mellanneolitiska gravar (Stockholm 1974).

Jennbert 2003: K. Jennbert, Animal Graves: Dog, Horse and Bear. Current Swedish Archaeology 11, 2003, 139–152.

Jordan 2003: P. Jordan, Material Culture and Sacred Landscape: The Anthropology of the Siberian Khanty (Walnut Creek 2003).

Kimura 1999: T. Kimura, Bearing the 'Bare Facts' of Ritual: A Critique of Jonathan Z. Smith's Study of the Bear Ceremony Based on a Study of the Ainu Iyomante. Numen 46(1), 1999, 88–114.

Kirkinen 2017: T. Kirkinen, "Burning Pelts": Brown Bear Skins in the Iron Age and Early Medieval (0–1300 AD) Burials in South-Eastern Fennoscandia. Estonian Journal of Archaeology 21(1), 2017, 3–29.

Kitagawa 1961: J. M. Kitagawa, Ainu Bear Festival (Iyomante). History of Religions 1(1), 1961, 95–151.

Klassen et al. 2020a: L. Klassen/T. D. Price/K.-G. Sjögren/L. Wincentz/B. Philippsen, Strontium and Lead Isotope Studies of Faunal and Human Remains from Kainsbakke and Kirial Bro. In: L. Klassen (ed.), The Pitted Ware Culture on Djursland: Supra-regional Significance and Contacts in the Middle Neolithic of Southern Scandinavia (Aarhus 2020) 407–446.

Klassen et al. 2020b: L. Klassen/R. Iversen/N. Nørkjær Johanssen/U. Rasmussen/O. B. Poulsen, The Pitted Ware Culture on Djursland in the Neolithic World. In: L. Klassen (ed.), The Pitted Ware Culture on Djursland: Supra-regional Significance and Contacts in the Middle Neolithic of Southern Scandinavia (Aarhus 2020) 451–489.

Kosintsev et al. 2018: P. A. Kosintsev/O. P. Bachura/V. S. Panov, Remains of Brown Bear (Ursus arctos L.) from the Kaninskaya Cave Sanctuary in the Northern Urals. Archaeology, Ethnology and Anthropology of Eurasia 46(2), 2018, 131–139.

Kwon 1999: H. Kwon, Play the Bear: Myth and Ritual in eastern Siberia. History of Religions 38(4), 1999, 373–387.

Larsson 2009: Å. M. Larsson, Breaking and Making Bodies and Pots: Material and Ritual Practices in Sweden in the Third Millennium BC (Uppsala 2009).

Lepiksaar 1974: J. Lepiksaar, Djurrester från den mellanneolitiska (gropkeramiska) boplatsen vid Äs, Romfartuna sn, Västmanland. In: L. Löfstrand, Yngre stenålderns kustboplatser: Undersökningarna vid Äs och studier i den gropkeramiska kulturens kronologi och ekologi (Uppsala 1974) 141–156.

Lindholm/Ljungkvist 2016: K.-J. Lindholm/J. Ljungkvist, The Bear in the Grave: Exploitation of Top Predator and Herbivore Resources in First Millennium Sweden – First Trends from a Long-Term Research Project. European Journal of Archaeology 19(1), 2016, 3–27.

Lindqvist/Possnert 1997: C. Lindqvist/G. Possnert, The Subsistence Economy and Diet at Jakobs/Ajvide, Eksta Parish, and Other Prehistoric Dwellings and Burial Sites on Gotland in Long-Term Perspective. In: G. Burenhult (ed.), Remote Sensing 1. Theses and Papers in North-European Archaeology 13:a. (Stockholm 1997) 29–90.

Lindström 2020: T. Lindström, Retrieving, Curating and Depositing Skulls at Pitted Ware Culture sites. Current Swedish Archaeology 28, 2020, 147–179.

Losey et al. 2013: R. J. Losey/V. I. Bazaliiskii/A. R. Lieverse/A. Waters-Rist/K. Faccia/A. W. Weber, The Bear-Able Likeness of Being: Ursine Remains at the Shamanka II Cemetery, Lake Baikal, Siberia. In: C. Watts (ed.), Relational Archaeologies: Humans, Animals, Things (New York 2013) 65–96.

Magnell 2020: O. Magnell. Osteologisk analys av djurben från den gropkeramiska boplatsen Tråsättra. In: N. Björck/M. Artursson/K.-F. Lindberg (eds.), Tråsättra: Aspekter på säljägarnas vardag och symbolik. Arkeologisk undersökning. Stockholms län, Uppland, Österåkers kommun, Österåker socken, Tråsättra 1:14, Österåker 553 (Stockholm 2020).

Magnell/Iregren 2010: O. Magnell/E. Iregren, Veistu hvé blóta skal? The Old Norse Blót in the Light of Osteological Remains from Frosö Church, Jämtland, Sweden. Current Swedish Archaeology 18, 2010, 223–250.

Makarewicz/Pleuger 2020: C. A. Makarewicz/S. Pleuger, Herder-Hunter-Fishers and Agricultural Contacts: Zooarchaeological Perspectives on Pitted Ware animal Exploitation Strategies from Djursland. In: L. Klassen (ed.), The Pitted Ware Culture on Djursland: Supra-regional Significance and Contacts in the Middle Neolithic of Southern Scandinavia (Aarhus 2020) 279–340.

Malmer 2002: M. P. Malmer, The Neolithic of South Sweden: TRB, GRK and STR (Stockholm 2002).

Mannermaa 2008: K. Mannermaa, Birds and Burials at Ajvide (Gotland, Sweden) and Zvejnieki (Latvia) about 8000–3900 BP. Journal of Anthropological Archaeology 27, 2008, 201–225.

Masuda et al. 2006: R. Masuda/T. Tamura/O. Takahashi, Ancient DNA Analysis of Brown Bear Skulls from a Ritual Rock Shelter Site of the Ainu Culture at Bihue, Central Hokkaido, Japan. Anthropological Science 114, 2006, 211–215.

Molnar Appelblom 1997: P. Molnar Appelblom, Arbete, förslitningar och könsroller: Om osteobiografins möjligheter att tolka forntida livssituationer. In: G. Burenhult (ed.), Ajvide och den moderna arkeologin (Stockholm 1997) 85–94.

Nadasdy 2007: P. Nadasdy, The Gift in the Animal: The Ontology of Hunting and Human–Animal Sociality. American Ethnologist 34(1), 2007, 25–43.

Norderäng 2008: J. Norderäng, Ajvideboplatsen. Rapport från arkeologisk undersökning 2007 av fornlämning nr. 171 på fastigheten Ajvide 2:1 i Eksta socken, Gotland (Visby 2008). http://urn.kb.se/resolve?urn=urn:nbn:se:hgo:diva-193.

Nuñez 1986: M. Nuñez, Clay Figurines from the Åland Islands and Mainland Finland. Fennoscandia Archaeologica 3, 1986, 17–34.

Oehrl 2013: S. Oehrl, Bear Hunting and its Ideological Context (as a Background for the Interpretation of Bear Claws and other Remains of Bears in Germanic Graves of the 1st Millennium AD). In: O. Grimm/U. Schmölcke (eds.), Hunting in Northern Europe until 1500 AD. Old Traditions and Regional Developments, Continental Sources and Continental Influences. Papers presented at a workshop organised by the Centre for Baltic and Scandinavian Archaeology, Schleswig, June 16th and 17th, 2011. Schriften des Archäologischen Landesmuseums, Ergänzungsreihe 7 (Neumünster 2013) 297–332.

Olson et al. 2011: C. Olson/N. Björck/J. Storå, Huts and Deposition of Refuse at Fräkenrönningen, a Neolithic Coastal Dwelling Site in Eastern Middle Sweden. International Journal of Osteoarchaeology 21, 2011, 173–186.

Overton 2016: N. J. Overton, More than Skin Deep: Reconsidering Isolated Remains of 'Fur-Bearing Species' in the British and European Mesolithic. Cambridge Archaeological Journal 26(4), 2016, 561–578.

Ozaki 1913: Y. Ozaki, Japan. In: W. Hutchinson (ed.), Customs of the World: A Popular Account of the Manners, Rites and Ceremonies of Men and Women in all Countries 1 (London 1913) 380–408.

Paisley/Saunders 2010: S. Paisley/N. J. Saunders, A God Forsaken: The Sacred Bear in Andean Iconography and Cosmology. World Archaeology 42(2), 2010, 245–260.

Pavlik 1997: S. Pavlik, The Role of Bears and Bear Ceremonialism in Navajo Orthodox Traditional Lifeway. The Social Science Journal 34(4), 1997, 475–484.

Pedersen 2001: M. A. Pedersen, Animism and North Asian Indigenous Ontologies. The Journal of the Royal Anthropological Institute 7(3), 2001, 411–427.

Pentikäinen 2005: J. Pentikäinen, Karhun kannoilla: Metsänpitäjä ja mies (Helsinki 2005).

Pilaar Birch 2018: S. E. Pilaar Birch, Introduction. In: S. E. Pilaar Birch (ed.), Multispecies Archaeology (Abingdon 2018) 1–7.

Rosengren 2018: E. Rosengren, A Partial Catalogue of the Mammalian Archaeofaunal Material in the Collection at Lund University Historical Museum (LUHM) (Lund 2018). http://lup.lub.lu.se/record/95548efc-8a0e-4dba-a38f-18c9d3b81ec6.

Runeson/Kihlstedt 2018: H. Runeson/B. Kihlstedt, Åby. En klassisk gropkeramisk lokal i det inre av Bråviken. Rapport 2017:28 (Stockholm 2018).

Rydving 2010: H. Rydving, The 'Bear Ceremonial' and Bear Rituals among the Khanty and the Sami. Temenos 46(1), 2010, 31–52.

Segerberg 1999: A. Segerberg, Bälinge mossar: Kustbor i Uppland under yngre stenåldern (Uppsala 1999).

Sjöling 2000: E. Sjöling, Benartefakterna från Korsnäs. En arkeoosteologisk analys av ben-, och tandartefakter från en gropkeramisk lokal i Södermanland. Unpubl. Master's thesis in osteoarchaeology, Stockholm University (Stockholm 2000).

Stenbäck 2003: N. Stenbäck, Människorna vid havet: Platser och keramik på Ålandsöarna perioden 3500–2000 f. Kr (Stockholm 2003).

Thomas 1954: S. E. Thomas, Sjöholmen, Site 179: A Re-Examination. In: C. A. Althin (ed.), The Chronology of Stone Age Settlement in Scania, Sweden. The Mesolithic Settlement (Lund 1954) 169–187.

Tylor 1871: E. B. Tylor, Primitive Culture: Researches into the Development of Mythology, Philosophy, Religion, Language, Art, and Custom (London 1871).

Wallin 2015: P. Wallin, Perfect Death: Examples of Pitted Ware Ritualisation of the Dead. In: R. Peyroteo Stjerna/K. von Hackwitz (eds.), Ancient Death Ways. Occasional Papers in Archaeology 59 (Uppsala 2015) 47–64.

Wallin 2017: P. Wallin, Analyser och tolkningar av begravningspraktiker hos mellanneolitisk gropkeramisk kultur på Gotland: Med Ajvidelokalen i fokus. In: P. Wallin/H. Martinsson-Wallin (eds.), Arkeologi på Gotland 2: Tillbakablickar och nya forskningsrön (Visby 2017) 117–128.

Westman Kuhmunen 2015: A. Westman Kuhmunen, A Female Perspective on Sami Bear Ceremonies. Journal of Northern Studies 9(2), 2015, 73–94.

Willerslev 2013: R. Willerslev, Taking Animism Seriously, but Perhaps Not Too Seriously. Religion and Society: Advances in Research 4, 2013, 41–57.

Willerslev/Pedersen 2010: R. Willerslev/M. A. Pedersen, Proportional Holism: Joking the Cosmos into the Right Shape in North Asia. In: T. Otto/N. Bubandt (eds.), Experiments in Holism: Theory and Practice in Contemporary Anthropology (Oxford 2010) 262–278.

Wincentz 2020: L. Wincentz, Kainsbakke and Kirial Bro: The Two Main Sites of the Pitted Ware Culture on Djursland. In: L. Klassen (ed.), The Pitted Ware Culture on Djursland: Supra-regional Significance and Contacts in the Middle Neolithic of Southern Scandinavia (Aarhus 2020) 35–140.

Wyszomirska 1979: B. Wyszomirska, A Double Grave with Yellow Ochre at Nymölla in Scania. Fornvännen 74, 1979, 73–84.

Wyszomirska 1984: B. Wyszomirska, Figurplastik och gravskick hos nord- och nordösteuropas neolitiska fångstkulturer (Lund 1984).

Yamada 2018: T. Yamada, The Ainu Bear Ceremony and the Logic behind Hunting the Deified Bear. Journal of Northern Studies 12(1), 2018, 35–51.

Zachrisson/Iregren 1974: I. Zachrisson/E. Iregren, Lappish Bear Graves in Northern Sweden. An Archaeological and Osteological Study (Stockholm 1974).

Zolotarev 1937: A. M. Zolotarev, The Bear Festival of the Olcha. American Anthropologist 39, 1937, 113–130.

Zvelebil 1997: M. Zvelebil, Hunter-Gatherer Ritual Landscapes: Spatial Organisation, Social Structure and Ideology among Hunter-Gatherers of Northern Europe and Western Siberia. In: A. V. Gijn/M. Zvelebil (eds.), Ideology and Social Structure of Stone Age Communities in Europe. Analecta Praehistorica Leidensia 29 (Leiden 1997) 33–50.

Zvelebil 2003: M. Zvelebil, People Behind the Lithics: Social Life and Social Conditions of Mesolithic Communities in Temperate Europe. In: L. Bevan/J. Moore (eds.), Peopling the Mesolithic in a Northern Environment (Oxford 2003) 1–26.

Österholm 1989: I. Österholm, Bosättningsmönstret på Gotland under stenåldern (Stockholm 1989).

Österholm 1991: I. Österholm, Resultat av gotländsk stenåldersforskning under 1980-talet. In: I. Österholm (ed.), Ur den gotländska jorden: Glimtar från arkeologiska undersökningar (Visby 1991) 7–19.

PhD student Tobias Lindström
Department of Archaeology and Classical Studies
Stockholm University
Stockholm
Sweden
tobias.lindstrom@ark.su.se

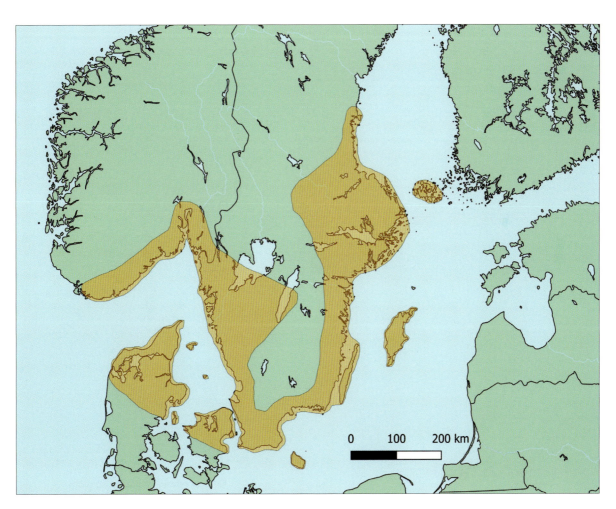

Fig. 1. Map of the approximate range of sites attributed to the Pitted Ware culture (map T. Lindström, with QGIS 3.10 using the Natural Earth Dataset).

Fig. 2. Bear skulls from Ainu bear ceremonies put up for display. Note the unilateral holes visible on several cranial vaults (photo J. Revilliod, after OZAKI 1913, 405).

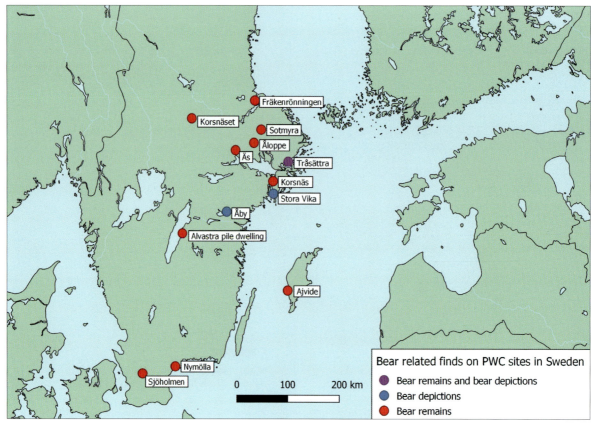

Fig. 3. Map of the Swedish Pitted Ware culture sites with bear remains and alleged bear depictions, as well as sites with both (map T. Lindström, with QGIS 3.10 using the Natural Earth Dataset).

Fig. 4. Tool made of a bear ulna from the Pitted Ware culture, found at Åloppe. Some have interpreted the artefact as an elk sculpture (photo T. Lindström, Swedish History Museum, CC BY 4.0).

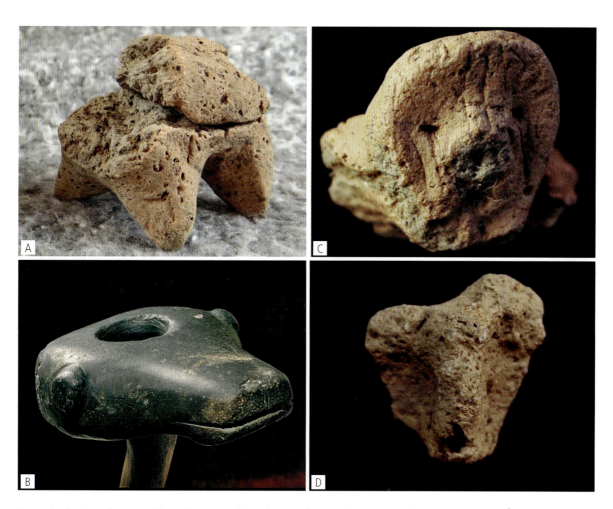

Fig. 5. A selection of suggested bear depictions from the Pitted Ware culture area; a: Clay figurine from Åby (photo courtesy of H. Runeson, Stiftelsen Kulturmiljövård); b: Stone club from Stora Vika (photo Swedish History Museum, CC BY-SA 2.5); c–d: Clay figurines from Tråsättra (after BJÖRCK et al. 2020, figs. 101–102; photos N. Björck, Arkeologerna/Swedish History museum, CC BY 4.0).

Table 1. Finds of bear remains at Pitted Ware culture sites in Sweden.

Site	No. of fragments	Bone element	Find context	Notes	References
Ajvide	1 (minimum)	Phalanx	Cultural layer	At least one bone, unclear documentation	Burenhult 1997, 18; Norderäng 2008, 5
Alvastra pile dwelling	4	Tooth, ulna	Deposit	Ulna found next to hearth	During 1986, 74
Fräkenrönningen	2	Undetermined	Cultural layer	Highly fragmented	Björck 1998, 43; Olson et al. 2011
Korsnäs	3	Metatarsal, metacarpal, tooth	Cultural layer	Tooth pendant	Aaris-Sørensen 1978; Sjöling 2000
Korsnäset	8	Cranium and „meat-poor parts"	Cultural layer		Ericson 1980, tab. 5–6
Nymölla	22	Cranium, mandible, tooth, scapula, carpal, metacarpal, femur, vertebra, metatarsal, metapodium, phalanges	Cultural layer	Multi-period site	Berlin 1941; Rosengren 2018, 47
Sjöholmen	1	Metacarpal	Cultural layer	Multi-period site	Forssander 1930; Thomas 1954; Brännborn et al. 2007, 84
Sotmyra	2	Phalanx	Cultural layer		Segerberg 1999, 179
Träsättra	2	Phalanx, carpal	Deposit	Metatarsal found next to cooking-pit	Magnell 2020, tab. 3
Åloppe	5	Tooth, phalanx, sesamoid, ulna	Cultural layer	Tool made of ulna	Hedell 1905; Indreko 1955; Gummesson 2008, 64
Äs	12	Tooth, phalanx, maxilla	Cultural layer/deposit	Teeth inside clay pot	Lepiksaar 1974, 150–152

The Kainsbakke bears and changing patterns in the human-bear relationship through the Danish Mesolithic and Neolithic

By Lutz Klassen and Kristian Murphy Gregersen

Keywords: Bears, rituals, Mesolithic, Neolithic, Denmark

Abstract: With a point of departure in the major discovery of ritually deposited bear bones at the Pitted Ware culture site of Kainsbakke on Djursland, Denmark, this paper sets out to investigate the human-bear relationship in Denmark during the Mesolithic and Neolithic (c. 9500–2400 cal BC). Studies of the abundance of bear bones in the archaeological record demonstrate a decreasing tendency throughout the entire Mesolithic and earlier parts of the Neolithic, followed by a rise in the early 3rd millennium BC (i.e. the Pitted Ware culture). Additional investigations of the treatment of the bones, the selection of specific bone elements and the contexts in which they were found indicate changing patterns in the human-bear relationship. These are interpreted as being due to a combination of environmental change and shifting contacts and interactions with societies in especially the eastern Baltic region.

The Kainsbakke bears

In the early 1980s, a number of brown bear (*Ursus arctos*) bones were excavated at the Neolithic site of Kainsbakke (Richter 1986; Makarewicz/Pleuger 2020, 303–305). Kainsbakke is situated on the Djursland peninsula in eastern Jutland; it is the largest settlement of the Pitted Ware culture (PWC) in Denmark (Wincentz Rasmussen 1984; 1991; Wincentz 2020) and is dated to the early 3rd millennium BC (Philippsen et al. 2020). J. Richter (1986) viewed the Kainsbakke bear bones as the remains of ordinary hunting activities, but this interpretation is, for several reasons, hardly tenable: The number of bones (34, representing at least ten different individuals – by far the largest total from any Neolithic site in southern Scandinavia), the diversity of skeletal elements, and the presence of entire crania are all exceptional. Furthermore, all the remains were found within a single pit – A47. Based on its construction and content of other finds, which included ordered depositions of large numbers of mandibles from several species, this pit can be identified as a ritual feature (Wincentz 2020, 44–56). The bear bones did not appear randomly distributed throughout the pit but sorted to some degree by skeletal element in several concentrations (Fig. 1). This suggests a conscious act of deposition rather than random disposal of waste. It is also remarkable that one of the two intact crania had been placed on top of an aurochs horn (Fig. 2). Furthermore, the taphonomic analysis undertaken by C. Makarewicz and S. Pleuger shows that the fracture patterns of the Kainsbakke bear bones deviate from those observed on the huge numbers of other mammal bones recovered from Kainsbakke – that bears were treated and handled differently from other wild and domestic species at the site (Makarewicz/Pleuger 2020, 311). The various lines of evidence outlined above clearly imply that the deposition of bear bones at Kainsbakke was ritual in nature.

Moreover, the ethnographical, historical and archaeological/ethnoarchaeological evidence for bear rituals in hunter-gatherer societies in the northern hemisphere (e.g. Hallowell 1926; Zachrisson/Iregren 1974; Paproth 1976; Black 1998; Wamers 2009; 2015; Helskog 2012) reinforces the interpretation of the Kainsbakke depositions as the result of ritual activities involving bears in the Danish Neolithic, as such practices can be demonstrated to have been widespread and abundant, both geographically and chronologically. Several observations at Kainsbakke provide further specific evidence. Of the several thousand bones recovered from pit A47, not only those of bears are remarkable; there are also a few bones of elk (*Alces alces*) and two wing bones of a Dalmatian pelican (*Pelecanus crispus*, cf. Makarewicz/Pleuger 2020, 307–308; Klassen et al. 2020a, 455–464). The elk bones are some of the latest examples of this species hitherto found in Denmark (as is also true of the bear bones), and the Dalmatian pelican is an exceptionally rare find in the Danish Neolithic with only two other known records (cf. Nikulina/Schmölcke 2015). Strontium isotope analysis demonstrates that the elk are non-local, and the same is true for at least some of the bears. These foreign elements had probably been brought to Kainsbakke in the form of single, disarticulated bones intended for ritual activities (Klassen et al. 2020b, 430–436; Price et al. 2021). The most likely region of origin for both the elk and the bear remains is the coastal area of western Sweden, about 100 km away on the opposite side of the Kattegat (Klassen et al. 2020b, 432–436; Price et al. 2021). The PWC people on Djursland were in close contact with PWC groups there. This combination of exceptional skeletal elements from bears, elk, and a very rare waterbird reflects the performance of shamanistic rituals, as clearly indicated by the ethnographical and archaeological evidence from northeastern Europe and western Siberia, which reveals these species to be a shaman's most important animals (Zvelebil 2003a; b; Klassen et al. 2020a). While such a direct historical analogy should, in general, obviously be applied with caution, in this particular case, and in this area in question, there is indeed evidence for a remarkable stability of such beliefs through long periods of time (Jordan 2003, 130; see Klassen et al. 2020a, 455–464), which justifies the approach adopted here.

The emergence of the PWC in the Danish regions bordering the Kattegat reflects contacts with PWC groups to the east and northeast. The culture has its roots in the Pit-Comb Ware culture of northeastern Europe (Iversen 2010, 28–30). This is evident from its material culture and in the complete or partial reversion to hunting (especially of seals) as a subsistence base in the PWC. The identification at Kainsbakke of shamanistic rituals (including those involving bears) typical of northeast European/western Siberian hunter-gatherer societies therefore fits well into the overall picture of the cultural development in the Kattegat region during the early 3rd millennium BC. At the same time, it poses several questions about the human-bear relationship during the earlier parts of prehistory in Denmark, which will be addressed below.

Was the performance of bear rituals by Neolithic communities in Denmark specific to the PWC, with its pronounced contacts with hunter-gatherer societies? Or is it also possible to identify traces of bear rituals in the Funnel Beaker culture (FBC), which had its roots in the central European Neolithic? Moreover, is there any evidence for the ritual significance of bears in hunter-gatherer communities earlier in southern Scandinavian prehistory? The latter question is relevant in the light of the abundant evidence of ritual depositions of elk bones/carcasses in the Early Mesolithic Maglemose culture (MMC), indicating shamanistic traditions that were at least related to those evident at Kainsbakke. As noted by K. Buck Pedersen and E. Brinch Petersen in a recent paper dealing with the elk depositions, the absence of bear counterparts from the same period is striking (Buck Pedersen/Brinch Petersen 2017, 251).

To provide answers to these questions, some simple studies were undertaken to evaluate the frequency of records of bear remains from the various cultures of the Danish Mesolithic and Neolithic (constancy and relative importance of bears), followed by qualitative assessments of the types of bones represented, finds contexts, modes of deposition and treatment/handling of bones. All the available evidence is then assessed and interpreted in a concluding summary.

Geographical, chronological and methodological framework

In the following, the question of the human-bear relationship will be addressed for the entire Mesolithic (i.e. MMC, Kongemose [KMC] and Ertebølle [EBC] cultures) and the Neolithic (i.e. FBC and PWC). The date of the extinction of the brown bear as an indigenous species in Denmark is not entirely clear. Bear bones have been found with some regularity in PWC contexts up to *c.* 2700 cal BC, while there is only one later record from the Late Single Grave culture (SGC) site of Holme Skanse, dated to *c.* 2400 cal BC (Andersen 1983, 78; Richter 1986, 126). Archaeological sites with preserved faunal material are, however, exceptionally rare for at least the next 1,000 years, and bears might therefore still have been present (in Jutland) in the early 2nd millennium BC. Post-Neolithic finds predominantly comprise distal phalanges, which occur primarily in Iron Age graves, probably representing imported bear skins (Møhl 1978; Aaris-Sørensen 2009, 18; Henriksen 2009). In absolute terms this study therefore covers the time between *c.* 9500 and 2400 cal BC.

From a geographical point of view, the study concentrates on the area of present-day Denmark, but finds from the adjacent regions of southern Sweden (Halland, Scania, Blekinge) and northern Germany (Schleswig-Holstein, Mecklenburg-Western Pomerania) are also considered. While this focus on the area of a modern state is potentially misleading in a general prehistoric context, in the present case there is some justification. Bears lived in northern Germany until the Early Middle Ages (cf. Schmölcke, this volume), and the same is true for at least parts of southern Sweden (see Magnell, this volume; Iregren 1988). In Denmark, on the other hand, the bear disappears from the faunal record, as indigenous species, in the mid-3rd millennium BC in the western parts of the country (Jutland), and it became extinct on the islands constituting the central and eastern parts of Denmark considerably earlier than this, around 6000 cal BC (Aaris-Sørensen 1980; 1985; 1990). The present study therefore encompasses all the archaeological evidence for bears from the period when the species actually lived in Denmark. The subfossil record indicates that the first bears were present slightly earlier, during the Allerød period (Aaris-Sørensen 2009, 18), but these finds derive exclusively from natural deposits and consequently do not permit any conclusions to be drawn about the human-bear relationship. Such evidence may, however, be represented by the difficult-to-date amber figurines dealt with by Gross/Vang Petersen (this volume).

During the time in question, the bears of northern Germany and southern Sweden had direct contacts with much larger bear populations in the vast areas of adjacent land. The same can be said of the human populations, who may well have been influenced (also in their relationship with bears) by these contacts. Denmark, on the other hand, developed during the Atlantic period into an area characterised by proximity to the sea (less then *c.* 50 km from any point) and an extremely long and complex coastline (*c.* 6500–3900 cal BC). These circumstances have shaped human identity in the region ever since: Depictions of ships and boats on Bronze Age bronzes and rock art, Iron Age burials formed like a ship or boat, and a comprehensive archaeological record relating to ships and to waterborne trade and warfare in the Viking Age constitute just a few examples from prehistory. Consequently, they may also have been important for the human-bear relationship during large parts of the Mesolithic and Neolithic, as indicated by ethnographical evidence which identifies local ecological conditions as a factor which influences bear-related beliefs (Black 1998, 346). Within the timespan under investigation here, Denmark was only part of a larger landmass during the MMC, in the Preboreal (*c.* 9500–8200 cal BC) and Boreal (*c.* 8200–6500 cal BC) periods.

Table 1 presents the basic data for all records of bear remains from archaeological contexts in the regions and periods under investigation here. The dating is based on stratigraphic evidence and ^{14}C dates (Table 2), which in most, but not all, cases permit the bear remains to be assigned to a single archaeological culture; this also applies to finds from multi-period sites. There are 32 records from Denmark, 23 from southern Sweden, and 13 from northern Germany, and their distribution is shown in Figures 3–6. The general scarcity or absence of finds from the Danish islands from the KMC onwards is the result of the previously mentioned extinction of bears here in the Early Atlantic: The few records from these areas represent imported items. The total absence of finds from the KMC and MMC in western Denmark is due to the acidic nature of Jutland's bogs, which has resulted in dissolution of all bones from these periods, and therefore does not reflect an absence of bears and bear hunting. Coastal settlements of these periods are either inaccessible (MMC) or have not been excavated to any significant degree in Jutland (KMC: only two of the 58 definite or possible Danish assemblages from the KMC containing bones are from Jutland).

Records from the EBC and FBC in Jutland are almost all from coastal or near-coastal sites: The bones here have been preserved in either kitchen middens or natural chalk-rich sediments.

Bird and/or fish bones have also been found at most Danish and Swedish sites that have yielded bear finds. This shows that the presence or absence of specific (especially smaller) bear bones at these sites is due neither to adverse conditions of preservation nor excavation technique but is in general likely to represent the prehistoric situation to the same degree as this is reflected in the overall extent of the bone assemblages. The same can be said of coastal sites in northern Germany, while there is a lack of fragile bones at some of the inland sites in Mecklenburg.

The chronologically isolated record of bear bones from the Late SGC site of Holme Skanse has been excluded from the present study, because a single assemblage does not permit any conclusions to be drawn about a specific culture or period.

Quantitative assessment

Methodological considerations

The methods employed below to assess the quantitative importance of bears during the Danish Stone Age follow, with some adjustments, those used by U. Schmölcke (this volume) in his contribution on the history of bears in the northern part of central Europe. Two quantitative factors are examined: Constancy and the relative abundance of bear bones. Constancy is calculated as the number of sites of a given culture yielding bear finds relative to the total number of known sites with preserved bones from that culture, while the relative abundance is expressed as the percentage of bear bones in the total assemblage of mammal bones from a given site. For a true evaluation of the outcomes of both calculations it is important to bear in mind several methodological conditions and limitations imposed by the nature of the available source material.

In his calculations of bear abundance, Schmölcke (this volume) only uses counts of wild mammals for the Neolithic assemblages. Differentiation of wild and domestic forms of *Bos* and *Sus* is, however, problematic. Furthermore, the percentage of bears relative to other wild meat animals logically increases in the Neolithic, as a substantial part of the meat supply comes from domesticated animals. Employing calculations based solely on wild mammals therefore hinders a direct comparison between Neolithic and Mesolithic assemblages when, as in the present study, the overall importance of bears for various human groups is to be evaluated. The calculations undertaken here are therefore based on the total number of identified specimens (NISP) of both wild and domesticated mammals.

The total assemblage size is important in calculations of abundancy, especially for a rare species like the bear. From a statistical point of view, it is necessary to investigate 300 bones to obtain a 95 % probability of finding at least one bone of an animal that constitutes 1 % of the total assemblage:

$$P_X(0) = \binom{300}{0} 0{,}01^0 (1 - 0{,}01)^{300-0} \approx 0.049$$

$$1 - 0.049 = 0.951 \approx 95\,\%$$

Calculations based on small assemblages are therefore in danger of yielding misleading results. Schmölcke (this volume) omitted assemblages with a total bone count of less than 200 from his calculations. In this study, the cut-off point has been raised to 300 bones, because Figure 7 (see below) suggests that unrealistically high bear frequencies (>2 %) are obtained for assemblages containing 200–300 bones.

One weakness associated with calculations of constancy when used to compare values for different periods is the underlying assumption that bone assemblages have had the same size distribution through time. This will hardly ever have been the case, and the calculations may therefore give skewed results with respect to constancy for rare species such as the bear, as the bones are more likely to be found in larger assemblages than in smaller ones.

Mixed assemblages represent a challenge to all the following calculations. Minor admixtures in assemblages from specific cultures must be assumed at most of the sites investigated here and do not present a major problem. The majority of the EBC records are, however, from sites that were also occupied during the Early FBC, in particular. The bone assemblages from these sites have typically been investigated/reported with no differentiation between these two cultural/chronological entities, while the bear bones have in almost all cases been assigned to the EBC occupation, either by stratigraphic observations or direct ^{14}C-dating. When comparing the percentage of bears in the EBC to that of other cultures, it must therefore be remembered that the numbers obtained are typically too low, due to the inevitable inclusion of FBC mammal bones in the calculations. The EBC part of the total assemblage is typically much greater than that of the FBC (due to several factors which affect and impact the upper layers of sites more than the lower ones, thereby leading to a greatly reduced FBC assemblage size). Deviations from the true values for the EBC can therefore generally be considered to be modest.

Mixed assemblages also present a challenge when enumerating the total number of bone assemblages from a specific culture employed in the calculations of the constancy of bear finds. The totals employed below are derived from the database of the Zoological Museum in Copenhagen (Table 3). Only finds that can definitely be assigned to either a single culture or to two, usually chronologically adjacent, cultures have been included. The latter have been weighted 50:50 in the site totals (e.g. a bone assemblage that can only be dated to either the EBC or the FBC contributes 0.5 points to the total for each of these two cultures). Uncertain finds and those with broad dating ranges have been excluded.

Finds complexes yielding imported bear bones have been omitted from the calculations, apart from Kainsbakke, where only some of the bear bones were imports. For both practical and methodological reasons, NISP (the number of identified specimens) is used instead of MNI (the minimum number of individuals) as a quantitative measure in the following. The MNI of bears in the assemblages recorded in Table 1 has only been determined in a few exceptional cases, while NISP is the only measure that is available for almost all assemblages. The decision to use NISP instead of MNI data also concurs with the methodological considerations of Makarewicz/Pleuger (2020, 281–282) with respect to the Kainsbakke assemblage.

Constancy of bear records

Constancy is presented as a range of values in Table 4, where these were calculated as a minimum (only sites with bear remains that can be ascribed to a single culture) and a maximum (including sites which may belong to either of two cultures).

Bear remains have been found at approximately one in six MMC settlements with preserved bones. There are no finds from Jutland and Funen (Fig. 3), but there is also a virtual absence of assemblages with preserved bones (only two are known from Funen, and none from Jutland). The constancy value given in Table 4 has therefore been calculated using data for Zealand only, but can nevertheless be assumed to be valid for the entire country.

A drastic decline is evident in the KMC, which is certainly at least partly due to the ongoing or possibly virtually complete extinction of bears on the Danish islands, the adverse conditions for the preservation of bones in Jutland, and the inadequate number of excavated coastal sites in the latter region: It does not necessarily indicate a situation drastically different from that during the MMC in those areas of Denmark where bears were still common. This conclusion is corroborated by the evidence from southern Sweden, where there are many finds of bears from KMC sites – both at coastal and inland locations (Table 1). It is not possible to calculate a reliable constancy rate for the parts of Denmark where bears must be assumed to have still been common (Jutland).

Due to the extinction of bears on the Danish islands, constancy for finds from the EBC has only been calculated for sites in Jutland. Only exceptional imports (Vedbæk-Bøgebakken, Syltholm) are known from the Danish islands. In Jutland, bear bones appear at one in eight to nine EBC sites with preserved bones. In the FBC, the same considerations regarding the regional extinction of bears apply as in the EBC, and the constancy has again only been calculated for Jutland. No less than four (or five) finds (nos. 9–12, and possibly no. 7, in Table 1) from islands close to Jutland must be perceived as imports, and they have therefore not been included in the calculations, leaving only between three and six sites in Jutland. Based on these numbers, constancy in the FBC is slightly lower than in the EBC, with bear bones appearing at one in seven to 13 sites. The true numbers are possibly somewhat lower, as a large percentage of the sites with finds potentially belonging to the FBC have been excluded from the calculations due to their very mixed character. The percentage of FBC sites considered is lower than that for all the other cultures (Table 4).

The constancy for bear finds in the PWC is exceptionally high, with the species being represented in 60–80 % of all known assemblages. Total numbers are small, and there is consequently a rather large statistical uncertainty. Four of the five assemblages from the Danish PWC with preserved bones are, however, either small or very small, but bear is still present in two of them (sites 2, 8). In Denmark, the PWC proper is only represented in Jutland, from where all known bone assemblages (with and without bear remains) derive.

While it is not possible to obtain any reliable numbers for the KMC, it appears that constancy declines continuously from the MMC to the FBC and then increases (possibly dramatically) in the PWC.

Abundance of bear remains

The relative abundance of bear bones at the individual sites is shown in Table 5, while Figure 7 gives a graphical representation. Table 6 presents averaged values for the individual cultures. Data for southern Sweden and northern Germany are included, because the number of Danish finds with available data is too small to give meaningful results.

The highest value is evident in the MMC, while a decreasing tendency is apparent in the succeeding Mesolithic periods. Values rise again in the FBC, and the second highest figure is seen in the PWC. But, as discussed above, the numbers estimated for the EBC are probably too low. Furthermore, a closer look at the FBC finds reveals that the three highest values were all obtained for small assemblages from Mecklenburg-Western Pomerania. It therefore appears that the FBC values for this region are not representative for Denmark. When the three assemblages in question are omitted from the calculation, the average abundance of bears at FBC sites is only 0.16. It therefore seems likely that the relative abundance of bears, at least at Danish sites, decreases from the MMC and throughout the entire Mesolithic, reaching its lowest values in the FBC, after which it increases again markedly in the PWC. Due to the lack of data for Denmark, the assessment of the KMC is based on data from southern Sweden.

Quantity of bear finds: Concluding summary
Due to the methodological uncertainties outlined above, as well as the (in some cases) statistically weak nature of the database, not too much emphasis should be placed on either of these two results when viewed in isolation. Nevertheless, it is remarkable that the two quantitative approaches yielded identical results, indicating a decreasing number of bears in the archaeological record from the MMC up to, and including, the FBC, followed by a possibly steep increase in the PWC.

The decreasing tendency throughout the Mesolithic and early parts of the Neolithic probably results from a combination of several factors. The natural environment during the MMC was well-suited as bear habitat, and only inland sites from this period are known. In the EBC, on the other hand, the economic focus had shifted to the marine environment created by Atlantic transgressions, and almost all the EBC assemblages with preserved bones are from coastal sites. Furthermore, bears had become extinct on the Danish islands, and their numbers may also have begun to decrease in the forests of Jutland, relative to the situation in the Boreal and Preboreal periods. In the subsequent Neolithic, settlements remained focused in coastal and near-coastal areas, with few known bone assemblages from inland sites. Moreover, Neolithic farmers destroyed bear habitats by clearing woodland for arable land and pasture and thereby probably contributed to a further decline in bear populations. The high constancy and relative abundance of bear finds in the PWC is at odds with all these prevailing factors, as woodland destruction was already well advanced at this point in time (at least in parts of Jutland), while the economic focus had at least partly shifted back to the marine environment, and all known bone assemblages derive from coastal sites.

QUALITATIVE ASSESSMENT

Skeletal element representation
Table 7 shows the presence of various bear skeletal elements at Mesolithic and Neolithic sites in Denmark and adjacent regions. Sites with uncertain or extremely broad dates have been omitted. In addition, several sites listed in Table 1 could not be included due to a lack of published or otherwise accessible information. The remaining 55 sites have been ordered chronologically to enable any potential development through time to be traced.

Two groups of skeletal elements must be discussed separately from the rest: teeth and third phalanges. Teeth are by far the most frequent finds, with a total of 78 recorded examples. They are abundant throughout the entire Mesolithic and Neolithic, being found at settlements, ritual sites and in graves, and no differences between the individual cultures are apparent. Teeth have been used as ornaments, either sewn onto clothes or perforated and strung on necklaces etc., and these items have also been exchanged with regions where bears no longer lived, for example the island of Zealand during the EBC. In general, it can be stated that the great frequency of teeth in the record demonstrates that bears were highly valued throughout the entire period under investigation here. This is true of both the area of present-day Denmark and the adjacent regions.

Bear phalanges are usually interpreted as indicating the presence of bear pelts rather than entire bears. This question has been discussed intensively in relation to Iron Age grave finds (predominantly cremations, but also inhumation graves) in the area of present-day Denmark (see above), as well as more northerly parts of Scandinavia (e.g. Petré/Wigardt 1973; Petré 1980; Grimm 2013) and central Europe (e.g. Kühl 1984; Beermann 2016). There is, however, also at least one record of distal bear phalanges from a Neolithic cremation grave in northwest Germany, indicating the incineration of a human corpse and a bear skin (Bärenfänger 2009). The evidence referred to above demonstrates that only the third phalanges remained with the pelt on skinning. Table 7 shows that distal phalanges are rather rare among the records of bear remains. This scarcity cannot be explained by their small size, as the first and second phalanges, which are of similar size, are rather (second phalanx) or considerably (first phalanx) more frequent in the record. Furthermore, given their distinctive, claw-like shape, third phalanges are easily recognisable when excavated. Records of isolated third phalanges (i.e. without either the first or second phalanges) are only known from the Early Mesolithic and the Neolithic, with records from three sites in each case. A closer look at the Neolithic evidence shows that one record is from the extreme southern periphery of the study area (Mecklenburg), while another is a perforated example from a passage grave (Dræby Mark). The latter indicates an (exchanged) ornament rather than a bear skin. This leaves two examples from the Kainsbakke site as the only evidence for bear skins in the entire Danish Neolithic. In many societies in the northern hemisphere where ethnographical studies have been undertaken, the bear's skin is one of the ritually preserved body parts (Black 1998, 346), and this may provide an explanatory model for the occurrence of bear skins at Kainsbakke. It can be concluded that the Early Mesolithic was apparently the only period when such skins entered the archaeological record from ordinary settlements with some regularity.

As only the third phalanges potentially bear witness to the presence of bear skins, the first and second phalanges can be assumed to indicate the presence of bear feet, and they will therefore be discussed together with the metapodials and other bones making up the paws. In general, the first and second phalanges are frequently encountered at settlements in all the periods and in all the regions under investigation here – with the remarkable exception of MMC settlements, where they are extremely rare. This absence is hardly coincidental, as carpals/metacarpals, as well as tarsals/metatarsals, have frequently been found at MMC settlements, and several have yielded extremely large assemblages. Most of the MMC sites discussed here were excavated in the first half of the 20th century, and the observed lack of first and second phalanges may therefore be a consequence of the excavation technique employed at that time. Third phalanges of bears, as well similarly sized or smaller fish and bird bones, have, however, been recovered from some of these settlements. It is therefore possible that the absence of the first and second phalanges either result from a period- or culture-specific technique employed in skinning and dismembering bear carcasses, whereby these elements of no nutritional value were possibly left at the kill site. Alternatively, it may reflect a taboo or a specific, possibly ritualised, treatment of this part of the bear feet, which prevented this element from entering the archaeological record for settlements.

When looking at the representation of fore- and hind feet, some clear differences are evident in the relative numbers, together with some obvious chronological variations. Forefeet are rare at Neolithic sites but much more frequent at those of the Mesolithic, while the reverse is true of hind feet, being rather common in the Neolithic, and clearly more frequent than forefeet, while the opposite is the case in the Mesolithic.

Distinct differences can also be observed in the presence of skeletal elements representing the hind- and forelimbs. Hindlimbs are, in general, extremely rare at settlements of the Neolithic and the EBC. The PWC may represent an exception to this general picture, as bones from the hindlimbs have been encountered at two of four/five sites. These sites are either geographically peripheral (Nymölla II) or ritual in nature (Kainsbakke). Bones representing the hindlimbs are moderately abundant (Swedish

KMC) or even frequent (MMC) in earlier parts of the Mesolithic, while the picture for forelimbs is quite different. The latter are moderately abundant or frequent in the Neolithic (especially the PWC) and the EBC, and also well represented in the earlier parts of the Mesolithic. It therefore seems that all meat-bearing parts of the animal were taken back to the settlements in the earlier parts of the Mesolithic, while some form of selection took place in the Late Mesolithic and Neolithic.

When the evidence for the presence of feet and limbs is combined, some specific patterns emerge. Bones from the forefeet are extremely rare during the Neolithic and EBC, while those of the forelimbs are rather frequent. The situation regarding the hindlimbs and feet is, on the other hand, the direct opposite: Bones from the hind feet are frequent, while those representing the hindlimbs are virtually absent. The finds from FBC and EBC sites show a similar pattern, while the evidence from the PWC possibly deviates somewhat, even though it retains a basic similarity. The almost mutually exclusive occurrence of bones from the forefeet and forelimbs and from the hind feet and hindlimbs during the Late Mesolithic and Neolithic is striking and probably reflects a ritually guided selection of which specific parts of the bear were permitted or not permitted at the settlements. The specific importance of the paws in ritual activities is especially well documented ethnographically (see Mansrud, this volume; Hallowell 1926 for numerous examples). In the earlier parts of the Mesolithic, bears were obviously treated differently with respect to their limbs and paws. Both forelimbs and forefeet are roughly evenly represented and abundant, and the same is true to some degree of the hindlimbs and feet, although the extremely high frequency of hindlimb bones at MMC settlements is exceptional. There is no direct evidence for these body parts having a ritual significance.

Records of bones from the bear's torso are extremely rare in the Neolithic and the EBC, while they are moderately frequent in the earlier parts of the Mesolithic.

Records of cranial bones and mandibles are infrequent in the Neolithic, especially when two FBC records from inland sites in Mecklenburg, i.e. on the periphery of the study area, are disregarded. The deposition of these bones may have been guided by factors other than those operating in southern Scandinavia. Cranial elements and/or mandibles have only been found at ritual sites in Denmark: Kainsbakke (see above) and Hygind (Andersen 1988; 1989). The Hygind causewayed enclosure is located on the island of Funen, and the bear bones (exclusively cranial fragments and a tooth) were therefore imported items. It can also be demonstrated that some of the cranial parts found at Kainsbakke were imported (Klassen et al. 2020b; Price et al. 2021). An atlas bone found at a third Neolithic ritual site at Ginnerup, a few kilometres distant from Kainsbakke, can possibly be added to this list, as the atlas is the first vertebra in the vertebral column and directly adjacent to the skull. Furthermore, the ring-shaped morphology of this bone is distinctively different from all other vertebrae. There is therefore clear evidence for the ritual importance of bear heads in the Danish Neolithic, which concurs with the general ethnographical information (Black 1998, 346). A similar situation in the Swedish PWC is evident from the deposition of bear cranial bones in a pottery vessel at Äs, outside the study area (Lepiksaar 1974, 150), and the final record of Neolithic cranial remains from the Swedish PWC site of Nymölla/Möllehusen II, listed in Table 7, therefore probably also has ritual connotations. The EBC site of Ringkloster is remarkable in that it yielded an entire bear cranium (Fig. 8). Apart from those found at Kainsbakke, this is the only example of its kind in the entire archaeological record. The second record of cranial bones in an EBC context is from the Virksund kitchen midden. These form part of a tiny bone assemblage from a very early (1860s) excavation and are therefore difficult to evaluate. Nevertheless, in general it can be concluded that cranial bones, together with bones from the limbs and paws, probably constitute the second group of skeletal elements indicating related ritual ideas in the Late Mesolithic and Neolithic.

Cranial bones and mandibles are more frequent in the earlier parts of the Mesolithic, and even common in the MMC. This shows that the bear heads were either less ritually important in the earlier Mesolithic than in the Late Mesolithic and Neolithic or, perhaps more likely in the light of ethnographical evidence, their ritual meaning and treatment differed from the situation in the later parts of the study period.

Finds contexts, modes of deposition and treatment of bones
All records of bear bones from the MMC derive from ordinary settlements, and there are no reports of any unusual modes of deposition. A mandible from the Øgaarde site (DEGERBØL 1943, 182–183 fig. 93) had been marrow-fractured, but there are no other reports of marrow-fracturing in the material addressed here. This underlines the conclusion that bear mandibles were treated differently in the MMC relative to all the other periods.

Due to the almost complete lack of finds, it is not possible to make any statements about the handling of bear bones in the Danish KMC. The southern Swedish KMC records are all from settlements, and no unusual modes of deposition have been reported. From the time of the later KMC there is, however, a ritual site with clear evidence for the ritual importance of bears located at Kanaljorden, Motala (Östergötland), approximately 200 km north of the study area (HALLGREN/ FORNANDER 2016; GUMMESSON et al. 2018). In addition to human skulls mounted on poles, a number of faunal remains were found on a stone packing located in shallow water in a small lake. Of these, the bear remains attract particular attention due to their distinctive character. Not only were they especially numerous at this site, but these bones were also deposited spatially apart from those of the other species. The only parallels to the remarkably large number of ribs, vertebrae, and mandibles in this combination are found at settlements of the MMC, from which the Kanaljorden deposition does, however, differ due to the almost total absence of elements from the hindlimbs.

The evidence for the EBC is more diverse. In addition to actual settlement finds, there are several examples of the ritual treatment and demonstrated social importance of bear bones. The cranium found at Ringkloster in eastern Jutland has already been mentioned, and a bear tarsal was found in grave X at the Skateholm I cemetery in Scania. A tooth pendant found in a grave at Vedbæk-Bøgebakken (AARIS-SØRENSEN 1980) is the earliest recorded instance of the exchange of bear elements. A bear fibula found at Syltholm on Lolland (unpublished; information courtesy of Museum Lolland-Falster) may constitute a second example of exchange during the EBC, but this bone may also date from the Early FBC.

There is increasing evidence for the exchange of bear bones in the FBC. Examples (apart from the possible find at Syltholm) comprise teeth (Spodsbjerg), an ornament or amulet from a grave (perforated third phalanx from the Dræby Mark passage grave), at least one skull or parts of it (Hygind), tarsals (Lindø), and forelimb elements (Bundsø). The latter may represent the exchange of a rare and prestigious joint of meat, but the context in which it was found, a settlement located on the site of a causewayed enclosure, possibly draws parallels with the cranial elements from Hygind. The perforated metatarsus from Heidmoor-Wolkenwehe in northern Germany (EWERSEN 2007, 293, 305 Taf. 2.3) belongs typologically to a group of metapodial-pendants that have typically been produced from the bones of other species (e.g. LEHMKUHL 1987). It is, however, an outlier in this group, with respect to both animal species and where it was found, as the main distribution area for metapodial-pendants is Switzerland and east-central Germany.

The abundant evidence for the ritual character of the PWC bear bones found at Kainsbakke, and the fact that the possible PWC atlas bone from Ginnerup was also found at a ritual site, has already been mentioned. The two remaining Danish records from this culture appear to be from actual settlements and the same is true for the remains from the Swedish site of Möllehusen/Nymölla II.

Changing patterns in the human-bear relationship during the Danish Mesolithic and Neolithic

As is evident from the above overview, there are considerable differences between the individual Mesolithic and Neolithic cultures in Denmark with regard to the representation of bear bones in the archaeological record. These differences doubtlessly reflect, at least in part, changing patterns in the human-bear relationship through time. Nevertheless, changes in the natural environment, the early extinction of bears in parts of the study area (probably at least partly due to contact between humans and bears) and the uneven representation of various types of sites in time and space do, however, create a complex background against which the observed variations must be evaluated. In the following, an attempt will be made to paint a picture, using a broad brush, of the changes in the human-bear relationship through the Danish Mesolithic and Neolithic, while the evidence from adjacent regions will primarily be used to qualify this picture.

The archaeological record for the MMC represents groups of forest dwellers with an economy adapted to that specific environment and to the opportunities offered by bodies of fresh water. Bears were hunted on rare occasions and in very modest numbers, compared to other game. Apart from the first and second phalanges, all bear body parts are represented in the archaeological record, with no obvious signs of ritual selection or other related activities. The only possible exception to this is a single marrow-fractured mandible (one of ten known mandibles from MMC context). Given the enormous body of ethnographical and historical evidence about the ritual significance of bears in the hunter-gatherer societies of the northern hemisphere, it would certainly be incorrect, based on the present evidence, to conclude that bears were viewed simply as ordinary (though rare) game by MMC people. There is every reason to believe that specific rituals were performed before, during and after the bear hunt. The relative ritual importance of the bear may, however, have been less than that evident from the archaeological record for later cultures. Especially in relation to Neolithic groups, this may be explained by the fact that hunting and gathering were the only sources of food during the MMC, and that the obviously ritually important elk (Buck Pedersen/Brinch Petersen 2017) was, according to the ethnographical record for northeastern Europe, often connected with beliefs associated with the reproduction of game animals (Zvelebil 2003a; b). A second possibility is that the bear figurines (cf. Gross/Vang Petersen, this volume) which, at least potentially, might date from the time of the MMC, may reflect specific beliefs and bear rituals. Finally, it is possible that bear rituals could have been performed at specific ritual sites in natural settings and at locations spatially distinct from those of the settlements – sites that may not yet have been discovered. The potential existence of such sites is indicated by the KMC locality at Kanaljorden mentioned above.

Due to the lack of evidence, it is not possible to evaluate directly the human-bear relationship during the Danish KMC. In southern Sweden, the record for this period resembles that of the preceding MMC. This is an interesting observation because the finds derive partly from ecological settings that differ completely from those encountered during the MMC (i.e. coastal settlements). The evidence from these KMC coastal settlements is notably dissimilar to that from the coastal sites of the subsequent EBC and FBC. This can be cautiously interpreted as indicating a certain degree of continuity of MMC traditions with respect to the treatment of bears in the KMC. There are no obvious signs of ritual activities associated with bears in the faunal record for the study area (but see remarks above regarding the Kanaljorden site).

While there are various indications of continuity from the earlier parts of the Mesolithic, different rules appear to have governed the selection and treatment of bear bones in the EBC. Definite signs of bear rituals are now detectable for the first time in the Danish faunal record, especially with regard to the treatment of crania, limbs and paws. These indicate changes in the human-bear relationship during the EBC, which may reflect several factors. The strong marine orientation of the EBC following drastic changes to the natural environment may well have prompted changes in attitude towards woodland game. Furthermore, bears had completely disappeared from eastern Denmark, and encounters with these animals in Jutland may have become increasingly rare. There is also a third factor that should not be overlooked. Distinct elements of cultural traditions from the eastern/northeastern Baltic, such as pottery, elk-antler hammers and specific burial customs, became incorporated into the EBC (KLASSEN 2004, 109–120, 136–139). As demonstrated by the archaeological record for the PWC (see below), specific beliefs in relation to bears which originated in these regions may well have accompanied these cultural influences and may therefore be responsible for the observed changes in the treatment of bears.

In general, the evidence for the human-bear relationship during the FBC is characterised by continuity from the EBC, i.e. the selection/rejection of specific body parts and the ritual emphasis placed on bear heads and their treatment. This demonstrates that beliefs rooted in hunter-gatherer societies lived on in a (semi-)agricultural society, thereby possibly indicating a considerable degree of population continuity across the Mesolithic-Neolithic transition. Bears are even less common in FBC contexts, and, at the same time, there is strong evidence for an increased social valorisation of bear remains. While the exchange of such remains in the EBC can only be demonstrated in one or possibly two cases, there is evidence in the FBC for exchange of skulls, teeth, ornaments made from third phalanges, the bones of the hind feet, and possibly fibulas. Further hints may be provided by the peculiar geographical distribution of the FBC bear remains. These appear in marked concentrations on the Djursland peninsula, in southeastern Jutland and adjacent islands, and in eastern Holstein (Fig. 6). All three regions are hotspots for Neolithic research, which may well explain the observed concentrations, just as the small number of FBC sites with preserved bones in Jutland may explain the absence of bear remains from other parts of the peninsula. However, the constancy of bear finds on Djursland and in southeastern Jutland and adjacent islands appears to be extremely high when compared to the general value calculated for Jutland. There are three records of bear remains from the islands south of Funen, out of 11 sites with preserved bones (27 %) in the region (NYEGAARD 1985), whereas on Djursland there are three or four FBC sites with bear remains, out of 13 sites with preserved bones (23–31 %). The concentration of bear finds, at least in these two distinct regional concentrations, appears therefore to be not solely a result of research intensity and favourable conditions for preservation. This conclusion is possibly supported by the distribution pattern for southeastern Jutland and adjacent islands (bear records from the Middle Neolithic FBC) and that for eastern Holstein and adjacent islands (Early Neolithic FBC or possibly Middle/Late EBC). The distribution pattern for bear bones in the northernmost of these two regions, which includes many instances of imported bones on the islands, resembles that for imported, socially valorised exotic items (copper flat axes and various types of central European stone artefacts; cf. Fig. 9a) which circulated here and in adjacent parts of northern Germany during the later parts of the EBC and Early FBC. It is therefore possible that the distribution of bear bones in the region indicates the continued existence of a probably ritualised exchange system and that these bones were treated like other socially valorised items and circulated between coastal groups in the area. In the same way, the distribution of Early Neolithic or Late EBC bear bones in eastern Holstein and the adjacent island of Lolland resembles that of specific types of imported Late Ertebølle/Early Neolithic stone tools (Fig. 9b) across the Femern Belt. Irrespective of the question of the existence of exchange systems, it is obvious from the comparatively large number of sites yielding imported bones or teeth that bears were subjected to special attention during the FBC in the region, even though this was possibly only done in specific groups.

The large number of bones found at coastal sites, despite the probable scarcity of bears (in Denmark), especially in the densely populated coastal regions, the efforts made to obtain bear bones from other areas and other aspects, especially at Kainsbakke, indicate a marked shift in, or intensification of, the human-bear relationship during the genesis of the PWC in Denmark. As already mentioned, the ritual significance of bears is also evident in the eastern Swedish distribution area of the PWC, and further evidence from Norway (amber figurine from Linnes, cf. Gross/Vang Petersen, this volume) can probably be added to this. The ritual focus on bears during the PWC must, without doubt, derive from the culture's roots in the Pit-Comb Ware culture complex of northeastern Europe, from which there are numerous depictions and figurines of bears (Wyszomirska 1984; cf. also Gross/Vang Petersen, this volume). The combination of bears, elks, and a rare waterbird found in the ritual assemblage at Kainsbakke points in the same direction as indicated by both the archaeological (rock art: Tansem/Johansen 2008; Helskog 2012) and ethnographical data (see above).

In conclusion, it appears that the human-bear relationship in Denmark may have passed through several stages. In an early phase, represented by the MMC and possibly at least partly by the KMC, the bear appears to have been a rare but still regularly hunted animal. There is no direct evidence (from Denmark) indicating that it had any special ritual role, but in the light of abundant ethnographical evidence it seems likely that the bear hunt, and use of the animal in general, was strictly regulated by complex sets of beliefs and rituals.

The first definite archaeological evidence for specialised bear rituals appears in the EBC. At the same time, finds from Zealand and Scania of "hammers" made of elk antler, interpreted as the drumsticks of shamans, indicate a distinct influence on southern Scandinavia from the shamanistic traditions of the eastern Baltic (Vang Petersen 1990, 19–20; 1998, 95ff; Timofeev 1998a, 228–230; 1998b, 44–46; Klassen 2004, 119–120). It is possibly no coincidence that the marked emphasis on bear rituals evident in the PWC in Denmark also coincides with clearly detectable influences from the eastern Baltic, including shamanistic rituals involving the elk. The changes in the human-bear relationship in Denmark through time therefore probably reflect a combination of ecological change, human impact on bear habitats and shifting cultural impact from neighbouring regions, especially the eastern Baltic. Whether contacts to the south had the same impact, as possibly indicated by the incorporation of bear remains in the circulation of imported, socially valorised copper and stone artefacts, remains to be seen.

Acknowledgements

This paper draws on information collected in course of the CONTACT project (East Jutland Museum 2014–2019) founded by the VELUX foundation. Thanks to Søren H. Andersen for permitting publication of the bear cranium from Ringkloster, for providing important information on several other bear finds from Danish Ertebølle sites, and for valuable discussions. Museum Lolland-Falster is thanked for permission to mention the recently excavated find from Syltholm. Further information on Danish and Swedish finds of bears was provided by Leif Jonsson and Kristoffer Buck Pedersen. Jacob Kveiborg, Søren A. Sørensen, Oliver Grimm, Daniel Groß, and Torsten Madsen made valuable comments on earlier versions of this paper. Kasper Funch Dubery provided important assistance with probability statistics. The English language of the paper was revised by David E. Robinson.

Bibliography

Aaris-Sørensen 1976: K. Aaris-Sørensen, A zoological investigation of the bone material from Sværdborg I – 1943. In: B. Bille Henriksen, Sværdborg I. Excavations 1943–44. A Settlement of the Maglemose Culture. Arkæologiske Studier III (København 1976) 137–148.

Aaris-Sørensen 1980: K. Aaris-Sørensen, Depauperation of the Mammalian Fauna of the Island of Zealand during the Atlantic Period. Videnskabelige Meddelelser fra Dansk Naturhistorisk Forening 142 (København 1980).

Aaris-Sørensen 1985: K. Aaris-Sørensen, Den terrestriske pattedyrsfauna i det sydfynske øhav gennem Atlantikum og Tidlig Subboreal. In: J. Skaarup, Yngre Stenalder på øerne syd for Fyn. Meddelelser fra Langelands Museum (Rudkøbing 1985) 458–466.

Aaris-Sørensen 1990: K. Aaris-Sørensen, Danmarks Forhistoriske Dyreverden (København 1990).

Aaris-Sørensen 2009: K. Aaris-Sørensen, Diversity and dynamics of the mammalian fauna in Denmark throughout the last glacial-interglacial cycle, 115-0 kyr BP. Fossils and Strata 57, 2009, 1–59.

Andersen 1961: K. Andersen, Verupbopladsen. En Maglemoseboplads i Åmosen. Aarbøger for Nordisk Oldkyndighed og Historie 1960, 1961, 118–151.

Andersen 1983: S. H. Andersen, Kalvø – A Coastal Settlement of the Single Grave Culture. Journal of Danish Archaeology 2, 1983, 71–80.

Andersen 1988: N. H. Andersen, Hygind. Arkæologiske Udgravninger i Danmark 1987 (København 1988) 123 nr. 120.

Andersen 1989: N. H. Andersen, Hygind. Arkæologiske Udgravninger i Danmark 1988 (København 1989) 1212 nr. 126.

Bärenfänger 2009: R. Bärenfänger, Die bislang ältesten Brandbestattungen Niedersachsens. Die neolithischen Gräber in Westhammrich bei Leer. Archäologie in Niedersachsen 12, 2009, 33–35.

Beermann 2016: S. Beermann, Bärenkrallen und Bärenfelle in Brand- und Körpergräbern der vorrömischen Eisenzeit bis Völkerwanderungszeit in Mittel- und Nordeuropa. Universitätsforschungen zur Prähistorischen Archäologie aus dem Seminar für Ur- und Frühgeschichte der Universität Göttingen 279 (Bonn 2016).

Berlin 1941: H. Berlin, Benfynden från stenåldersboplatsen I Gualöv. Meddelanden Från Lunds Universitets Historiska Museum 1941, 151–152.

Black 1998: L. T. Black, Bear in Human Imagination and Ritual. Ursus 10, 1998, 343–347.

Brännborn et al. 2007: P. Brännborn/J. Jardfelt/A. Karlsson, Fågel, fisk och mitt i mellan-osteologisk analys af materialet från mesolitiska-neolitiskka boplatsen Sjöholmen. C-uppsats, Historisk osteologi, Institutionen för Arkeologi och Antikens Historia vid Lunds Universitet (Lund 2007).

Bray et al. 2013: S. C. E. Bray/J. J. Austin/J. L. Metcalf/K. Østbye/E. Østbye/S.-E. Lauritzen/K. Aaris-Sørensen/C. Valdiosera/C. J. Adler/A. Cooper, Ancient DNA identifies post-glacial recolonisation, not recent bottlenecks, as the primary driver of contemporary mtDNA phylogeography and diversity in Scandinavian brown bears. Diversity and Distributions 19, 2013, 245–256.

Broholm 1928: H. C. Broholm, Langøfundet. En boplads fra ældre stenalder paa Fyn. Aarbøger for Nordisk Oldkyndighed og Historie 1928, 129–190.

Buck Pedersen/Brinch Petersen 2017: K. Buck Pedersen/E. Brinch Petersen, Bringing Home the Elk: Preboreal Elk Depositions from southern Scandinavia. In: M. Sørensen/K. Buck Pedersen, Problems in Palaeolithic and Mesolithic Research. Arkæologiske Studier 12 (København 2017) 237–256.

Ceglielka et al. 1995: S. Ceglielka/M. Eriksson/J. Ingvald/O. Magnell/O. Nilsson/P. Nilsson, Jakten på det levande – en analys av det osteologiska materialet från den senmesolitiska boplatsen Bökeberg III. C-uppsats vid Arkæologiske Institutionen, Lunds Universitet (Lund 1995).

Degerbøl 1933: M. Degerbøl, Danmarks pattedyr i fortiden i sammenligning med recente former. Videnskabelige Meddelelser fra Dansk Naturhistorisk Forening i København 96, 1933, 357–641.

Degerbøl 1939: M. Degerbøl, Bundsø, en yngre Stenalders Boplads paa Als. IV: Dyreknogler. Aarbøger for Nordisk Oldkyndighed og Historie 1939, 85–198.

Degerbøl 1942: M. Degerbøl, Et knoglemateriale fra Dyrholm-bopladsen, en ældre stenalder-køkkenmødding. Med særlig Henblik paa uroksens køns-dimorphisme og paa kannibalisme i Danmark. In: T. Mathiassen/M. Degerbøl/J. Troels-Smith, Dyrholmen, En Stenalderboplads Paa Djursland. Det Kongelige Danske Videnskabernes Selskab, Arkæologisk-Kunsthistoriske Skrifter 1,1 (København 1942) 79–135.

Degerbøl 1943: M. Degerbøl, Om dyrelivet i Aamosen i stenalderen. In: T. Mathiassen, Stenalderbopladser i Aamosen. Nordiske Fortidsminder 3 (København 1943) 165–206.

Degerbøl/Krog 1951: M. Degerbøl/H. Krog, Den europæiske Sumpskildpadde (*Emys orbicularis* L.) i Danmark. En zoologisk og geologisk undersøgelse af danske postglaciale fund og deres betydning for bedømmelsen af temperaturforholdene i forhistorisk tid. Danmarks Geologiske Undersøgelse II, Række nr. 78, 1951, 1–130.

Eriksson/Magnell 2001: M. Eriksson/O. Magnell, Det djuriska Tågerup. Nya rön Kring Kongemose-Ertebøllekulturens jakt och fiske. In: P. Karsten/B. Knarrstöm (eds.), Tågerup. Specialstudier. Skånska spår. Arkeologi längs Västkustbanan (Lund 2001) 156–237.

Ewersen 2007: J. Ewersen, Die Tierknochen aus der neolithischen Siedlung Heidmoor, Kr. Segeberg. Untersuchungen zur Fauna eines jungsteinzeitlichen Fundplatzes unter besonderer Berücksichtigung wirtschaftshistorischer Aspekte. Untersuchungen und Materialien zur Steinzeit in Schleswig-Holstein aus dem Archäologischen Landesmuseum der Stiftung Schleswig-Holsteinische Landesmuseen Schloss Gottorf 4 (Neumünster 2007).

Forssander 1930: F. E. Forssander, En märklig stenåldersboplats vid Sjöholmen. Meddelanden från Lunds Universitets Historiska Museum 1929/30, 1930, 25–42.

Friis Johansen 1919: K. Friis Johansen, En boplads fra den ældste Stenalder i Sværdborg Mose. Aarbøger for Nordisk Oldkyndighed og Historie 1919, 106–235.

Gehl 1973: O. Gehl, Die Jagd- und Haustiere der steinzeitlichen Siedler von Basedow, Kreis Malchin. Jahrbuch Bodendenkmalpflege Mecklenburg 1973, 67–87.

Gehl 1975: O. Gehl, Die steinzeitliche Siedlung Stinthorst bei Waren/Müritz im Spiegel des Säugetierfundgutes. Jahrbuch Bodendenkmalpflege Mecklenburg 1975, 39–53.

Gehl 1979: O. Gehl, Die Nutzung von Haus- und Wildtieren nach dem Knochenfundgut der neolithischen Siedlung bei Glasow an der Randow, Kreis Pasewalk. Jahrbuch Bodendenkmalpflege Mecklenburg 1979, 39–48.

GLYKOU 2016: A. GLYKOU, Neustadt LA 156. Ein submariner Fundplatz des späten Mesolithikums und des frühesten Neolithikums in Schleswig-Holstein. Untersuchungen zur Subsistenzstrategie der letzten Jäger, Sammler und Fischer an der norddeutschen Ostseeküste. Untersuchungen und Materialien zur Steinzeit in Schleswig-Holstein und im Ostseeraum 7 (Kiel, Hamburg 2016).

GRIMM 2013: O. GRIMM, Bear-skins in northern European burials and some remarks on other bear-related furnishings in the north and middle of Europe in the 1st millennium AD. In: O. Grimm/U. Schmölcke (eds.), Hunting in northern Europe until 1500 AD. Old traditions and regional developments, continental sources and continental influences. Papers presented at a workshop organized by the Centre for Baltic and Scandinavian Archaeology (ZBSA) Schleswig, June 16th and 17th, 2011. Schriften des Archäologischen Landesmuseums, Ergänzungsreihe 7 (Neumünster 2013) 277–296.

GUMMESSON et al. 2018: S. GUMMESSON/F. HALLGREN/A. KJELLSTRÖM, Keep your head high: skulls on stakes and cranial trauma in Mesolithic Sweden. Antiquity 92, 2018, 74–90.

HALLGREN/FORNANDER 2016: F. HALLGREN/E. FORNANDER, Skulls on stakes and skulls in water. Mesolithic mortuary rituals at Kanaljorden, Motala, Sweden, 7000 BP. In: J. M. Grünberg/B. Gramsch/L. Larsson/J. Orschiedt/H. Meller (eds.), Mesolithic burials – Rites, symbols and social organisation of early postglacial communities. International Conference Halle (Saale), Germany, 18th–21st September 2013. Tagungen des Landesmuseum für Vorgeschichte 13.1 (Halle [Saale] 2016) 161–174.

HALLOWELL 1926: A. I. HALLOWELL, Bear ceremonialism in the northern hemisphere. American Anthropologist 28(1), 1926, 1–175.

HEINRICH 1999: D. HEINRICH, Die Tierknochen des frühneolithischen Wohnplatzes Wangels LA 505. Ein Vorbericht. Offa 54/55, 1997/98, 43–49.

HELSKOG 2012: K. HELSKOG, Bears and Meaning among Hunter-fisher-gatherers in northern Fennoscandia 9000–2500 BC. Cambridge Archaeological Journal 22(2), 2012, 209–236.

HENRIKSEN 2009: M. B. HENRIKSEN, Brudager mark – en romertidsgravplads nær Gudme på Sydøstfyn. Fynske Jernaldergrave 6,1. Fynske Studier 22,1 (Odense 2009).

IREGREN 1988: E. IREGREN, Finds of Brown Bear (*Ursus arctos*) in Southern Scandinavia – Indications of local hunting or trade? In: B. Hårdh/L. Larsson/D. Olausson/R. Petré (eds.), Trade and Exchange in Prehistory. Studies in Honour of Berta Stjernquist. Acta Archaeologica Lundensia Series in 8°, N°16 (Lund 1988) 295–308.

IREGREN et al. 1990: E. IREGREN/B. RINGBERG/A.-M. ROBERTSSON, The brown bear (*Ursus arctos* L.) find from Ugglarp, southernmost Sweden. The skeleton, its age and environment. Sveriges Geologiska Undersökning, Serie C, Nr. 824 (Uppsala 1990).

IVERSEN 2010: R. IVERSEN, In a World of Worlds. The Pitted Ware Complex in a large-scale Perspective. Acta Archaeologica 81, 2010, 5–43.

JESSEN 1929: K. JESSEN, Bjørnen (*Ursus arctos* L.) i Danmark. Meddelelser fra Dansk Geologisk Forening 7, 1929, 273–286.

JONSSON 1988: L. JONSSON, The vertebrate faunal remains from the Late Atlantic settlement Skateholm in Scania, South Sweden. In: L. Larsson (ed.), The Skateholm Project. I. Man and Environment. Acta Regiae Societatis Humaniorum Litterarum Lundensis 79 (Stockholm 1988) 56–88.

JORDAN 2003: P. D. JORDAN, Investigating Post-Glacial hunter gatherer landscape enculturation: ethnographic analogy and interpretative methodologies. In: L. Larsson/H. Kindgren/K. Knutsson/D. Loeffler/A. Åkerlund (eds.), Mesolithic on the Move. Papers presented at the Sixth International Conference on the Mesolithic in Europe, Stockholm 2000 (Oxford 2003) 128–138.

KLASSEN 2004: L. KLASSEN, Jade und Kupfer. Untersuchungen zum Neolithisierungsprozess im westlichen Ostseeraum unter besonderer Berücksichtigung der Kulturentwicklung Europas 5500–3500 BC. Jysk Arkæologisk Selskabs Skrifter 47 (Højbjerg 2004).

KLASSEN et al. 2020a: L. KLASSEN/R. IVERSEN/N. NØRKJÆR-JOHANNSEN/U. RASMUSSEN/O. B. POULSEN, The Pitted Ware culture on Djursland in the Neolithic world. In: L. Klassen (ed.), The Pitted Ware Culture on Djursland. Supra-regional significance and contacts in the Middle Neolithic of southern Scandinavia. East Jutland Museum Publications 5 (Aarhus 2020) 451–489.

KLASSEN et al. 2020b: L. KLASSEN/T. D. PRICE/K.-S. SJÖGREN/L. WINCENTZ/B. PHILIPPSEN, Strontium and lead isotope studies of faunal and human remains from Kainsbakke and Kirial Bro. In: L. Klassen (ed.), The Pitted Ware Culture on Djursland. Supra-regional significance and contacts in the Middle Neolithic of southern Scandinavia. East Jutland Museum Publications 5 (Aarhus 2020) 407–446.

KÜHL 1984: I. KÜHL, Animal Remains in Cremations from the Bronze Age to the Viking Period in Schleswig-Holstein, North Germany. In: C. Grigson/J. Clutton-Brock (eds.), Animals and Archaeology 4. Husbandry in Europe. BAR International Series 227 (Oxford 1984) 209–220.

KURCK 1917: C. KURCK, Den forntida utbredningen af kärrsköldpaddan *Emys orbicularis* (Lin.) i Sverige, Danmark och angränsande länder. Lunds Universitets Årsskrift N.F. 13, 1917, 1–128.

LAGERGREN-OLSSON 2001: A. LAGERGREN-OLSSON, Våtmarkslager och mesolitiska lämningar i Önsvala, Skåne, Nevishögs socken, Mölleberga 2:18. Arkeologisk Förundersökning. UV Syd Rapport 2001:42 (Lund 2001).

LARSSON 1985: L. LARSSON, En kustboplats från tidig bondestenålder i södra Skåne. Ale 1985(4), 1–14.

LARSSON 1986: M. LARSSON, Bredasten – An early Ertebølle Site with a Dwelling Structure in South Scania. Meddelanden från Lunds Universitets Historiska Museum 1985/86, 1986, 25–49.

LEHMKUHL 1987: U. LEHMKUHL, Archäozoologische und typologische Untersuchungen an Metapodia-Anhängern des Neolithikums und der Bronze- bis frühen Eisenzeit in der DDR. Jahrbuch für Bodendenkmalpflege Mecklenburg 1986, 1987, 19–38.

Lehmkuhl 1988: U. Lehmkuhl, Zur Kenntnis der Fauna vom mesolithischen Fundplatz Tribsees, Kreis Stralsund. Bodendenkmalpflege in Mecklenburg, Jahrbuch 1987, 1988, 47–82.

Lehmkuhl 1989: U. Lehmkuhl, Erste Ergebnisse der Tierknochenuntersuchungen von der neolithischen Siedlung Parchim (Löddigsee). Bodendenkmalpflege in Mecklenburg, Jahrbuch 1988, 1989, 47–83.

Lehmkuhl 2010: U. Lehmkuhl, Die Tierbeigaben aus den Gräbern von Ostorf (Mecklenburg) unter besonderer Berücksichtigung der durchbohrten Tierzähne. Mit einem Beitrag zu den Tierbeigaben aus den Gräbern von Tangermünde bei Stendal. Bericht der Römisch-Germanischen Kommission 88, 2007 (2010), 385–414.

Lepiksaar 1974: J. Lepiksaar, Djurrester från den mellanneolitiska (gropkeramiska) boplatsen vid Äs, Romfartuna sn, Västmanland. In: L. Löfstrand, Yngre stenålderns kustboplatser: undersökningarna vid Äs och studier i den gropkeramiska kulturens kronologi och ekologi. Archaeological Studies Uppsala Universitet 1, AUN 1 (Uppsala 1974) 140-156.

Lepiksaar 1978: J. Lepiksaar, Bone remains from the Mesolithic settlements Ageröd I:B and Ageröd I:D. In: L. Larsson, Ageröd I:B – Ageröd I:D. A Study of Early Atlantic Settlement in Scania. Acta Archaeologica Lundensia, Series in 4°, N°12 (Bonn, Lund 1978) 234–258.

Lepiksaar 1982: J. Lepiksaar, Djurrester från den tidigatlantiska boplatsen vid Segebro nära Malmö i Skåne (Sydsverige). In: L. Larsson, Segebro. En tidigatlantisk boplats vid sege ås mynning. Malmöfynd 4 (Malmö 1982) 105–128.

Lepiksaar 1983: J. Lepiksaar, Animal remains from the Mesolithic bog site at Ageröd V in central Scania. In: L. Larsson, Ageröd V. An Atlantic Bog Site in Central Scania. Acta Archaeologica Lundensia, Series in 8°, N°12 (Lund 1983) 159–167.

Liljegren 1975: R. Liljegren, Subfossila vertebratfynd från Skåne. University of Lund, Department of Quaternary Geology Report 8 (Lund 1975).

Lüttschwager 1954: J. Lüttschwager, Studien an vorgeschichtlichen Wirbeltieren Schleswig-Holsteins. Schriften des Naturwissenschaftlichen Vereins für Schleswig-Holstein 27, 1954, 22–33.

Lüttschwager 1967: J. Lüttschwager, Kurzbericht über Tierfunde aus meso- und neolithischen Moorsiedlungen in Schleswig-Holstein. Schriften des Naturwissenschaftlichen Vereins für Schleswig-Holstein 37, 1967, 53–64.

Madsen et al. 1900: A. P. Madsen/S. Müller/C. Neergaard/C. G. J. Petersen/E. Rostrup/K. J. V. Steenstrup/H. Winge, Affaldsdynger Fra Stenalderen i Danmark. Undersøgte for Nationalmuseet (Paris, København, Leipzig 1900).

Magnell 2006: O. Magnell, Tracking Wild Boar and Hunters. Osteology of Wild Boar in Mesolithic South Scandinavia. Studies in Osteology 1. Acta Archaeologica Lundensia Series in 8°, N°51 (Lund 2006).

Magnell 2007: O. Magnell, Djuren och människan. In: M. Andersson (ed.), Kustslättens mötesplatser (Stockholm 2007) 51–86.

Makarewicz/Pleuger 2020: C. Makarewicz/S. Pleuger, Herder-hunter fishers and agricultural contacts. Zooarchaeological perspectives on Pitted Ware animal exploitation strategies from Djursland. In: L. Klassen (ed.), The Pitted Ware Culture on Djursland. Supraregional significance and contacts in the Middle Neolithic of southern Scandinavia. East Jutland Museum Publications 5 (Aarhus 2020) 279–339.

Marseen 1953: O. Marseen, Fangstfolk på Selbjerg. Kuml 1953, 102–120.

Møhl 1978: U. Møhl, Bjørnekløer og brandgrave. Dyreknogler fra germansk jernalder i Stilling. Kuml 1977, 119–129.

Nikulina/Schmölcke 2015: E. Nikulina/U. Schmölcke, First archaeogenetic results verify the mid-Holocene occurrence of Dalmatian pelican *Pelecanus crispus* far out of present range. Journal of Avian Biology 46, 2015, 344–351.

Noe-Nygaard 1983: N. Noe-Nygaard, A new Find of Brown Bear (*Ursus arctos*) from Star Carr and Other Finds in the Late Glacial and Post Glacial of Britain and Denmark. Journal of Archaeological Science 10, 1983, 317–325.

Nyegaard 1985: G. Nyegaard, Faunalevn fra yngre stenalder på øerne syd for Fyn. In: J. Skaarup, Yngre Stenalder på øerne syd for Fyn. Meddelelser fra Langelands Museum (Rudkøbing 1985) 426–457.

Paproth 1976: H.-J. Paproth, Studien über das Bärenzeremoniell I. Bärenjagdriten und Bärenfeste bei den tungusischen Völkern. Skrifter utgivna av religionshistoriska institutionen i Uppsala 15 (Uppsala 1976).

Petré 1980: M. Petré, Björnfällen i begravningsritualen – statusobjekt speglande regional skinnhandel? Fornvännen 75, 1980, 5–14.

Petré/Wigardt 1973: B. Petré/M. Wigardt, Björnklor i svenska gravfynd från jernalderen. Trebetygsuppsats framladg i Seminariet för arkeologi, särskilt nordeuropeisk vid Stockholms Universitet (Stockholm 1973).

Philippsen et al. 2020: B. Philippsen/R. Iversen/L. Klassen, The Pitted ware culture chronology on Djursland. New evidence from Kainsbakke and other sites. In: L. Klassen (ed.), The Pitted Ware Culture on Djursland. Supra-regional significance and contacts in the Middle Neolithic of southern Scandinavia. East Jutland Museum Publications 5 (Aarhus 2020) 257–277.

Price et al. 2021: T. D. Price/L. Klassen/ K.-G. Sjögren, Pitted Ware Culture: Isotopic Evidence for Contact between Sweden and Denmark across the Kattegat in the Middle Neolithic, ca. 3000 BC. Journal of Anthropological Archaeology 61, 2021. https://doi.10.1016/j.jaa.2020.101254.

Reichstein 1985: H. Reichstein, Die Tierknochen vom mittelneolithischen Fundplatz Neukirchen Bostholm, Kreis Schleswig-Flensburg. Offa 42, 1985, 331–345.

Richter 1986: J. Richter, Brown Bear (*Ursus arctos*) from Kainsbakke, East Jutland. Journal of Danish Archaeology 5, 1986, 125–134.

Rosengren 2018: E. Rosengren, A partial Catalogue of the Mammalian Archaeofaunal Material in the Collection at Lund University Historical Museum (LUHM) (Lund 2018).

Rosenlund 1980: K. Rosenlund, Knoglematerialet fra bopladsen Lundby II. In: B. Bille Henriksen, Lundbyholmen. Pladser af Maglemose-type i Sydsjælland. Nordiske Fortidsminder Serie B6 (København 1980) 128–137.

Rowley-Conwy 1998: P. Rowley-Conwy, Meat, furs and skins: Mesolithic animal bones from Ringkloster, a seasonal hunting camp in Jutland. Danish Journal of Archaeology 12, 1994/95 (1998), 87–98.

Salomonsson 1971: B. Salomonsson, Malmötraktens Förhistoria. In: O. Bjurling (ed.), Malmö Stads Historia 1 (Malmö 1971) 15–170.

Sarauw 1903: G. F. L. Sarauw, En Stenalders Boplads i Maglemose ved Mullerup, sammenholdt med beslægtede Fund. Bidrag til Belysning af Nystenalders Begyndelse i Norden. Aarbøger for Nordisk Oldkyndighed og Historie 1903, 148–315.

Skousen 2008: H. Skousen, Arkæologi i lange baner. Undersøgelser forud for anlæggelsen af motorvejen nord om Århus 1998–2007 (Højbjerg 2008).

Tansem/Johansen 2008: K. Tansem/H. Johansen, The World Heritage Rock Art in Alta. Adoranten 2008, 65–84.

Thomas 1954: S. Thomas, Sjöholmen, Site 179. A re-examination. In: C. A. Althin (ed.), The chronology of stone age settlement in Scania, Sweden I. The Mesolithic settlement. Acta Archaeologica Lundensia Series in 4°, N°1 (Lund 1954) 169–187.

Thomsen/Jessen 1906: T. Thomsen/A. Jessen, Brabrandfundet fra den ældre stenalder, arkæologisk of geologisk behandlet. Aarbøger for Nordisk Oldkyndighed og Historie 1906, 1–74.

Timofeev 1998a: V. I. Timofeev, The beginnings of the Neolithic in the eastern Baltic. In: M. Zvelebil/L. Domanska/R. Denell (eds.), Harvesting the Sea, Farming the Forest. The Emergence of Neolithic Societies in the Baltic Region. Sheffield Archaeological Monographs 10 (Sheffield 1998) 225–236.

Timofeev 1998b: V. I. Timofeev, The east-west relations in the Late Mesolithic and Neolithic in the Baltic region. Baltic-Pontic Studies 5, 1998, 44–58.

Vang Petersen 1990: P. Vang Petersen, Eksotiske faunarester i Kongemose- og Ertebølletid – et resultat af udveksling? Hikuin 16, 1990, 17–30.

Vang Petersen 1998: P. Vang Petersen, Rav, hjortetak og mesolitisk magi. Danefæ fra jægerstenalderen. Nationalmuseets Arbejdsmark 1998, 87–100.

Wamers 2009: E. Wamers, Von Bären und Männern. Berserker, Bärenkämpfer und Bärenführer im frühen Mittelalter. Zeitschrift für Archäologie des Mittelalters 37, 2009, 1–46.

Wamers 2015: E. Wamers, „Steh nun auf, mein lieber Bär, deine Gäste zu empfangen!" Zum Bärenkult zirkumpolarer Jägervölker. In: E. Wamers (ed.), Bärenkult und Schamanenzauber. Rituale früher Jäger (Regensburg 2015) 41–83.

Wincentz 2020: L. Wincentz, Kainsbakke and Kirial Bro: the two main sites of the Pitted Ware culture on Djursland. In: L. Klassen (ed.), The Pitted Ware Culture on Djursland. Supra-regional significance and contacts in the Middle Neolithic of southern Scandinavia. East Jutland Museum Publications 5 (Aarhus 2020) 35–140.

Wincentz Rasmussen 1984: L. Wincentz Rasmussen, Kainsbakke A 47. A Settlement Structure from the Pitted Ware Culture. Journal of Danish Archaeology 3, 1984, 83–98.

Wincentz Rasmussen 1991: L. Wincentz Rasmussen, Kainsbakke, en kystboplads fra yngre stenalder. In: L. Wincentz Rasmussen/J. Richter (eds.), Kainsbakke (Grenå 1991) 1–69.

Winge 1904: H. Winge, Om jordfundne pattedyr fra Danmark. Videnskabelige Meddelelser fra Dansk Naturhistorisk Forening i København 56, 1904, 193–304.

Wyszomirska 1984: B. Wyszomirska, Figuralplastik och gravskick hos Nord- och Nordösteuropas neolitiska fångst-kulturer. Acta Archaeologica Lundensia Series in 4°, N°18 (Lund 1984).

Wyszomirska 1988: B. Wyszomirska, Ekonomisk stabilitet vid kusten: Nymölla III, en Tidigneolitisk bosättning med fångstekonomi i nordöstra Skåne. Acta Archaeologica Lundensia, Series in 8°, N°17 (Stockholm 1988).

Zachrisson/Iregren 1974: I. Zachrisson/E. Iregren, Lappish Bear Graves in Northern Sweden. Archaeological and Osteoarchaeological Study. Early Norrland 5 (Stockholm 1974).

Zvelebil 2003a: M. Zvelebil, People behind the lithics: Social life and social conditions of Mesolithic communities in temperate Europe. In: J. Moore/L. Bevan (eds.), Peopling the Mesolithic in Northern Environment. BAR International Series 1157 (Oxford 2003) 1–26.

Zvelebil 2003b: M. Zvelebil, Enculturation of Mesolithic Landscapes. In: L. Larsson/H. Kindgren/K. Knutsson/D. Loeffler/A. Åkerlund (eds.), Mesolithic on the Move. Papers presented at the Sixth International Conference on the Mesolithic in Europe, Stockholm 2000 (Oxford 2003) 65–73.

Dr. Lutz Klassen
Museum Østjylland
Randers
Denmark
lk@museumoj.dk

Associate Prof. Kristian Murphy Gregersen
Det Kongelige Akademi
Institut for Konservering
København
Denmark
kgre@kadk.dk

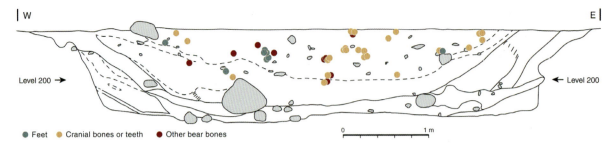

Fig. 1. The distribution of bear bones in pit A47 at the Kainsbakke site on the Djursland peninsula in Denmark, dated to the early 3rd millennium BC (after WINCENTZ 2020, 51 fig. 18).

Fig. 2. Kainsbakke (Djursland, Denmark). Cranium of bear deposited on top of an aurochs horn core in the early 3rd millennium BC (photo L. Wincentz, after WINCENTZ 2020, 50 fig. 17).

Fig. 3. Maglemose culture bear records from Denmark, southern Sweden, and northern Germany. Numbers refer to Table 1. Records that may belong to a chronologically adjacent culture are indicated by a lighter signature.

Fig. 4. Kongemose culture bear records from Denmark, southern Sweden, and northern Germany. Numbers refer to Table 1. Records that may belong to a chronologically adjacent culture are indicated by a lighter signature.

Fig. 5. Ertebølle culture bear records from Denmark, southern Sweden, and northern Germany. Numbers refer to Table 1. Records that may belong to a chronologically adjacent culture are indicated by a lighter signature.

Fig. 6. Neolithic bear records from Denmark, southern Sweden, and northern Germany. Numbers refer to Table 1. Red symbols: Funnel Beaker culture; blue symbols: Pitted Ware culture. Uncertain records or Funnel Beaker records that may belong to a chronologically adjacent culture are indicated by a lighter signature. The Holme Skanse site from the Late Single Grave culture is indicated by a yellow signature.

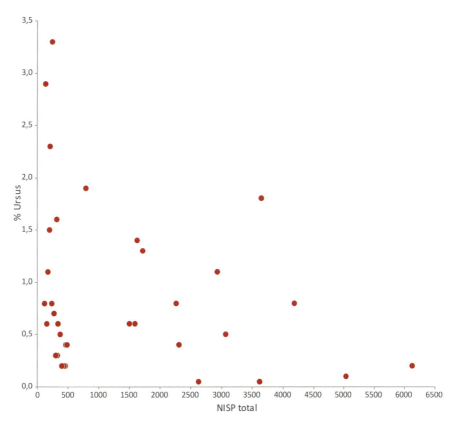

*Fig. 7. Relative abundance of bear (*Ursus*) bones at individual sites in Denmark, southern Sweden, and northern Germany. Data from Table 5. The Virksund site (% Ursus >16) has been omitted.*

Fig. 8. Bear cranium from the Late Mesolithic (Ertebølle culture) site of Ringkloster in eastern Jutland, dated to c. 4600 BC (courtesy of S. H. Andersen, photo Moesgaard Museum).

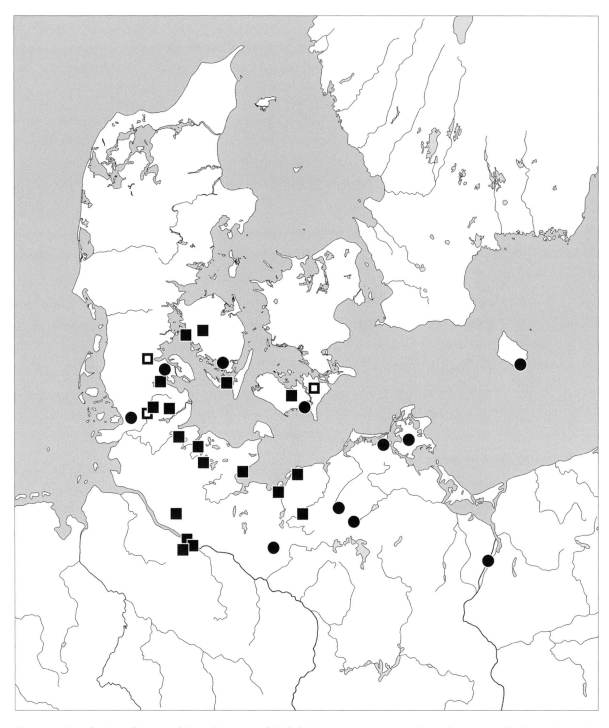

Fig. 9. a: Distribution of imported Danubian axes of Erikshale type (squares) and the earliest copper finds (circles) in the southwestern part of the Baltic, demonstrating the circulation of socially valorised items among coastal groups in the area during the Late Mesolithic/Early Neolithic. Open signatures indicate finds with imprecisely known find locations (after KLASSEN *2004, 319 fig. 153).*

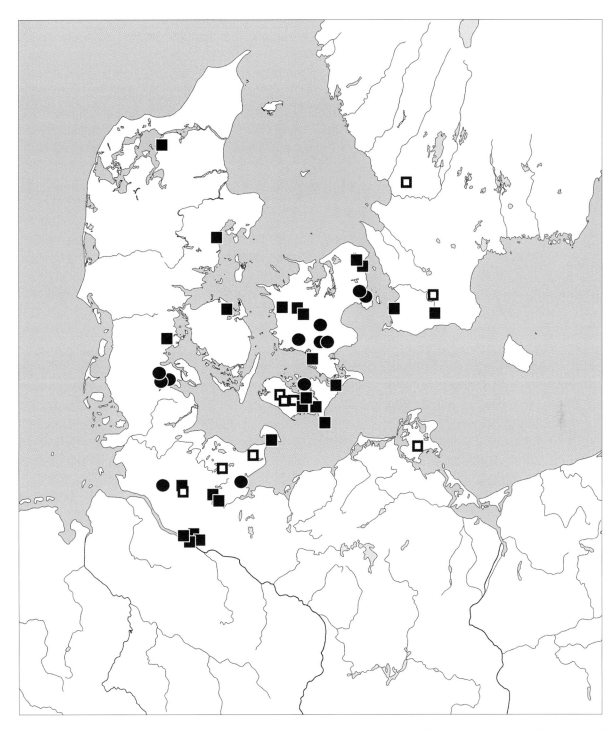

Fig. 9. b: Distribution of imported Danubian axes of the Böken, Neversdorf and Skeldby types (squares), as well as of axes of alpine rock (circles) in the southwestern Baltic, demonstrating an exchange across the Femern Belt during the Late Mesolithic/ Early Neolithic. Open signatures indicate finds with imprecisely known find locations (after KLASSEN 2004, 319 fig. 154).

Table 1. Basic data for records of bear bones from archaeological contexts in Denmark, southern Sweden, and northern Germany dating from the Mesolithic and Neolithic (c. 9500–2600 BC). MMC = Maglemose culture; KMC = Kongemose culture; EBC = Ertebølle culture; FBC = Funnel Beaker culture; PWC = Pitted Ware culture; BAC = Battle Axe culture; SGC = Single Grave culture; MN = Middle Neolithic; EN = Early Neolithic; det. = determined by.

Site no.	Site	Province	Country	Type of site	Culture	Date BC cal	NISP Ursus	Site Location	Reference
1	Kainsbakke	Djursland	Denmark	Ritual site	PWC	3000–2800 BC	34	Coastal	Richter 1986; Makarewicz/Pleuger 2020
2	Kirial Bro	Djursland	Denmark	Settlement	PWC	3000–2800 BC	1	Coastal	Makarewicz/Pleuger 2020
3	Kolind	Djursland	Denmark	Settlement	FBC MN A	3330–2930 BC	1	Coastal	Degerbøl 1942; Klassen et al. 2020b
4	Ørum Aa	Djursland	Denmark	Settlement	FBC EN II	3630–3370 BC	2	Coastal	Madsen et al. 1900; Degerbøl 1939; Klassen et al. 2020b
5	Ginnerup	Djursland	Denmark	Ritual site	FBC–PWC	3200–3000 BC	1	Coastal	unpublished; Jakob Kveiborg det.
6	Lindegårds Mose	East Jutland	Denmark	Settlement	FBC EN	3800–3300 BC	3	Coastal	Skousen 2008
7	Bundsø	South Jutland	Denmark	Ritual site/Settlement	FBC MN A	3340–2950 BC	4	Near coastal	Degerbøl 1939; Nyegaard 1985; Klassen et al. 2020b
8	Selbjerg	North Jutland	Denmark	Settlement	PWC	2880–2630 BC	1	Coastal	Marsen 1953; Klassen et al. 2020b
9	Hygind	Funen	Denmark	Ritual site	FBC EN II/MN A	3500–2800 BC	4	Near coastal	unpublished
10	Draby Mark	Funen	Denmark	Grave	FBC MN A	3300–2800 BC	1	Near coastal	Broholm 1928; Jessen 1929; Degerbøl 1939
11	Spodsbjerg	Langeland	Denmark	Settlement	FBC MN A V	3000–2800 BC	1	Coastal	Nyegaard 1985
12	Lindø	Langeland	Denmark	Settlement	FBC MN A	3300–2800 BC	2	Near coastal	Degerbøl 1939; Nyegaard 1985
13	Øgaarde	Zealand	Denmark	Settlement	MMC	9000–6000 BC	12	Inland	Degerbøl 1943; Degerbøl/Krogh 1951
14	Dyrholmen	Djursland	Denmark	Settlement	Early EBC	5200–5000 BC	1	Coastal	Degerbøl 1942
15	Vedbæk-Bøgebakken	Zealand	Denmark	Grave	Early EBC	5300–4600 BC	1	Coastal	Aaris-Sørensen 1980
16	Virksund	North Jutland	Denmark	Settlement	Late EBC	4300–4000 BC	8	Coastal	Winge 1904; Bray et al. 2013
17	Bjørnsholm	North Jutland	Denmark	Settlement	Late EBC	4400–4150 BC	1	Coastal	Møhl 1978
18	Ringkloster	East Jutland	Denmark	Settlement	Middle/Late EBC	4600–4400 BC	5	Inland	Rowley-Conwy 1998; Glykou 2016; S.H. Andersen pers. comm.
19	Flynderhage	East Jutland	Denmark	Settlement	Middle/Late EBC	4600–4000 BC	1	Coastal	Aaris-Sørensen 1980; S.H. Andersen pers. comm.
20	Brabrand	East Jutland	Denmark	Settlement	Late EBC	4400–4000 BC	2	Coastal	Thomsen/Jessen 1906; Jessen 1929
21	Øster Jølby	North Jutland	Denmark	Settlement	EBC/FBC EN	5300–3300 BC	1	Coastal	S.H. Andersen pers. comm.
22	Kolding Fjord	East Jutland	Denmark	Settlement (?)	EBC/FBC	5300–2800 BC	1	Coastal	Jessen 1929

Cont. Tab. 1.

Cont. Table 1.

Site no.	Site	Province	Country	Type of site	Culture	Date BC cal	NISP *Ursus*	Site Location	Reference
23	Kolind	Djursland	Denmark	Settlement	Early EBC	5300–4000 BC	1	Coastal	Degerbøl 1942
24	Mullerup	Zealand	Denmark	Settlement	MMC	9000–6000 BC	10	Inland	Sarauw 1903
25	Sværdborg Mose I	Zealand	Denmark	Settlement	MMC	9000–6000 BC	31	Inland	Friis Johansen 1919; Aaris-Sørensen 1976
26	Lundby I	Zealand	Denmark	Settlement	MMC	9000–6000 BC	2	Inland	Degerbøl 1933
27	Verup	Zealand	Denmark	Settlement	MMC	9000–6000 BC	2	Inland	Degerbøl/Krogh 1951; Andersen 1961
28	Holmegaard V	Zealand	Denmark	Settlement	MMC	9000–6000 BC	1 (?)	Inland	Iregren et al. 1990; K. M. Gregersen, pers. comm.
29	Gislinge Lammefjord	Zealand	Denmark	Settlement	KMC	6000–5300 BC	?	Coastal	Aaris-Sørensen 1980
30	Rävgrav	South Scania	Sweden	Settlement	FBC MN	3300–2800 BC	1	Coastal	Larsson 1985; Iregren 1988; L. Jonsson, pers. comm.
31	Nymölla/Möllehusen II	East Scania	Sweden	Settlement	PWC	3200–2500 BC	22	Coastal	Berlin 1941; Iregren 1988; Rosengren 2018
32	Hunneberget	East Scania	Sweden	Settlement	FBC MN ?	3300–2800 BC	1	Inland	Magnell 2007; Rosengren 2018
33	Sjöholmen	Central Scania	Sweden	Settlement	EBC/FBC EN	5300–3300 BC	16	Inland	Forssander 1930; Thomas 1954; Rosengren 2018
34	Skateholm I	South Scania	Sweden	Settlement (2), Grave (2)	EBC	5300–4000 BC	4	Coastal	Jonsson 1988; Rosengren 2018
35	Skateholm II	South Scania	Sweden	Grave	EBC	5300–4000 BC	1	Coastal	Jonsson 1988; Rosengren 2018
36	Limhamn	South Scania	Sweden	Settlement	EBC	5300–4000 BC	1	Coastal	Liljegren 1975; Salomonsson 1971
37	Bökeberg	Central Scania	Sweden	Settlement	EBC	5300–4000 BC	9	Inland	Ceglielka et al. 1995; Eriksson/Magnell 2001; Rosengren 2018
38	Önsvala	Central Scania	Sweden	Settlement	EBC/FBC EN/MN	5300–2800 BC	1	Inland	Lagergren-Olsson 2001; Rosengren 2018
39	Hyllie Fattiggård	South Scania	Sweden	Settlement	EBC/FBC EN	5300–3300 BC	1	Coastal	Salomonsson 1971
40	Nymölla/Möllehusen III	East Scania	Sweden	Settlement	EBC/FBC EN	5300–3300 BC	1	Coastal	Wyszomirska 1988; Rosengren 2018
41	Sjöholmen	Central Scania	Sweden	Settlement	EBC-BAC	5300–2400 BC	1	Inland	Brännborn et al. 2007; Rosengren 2018
42	Tågerup, fase III	West Scania	Sweden	Settlement	EBC	5300–4000 BC	2	Coastal	Eriksson/Magnell 2001; Rosengren 2018

Cont. Tab. 1.

Cont. Table 1.

Site no.	Site	Province	Country	Type of site	Culture	Date BC cal	NISP *Ursus*	Site Location	Reference
43	Tågerup, fase II	West Scania	Sweden	Settlement	KMC-EBC	5600–5000 BC	8	Coastal	Eriksson/Magnell 2001; Rosengren 2018
44	Tågerup, fase I	East Scania	Sweden	Settlement	KMC	6000–5300 BC	23	Coastal	Eriksson/Magnell 2001; Rosengren 2018
45	Ringsjöholm	Central Scania	Sweden	Settlement	KMC	6000–5300 BC	16	Inland	Rosengren 2018
46	Arlöv I	South Scania	Sweden	Settlement	KMC	6000–5300 BC	2	Coastal	Salomonsson 1971; Lepiksaar 1983; Rosengren 2018
47	Ageröd V	Central Scania	Sweden	Settlement	KMC	6000–5300 BC	1	Inland	Rosengren 2018
48	Ageröd I:D	Central Scania	Sweden	Settlement	MMC-KMC	6400–5800 BC	2	Inland	Lepiksaar 1978; Rosengren 2018
49	Ageröd I:B	Central Scania	Sweden	Settlement	MMC-KMC	6400–5800 BC	5	Inland	Lepiksaar 1978; Rosengren 2018
50	Bredasten	South Scania	Sweden	Settlement	EBC	5300–4000 BC	10	Inland	Larsson 1986; Magnell 2006; Rosengren 2018
51	Ageröd I:HC	Central Scania	Sweden	Settlement	MMC	9000–6000 BC	65	Inland	Rosengren 2018
52	Segebro	South Scania	Sweden	Settlement	KMC	6000–5300 BC	17	Coastal	Lepiksaar 1982
53	Neukirchen-Bostholm	Schleswig	Germany	Settlement	FBC MN A III/IV	3200–3100 BC	1	Coastal	Reichstein 1985
54	Wolkenwehe-Heidmoor	Holstein	Germany	Settlement	FBC MN A V	3000–2800 BC	10	Inland	Ewersen 2007
55	Wangels	East Holstein	Germany	Settlement	FBC EN	4100–3800 BC	1	Coastal	Heinrich 1999
56	Siggeneben-Süd	East Holstein	Germany	Settlement	FBC EN	3800–3500 BC	1	Coastal	Heinrich 1999; U. Schmölcke pers. comm.
57	Neustadt i.H.	East Holstein	Germany	Settlement	FBC EN	4100–3500 BC	1	Coastal	Glykou 2016
58	Basedow	South Mecklenburg	Germany	Settlement	FBC MN A	3300–2800 BC	2	Inland	Gehl 1973
59	Waren-Stinthorst	South Mecklenburg	Germany	Settlement	FBC MN A (?)	3300–2800 BC	5	Inland	Gehl 1975
60	Glasow	Western Pomerania	Germany	Settlement	FBC MN A	3300–2800 BC	2	Inland	Gehl 1979
61	Ostorf	Mecklenburg	Germany	Graves	FBC MN A	3300–3000 BC	21	Inland	Lehmkuhl 2010
62	Hohen Viecheln	Mecklenburg	Germany	Settlement	MMC	9000–7000 BC	9	Inland	Schmölcke this volume
63	Bondebrück	Schleswig	Germany	Settlement	EBC	5300–4000 BC	1	Inland	Lüttschwager 1967

Cont. Tab. 1.

Cont. Table 1.

Site no.	Site	Province	Country	Type of site	Culture	Date BC cal	NISP Ursus	Site Location	Reference
64	Tribsees	Western Pomerania	Germany	Settlement	MMC	9000–6000 BC	2	Inland	LEHMKUHL 1988
65	Kiel Ellerbek	Holstein	Germany	Settlement	EBC	5300–4000 BC	1	Coastal	KURCK 1917; LÜTTSCHWAGER 1954
66	Lundby II	Zealand	Denmark	Settlement	MMC	8900–6400 BC	15	Inland	ROSENLUND 1980
67	Holme Skanse	Djursland	Denmark	Settlemnt	Late SGC (?)	2400 BC (?)	4	Coastal	ANDERSEN 1983; RICHTER 1986; S.H. Andersen, pers. comm.
68	Syltholmen	Lolland	Denmark	Settlement	EBC/FBC	4700–3500 BC	1	Coastal	Unpublished; information courtesy Museum Lolland-Falster

Table 2. ^{14}C dates for remains of bears from archaeological contexts in Denmark. Dating attempts on several other finds (Lindegårds Mose, Ørum Aa, Hygind, Holme Skanse) failed due to lack of preserved collagen in the samples submitted for dating.

Site no.	Site name	Lab. no.	Date bp	Date BC cal (2δ)	Reference
3	Kolind	AAR-21420	4441 ± 29	3390–2930	Klassen et al. 2020b
4	Ørum Aa	AAR-21421	4450 ± 30	3630–3370	Klassen et al. 2020b
7	Bundsø	AAR-21416	4686 ± 29	3340–2950	Klassen et al. 2020b
8	Selbjerg	AAR-21415	4148 ± 29	2880–2630	Klassen et al. 2020b
14	Dyrholmen	AAR-21418	6165 ± 31	5220–5020	unpublished
16	Virksund	CURL-10287	5310 ± 20	4240–4050	Bray et al. 2013
17	Bjørnsholm	AAR-21422	5533 ± 31	4450–4340	unpublished

Table 3. Number of assemblages containing preserved bones, number of assemblages included after removal of mixed localities, site totals employed in calculations of bear constancy and abundance, and percentage of selected assemblages in relation to all known sites (including very mixed assemblages), based on records held at the Zoological Museum in Copenhagen. For abbreviations, see Table 1.

	MMC	KMC	EBC	FBC	PWC
All sites	58	65	230	430	5
Selected sites	46	58	208	286	5
Site total	43.5	46	181.5	213.5	5
Percentage of selected sites	79.3	89.2	90.4	66.5	100

Table 4. Constancy of bear records in the Danish Stone Age. For abbreviations, see Table 1.

	Sites with bear bones	Sites with preserved bones	Constancy
PWC	3 (4) (Jutland only)	5 (Jutland)	60–80 % (Jutland)
FBC	3 (6)	39 (Jutland)	7.7–15.4 % (Jutland)
EBC	7 (9) (Jutland only)	65.5 (Jutland)	10.7–13.7 % (Jutland)
KMC	1 (Zealand only)	43.5 (Zealand)	2.3 % (Zealand)
MMC	7 (Zealand only)	43.5 (Zealand)	16.1 % (Zealand)

Table 6. Average abundance of bear bones in the individual cultures based on the data in Table 5. For abbreviations, see Table 1.

	MMC	KMC (Sweden)	EBC	FBC incl. (excl.) Mecklenburg	PWC
n	4	5	6	8 (5)	3
% Ursus	1.34	0.69	0.32	0.41 (0.16)	0.80

Table 5. Number of bear bones (NISP Ursus), assemblage size (NISP total) for mammal bones and abundance of bears (% Ursus) for sites in Denmark, southern Sweden, and northern Germany.

Site no.	Site	Culture	NISP Ursus	NISP total	% Ursus
1	Kainsbakke	PWC	34	4,178	0.8
2	Kirial Bro	PWC	1	329	0.3
8	Selbjerg	PWC	1	<125	>0.8
31	Nymölla/Möllehusen II	PWC	22	1,717	1.3
11	Spodsbjerg	FBC MN	1	3,896	0.05
12	Lindø	FBC MN	3	>779	<0.4
54	Wolkenwehe-Heidmoor	FBC MN	10	6,125	0.2
58	Basedow	FBC MN	2	477	0.4
53	Neukirchen Bostholm	FBC MN	1	403	0.2
60	Glasow	FBC MN	2	339	0.6
59	Waren-Stinthorst	FBC MN	5	316	1.6
9	Hygind	FBC EN/MN	4	812	0.5
30	Rävgrav	FBC EN	1	2,630	0.05
6	Lindegårds Mose	FBC EN	3	205	1.5
57	Neustadt i.H.	FBC EN	1	>3,617	0.05
55	Wangels	FBC EN	1	446	0.2
56	Siggeneben-Süd	FBC EN	1	155	0.6
37	Bökeberg	EBC	9	1,594	0.6
50	Bredasten	EBC	10	2,309	0.4
34	Skateholm I	EBC	4	138	2.9
42	Tågerup phase III	EBC	2	463	0.4
55	Bondebrück	EBC	1	302	0.3
17	Bjørnsholm	EBC	1	>100	<1.0
20	Brabrand	EBC	2	>100	>2.0
18	Ringkloster	EBC	5	5,036	0.1
16	Virksund	EBC	8	<50	>16.0
14	Dyrholmen	EBC	1	>1,000	<0.1
43	Tågerup phase II	KMC	8	244	3.3
44	Tågerup phase I	KMC	23	1,627	1.4
45	Ringsjöholm	KMC	16	3,064	0.5
46	Arlöv I	KMC	2	371	0.5
47	Ageröd V	KMC	1	441	0.2
52	Segebro	KMC	17	2,262	0.8
49	Ageröd I:B	KMC	5	211	2.3
48	Ageröd I:D	KMC	2	179	1.1
66	Lundby II	MMC	15	791	1.9
25	Sværdborg I	MMC	31	2,923	1.1
27	Verup	MMC	2	240	0.8
28	Holmegaard V	MMC	few bones	3,452	?
51	Ageröd I:HC	MMC	65	3,649	1.8
64	Tribsees	MMC	2	277	0.7
62	Hohen Viecheln	MMC	9	1,498	0.6

Table 7. Distribution of individual skeletal elements. Sites are listed in approximate chronological order. Imprecisely dated sites and sites lacking information have been omitted. Note that the sum of bear bones for the individual sites is not necessarily identical to the number of bones given in Table 1 because the accessible bone identifications were, in some cases, inadequate. Some "phalanges" and "metapodials" included in the totals in Table 1 are missing. For abbreviations, see Table 1.

Site no.	Site	Culture	Head			Forelimb			Hindlimb	
			Cranium	Mandible	Tooth	Scapula	Humerus	Ulna	Radius	Pelvis
67	Holme Skanse	Late SGC								
8	Selbjerg	PWC							1	
1	Kainsbakke	PWC	9	5	5	1	1	2		2
2	Kirial Bro	PWC								
31	Nymölla/Möllehusen II	PWC	1	1	1	1				
5	Ginnerup	FBC/PWC								
11	Spodsbjerg	FBC MN			1					
54	Wolkenwehe-Heidmoor	FBC MN			2	1			2	
53	Neukirchen-Bostholm	FBC MN								
3	Kolind	FBC MN								
7	Bundsø	FBC MN				1	1	1		
12	Lindø	FBC MN								
10	Dræby Mark	FBC MN								
58	Basedow	FBC MN			2					
59	Stinthorst	FBC MN	1		1					
60	Glasow	FBC MN	1							
61	Ostorf	FBC MN			21					
9	Hygind	FBC EN/MN	3		1					
30	Rävgrav	FBC EN/MN								
4	Ørum Aa	FBC EN/MN			1		1			
6	Lindegårds Mose	FBC EN								
56	Siggeneben-Süd	FBC EN	1							
68	Syltholm	FBC EN/EBC								
57	Neustadt i.H.	EBC/FBC EN							1	
17	Bjørnsholm	EBC/FBC EN						1		
16	Virksund	EBC	2	1	3					
20	Brabrand	EBC		1	1					
18	Ringkloster	EBC	1							
23	Kolind	EBC				1				
37	Bökeberg	EBC							1	
63	Bondebrück	EBC				1				
15	Vedbæk-Bøgebakken	EBC				1				
14	Dyrholmen	EBC								
50	Bredasten	EBC								
34	Skateholm I	EBC			2				1	
35	Skateholm II	EBC								
42	Tågerup fase III	EBC								

	Hindlimb				Forefoot	Hindfoot	Torso		Phalanges		
Femur	Patella	Tibia	Fibula	Carpal/Metacarpal	Tarsal/Metatarsal	Vertebra	Rib	1st phalanx	2nd phalanx	3rd phalanx
			1	2						
	1				1			2	3	2
									1	
1				3	3	1		6	3	
						1				
			1		1	3				
								1		
					1					
					2			1		
										1
				2						
									1	1
					1					
					3					
			1							
				2						
		1		1	1			1		
				1	1	2	1	2	1	
						1				
				2	3	2				
								1		
					1					
								1	1	

Cont. Tab. 7. ▶

Cont. Tab. 7.

				Head			Forelimb			Hindlimb	
Site no.	Site	Culture	Cranium	Mandible	Tooth	Scapula	Humerus	Ulna	Radius	Pelvis
43	Tågerup fase II	EBC/KMC			6					
44	Tågerup fase I	KMC		1	6	2		1		
45	Ringsjöholm	KMC			4	1	1	2		1
46	Arlöv I	KMC		1						
47	Ageröd V	KMC								
52	Segebro	KMC		2			4			2
49	Ageröd I:B	KMC/MMC			4					
48	Ageröd I:D	KMC/MMC					1			
26	Lundby I	MMC								
66	Lundby II	MMC	8							1
13	Øgaarde	MMC		1	2				2	1
28	Holmegaard V	MMC								
25	Sværdborg I	MMC		1	3	2		2	1	
24	Mullerup	MMC	1	1	1		1	1	2	
27	Verup	MMC								
51	Ageröd I:HC	MMC	4	6	8	1	2	6		1
64	Tribsees	MMC			1					
62	Hohen Viecheln	MMC		1			2			1

	Hindlimb				Forefoot	Hindfoot	Torso		Phalanges		
	Femur	Patella	Tibia	Fibula	Carpal/ Metacarpal	Tarsal/ Metatarsal	Vertebra	Rib	1st phalanx	2nd phalanx	3rd phalanx
					1					1	
					5	4		1	2		
					2	3	3		1		
	1										
									1		
					2	3	1		1		
									1		
					1						
	2										
						1					
			1								
	2	1	2		3						3
					2						2
							2				
	1	1	4		7	3	7	8		3	1
			1								
	1				2		1				

385

Bears and the Viking Age transition in Sweden

By John Ljungkvist and Karl-Johan Lindholm

Keywords: Sweden, brown bear, Ursus arctos, Iron Age, Vendel period, Viking Age, The Bear in the Grave (BiG) database

Abstract: The main objective of this paper is to generate a deeper understanding of bear and human relationships in the 1st millennium AD. This will primarily be achieved by the analysis of a detailed chronology of bear phalanges from Iron Age burials. The aim is to note changes in the deposition patterns of bear remains in burials in order to identify human impact – which we consider significant for hunting pressure – on the Scandinavian bear populations of the 1st millennium AD. The authors suggest that bear hunting can be considered as part of the larger processes of intensified exploitation of the boreal forest in the Iron Age that contributed to the formation of interregional trade networks. It is suggested that this exploitation affected the bear population to such an extent that overused animal resources can be understood as one of several contributing factors behind the Viking expansion outside Scandinavia. An understanding of bear and human relationships will contribute to a better knowledge of the cultural history of Scandinavia's forested region and of interregional contacts with the central agricultural regions.

Introduction

More than half of Scandinavia's land area consists of boreal forest composed of coniferous tree species, such as pine and spruce, mixed with broadleaf species, such as aspen and birch. The boreal forest covers a hilly and undulating topography interspersed by numerous lakes, rivers, streams and mires, with a climate characterised by long winters and short growing seasons. Archaeology and palaeoecology have discovered that the forested landscapes of inland Scandinavia contain a diverse and fairly repetitive record of archaeological sites related to the use of forest resources, such as game, fish, pasture, wood, fuel and minerals (see for example EMANUELSSON et al. 2003; KRESTEN 2008; ASHBY et al. 2015; LOFTSGARDEN 2015; HENNIUS 2018; BAUG et al. 2019). In general, the forested inland has been considered to have negligible importance for understanding the larger societal developments of Scandinavia's past, even if research has repeatedly noted connections between the inland central agricultural plains and the coastal areas. An explanation for this view is the less accentuated social elites in the archaeological record of the forested region and the small amount of written documentation prior to AD 1500. Moreover, the first permanent agrarian settlements have generally been perceived as belonging to the Viking Age (*c.* AD 750–1050) or the Middle Ages (*c.* AD 1050–1520). This is generally explained as the result of population growth and new technologies that facilitated farming in forested areas.

In the Viking Age, archaeological and palaeoecological observations suggest that the craft and trading systems of the forested inland region changed. The often very long and large systems of hunting-pits, primarily used for hunting elk and reindeer, must have been labour-intensive enterprises requiring large bodies of people and coordination – at least seasonally – in terms of construction, maintenance and use (Lindholm/Ljungkvist 2016). In the Viking Age they seem to have played a lesser role in several areas, although in the Middle Ages they became more actively used again (Hennius 2020a). Moreover, the reduced use of the pits coincides with growing indications of agriculture and more distinct shieling systems in the pollen diagrams, if compared with the earlier phase (Svensson 1998; Emanuelsson 2001; Emanuelsson et al. 2003; Karlsson et al. 2010; Dögg Eddudóttir et al. 2021; Larsson 2021).

Direct archaeological evidence for the Early Iron Age settlements still has to be retrieved from the inland region. We suggest that the distribution of pitfall systems can be considered as proxy data for areas with more intense exploitation of wild faunal resources, as well as larger groups of people cooperating within a permanent field-and-meadow land organisation in the boreal zone (Lindholm et al. 2013; Lindholm/Ljungkvist 2016; Hennius 2020a). Even if most animals found in burial contexts were not hunted using pitfall traps, we consider the features indicative of a socio-economic system that incorporated forest resources and hunting (Fig. 1).

Lindholm/Ljungkvist (2016, fig. 7) have noted additional processes that can be associated with the shift in the Viking Age, and the observed pattern is further strengthened by recent research. In the Roman Iron Age, tar was associated with farmsteads, but in the Viking Age, and perhaps earlier, it was moved into the actual resource areas, i.e. the forested outland where considerably larger volumes were produced at larger production sites (Hennius 2018). This, in turn, necessitated the reorganisation of labour involving entire communities, which also had to engage in long-term forest management to provide adequate fuel for the tar production sites (ibid.). A similar process appeared in Norway, where an expanded need for iron resulted in the establishment of more sophisticated and reusable furnaces in order to produce large quantities of iron bloom. This intensification was based on decentralised and farm-based iron production, and it was initiated long before the establishment of industrial blast furnaces (Rundberget 2013; Indrelid et al. 2015; Loftsgarden 2017).

Based on this background, a more detailed analysis of bear phalanx contexts might help in gaining a more specific date for the shift, and thereby help in acquiring a better understanding of the changing interdependencies of the Iron Age ecological globalisation. In our analysis, we will consider chronological trends, such as an increase in bear claws in burials, as a reflection of increased hunting. Reduced numbers could, on the other hand, be related to changes in fashion or in the burial ritual, but could also be seen as signs of overexploitation, especially if the trend does not coincide with signs of lessened pressure on outland/forest resources. The presumption of this paper is that claws mainly represent bear furs. A critique of this assumption is presented by Grimm (2013; cf. Jordahl et al., this volume), who shows that claws could also represent amulets. Therefore, it is important to raise a number of source critical issues particularly relating to the material from mainland Sweden, but also from Gotland and Öland, where the vast majority of the graves are cremation burials. In addition, almost all of the relatively few inhumation burials have usually been plundered, making it difficult to analyse how claws were distributed in the burial chambers – and thus making it difficult to assess whether furs were deposited in graves or not. One way to approach this topic is to register the number of claws from each burial and thereafter register the minimum number of paws from a burial. This has not been done in a consistent way for this study due to the very low number of intact and well-preserved inhumation burials. Among the cremation burials, we are faced with the problem that many finds are derived from older excavations. This means that the bone material has not been retrieved in a consistent way during excavation, e.g. the material has not been sieved, and in many cases the bone material has not been analysed by osteologists, meaning that it can be assumed that claw fragments often remained undiscovered. Nonetheless, we know from Rösta in Ås and a few Gotlandic cases that inhumation burials can contain, for example, a few claws in a pouch or placed in a container by the head of the deceased (see JORDAHL et al., this volume). There are also cases of cremation burials with more than 10 claws, representing probably at least two paws and therefore with a higher probability of a fur or parts of such. However, even if it is crucial to have insights into these source critical issues, it does not directly affect the main topic and results of this paper, since bears have to be hunted to get access to either amulets, furs or even canine teeth (for the latter see MAGNELL/IREGREN 2010; cf. MAGNELL, this volume, on Frösö).

THE BEAR IN THE GRAVE DATABASE

The paper's main method is a renewed analysis of the BiG database, which is now primarily aimed at a chronological analysis with a higher temporal resolution. This analysis will enable the identification of notable changes between sub-phases of the main periods of the Iron Age. The analysis also includes additional animal species, such as lynx and birds of prey. Since the previous study, a number of new burials containing bear claws have been registered. With the new registrations, also including those presented by JORDAHL et al. in this volume, the Swedish finds currently number 407, of which 147 come from the island of Gotland. Five finds come from another Baltic Sea island, Öland, and the remaining originate from the mainland (Fig. 1). In comparison with LINDHOLM/LJUNGKVIST (2016), based on 323 contexts, the current database represents an increase of about 26 %.

This increase does not change the geographical aspect of bear finds from burial contexts in any substantial way. Besides the huge concentration on Gotland, two other areas have significant concentrations of burials with bear phalanges. One is the Lake Mälaren region, constituted by the counties (Swedish: *landskap*) of Uppland, Södermanland, and Västmanland. In the third area, the Norrland/Bothnian counties of Medelpad and Ångermanland, the actual number of contexts is considerably lower, but if the actual population density and the few excavations undertaken in this region are taken into consideration, the number of contexts is high. The reasons behind the varying frequency of bear claws cannot be attributed to one single factor. There is a strong correlation between a large number of excavated graves and bear claw finds, particularly in the Mälaren region and on Gotland.

Another factor is varying burial practices between different periods and areas. Once again, Gotland in particular and, to some degree, the Mälaren region stand out with many well-preserved burial grounds containing substantial amounts of grave goods from more than one archaeological period. The burial practice itself does, however, show major differences between mainland Sweden and Gotland. As mentioned above, mainland Sweden is completely dominated by cremation burials while, on Gotland, there is a significantly higher number of inhumation burials, even though the cremation burial practice dominates in most periods. The limestone bedrock of Gotland also provides favourable preservation conditions for both human skeletons and bear claws in the inhumation burials. Some regions, such as the island of Öland, are archaeologically very rich, with many excavated graves but with surprisingly few bear claw finds. One reason might be the number of Migration to Viking period burials that are not so well represented as those from earlier periods. For example, if there was a strong Migration period increase in bear claws on Öland, it cannot be observed in the present burial record. In south Scandinavia in general, a considerably lesser number of graves has been excavated and, in comparison with Middle Sweden and Gotland, there is also a clear trend toward more sparse grave goods from the Migration period and onwards into the Viking Age. The sparse number of bear claws in these areas are not direct evidence of reduced imports, but rather a sign of different burial and deposition patterns.

One basic source-critical issue, closely related to hunting pressure and the export of products from bears and other large mammals, is the extent to which the archaeological finds represent a local resource or imports from other regions. This question is related to the size of the bear population in Scandinavia during the Iron Age. The factual situation is uncertain, but for sure bears have never lived on the large Baltic islands. In the agricultural regions of south Scandinavia, they have probably been extinct for a very long time (see below, and also MAGNELL, this volume, on bears and humans in Sweden). The situation is more complex further north. For example, the medieval law texts do mention bears in, for example, the landscapes of Uppland and Östergötland, but whether their presence was sporadic or well established is uncertain (see summaries in KARLSSON 2016, 64–66). Surely, it is possible that local bear populations existed in forested areas on the fringes of the settlements in the densely inhabited areas. On the other hand, an argument against this notion is the general absence of predators, including wolf and bear, as well as bones from their natural prey (LINDHOLM/LJUNGKVIST 2016, 9). In addition, hunting for bears, as well as wolves and foxes, tends to have been free in the Middle Ages since they were all seen as a threat to livestock (KARLSSON 2016, 65, 66). Settlement layers in agrarian or urban settlement contexts in Middle Sweden contain no, or very few, bones from large wild mammals (LINDHOLM/LJUNGKVIST 2016). Animals such as elk, roe deer or red deer are very rare in the archaeological records, which means that regionally low or very low populations of these animals can be reasonably assumed. They can be seen as a proxy for bear and lynx populations, based on the notion that populations of larger mammals live side-by-side, no matter if they are grazers or predators. Consequently, we consider the vast majority of bear claw finds as imported, either from the boreal regions of Scandinavia or from areas beyond the eastern and southern shores of the Baltic Sea.

Of BiG's current 407 burials with bear claws, 248 graves have been dated to a specific archaeological period (Table 1). Some graves are, for various reasons, difficult to date to a specific phase within the time period, primarily due to poor preservation conditions and disturbed grave contents because of plundering or damage. Vague datings are also to some degree related to whether the graves belong to a transition phase or if they lack sufficient chronological data. Despite these source-critical points, the trends of the statistics are still so clear that these uncertainties do not affect the general basis for the analysis. As already noted, the number of dated graves represents a 26 % increase in comparison with the previous study (LINDHOLM/LJUNGKVIST 2016, 12 table 2). In comparison with this, the increase has not changed the overall statistical outlook of tendencies in the record significantly, except for the increased number of finds related to graves dated to the Roman Iron Age, primarily on Gotland, which see a more than double increase from eight to 18 burials. Only a few Roman Iron Age burials contain bear claws, although their geographical distribution shows that they are clearly associated with areas outside the natural distribution area for bears, which is particularly apparent through the concentration associated with the island of Gotland (see JORDAHL et al., this volume). In the Migration period, the numbers increase dramatically, while they become considerably lower during the Vendel and Viking periods (see Table 1).

In the previous study, it was mentioned but not further elaborated upon that a significant change in the frequency of graves with bear claws took place in the Vendel period (LINDHOLM/LJUNGKVIST 2016). In this study, a larger number of graves were dated to specific phases of the Vendel period, based on a method for find combinations that originally developed for the Valsgärde burial ground (Table 2). This is founded on a detailed chronology, based on studies by other authors. A total of 62 more precisely dated Vendel period burials are presented in Table 3. Statistically, this is not a significant number of graves. But the strong contrast between the two halves of the Vendel period is still apparent and noteworthy, and the shift reveals a dramatic drop in the number of bear claw burials after the mid-7th century.

In order to test if this shift could be observed in the appearance of other exclusive animals, a compilation of burial records of lynx claws and birds of prey were added to the BiG database in order to serve as a comparison (Table 4a; b). The total number of finds from these animals is significantly lower but, nonetheless, clear patterns can be observed. For lynx, the drop seems to be more or less similar to what we see among the bear claws. For the birds of prey, the pattern is not as dramatic, even if a clear change can be seen. Something that especially shows up in the frequency of the birds of prey is that the most exclusive birds, i.e. peregrine falcons, are almost completely absent in burials after *c.* AD 700. They appear in eight to nine burials in phase VET 1 to VET 2–3, but only in one burial in the interval VET 4 to VIT 4.

A DECLINE AND ITS CAUSES

It is difficult to divide the Migration period records into chronological phases in the same way as the Vendel period. Still, the overall pattern reveals a major change in the burial deposition pattern of exclusive animals around or slightly after AD 650, and it seems necessary to discuss possible reasons behind the change. To begin with, it may be relevant to first inquire into how widespread the custom of depositing bear remains was at its peak. Therefore, in order to enable this discussion, the renewed analysis of the BiG database has included a sex determination of the burials, primarily based on find combinations. This has been done because the osteological data from cremation burials are more fragmentary and more affected by source-critical uncertainties. Discussions concerning gender are usually related to the analysis of gendered social relations, but, in this case, we suggest that gender associations and related frequencies of animal remains can also be used for insights into how common bear furs were in societies. Further, such an analysis can help to identify the impact of hunting on bear populations. From a general perspective, without considering potential regional differences in Sweden, it is possible to conclude that bear claws are more frequent in male burials. However, as PETRÉ (1980) has already noted, the proportion of females is also high and the occurrence of bear claws cannot be seen as a strictly gender-related attribute (Table 5). This pattern is in clear contrast to lynx claws, which seem to be almost entirely associated with women (see ZACHRISSON/KRZEWIŃSKA 2019).

Hence, bear remains seem to be equally related to both male and female burials. In certain families, particularly high-status families of the community, it is possible that at least two members possessed bear-related items, whether it was furs or not (see JORDAHL et al., this volume; GRIMM, this volume). The relative distribution of bear-related items between female and male burials is in itself a delicate topic. The gender relation of bear finds/claws is probably a matter of regional patterns. Without presenting numbers, it seems that bear remains are more common in female burials on Gotland. However, in the Lake Mälaren region they are, for example, in a minority on Helgö. Of 20 burials with bear claws, only one is considered a more or less certain female burial, and five are uncertain or double burials. But on the Lunda and Viken burial sites on the adjacent island of Lovön three out of six burials that contain bear claws are female. In addition, based on find combinations, it seems that bear furs were associated with burials at more than one level of the society, to such extent that they can be considered a "middle-class phenomenon" between *c.* AD 400 and 650. Moreover, at the Barshalder burial ground on Gotland (which has been osteologically examined in detail) 13 % of the graves, both male and female, contained bear claws (RUNDKVIST 2003). It is far from certain whether Barshalder is representative of other sites on Gotland, but the number of finds from the island suggests it is not unique. From Broa in Halla, an excavation of 120 burials without a detailed bone analysis has generated at least 12 with bear claws (LANGHAMMAR 2012; Statens Historiska Museum [SHM] bone database Inv. No. 35335). If these burial grounds are representative of average grave fields on

Gotland, it must mean that an extraordinary number of furs and other bear-related items with claws circulated in this period. For the time around AD 500, it has been estimated that the number of farms on Gotland was almost 2,000 (Svedjemo 2014).

If this reasoning can be considered a valid reflection of hunting pressure during the time period, it raises concerns as to what extent the bear populations became affected by hunting. Moreover, the extent to which the decrease in the middle part of the Vendel period should be understood as related to changed exchange routes, the alteration of the burial rituals, or the scarcity of these animals/products, together with increased prices, can be questioned. Possible reasons may of course intersect; for example, a change in a funeral ritual can be the result of economic factors, such as the reduced ability of community members to deposit items that have suddenly become scarce. The scarcity in turn could have been caused by increased distances to trade routes, as well as a general scarcity of bears due to overexploitation.

Recent findings in genetics can help with some insights into the issue (Xenikoudakis et al. 2015). The brown bears on the Scandinavian peninsula can be separated into two larger groups based on distinctive lineages (or clades) characterised by their mitochondrial DNA haplotypes. These clades reflect the Scandinavian bears' phylogeography, which is shaped by faunal history and the colonisation of the area from the south and northeast after the last glaciation (Taberlet et al. 1995; Bray et al. 2013). A contact zone in a distinct latitudinal division has been identified in the central Scandinavian peninsula, in the regions of northern Trøndelag in Norway and northern Jämtland in Sweden (Fig. 1). Since it is exclusively females who pass on mtDNA, the division is reflective of the maternal inheritance and female bears. Modelling of the genetic substructure and variation by computational statistics, also using more genetic data, implies a demographic bottleneck and a general decline in effective population size in both the southern and northern parts of the Scandinavian brown bear population, as well as in the contact zone (see also Kopatz, this volume). The loss in genetic diversity and the current genetic structure may, however, have been caused by long term historical-ecological processes preceding the major decline in bear population size caused by intense hunting approximately 100 years ago. Hence, the structure of the bear population seems to have an ancient origin that has remained largely unaffected over the last centuries. The timing for this process remains to be outlined (Xenikoudakis et al. 2015). Still, the genome of current bear populations has a strong relationship to hunting pressure and habitation loss due to human expansion. The question is, when did these changes in the genome take place? We would suggest that the chronology of the burial record with associated bear remains might present a temporal reflection of this process.

Moreover, in order to understand the shift in the Vendel period, it could be useful to compare the frequencies of various animal remains associated with the different phases of the Viking Age. It is clear that there are fewer claws of bear and lynx, and the diversity of birds of prey is consistently lower from the 7th to the 10th century in comparison with the phase of *c.* AD 560–650. To illustrate this process we can look at Helgö, which served as a considerably earlier economic centre from at least the Migration period. During its later stages, Helgö's role as a production site seems to have diminished, but it was still a place with considerable wealth well into the Viking Age. Here, there are bear claws recorded from 22 out of a total of 123 Migration to Viking Age burials containing analysed bones. This is 18 % of the total (Melin/Sigvallius 2001, 123–132). A total of 16 originate from the Migration or Vendel periods. At Birka, which was a later economic centre from *c.* AD 750–980, the number of bear claws is considerably lower, with currently only six out of more than 1,000 excavated graves registered. This is noteworthy since this town served as a hub for the importation of furs and antlers. Birka seems to follow the broader trend of the Viking Age, with a low number of bears and birds of prey finds, and a virtual absence of lynx claws (Lindholm/Ljungkvist 2016, 13). Nevertheless, an increase in bear claws in burials can be seen in the late 9th and early to mid-10th century (VIT 3–4), but the levels are far from those of the AD 400–650 interval. Interestingly, the pattern observed at Birka has parallels with the one observed for the Baltic island of Åland, and one can ask if it coincides with a markedly higher level of eastern and oriental influences on the material culture in eastern Sweden (see Gustavsson/Ljungkvist, this volume).

In addition, some other differences can be noted by comparing burials from the 7th century with those of the 9th century. During an earlier stage, *c.* 560–750, the burials contained more animal remains, and particularly the monumental burial mounds were on average larger. The 8th century is a transition period, in which domestic animals in the burials and the amount of labour on the actual grave monuments seem to decrease. This is clearly visible in the monumental burials, but similar changes have also been observed at large cremation burial grounds (Bratt 2008; Magnell et al. 2017, 222). A trend that can be clearly observed among the elite boat burials is a decrease in certain prestigious objects. Some of the Vendel period burials from *c.* 560–700 contain up to four swords, two long swords and skramasaxes, respectively, and, furthermore, up to three shields. In comparison, Viking Age burials normally contain only one sword and one shield (see Stolpe/Arne 1912; Lundström 1980; Schönbäck/Thunmark-Nylén 2002). These changes correlate with the decreased frequency of birds of prey in the burials (see Tables 3–4). The differences in the frequencies of bear remains between the regions of present-day Sweden also seem to appear at this time, which other chapters in this volume clearly help to demonstrate (see Gustavsson/Ljungkvist, this volume; Jordahl et al., this volume).

Discussion

In the final section of the paper, we want to discuss the implications of the renewed analysis of the BiG database by investigating other areas. To begin with, settlement studies in the Lake Mälaren region, i.e. the large and densely populated basin in southeastern Sweden, which borders the boreal forest, suggest that the settlement concentration was almost as large in the Roman Iron Age as in the Viking Age (Göthberg 2000, 208). This in turn suggests a several-century-long continuity of heavy landscape utilisation that would have created a general lack of valuable animal resources in the nearest area. For southern Sweden, it is possible to follow a process of increased utilisation of large mammal resources over a long time span. This continuity probably goes back to the Bronze Age or even to the Neolithic, when high hunting pressure in, for example, Denmark and southernmost Sweden had already led to a sharp decrease in, and even an extinction of, large mammals such as bear, lynx, and elk in the Mesolithic (Aaris-Sørensen 2009; Magnell, this volume, on bears and humans in Sweden). Due to a weaker burial record in southern Scandinavia (see above), it is more difficult to estimate how high the demand was for bears, their claws, fur and teeth in this area. However, we do know that some demand existed as there is evidence for a bear being imported to the royal court in Lejre, Denmark, as well as finds of claws and canine teeth from Haithabu culture layers and claws from the sequence of halls in Uppåkra (Christensen 2015, 168–169; Schietzel 2018, 304; cf. Magnell, this volume, on Frösö church). It is yet not possible to define how many of these bear remains originated from northern Scandinavia or eastern Europe, respectively, although we suspect that the majority of these south Scandinavian finds originate from present-day Finland, Poland, the Baltic states, and regions further east. This relates to the view of Iregren (1988) and the conclusion of this paper that the few finds of bears in south Scandinavia reflect trading networks positioned to large degree in central Sweden and on Gotland.

Nevertheless, from the Roman Iron Age, it is possible to identify a broad demand for antlers, primarily used for manufacturing combs, and furs or other objects with attached claws of bears and lynx in the central agricultural regions of Scandinavia. We suggest that this fashion required the inhabitants of these regions to import either finished products or the raw material from areas that still maintained viable animal populations, i.e. the forested region of inland Scandinavia or the eastern parts of Europe. This demand increased dramatically in the Migration period – possibly a result of intensified trade networks – when a very high number of bear claws, and to some degree lynx claws, appeared within a rather broad social spectrum of the society, at least on Gotland and in the Lake Mälaren region. This trend goes hand in hand with increased numbers of composite antler combs in the burials, which eventually became one of the most common finds in Iron Age burials, which in turn reflects a strong demand for forest products in Iron Age society. For example, at the Lunda 27 burial site, located just west of Stockholm, 72 % of the 155 burials that date to between the Migration and the Early Viking periods contained antler combs (PETRÉ 1984, 70). Thus, when we reach the 6th century, a high percentage of the population seems to have owned at least one item produced from resources that had their origins in the forested regions, and these patterns stayed the same throughout the Viking Age. In the late 6th-century burial record a new, expensive sport, falconry, appears in the shape of buried birds of prey (VRETEMARK 2013, table 1). So far, 44 graves containing raptors have been registered in the BiG database. Goshawk and eagle owl are the dominant species, but the more exclusive birds such as peregrine falcons appear in eight of the records. The appearance of birds of prey in burials coincides to some degree with a massive importation of whalebone gaming pieces, most likely originating from arctic Norway (HENNIUS et al. 2018; LJUNGKVIST/HENNIUS 2020). In comparison with other phases, particularly with the phase AD 560–650, this is a period of considerably higher investments and effort in the burial goods, the number of sacrificed animals, and the size of the burial monuments (see for example LJUNGKVIST 2006; BRATT 2008). Antler combs, bear and lynx claws, and birds of prey are all separate indications of outland exploitation that seems to increase considerably at this time. The question is whether this fashion was ultimately sustainable.

In order to discuss the question of sustainable hunting, it might be illustrative to use an analogy with early modern to more recent cases where top predators and large mammal populations, birds and fishes have reached tipping points, i.e. the population has not been able to sustain a high demand and hunting pressure. In such cases, the valued animal populations have become associated with a strong relation to market value, since diminishing numbers resulted in increasingly higher prices of the animal products and a potentially more exclusive status of the product for those who lived outside the exploitation areas. Such a relationship has, for example, been recorded in the cod and bluefin tuna industries during the last century (AINSWORTH/U'SUMALIA 2011; RODRIGUES et al. 2021). Almost all stocks of these species dropped dramatically during and after the 1980s. This resulted in considerably higher prices for the consumers but did not immediately lead to sufficiently managed stocks. Today, the situation for bluefin tuna is more stable, while many or most Atlantic cod stocks remain in a disastrous state. The primary illustration of how certain animal products could be valued in medieval times, and probably earlier, were the high prices paid for Norwegian falcons in England and the far greater amounts paid in the eastern Mediterranean (ANDERSEN 2019, 50; CHIESA 2021, 10). So, the dramatic decline in bear claws, lynx, and peregrine falcons in burials in the late 7th century can be considered as a sign of something happening with the animal populations, and/or their availability. If a globally expanding market resulted in high demands and high hunting pressures on certain animals, it is perhaps not surprising if fewer exclusive birds and perhaps also bear furs ended up in Scandinavian graves during the Viking Age.

Here, we would also like to follow another line of argumentation for understanding the shift. There are indications that the interaction between eastern Sweden and the regions on the opposite side of the Baltic increased about 50 years after the decline in the mid-7th century. These contacts are particularly visible through the AIII-pottery, particularly from Finland and Estonia (CALLMER 2017). Additionally, more conflict-related evidence comes from the Saaremaa boat graves in Estonia, indicating a Middle Swedish military presence outside their homelands (PEETS et al. 2011; 2013; PRICE et al. 2016). An early example of increasing Gotlandic interest in the Baltic region during this period (and probably earlier) can be found in Grobina in Latvia (NERMAN 1958). As Callmer, among others, has stated, the largest Scandinavian motivation for the trade in the east was fur products (CALLMER 2017). The Scandinavians were likely aware of these eastern resources before the Viking Age, since there had been long-lasting contacts over the Baltic, but still, evidence from the material culture suggests these contacts increased in the early 8th century. In the 9th century, a strong Scandinavian presence built up along Russian river and lake systems, from the Ladoga-Ilmen-Peipus triangle to Kiev, and their activities stretched into the Black Sea. Further, there are increased interactions between Scandinavians and communities in the southern Baltic, which in turn connected with other large forest areas of the boreal region (see, for example, CALLMER 2017).

The causes of the Viking phenomenon have been discussed on numerous occasions, and there is probably not just a single one, not least since the mobility and expansion of the Scandinavians and their interaction with people from other regions had so many different expressions (see for example ASHBY 2015; BARRETT 2015). A shortage of bear fur is not in itself an explanation for the major changes in Scandinavia, but still, the bear claws are among the most substantial pieces of evidence we have of resource exploitation that could have reached unsustainable levels, at least in Scandinavia. Walrus hunting is another example. Recent research in the North Atlantic has shown how Scandinavians could kill off large mammal populations within relatively short periods during the late Viking to Middle Ages, and, furthermore, how the hunters expanded and made use of new killing grounds as soon as the previous ones had been depleted (STAR et al. 2020). Such processes of "resource colonisation", by which the quest for valuable natural resources resulted in intensive exploitation and production landscapes (LINDHOLM et al. 2021), constituted a process that has recently been conceptualised as ecological globalisation, creating interdependencies between animal populations, the resource-extracting communities and distant centres of consumption (BARRETT et al. 2020). The bears, lynx, and raptors were probably not primarily important as animal trading resources, but perhaps they can, as the walrus, serve as proxies for intense resource use and coupled relations between animal population dynamics and human mobility. Either people moved to find new hunting grounds for valued resources or they tried to control the trade and trade routes in other regions – an important drive for the expansion of Scandinavians to other regions.

In conclusion, current research proposes a settlement expansion in the Middle Iron Age (*c.* AD 200–650). This seems to have been driven by diversified outland use through which resources from the boreal forest environments were transformed into commodities for trade and exchange, and arguably one significant trade item was bear furs and other bear-related items, such as claws and canine teeth, as is reflected in the burial records presented in this paper. In the Viking Age, on the other hand, the trading systems of the forested inland region seem to have changed without direct signs of decreased exploitation of the forests. Instead, it is possible to note increased indications of agriculture and more distinct shieling systems in pollen diagrams, as well as a regionally constituted intensification of pastoralism, as well as tar and iron production. We think this is related to the opening of new facets, or niches, in the multifunctional boreal environments gradually putting a larger focus on resources other – for both subsistence and trade – than wild animals. This paper proposes that one explanation for this shift was a depletion of animals such as the bear in the Scandinavian inland region, which is also manifested in the decreased frequencies of bear claws in the BiG database (Table 4). The increase in bear claws from the 10th century onwards shows that these were possibly, at least partially, from bears that came from forests in the east, which coincides with similar cases on the Baltic Åland islands (see GUSTAVSSON/LJUNGKVIST, this volume). Our model is straightforward and carries several source-critical issues, but, at the same time, it could serve as a hypothesis for future interdisciplinary research committed to the boreal forests of Scandinavia and their contacts with the worlds outside.

Acknowledgements

John Ljungkvist's work on this paper was undertaken within the "The Viking Phenomenon" project at Uppsala University, funded by the Swedish Research Council (VR 2015-00466).

Karl-Johan Lindholm's work was undertaken within the project "Contesting marginality: The boreal forest of inland Scandinavia and the worlds outside, 1–1500 AD (UTMA)", supported by the Swedish Research Council (VR 2017-01483).

We would like to thank the editor of this volume, Oliver Grimm, for his immense support and patience. In addition, we want to acknowledge the reviewers and express our gratitude for their constructive comments.

Bibliography

Aaris-Sørensen 2009: K. Aaris-Sørensen, Diversity and dynamics of the mammalian fauna in Denmark throughout the last glacial-interglacial cycle, 115–0 kyr BP. Fossils and Strata 57, 2009, 1–59.

Ainsworth/U'Sumalia 2011: C. Ainsworth/R. U'Sumalia, Intergenerational valuation of fisheries resources can justify long-term conservation: A case study in Atlantic cod (*Gadus morhua*). Canadian Journal of Fisheries and Aquatic Sciences, April 2011, 1104–1110.

Aronsson 1994: K.-Å. Aronsson, Pollen evidence of Saami settlement and reindeer herding in the boreal forest of northernmost Sweden – an example of modern pollen rain studies as an aid in the interpretation of marginal human interference from fossil pollen data. Review of Paleobotany and Palynology 82, 1994, 37–45.

Andersen 2019: R. Andersen, Norske Rovfugler Brukt i Engelsk Falkonering i Høymiddelalderen. Master's thesis, Stavanger University (Stavanger 2019). http://hdl.handle.net/11250/2601397.

Ashby 2015: S. Ashby, What really caused the Viking Age. Archaeological Dialogues 22(1), 2015, 89–106.

Ashby et al. 2015: S. P. Ashby/A. Coutu/S. M. Sindbæk, Urban networks and Arctic outlands: Craft specialists and reindeer antler in Viking towns. European Journal of Archaeology 18, 2015, 679–704.

Barrett 2015: J. Barrett, What caused the Viking Age? Antiquity 82, 2015, 671–685.

Barrett et al. 2020: J. Barrett/S. Boessenkool/C. J. Kneale/T. C. O'Connel/B. Star, Ecological Globalisation, Serial Depletion and the Medieval Trade of Walrus Rostra. Quaternary Science Reviews 229, 2020. https://doi.org/10.1016/j.quascirev.2019.106122.

Baug et al. 2019: I. Baug/D. Skre/T. Heldal/J. Ø. Janssen, The beginning of the Viking Age in the West. Journal of Maritime Archaeology 14, 2019, 43–80. https://doi.org/10.1007/s11457-018-9221-3.

Bergman et al. 2013: I. Bergman/O. Zachrisson/L. Lindgren, From Hunting to Herding: Land Use, Ecosystem Processes, and Social Transformation among Sami AD 800–1500. Arctic Anthropology 50, 2013, 25–39.

Björklund 2013: I. Björklund, Domestication, Reindeer Husbandry and the Development of Sámi Pastoralism. Acta Borealia 30, 2013, 174–189.

Bratt 2008: P. Bratt, Makt uttryckt i jord och sten: stora högar och maktstrukturer i Mälardalen under järnåldern. Stockholm Studies in Archaeology 46 (Stockholm 2008).

Bray et al. 2013: S. C. E. Bray/J. J. Austin/J. L. Metcalf/K. Østbye/E. Østbye/S.-E. Lauritzen/K. Aaris-Sørensen/C. Valdiosera/C. J. Adler/A. Cooper, Ancient DNA identifies post-glacial recolonisation, not recent bottlenecks, as the primary driver of contemporary mtDNA phylogeography and diversity in Scandinavian brown bears. Diversity and Distributions 19, 2013, 245–256.

Callmer 1997: J. Callmer, Beads and bead production in Scandinavia and the Baltic Region c. AD 600–1100: a general outline. In: U. von Freeden/A. Wieczorek (eds.), Perlen. Archäologie, Techniken, Analysen (Bonn 1997) 197–202.

Callmer 2017: J. Callmer, The Rise of the Dominion of the ar-Rus in the Northern Parts of Eastern Europe, Seventh to Ninth Centuries A.D.: A Case of Culture Construction. In: J. Callmer/I. Gustin/M. Roslund (eds.), Identity Formation and Diversity in the Early Medieval Baltic and Beyond Communicators and Communication. The Northern World 75 (Leiden 2017) 136–167.

Chiesa 2021: P. Chiesa, Marckalada: The First Mention of America in the Mediterranean Area (c. 1340). Terra Incognitae 53, 2021, 88–106. https://doi.org/10.1080/00822884.2021.1943792.

Christensen 2015: T. Christensen, Lejre bag myten (Aarhus 2015).

Dögg Eddudóttir et al. 2021. S. Dögg Eddudóttir/E. Svensson/S. Nilsson, The history of settlement and agrarian land use in a boreal forest in Värmland, Sweden, new evidence from pollen analysis. Vegetation History and Archaeobotany 30, 2021, 759–771. https://doi.org/10.1007/s00334-021-00829-y.

Emanuelsson 2001: M. Emanuelsson, Settlement and Land-Use History in the Central Swedish Forest Region: The Use of Pollen Analysis in Interdisciplinary Studies. PhD thesis, Swedish University of Agricultural Sciences, Umeå (Umeå 2001).

Emanuelsson et al. 2003: M. Emanuelsson/A. Johansson/S. Nilsson/S. Pettersson/E. Svensson, Settlement, Shieling and Landscape: The Local History of a Forest Hamlet (Stockholm 2003).

Eriksson 2020: O. Eriksson, Origin and Development of Managed Meadows in Sweden: A Review. Rural Landscapes: Society, Environment, History 7(1), 2020. https://doi.org/10.16993/rl.51.

Feveile/Jensen 2000: C. Feveile/S. Jensen, Ribe in the 8th and 9th century: A contribution to the Archaeological Chronology of North Western Europe. Acta Archaeologica 71, 2000, 9–24.

Göthberg 2000: H. Göthberg, Bebyggelse i förändring. OPIA 25 (Uppsala 2000).

Grimm 2013: O. Grimm, Bear-skins in northern European burials and some remarks on other bear-related furnishings in the north and middle of Europe in the 1st millennium AD. In: O. Grimm/U. Schmölcke eds.), Hunting in Northern Europe until 1500 AD – Old traditions and regional developments, continental sources and continental influences. Papers presented at a workshop organised by the Centre for Baltic and Scandinavian Archaeology, Schleswig, June 16th and 17th, 2011. Schriften des Archaologischen Landesmuseums, Ergänzungsreihe 7 (Neumünster 2013) 277–296.

Hatlestad et al. 2021: K. Hatlestad/J. Wehlin/K. J. Lindholm, Coping with Risk. A Deep-Time Perspective on Societal Responses to Ecological Uncertainty in the River Dalälven Catchment Area and Sweden. Land 10(8), 2021, 883. https://doi.org/10.3390/land10080883.

Hennius 2018: A. Hennius, Viking Age tar production and outland exploitation. Antiquity 92, 2018, 1349–1361.

Hennius 2020a: A. Hennius, Towards a Refined Chronology of Prehistoric Pitfall Hunting in Sweden. European Journal of Archaeology 23, 2020, 530–546. https://doi.org:10.1017/eaa.2020.8.

Hennius 2020b: A. Hennius, Outland exploitation and the emergence of seasonal settlements. Bebyggelsehistorisk tidskrift 79, 2020, 8–24.

Hennius 2021: A. Hennius, Outlanders? Resource colonisation, raw material exploitation and networks in Middle Iron Age Sweden. Occasional Papers in Archaeology 73 (Uppsala 2021).

Hennius et al. 2018: A. Hennius/R. Gustavsson/J. Ljungkvist/L. Spindler, Whale bone gaming pieces: Aspects of marine mammal exploitation in Vendel and Viking Age Scandinavia. European Journal of Archaeology 21, 2018, 612–631. https://doi.org/ 10.1017/eaa.2018.15.

Indrelid et al. 2015: S. Indrelid/K. L. Hjelle/K. Stene (eds.), Exploitation of outfield resources – Joint Research at the University Museums of Norway (Bergen 2015).

Iregren 1988: E. Iregren, Finds of brown bear (*Ursus arctos*) in Sothern Scandinavia: indicators of local hunting or trade? In: B. Hårdh/L. Larsson/D. Olausson/R. Petré (eds.), Trade and Exchange in Prehistory: Studies in Honour of Berta Stjernquist (Lund 1988) 295–308.

Jørgensen/Nørgård Jørgensen 1997: L. Jørgensen/A. Nørgård Jørgensen, Nørre Sandegård Vest: a cemetery from the 6th–8th centuries on Bornholm (Copenhagen 1997).

Karlsson 2016: J. Karlsson, Spill: om djur, hantverk och nätverk i Mälarområdet under vikingatid och medeltid. Theses and Papers in Osteoarchaeology 8 (Stockholm 2016).

Karlsson et al. 2010: H. Karlsson/M. Emanuelsson/U. Segerström, The history of a farm-shieling system in the central Swedish forest region. Vegetation History and Archaeobotany 19, 2010, 103–19.

Kresten 2008: P. Kresten, Geological analysis of selected stone artefacts from Helgö. In: H. Clarke/K. Lamm (eds.), Excavations at Helgö 17 (Stockholm 2008) 148–157.

Langhammar 2012: D. Langhammar, Rapportsammanställning Gotland. Halla socken. Broe 1:71. Unpublished report, ATA (Stockholm 2012).

Larsson 2021: P. I. Larsson, Continuous Presence: A Historical Ecology of Ängesviken, Jämtland. MA-thesis, Department of Archaeology and Ancient History, Uppsala University (Uppsala 2019).

Lindgren 2019: S. Lindgren, Mötesplatser i fångstmarken. MA-thesis, Department of Archaeology and Ancient History, Uppsala University (Uppsala 2019).

Lindholm/Ljungkvist 2016: K.-J. Lindholm/J. Ljungkvist, The bear in the grave – Exploitation of top predator and herbivore resources in first millennium Sweden – first trends from a long term research project. European Journal of Archaeology 19, 2016, 3–27.

Lindholm et al. 2013: K.-J. Lindholm/E. Sandström/A. K. Ekman, The archaeology of the commons. Journal of Archaeology and Ancient History 10, 2013, 1–49.

Lindholm et al. 2021: K.-J. Lindholm/E. Ersmark/A. Hennius/S. Lindgren/K. Loftsgarden/E. Svensson, Contesting Marginality: The Boreal Forest of Middle Scandinavia and the Worlds Outside. In: C. Symes (ed.), The Global North: Spaces, Connections, and Networks from the Iron Age to 1650. The Medieval Globe 7.1 (Leeds 2021) 9–35.

Ljungkvist 2006: J. Ljungkvist, En hiar atti rikR. Om elit struktur och ekonomi kring Uppsala och Mälaren under yngre järnålder (Uppsala 2006).

Ljungkvist 2008: J. Ljungkvist, The development and chronology of the Valsgärde cemetery. In: S. Norr (ed.), Valsgärde Studies: the Place and its People, Past and Present. OPIA 42 (Uppsala 2008) 13–55.

Ljungkvist/Hennius 2020: J. Ljungkvist/A. Hennius, Correlations and conflicts between 14C dates and find chronologies in a transition phase – exemplified by the Ottar's mound in Vendel. In: C. Hillerdal/K. Ilves (eds.), Re-imagining Periphery: Archaeology and Text in Northern Europe from Iron Age to Viking and Early Medieval Periods (Oxford 2020) 91–101.

Loftsgarden 2015: K. Loftsgarden, Kolgroper – gull eller gråstein? In: I. M. Berg-Hansen (red.), Arkeologiske Undersøkelser 2005–2006, Kulturhistorisk museum, Universitetet i Oslo (Oslo 2015) 142–154.

Loftsgarden 2017: K. Loftsgarden, Marknadsplassar omkring Hardangervidda – ein arkeologisk og historisk analyse av innlandets økonomi og nettverk i vikingtid og mellomalder. PhD thesis, Bergen University (Bergen 2017).

Lundström 1980: A. Lundström, Gravgåvorna i Valsgärde. In: A. Sandwall (ed.), Vendeltid (Stockholm 1980) 65–79.

Magnell/Iregren 2010: O. Magnell/E. Iregren, Veitstu Hvé Blóta Skal: The Old Norse Blót in the Light of Osteological Remains from Frösö Church, Jämtland, Sweden. Current Swedish Archaeology 18, 2010, 223–250.

Magnell et al. 2017: O. Magnell/S. Prata/E. Sjöling, Att behandlas som ett djur – vendel- och vikingatida djurben från gravar och gårdar. In: L. Beronius Jörpeland/H. Göthberg/A. Seiler/J. Wikborg (eds.), at Upsalum – människor och landskapande: Utbyggnad av Ostkustbanan genom Gamla Uppsala (Stockholm 2017) 209–224.

Melin/Sigvallius 2001: J. Melin/B. Sigvallius, Excavations at Helgö XIV (Stockholm 2001).

Nerman 1958: B. Nerman, Grobin-Seeburg. Ausgrabungen und Funde (Stockholm 1958).

Peets et al. 2011: J. Peets/R. Allmäe/L. Maldre, Archaeological investigations of Pre-Viking Age burial boat in Salme village at Saaremaa. Archaeological Fieldwork in Estonia 2010, 29–48.

Peets et al. 2013: J. Peets/R. Allmäe/L. Maldre/R. Saage/T. Tomek/L. Lõugas, Research results of the Salme ship burials in 2011–2012. Archaeological Fieldwork in Estonia 2012, 43–60.

Petré 1980: B. Petré, Björnfällen i begravningsritualen – Statusobjekt speglande regional skinnhandel? Fornvännen 75, 1980, 5–14.

Petré 1984: B. Petré, Arkeologiska undersökningar på Lovö 4 (Stockholm 1980).

Pilø et al. 2018: L. Pilø/E. Finstad/C. B. Ramsey/J. R. P. Martinsen/A. Nesje/B. Solli/V. Wangen/M. Callanan/J. H. Barrett, The chronology of reindeer hunting on Norway's highest ice patches. Royal Society Open Science 5(171738), 2018. https://doi.org/10.1098/rsos.171738.

Price et al. 2016: T. D. Price/J. Peets/R. Allmäe/L. Maldre/E. Oras, Isotopic provenancing of the Salme ship burials in Pre-Viking Age Estonia. Antiquity 90, 2016, 1022–1037. https://doi.org/10.15184/aqy.2016.106.

Ramqvist 2001: P. H. Ramqvist, Utbytessystem under det första årtusendet e.Kr. Idéer utgående från tre mellannorrländska älvar. Fornvännen 96, 2001, 1–21.

Ramqvist 2007: P. H. Ramqvist, Fem Norrland: om norrländska regioner och deras interaktion. Arkeologi i norr 10, 2007, 153–80.

Rodrigues et al. 2021: M. Rodrigues/D. Calvo-Dopico/E. Mourelle, Impact of stock health on fish prices: Evaluation and implications for food accessibility. PLoS ONE 16(12), 2021. https://doi.org/10.1371/journal.pone.0261580.

Rundberget 2013: B. Rundberget, Jernets dunkle dimension. Jernvinna i sørlige Hedmark. Sentral økonomisk faktor og premiss for samfundsutvikling c. AD 700–1300 (Oslo 2013).

Rundkvist 2003: M. Rundkvist, Barshalder 2. Studies of late Iron Age Gotland. PhD thesis (vol. 1–2), Stockholm University (Stockholm 2003).

Schietzel 2018: K. Schietzel, Spurensuche Haithabu (Schleswig 2018).

Schönbäck/Thunmark-Nylèn 2002: B. Schönbäck/L. Thunmark-Nylèn, De vikingatida båtgravarna vid Valsgärde – relativ kronologi. Fornvännen 97, 2002, 1–8.

Skibsted Klæsøe 1999: I. Skibsted Klæsøe, Vikingetidens kronologi – en nybearbejning af det arkæologiske material. Aarbøger for nordisk oldkyndighed og historie 1997, 189–242.

Star et al. 2018: B. Star/J. H. Barrett/A. T. Gondek/S. Boessenkool, Ancient DNA reveals the chronology of walrus ivory trade from Norse Greenland. Proceedings of the Royal Society, B 285: 20180978. https://doi.org/10.1098/rspb.2018.0978.

Stolpe/Arne 1912: H. Stolpe/T. J. Arne, Graffältet vid Vendel (Stockholm 1912).

Storli 1993: I. Storli, Sami Viking Age Pastoralism – or "The Fur Trade Paradigm" Reconsidered. Norwegian Archaeological Review 26, 1993, 1–20.

Svedjemo 2014: G. Svedjemo, Landscape dynamics: spatial analyses of villages and farms on Gotland AD 200–1700 (Uppsala 2014).

Svensson 1998: E. Svensson, Människor i utmark. PhD thesis, Lund University (Lund 1998).

Taberlet et al. 1995: P. Taberlet/J. E. Swenson/F. Sandegren/A. Bjärvall, Localization of a Contact Zone between Two Highly Divergent Mitochondrial DNA Lineages of the Brown Bear Ursus arctos in Scandinavia. Conservation Biology 9(5), 1995, 1255–1261 https://doi.org/10.1046/j.1523-1739.1995.951255.

Vretemark 2013: M. Vretemark, The Vendel Period royal follower's grave at Swedish Rickeby as starting point for reflections about falconry in Northern Europe. In: O. Grimm/U. Schmölcke (eds.), Hunting in northern Europe until 1500 AD. Old traditions and regional developments, continental sources and continental influences. Papers presented at a workshop organized by the Centre for Baltic and Scandinavian Archaeology, ZBSA. Schriften des Archäologischen Landesmuseums, Ergänzungsreihe 7 (Neumünster 2013) 379–386.

Xenikoudakis et al. 2015: G. Xenikoudakis/E. Ersmark/J. L. Tisson/L. Waits/J. Kindberg/J. E. Swenson/L. Dalèn, Consequences of a demographic bottleneck on genetic structure and variation in the Scandinavian brown bear. Molecular Ecology 24(13), 2015, 3441–3454.

Zachrisson 2010: T. Zachrisson, Vittnesbörd om pälshandel? Ett arkeologiskt perspektiv på romerska bronsmynt funna i norra Sverige. Fornvännen 105, 2010, 187–202.

Zachrisson/Krzewińska 2019: T. Zachrisson/M. Krzewińska, The «Lynx Ladies» – Burials furnished with lynx skins from the Migration and Merovingian periods found in present-day Sweden. In: M. Augstein/M. Hardt (eds.), Sächsische Leute und Länder. Benennung und Lokalisierung von Gruppenidentitäten im ersten Jahrtausend. 66. Internationales Sachsensymposion = Neue Studien zur Sachsenforschung 10 (Braunschweig 2019) 103–119.

Prof. Karl-Johan Lindholm
Department of Archaeology and Ancient History
Uppsala University
Uppsala
Sweden
karl-johan.lindholm@arkeologi.uu.se

Associate Prof. John Ljungkvist
Department of Archaeology and Ancient History
Uppsala University
Uppsala
Sweden
john.ljungkvist@arkeologi.uu.se

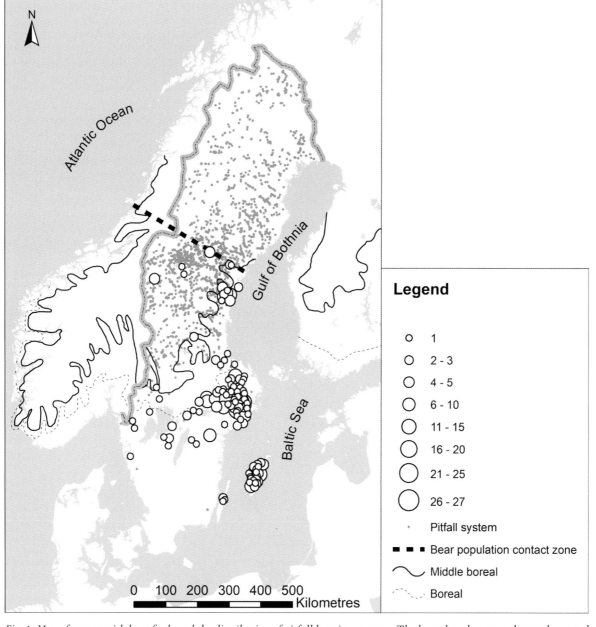

Fig. 1. Map of graves with bear finds and the distribution of pitfall hunting systems. The boundary between the southern and middle boreal zones is indicated on the map, as well as the contact zone between the southern and northern bear population (map K. J. Lindholm).

Table 1. Number of graves with bear claws from the main Iron Age periods in Sweden. Upper row shows the graves from the database with allocation to a specific time period. Contexts that cut through more than one phase due to uncertain dating circumstances are excluded. Lower row reveals the number of graves per year according to estimated length of the period. PRIA (Pre-Roman Iron Age), 400 years (400–1 BC); RIA (Roman Iron Age), 400 years (AD 1–400); MP (Migration Period), 160 years (AD 400–560); VET (Vendel Period), 215 years (AD 560–775); VIT (Viking Age), 275 years (AD 775–1050).

	PRIA	RIA	MP	VET	VIT
Find per phase	3	18	101	76	50
Find per year of phase	0.008	0.045	0.63	0.35	0.18

Table 2. A model of late Iron Age phases in present-day Sweden, based on different overlapping Scandinavian chronological studies (modified after LJUNGKVIST 2008).

Phase	Years	Key reference
VET 1	560/70–620/30	PETRÉ 1984 (bead horizon P3); JØRGENSEN/NØRGÅRD JØRGENSEN 1997 (1B1–1B2)
VET 2	620/30–660/70	PETRÉ 1984 (bead horizon P4); JØRGENSEN/NØRGÅRD JØRGENSEN 1997 (1C–1D2)
VET 3	660/700–710	JØRGENSEN/NØRGÅRD JØRGENSEN 1997 (1D2); PETRÉ 1984 (P4 late)
VET 4	710–760/70	FEVEILE/JENSEN 2000 (Phase B–C); JØRGENSEN/NØRGÅRD JØRGENSEN 1997 (1D2–2A); PETRÉ 1984 (bead horizon P5); CALLMER 1997 (Pl. 15A)
VIT 1	760/70–800	FEVEILE/JENSEN 2000 (Phase D–F); CALLMER 1997 (Pl. 16A); SKIBSTED KLÆSØE 1999 (per. 1)
VIT 2	800–850	FEVEILE/JENSEN 2000 (Phase G–H); CALLMER 1997 (Pl. 16B–C); SKIBSTED KLÆSØE 1999 (per. 1–2a1)
VIT 3	850–900	CALLMER 1997 (Pl. 17A); SKIBSTED KLÆSØE 1999 (per. 2a1–2a2)
VIT 4	900–950	CALLMER 1997 (Pl. 17B); SKIBSTED KLÆSØE 1999 (per. 2a1–2b).
VIT 5	950–1000	CALLMER 1997 (Pl. 18A–B)
VIT 6	1000–1050	CALLMER 1997 (Pl. 18C)

Table 3. Graves with bear claws dated to a specific phase of the Vendel period and divided into three blocks of the main period: early phase, transition phase, later phase.

	VET 1	VET 1–2	VET 2	VET 2–3	VET 3	VET 3–4	VET 4
Finds per phase	23	19	10	6	1	2	1
Numbers in larger phases			Early phase: 52 graves	Transition phase: 6 graves			Late phase: 4 graves

Table 4a. Graves containing bear, bird of prey, and lynx remains from the Vendel period, dated to a specific phase.

Phase	VET 1	VET 1–2	VET 2	VET 2–3	VET 3	VET 3–4	VET 4	VET 4–VIT 1
Bear claw	23	19	10	6	1	2	1	2
Lynx claw	6	4	2	4			1	1
Birds of prey	8	13	7	9	1	4	3	3

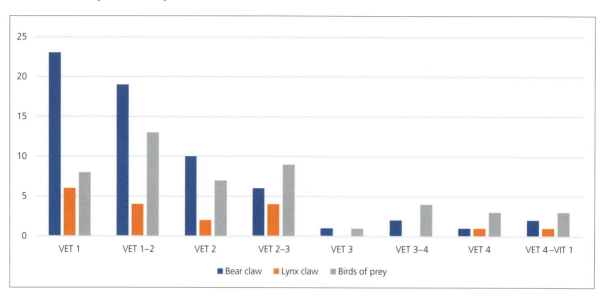

Table 4b. Visual representation of Table 4a.

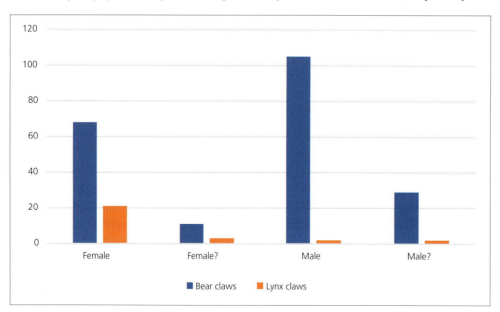

Table 5. Frequency of bear and lynx claws in graves with female and male attributes, respectively.